STUDENT'S SOLUTIONS MANUAL
TO ACCOMPANY

CALCULUS AND ANALYTIC GEOMETRY
by GEORGE B. THOMAS, JR. and ROSS L. FINNEY   PART I / SIXTH EDITION

KENNETH R. BALLOU
University of California, Berkeley

CHARLES SWANSON
University of Minnesota

ADDISON-WESLEY PUBLISHING COMPANY
Reading, Massachusetts • Menlo Park, California
Wokingham, Berkshire • Amsterdam • Don Mills, Ontario • Sydney

CONTENTS

Reproduced by Addison-Wesley from camera-ready copy supplied by the authors.

Copyright © 1984 by Addison-Wesley Publishing Company, Inc.

ISBN 0-201-16298-9
  BCDEFGHIJ-AL-8987654

CHAPTER 1

Article 1.1

From the diagram (or from the midpoint formula), we see that if P has coordinates (x,y), then Q has coordinates (x,−y), R has coordinates (−x,y), and S has coordinates (−x,−y). The coordinates of point T, however, are not so obvious.

Since PA = AT, PX = XA = AY = YT. Hence, the coordinates of T are (y,x).

We then have the following chart.

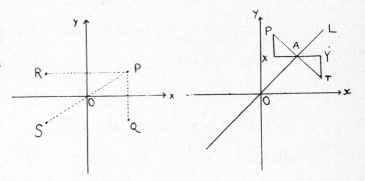

	P(x,y)	Q(x,−y)	R(−x,y)	S(−x,−y)	T(y,x)
1.	(1,−2)	(1,2)	(−1,−2)	(−1,2)	(−2,1)
2.	(2,−1)	(2,1)	(−2,−1)	(−2,1)	(−1,2)
3.	(−2,2)	(−2,−2)	(2,2)	(2,−2)	(2,−2)
4.	(−2,1)	(−2,−1)	(2,1)	(2,−1)	(1,−2)
5.	(0,1)	(0,−1)	(0,1)	(0,−1)	(1,0)
6.	(1,0)	(1,0)	(−1,0)	(−1,0)	(0,1)
7.	(−2,0)	(2,0)	(2,0)	(2,0)	(0,−2)
8.	(0,−3)	(0,3)	(0,−3)	(0,3)	(−3,0)
9.	(−1,−3)	(−1,3)	(1,−3)	(1,3)	(−3,−1)
10.	(/2,− /2)	(/2, /2)	(− /2,− /2)	(− /2, /2)	(− /2, /2)
11.	(−π,−π)	(−π,π)	(π,−π)	(π,π)	(−π,−π)
12.	(−1.5,2.3)	(−1.5,−2.3)	(1.5,2.3)	(1.5,−2.3)	(2.3,−1.5)
13.	(x,y)	(x,−y)	(−x,y)	(−x,−y)	(y,x)

15.

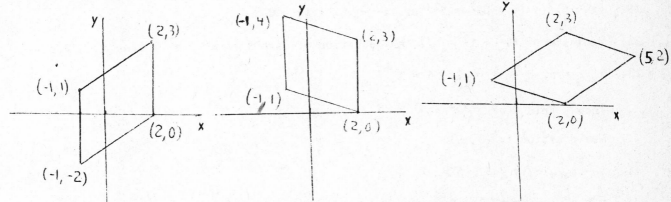

17. L = 3h, 2(L + h) = 56 => L = 21, h = 7. So the coordinates are A(−12,2), B(−12,−5), C(9,−5).

19. $\frac{1}{1} = \frac{b}{2}$ => b = 2

Article 1.2

1. a) $\Delta x = 57 - 0 = 57$; $\Delta y = 22 - 32 = -10$

    b) $\Delta x = 26 - 28 = -2$; $\Delta y = 6 - 18 = -12$

    c) $\Delta x = 40 - 39 = 1$; $\Delta y = 4 - 12 = -8$

3. a) $C(1,1)$ b) $\Delta x = 1 - (-1) = 2$

    c) $\Delta y = 2 - 1 = 1$ d) $AB = \sqrt{(\Delta x)^2 + (\Delta y)^2} = \sqrt{5}$

5. a) $C(-1,2)$ b) $\Delta x = -1 - (-3) = 2$

    c) $\Delta y = -2 - 2 = -4$ d) $AB = \sqrt{(\Delta x)^2 + (\Delta y)^2} = \sqrt{20}$

7. a) $C(-8,1)$ b) $\Delta x = -8 - (-3) = -5$

    c) $\Delta y = 1 - 1 = 0$ d) $AB = \sqrt{(\Delta x)^2 + (\Delta y)^2} = 5$

9. Since $x > 0$ and $y > 0$, P_2 lies above and to the right of P_1.

11. Since $x < 0$ and $y = 0$, P_2 lies to the left of P_1.

13. Since $x > 0$ and $y < 0$, P_2 lies below and to the right of P_1.

15. Since $x = 0$ and $y < 0$, P_2 lies below P_1.

17. $(x - 0)^2 + (y - 0)^2 = 5^2$ or $x^2 + y^2 = 25$

19. $(x - 5)^2 + (y - 0)^2 = 5^2$ or $x^2 - 10x + y^2 = 0$

21. radius $r = \sqrt{1^2 + 1^2} = \sqrt{2}$ => equation of circle is $x^2 + y^2 = 2$.

23. $x_{new} = x_{old} + x = -2 + 5 = 3$

    $y_{new} = y_{old} + y = 3 + (-6) = -3$

    New position is $(3,-3)$

25. $x = x_{new} - \Delta x = 3 - 5 = -2$

    $y = y_{new} - \Delta y = -3 - 6 = -9$

27. $x_{old} = x_{new} - \Delta x = u - h$

    $y_{old} = y_{new} - y = v - k$

    Starting position is $(u-h, v-k)$.

29. Since v is constant, the bases of all five triangles are of equal length; since

all five triangles have a common altitude, $A_1 = \ldots = A_5$.

Article 1.3

$$M_{AB} = \frac{Y_B - Y_A}{x_B - x_A} \quad (x_B \neq x_A);$$

$M_1 = -\dfrac{1}{M_{AB}}$ $(M_{AB} \neq 0)$. Also, if either M_{AB} or M_1 is zero, the other is undefined.

	M_{AB}	M_1			M_{AB}	M_1
1.	3	$-\frac{1}{3}$		9.	-1	1
2.	3	$-\frac{1}{3}$		10.	undefined	0
3.	$-\frac{1}{3}$	3		11.	2	$-\frac{1}{2}$
4.	$-\frac{1}{2}$	2		12.	$\frac{2}{3}$	$-\frac{3}{2}$
5.	-1	1		13.	$\frac{y}{x}$	$-\frac{x}{y}$
6.	0	undefined		14.	0	undefined
7.	0	undefined		15.	undefined	0
8.	-3	$\frac{1}{3}$		16.	$-\frac{b}{a}$	$\frac{a}{b}$

17. $M_{AB} = M_{CD} = 1$; $M_{BC} = M_{DA} = -1$. Hence ABCD is a rectangle.

19. $M_{AB} = M_{CD} = \frac{1}{3}$; $M_{BC} = M_{DA} = -3$. Hence ABCD is a rectangle.

21. $M_{AB} = M_{CD} = -1$; $M_{BC} = \frac{1}{2}$; $M_{DA} = 2$. Hence ABCD is a trapezoid.

23. $\tan \emptyset = \dfrac{R}{T} \Rightarrow \tan 40^\circ = \dfrac{R}{9}$

$\Rightarrow R = 9 \tan 40^\circ = 7.55$ inches.

25. a) Rate is approximately $\dfrac{68^\circ - 70^\circ}{0.35 \text{ in}}$ $\quad -5.7^\circ/\text{in}$

b) Rate is approximately $\dfrac{10^\circ - 68^\circ}{3.6 \text{ in}}$ $\quad -16.1^\circ/\text{in}$

c) Rate is approximately $\frac{5^{\circ} - 10^{\circ}}{0.5 \text{ in}} \approx -10^{\circ}/\text{in}$

NOTE: Student's answers will vary depending on how the student reads the graph. (Because of the thickness of the curve, the above numbers are at best approximate.)

27. $\frac{y - 0}{x - 0} = \frac{y}{x} = 2 \Rightarrow y = 2x$

$\frac{y - 0}{x - (-1)} = \frac{y}{x + 1} = 1 \Rightarrow y = x + 1 \Rightarrow 2x = x + 1$

$(x,y) = (1,2)$

29. The equation of the line is $y - 0 = -\frac{1}{2}(x - 0)$ or $y = -\frac{1}{2}x$.

31. $M_{AB} = 2$, $M_{BC} = 3 \Rightarrow$ A,B,C are not collinear.

33. $M_{AB} = -\frac{1}{2}$, $M_{BC} = -1 \Rightarrow$ A,B,C are not collinear.

35. $M_{AB} = \frac{1}{4}$, $M_{BC} = -4 \Rightarrow AB \perp BC$.

Because $\angle ABC$ is an inscribed right angle in the circumcircle, AC must be a diameter of the circumcircle. Hence, the center of the circumcircle is the midpoint of AC, or $(\frac{3}{2},\frac{3}{2})$. The radius of the circumcircle is $\sqrt{(\frac{3}{2})^2 + (\frac{3}{2} - 1)^2} = \frac{1}{2}\sqrt{34}$.

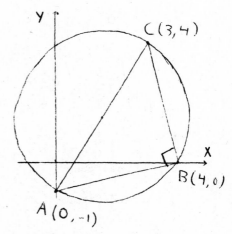

37. $y = 3 x$, $y = -2$ (measured from P_1 to the x-axis) $\Rightarrow x = -\frac{2}{3}$. So the line crosses the x-axis at $(\frac{1}{3},0)$.

Article 1.4

1. a) vertical line: $x = 2$
   b) horizontal line: $y = 3$
3. a) vertical line: $x = 0$
   b) horizontal line: $y = 0$
5. a) vertical line: $x = -4$
   b) horizontal line: $y = 0$
7. a) vertical line: $x = 0$
   b) horizontal line: $y = b$
9. $y - 1 = 1(x - 1)$ or $y = x$
11. $y - 1 = 1(x - (-1))$ or $y = x + 2$
13. $y - b = 2(x - 0)$ or $y = 2x + b$

15. $m = \frac{3 - 0}{2 - 0} = \frac{3}{2}$; equation of line is therefore

$y - 0 = \frac{3}{2}(x - 0)$ or $y - 3 = \frac{3}{2}(x - 2)$ or $y = \frac{3}{2}x$.

17. vertical line: $x = 1$

19. vertical line: $x = -2$

21. $m = -\frac{F_0}{T} \Rightarrow y - F_0 = -\frac{F_0}{T}(x - 0)$

or $y - 0 = -\frac{F_0}{T}x + F_0$.

23. vertical line: $x = 0$

25. $m = -\frac{1}{3} \Rightarrow y - 1.5 = -\frac{1}{3}(x + 0.7)$

or $y - 0.8 = -\frac{1}{3}(x - 1.4)$

27. $m = \frac{y_1 - y_0}{x_1 - x_0} \Rightarrow y - y_0 = \frac{y_1 - y_0}{x_1 - x_0}(x - x_0)$

or $y - y_1 = \frac{y_1 - y_0}{x_1 - x_0}(x - x_1)$

29. $y = mx + b = 3x - 2$

31. $y = x + \sqrt{2}$

33. $y = -5x + 2.5$

35. $y = mx + b = 3x + 5 \Rightarrow m = 3, b = 5$

37. $x + y = 2 \Rightarrow y = -x + 2 \Rightarrow m = -1, b = 2$

39. $x - 2y = 4 \Rightarrow y = \frac{x}{2} - 2 \Rightarrow m = \frac{1}{2}, b = -2$

41. $4x - 3y = 12 \Rightarrow y = \frac{4}{3}x - 4 \Rightarrow m = \frac{4}{3}, b = -4$

43. $\frac{x}{3} + \frac{y}{4} = 1 \Rightarrow y = -\frac{4}{3}x + 4 \Rightarrow m = -\frac{4}{3}, b = 4$

45. $\frac{x}{2} - \frac{y}{3} = -1 \Rightarrow y = \frac{3}{2}x + 3 \Rightarrow m = \frac{3}{2}, b = 3$

47. $1.05x - 0.35y = 7 \Rightarrow y = 3x - 20 \Rightarrow m = 3, b = -20$

49. $\frac{x}{a} + \frac{y}{b} = 1 \Rightarrow y = -\frac{b}{a}x + b \Rightarrow$ slope is $-\frac{b}{a}$, y-intercept is b.

To find the x-intercept, set $y = 0$: $\frac{x}{a} = 1 \Rightarrow x = a$.

CHAPTER 1 - Article 1.4

51. a) $y - 1 = 3(x - 2) \Rightarrow y = 3x - 5$

   b) $y = mx + b = -\dfrac{1}{\sqrt{3}}x + 1$

53. a) The line through P parallel to L has the same slope as L. The required line

   is therefore $y - 1 = x - 2$, or $y = x - 1$.

   b) The slope of the line perpendicular to L is $M_\perp = \dfrac{-1}{M_L} = -1$. Hence the

   required line is $y - 1 = -(x - 2)$ or $y = 3 - x$.

   c) As in Example 5, the line through P perpendicular to L is $y = 3 - x$. The

   intersection of this line and L is $y = 3 - x = x + 2 \Rightarrow x = \dfrac{1}{2}$

   $\Rightarrow y = \dfrac{5}{2}$. Then the distance from P to L is the distance from P to this

   point of intersection:

   $$d = \sqrt{(2 - \tfrac{3}{2})^2 + (1 - \tfrac{5}{2})^2} = \sqrt{(\tfrac{1}{2})^2 + (-\tfrac{3}{2})^2} = \frac{\sqrt{10}}{2}.$$

55. a) $m_L = \dfrac{-1}{\sqrt{3}} \Rightarrow y = \dfrac{-x}{\sqrt{3}}$

   b) $m_\perp = -\dfrac{1}{M_L} = \sqrt{3} \Rightarrow y = \sqrt{3}\,x.$

   c) Point of intersection of L and $L_\perp$ is $(\tfrac{3}{4}, \tfrac{3\sqrt{3}}{4})$

   $d = \sqrt{(\tfrac{3}{4})^2 + (\tfrac{3\sqrt{3}}{4})^2} = \dfrac{3}{2}$

57. a) $m_L = -2 \Rightarrow y = -2x - 2$

   b) $m_\perp = \dfrac{1}{2} \Rightarrow y = \dfrac{1}{2}x + 3$

   c) L and $L_\perp$ intersect at $(\tfrac{2}{5}, \tfrac{16}{5})$. So,

   $d = \sqrt{(\tfrac{2}{5} + 2)^2 + (\tfrac{16}{5} - 2)^2} = \dfrac{6}{5}\sqrt{5}$

59. a) $M_L = 2 \Rightarrow y = 2x - 2$

   b) $M_\perp = -1/M_L = -\dfrac{1}{2} \Rightarrow y = -\dfrac{1}{2}x + \dfrac{1}{2}$

   c) L and $L_\perp$ intersect at $(-\tfrac{3}{4}, \tfrac{4}{5}) \Rightarrow$ c

   $d = \sqrt{(1 + \tfrac{3}{5})^2 + (0 - \tfrac{4}{5})^2} = \dfrac{4}{5}\sqrt{5}$

61. a) L is vertical, so the equation of the line through (3,2) parallel to L is

   $x = 3$.

   b) The equation of the line $L_\perp$ through (3,2) perpendicular to L is $y = 2$.

   c) L and $L_\perp$ intersect at $(-5,2) \Rightarrow d = \sqrt{(3 + 5)^2 + (2 - 2)^2} = 8.$

63. a) L is vertical, so the equation of the line through (a,b) parallel to L is

   x = a.

   b) The equation of $L_\perp$ is y = b.

   c) L, $L_\perp$ intersect at (-1,b). Hence
   $$d = \sqrt{(a - (-a))^2 + (b - b)^2} = |a + 1|.$$

65. a) $M_L = -\frac{4}{3} \Rightarrow y = -\frac{4}{3}x + \frac{34}{3}$

   b) $M_\perp = -\frac{1}{M_L} = \frac{3}{4} \to y = \frac{3}{4}x + 3$

   c) L and $L_\perp$ intersect at $(\frac{12}{25}, \frac{84}{25}) \Rightarrow$
   $$d = \sqrt{(4 - \frac{12}{25})^2 + (6 - \frac{84}{25})^2} = \frac{110}{25} = 4.4$$

67. $\tan\phi = m = 1 \Rightarrow \phi = 45^\circ$

69. $\tan\phi = m = -\frac{1}{\sqrt{3}} \Rightarrow \phi = 150^\circ$

71. $\tan\phi = m = -2 \Rightarrow \phi = 116.6^\circ$

73. $\tan\phi = m = -\frac{4}{3} \Rightarrow \phi = 126.9^\circ$

75. $m = \tan 60^\circ = \sqrt{3} \Rightarrow y = \sqrt{3}x + (4 - \sqrt{3})$

77. $\phi = 90^\circ \Rightarrow$ the required line is vertical. So, the required line is x = -2.

79. Determine k: when d = 100, p = 10.94

   10.94 = 100k + 1 $\Rightarrow$ k = 0.0994

   At d = 50: p = (0.0994)(50) + 1 = 5.97 atm

81. $m = \frac{35.16 - 35}{135 - 65}$ 0.0023 $\Rightarrow$ S - 35 = 0.0023(t - 65) or

   S = 0.0023t + 34.85

83. If $0 < \alpha < 90^\circ$, then from the diagram

   shows that the x-intercept of L is $\frac{p}{\cos\alpha}$

   and that the y-intercept of L is

   $\frac{p}{\sin\alpha}$. Then the equation of L is

   $\frac{x}{p/\cos\alpha} + \frac{y}{p/\sin\alpha} = 1,$

   or $x\cos\alpha = y\sin\alpha = p.$

Article 1.5

1. $x + 4 \geq 0 \Rightarrow x \geq -4$. Domain is $-4 \leq x \leq \infty$.
3. $x - 2 \neq 0 \Rightarrow x \neq 2$. Domain is $x: -\infty < x < 2$ or $2 < x < \infty$
5. $x \geq 0$ if $\sqrt{x}$ is to be defined. Domain is $0 \leq x < \infty$.
7. y is defined for all x. Domain is $-\infty < x < \infty$.
9. y is defined for all x. Domain is $-\infty < x < \infty$.
11. y is defined for all x. Domain is $-\infty < x < \infty$.

13. a) y is defined for all x. Domain is $-\infty < x < \infty$.

    b) $x^2 \geq 0$ for all x. Range is $1 \leq y < \infty$.　　　c) answer in back of text.

15. a) y is defined for all x. Domain is $-\infty < x < \infty$.

    b) $x^2 \geq 0$ for all x, so $-x^2 \leq 0$. Range is $-\infty < y < \infty$.　　c) answer in back of text.

17. a) $x + 1 \geq 0 \Rightarrow x \geq -1$. Domain is $-1 \leq x < \infty$.

    b) By convention $\sqrt{x + 1} < 0$. Range is $0 < y < \infty$.　　c) answer in back of text.

19. a) $x \geq 0$. Domain is $0 \leq x < \infty$.

    b) $\sqrt{x} \geq 0$. Range is $1 \leq y < \infty$.　　c) answer in back of text.

21. a) For $\sqrt{2x}$ to be defined, we must have $x \geq 0$. Thus the domain is $0 \leq x < \infty$.

    b) $(\sqrt{2x})^2 \geq 0$ for all x in the domain. Hence the range is $0 \leq y < \infty$.　c) answer in back of text.

23. a) $x \neq 0 \Rightarrow$ Domain is $x: -\infty < x < 0$ or $0 < x < \infty$.

    b) The fraction takes on all values except zero. The range is therefore

    $y: -\infty < y < 0$ or $0 < y < \infty$.　　c) answer in back of text.

25. a) y is defined for all x. Domain is $-\infty < x < \infty$.

    b) $-1 \leq \sin 2x \leq 1$, so the range is $-1 \leq y \leq 1$.　　c) answer in back of text.

27. a) y is defined for all x. Domain is $-\infty < x < \infty$.

    b) $|\sin x| \leq 1$, so the range is $0 \leq y \leq 1$.　　c) answer in back of text.

29. a) y is defined for all x. Domain is $-\infty < x < \infty$.

b) $-1 \leq \sin x \leq 1$, so the range is $0 \leq y \leq 2$.   c) answer in back of text.

31. a) No, x cannot be negative (otherwise, $\sqrt{x}$ is undefined).

    b) No, x = 0 leads to division by zero.

    c) y is defined for all x > 0. Domain is $0 < x < \infty$.

33. a) No, otherwise $\frac{1}{x} - 1 < 0$.

    b) No, x = 0 leads to division by zero.

    c) No, $x > 1 \Rightarrow \frac{1}{x} < 1 \Rightarrow \frac{1}{x} - 1 < 0$.

    d) From a, b, and c, the possible domain is $0 < x \leq 1$. A quick check shows

    that $\frac{1}{x} - 1 \geq 0$ for these values of x. Hence, the domain is in fact

    $0 < x \leq 1$.

35. Graphs a, c, and d cannot be the graph of $y = 4x^2$. At x = 0, y = 0, only

    graph (b) passes through (0,0).

37. y(0) = 0, y(1) = 1, y(2) = 0.

39.

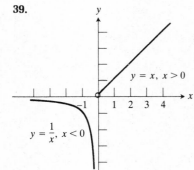

41.

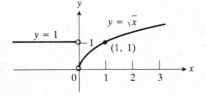

43. Domain(f) = $-\infty < x < \infty$, domain (g) = $1 \leq x < \infty$. The domains of f + g,

    f $-$ g, and f . g are domain (f) $\bigcap$ domain (g) = $1 \leq x < \infty$. In calculating

    domain (f/g) and domain (g/f) we must avoid division by 0. Hence, domain

    $(\frac{f}{g})$ = $1 < x < \infty$ and domain (g/f) = $1 \leq x < \infty$.

45. Domain $(f) = 0 \leq x < \infty$; domain $(g) = -1 \leq x < \infty$. Domain $(f+g) =$ domain $(f-g) =$ domain $(f \cdot g) =$ domain $(f) \cap$ domain $(g) = 0 \leq x < \infty$. In finding the domains of f/g and g/f, we must be careful to exclude division by zero. Hence, domain $(f/g) = 0 \leq x < \infty$, and domain $(g/f) = 0 < x < \infty$.

47. a) $h(-1) = 1 + \dfrac{5}{-1} = -4$.

　　b) $h(1/2) = 1 + \dfrac{5}{1/2} = 11$.

　　c) $h(5) = 1 + \dfrac{5}{5} = 2$.

　　d) $h(5x) = 1 + \dfrac{5}{5x} = 1 + \dfrac{1}{x} = \dfrac{x+1}{x}$.

　　e) $h(10x) = 1 + \dfrac{5}{10x} = 1 + \dfrac{1}{2x} = \dfrac{2x+1}{2x}$.

　　f) $h(\dfrac{1}{x}) = 1 + \dfrac{5}{1/x} = 1 + 5x$.

49. $f(1-x) = \dfrac{(1-x)-1}{1-x} = \dfrac{-x}{1-x}$. $f(x) \cdot f(1-x) = \dfrac{x-1}{x} \cdot \dfrac{-x}{1-x} = 1$.

51. $\dfrac{F(t+h) - F(t)}{h} = \dfrac{[4(t+h)-3] - [4t-3]}{h} = \dfrac{4h}{h} = 4$.

53. $f(x) = \sin^2 x$, $g(x) = \sqrt{x}$: $g(f(x)) = \sqrt{\sin^2 x} = |\sin x|$, $f(g(x)) = \sin^2(\sqrt{x})$.

55. $-\infty < x \leq -2$ or $2 \leq x < \infty$.

57. $x - 3 < -3$ or $x - 3 > 3$; equivalently, $x < 0$ or $x > 6$.

59. $|\dfrac{1}{x}| = \dfrac{1}{|x|} \leq 1 \Rightarrow |x| \geq 1 \Rightarrow x \leq -1$ or $x \geq 1$.

61. $-8 < x < 8 \Rightarrow |x| < 8$.

63. $-5 < x < 1 \Rightarrow -3 < x + 2 < 3 \Rightarrow |x + 2| < 3$.

65. $-a < y < a \Rightarrow |y| < a$.

67. $L - \varepsilon < y < L + \varepsilon \Rightarrow -\varepsilon < y < L < \varepsilon \Rightarrow |y - L| < \varepsilon$.

69. $x_0 - 5 < x < x_0 + 5 \Rightarrow -5 < x < x_0 < 5 \Rightarrow |x - x_0| < 5$.

71. Answer: g. $|x + 3| < 1 \Rightarrow -1 < x + 3 < 1 \Rightarrow -4 < x < -2$.

73. Answer: e. $\left|\dfrac{x}{2}\right| < 1 \Rightarrow |x| < 2 \Rightarrow -2 < x < 2$.

75. Answer: h. $|2x \cdot 5| < 1 \Rightarrow -1 < 2x - 5 < 1 \Rightarrow 4 < 2x < 6 \Rightarrow 2 < x < 3$.

77. Answer: b. $\dfrac{|x-1|}{2} < 1 \Rightarrow |x - 1| < 2 \Rightarrow -2 < x - 1 < 2 \Rightarrow -1 < x < 3$.

79. Answer: i. $|x^2 - 2| \leq 2 \Rightarrow -2 \leq x^2 - 2 \leq 2 \Rightarrow 0 \leq x^2 \leq 4 \Rightarrow -2 \leq x \leq 2$.

81. $|1 - x| = 1 - x$ if $1 - x \geq 0$ or $x \leq 1$.

$|1 - x| = x - 1$ if $1 - x \leq 0$ or $x \geq 1$.

83.

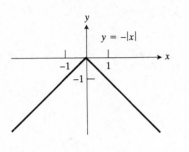

85.

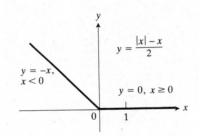

87.

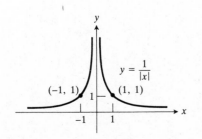

89. a) From the graph: minimum value is y = -2, no maximum value.

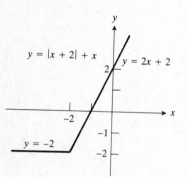

b) If x ≤ 0, y = -x + (1 - x) = 1 - 2x.

If 0 ≤ x ≤ 1, y = x + (1 - x) = 1.

If 1 ≤ x, y = x + (x - 1) = 2x - 1.

Minimum value is y = 1; there is no maximum value.

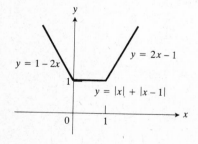

91.

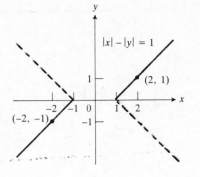

93. [x] = 0 for 0 ≤ x < 1.

95. From Example 22, $y = 4 \left(\dfrac{Th}{d}\right)^{1/2} = 9.29$ km. Thus, the Concorde should fly

9.3 km south of Nantucket.

97. $f(-x) = -x - (\dfrac{1}{-x}) = -x + \dfrac{1}{x} = -f(x)$. Hence f(x) is an odd function.

Article 1.6

1. m = 2x - 2
   m = 0 if x = 1

x	y	m
-2	5	-6
-1	0	-4
0	-3	-2
1	-4	0
2	-3	2
3	0	4
4	5	6

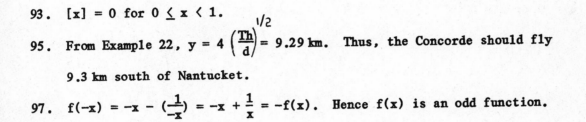

3. m = -2x

x	y	m
3	-5	-6
2	0	-4
1	3	-2
0	4	0
-1	3	2
-2	0	4
-3	-5	6

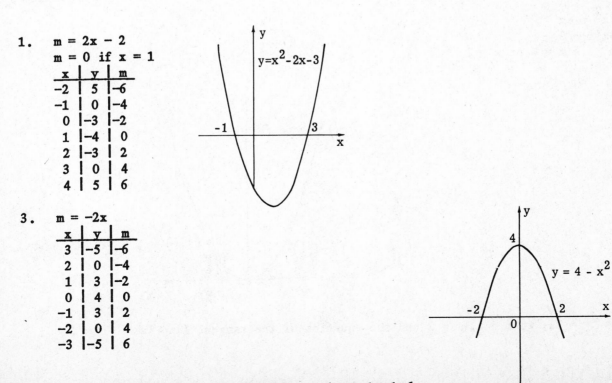

5. $y = x^2 - 4x + 4 \rightarrow \Delta y/\Delta x$

    $= [(x + \Delta x)^2 - 4(x + \Delta x) + 4 - x^2 + 4x - 4]/\Delta x \rightarrow$

    $\Delta y/\Delta x = [2x\Delta x + (\Delta x)^2 - 4\Delta x]/\Delta x = 2x - 4 + \Delta x \Rightarrow m = 2x - 4.$

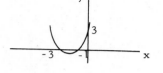

x	2	3	4	1	0
y	0	1	4	1	4
m	0	2	4	-2	-4

7. $m = 1 - 2x$, $m = 0$ if $x = 1/2$

9. $y = x^2 + 3x + 2$, $\Delta y = (x + \Delta x)^2 + 3(x + \Delta x) + 2 - x^2 - 3x - 2$

    $= 2x\Delta x + (\Delta x)^2 + 3\Delta x \rightarrow \Delta y/\Delta x = 2x + 3.$

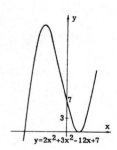

x	-3/2	0	-1	-2	-3
y	-1/4	2	0	0	2
m	0	3	1	-1	-3

11. $m = 6x^2 + 6x - 12$

    $= 6(x + 2)(x - 1)$

    $m = 0$ if $x = -2$ or $x = 1$

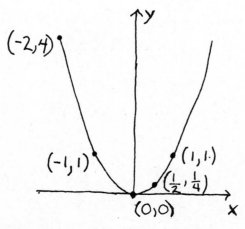

$y = 2x^2 + 3x^2 - 12x + 7$

13. $m = 3x^2 - 12$; $m = 0$ if $x \pm 2$.

15. $m = 3x^2 - 6x$

    $= 3x(x - 2)$

    $m = 0$ if $x = 0$ or $x = 2$

17. a)

(-2,4)

(-1,1)   (1,1)

$\left(\tfrac{1}{2}, \tfrac{1}{4}\right)$

(0,0)

b) $m = \dfrac{dy}{dx} = 2x$; at $x = 1$, $m = 2$ and the equation of the tangent line is

$y - 1 = 2(x - 1)$ or $y = 2x - 1$.

c) Through $(-2,4)$: $m = \dfrac{4 - 1}{-2 - 1} = -1$

Through $(-1,1)$: $m = \dfrac{-1 - 1}{-1 - 1} = 0$

Through $(0,0)$: $m = \dfrac{0 - 1}{0 - 1} = 1$

Through $(\frac{1}{2},\frac{1}{4})$: $m = \dfrac{1 - 1/4}{1 - 1/2} = \dfrac{3}{2}$.

d) $m = \dfrac{1 - (1 - \Delta x)^2}{1 - (1 - \Delta x)} = \dfrac{2\Delta x - (\Delta x)^2}{\Delta x} = 2 - \Delta x$

$\lim\limits_{x \to 0} m = 2 = $ slope of tangent line to the parabola at $x = 1$.

19. a) At Q: $x - h$, $y = h^3 - 3h + 3$

Slope of PQ is $m = \dfrac{(h^3 - 3h + 3) - 3}{n - 0} = \dfrac{h^3 - 3h}{h} = h - 3$.

b) $\lim\limits_{h \to 0} m = -3$.

c) Tangent line: $y - 3 = -3(x - 0)$ or $y = -3x + 3$.

Article 1.7

1. $f'(x) = \lim\limits_{h \to 0} \dfrac{f(x + h) - f(x)}{h}$

$= \lim\limits_{h \to 0} \dfrac{(x + h)^2 - x^2}{h} = \lim\limits_{h \to 0} \dfrac{2hx + h^2}{h}$

$= \lim\limits_{h \to 0} 2x + h = 2x$.

The slope of the tangent to the curve $y = f(x)$ at $x = 3$ is $f'(3) = 6$. Hence, the tangent line at $(3,9)$ is $y - 9 = 6(x - 3)$ or $y = 6x - 9$.

3. $f'(x) = \lim\limits_{h \to 0} \dfrac{f(x + h) - f(x)}{h} = \lim\limits_{h \to 0} \dfrac{[2(x + h) + 3] - (2x + 3)}{h}$

$= \lim\limits_{h \to 0} \dfrac{2h}{h} = \lim\limits_{h \to 0} 2 = 2$. So, $f'(x) = 2$.

For any curve $y = f(x)$, the slope of the tangent at $(3,f(3))$ is $f'(3)$. So, for $y = 2x + 3$, $f(3) = 9$, $f'(3) = 2$, and the tangent line at $(3,9)$ is $y - 9 = 2(x - 3)$ or $y = 2x + 3$. We note that the curve $y = 2x + 3$ is itself a straight line. So, it is reasonable in this case that the curve is its own tangent.

5. $f'(x) = \lim\limits_{h \to 0} \dfrac{1/(x + h) - 1/x}{h} = \lim\limits_{h \to 0} \dfrac{x - (x + h)}{hx(x + h)}$

$$= \lim_{h\to 0} \frac{-1}{x(x+h)} = -\frac{1}{x^2}.$$

The slope of the tangent at $(3,f(3))$ is $f'(3) = -\frac{1}{9}$. Hence the tangent line at $(3,\frac{1}{3})$ is $y = \frac{1}{3} = -\frac{1}{9}(x-3)$ or $y = -\frac{1}{9}x + \frac{2}{3}$.

7. $f'(x) = \lim_{h\to 0} \frac{1/(2(x+h)+1) - 1/(2x+1)}{h}$

$$= \lim_{h\to 0} \frac{-2h}{h(2x+1)[2(x+h)+1]} = \frac{-2}{(2x+1)^2}$$

Slope of tangent is $-\frac{2}{49}$; tangent is $y = -\frac{2}{49}x + \frac{13}{49}$.

9. $f'(x) = \lim_{h\to 0} \frac{f(x+h) - f(x)}{h}$

$$= \lim_{h\to 0} \frac{2(x+h)^2 - (x+h) + 5] - (2x^2 - x + 5)}{h}$$

$$= \lim_{h\to 0} \frac{2xh + h^2 - h}{h} = \lim_{h\to 0} 2x + h - 1 = 2x - 1.$$

Slope of tangent at $(3,f(3))$ is $f'(3) = 5$. Equation of tangent line at $(3,20)$ is $y - 20 = 5(x-3)$ or $y = 5x + 5$.

11. $f'(x) = \lim_{h\to 0} \frac{(x+h)^4 - x^4}{h} = \lim_{h\to 0} \frac{4x^3h + 6x^2h^2 + 4xh^3}{h} = 4x^3.$

Slope of tangent is $f'(3) = 108$. Equation of tangent line at $(3,81)$ is

$y - 81 = 108(x-3)$ or $y = 108x - 243$.

13. $f'(x) = \lim_{h\to 0} \frac{[x+h - 1/(x+h)] - (x - 1/x)}{h}$

$$= \lim_{h\to 0} \frac{h + h/((x+h)x)}{h} = 1 + \frac{1}{x^2} = \frac{x^2+1}{x^2}$$

The slope of the tangent line at $(3,\frac{8}{3})$ is $f'(3) = \frac{10}{9}$. The equation of the tangent line at $(3,\frac{8}{3})$ is thus $y - \frac{8}{3} = \frac{10}{9}(x-3)$ or

$y = \frac{10}{9}x - \frac{2}{3}$.

15. As in Example 5, we are going to use the method of rationalizing the numerator:

$$\frac{d}{dx}(\sqrt{2x}) = \lim_{h\to 0} \frac{\sqrt{2(x+h)} - \sqrt{2x}}{h}$$

$$= \lim_{h\to 0} \frac{\sqrt{2(x+h)} - \sqrt{2x}}{h} \cdot \frac{\sqrt{2(x+h)} + \sqrt{2x}}{\sqrt{2(x+h)} + \sqrt{2x}}$$

$$= \lim_{h\to 0} \frac{2}{\sqrt{2(x+h)} + \sqrt{2x}} = \frac{2}{2\sqrt{2x}} = \frac{1}{\sqrt{2x}}$$

Then, the slope of the tangent line to the curve $y = \sqrt{2x}$ at the point $(3, \sqrt{6})$

is $f'(3) = \dfrac{1}{\sqrt{6}} = \dfrac{\sqrt{6}}{6}$. The tangent line is, therefore, $y - \sqrt{6} = \dfrac{\sqrt{6}}{6}(x - 3)$

$(x - 3)$ or $y = \dfrac{\sqrt{6}}{6}x + \dfrac{\sqrt{6}}{2}$.

17. $f'(x) = \lim\limits_{h \to 0} \dfrac{\sqrt{2(x + h) + 3} - \sqrt{2x + 3}}{h}$

$= \lim\limits_{h \to 0} \dfrac{\sqrt{2(x + h) + 3} - \sqrt{2x + 3}}{h} \cdot \dfrac{\sqrt{2(x + h) + 3} + \sqrt{2x + 3}}{\sqrt{2(x + h) + 3} + \sqrt{2x + 3}}$

$= \lim\limits_{h \to 0} \dfrac{2}{\sqrt{2(x + h) + 3} + \sqrt{2x + 3}} = \dfrac{1}{\sqrt{2x + 3}}$

Slope of tangent line through $(3,3)$ is $\dfrac{1}{3}$; the equation of the tangent line

is $y - 3 = \dfrac{1}{3}(x - 3)$ or $y = \dfrac{1}{3}x + 2$.

19. $f'(x) = \lim\limits_{h \to 0} \dfrac{1/\sqrt{2(x + h) + 3} - 1/\sqrt{2x + 3}}{h}$

$= \lim\limits_{h \to 0} \dfrac{\sqrt{2x + 3} - \sqrt{2(x + h) + 3}}{h\sqrt{2x + 3}\sqrt{2(x + h) + 3}} \cdot \dfrac{\sqrt{2x + 3} + \sqrt{2(x + h) + 3}}{\sqrt{2x + 3} + \sqrt{2(x + h) + 3}}$

$= \lim\limits_{h \to 0} \dfrac{-2}{\sqrt{2x + 3}\,\sqrt{2(x + h) + 3}\,[\sqrt{2x + 3} + \sqrt{2(x + h) + 3}]}$

$= -(2x + 3)^{-3/2}$.

The slope of the tangent line through $(3,\dfrac{1}{3})$ is $-\dfrac{1}{27}$. The equation of

the tangent line is $y - \dfrac{1}{3} = -\dfrac{1}{27}(x - 3)$ or $y = -\dfrac{1}{27}x + \dfrac{4}{9}$.

21. The two graphs are the same. If $x > 0$, $|x| = x$, and

$y = \dfrac{|x|}{x} = \dfrac{x}{x} = 1$; if $x < 0$, $|x| = -x$, and

$y = \dfrac{|x|}{x} = \dfrac{-x}{x} = -1$.

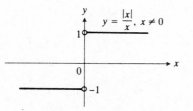

23. The derivative is greatest at $t = 20$ days, at which time the rabbit population
is 1,700. The derivative reaches its lowest value at $t = 65$ days, at which
time the rabbit population is 1,300.

25. a.

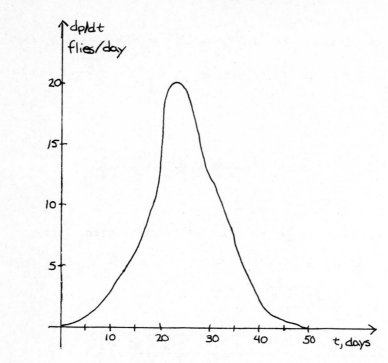

b. The fruit fly population seems to grow
fastest for 20 $\leq$ t $\leq$ 30 and slowest for 0
$\leq$ t $\leq$ 5, 45 $\leq$ t $\leq$ 50.

27. If x is between −1 and 0, f'(x) = −2. So,
y = −2x + b. Since (−1,1) is on the graph of
f, b = −1, and y = −2x − 1. This segment ends
at the point (0,−1). If x is between 0 and 1,
f'(x) = 0. So, here y = b = −1 (since (0,−1)
is on the graph of f). This line segment ends
at the point (1,−1). If x is between 1 and 4,
f'(x) = 1 and y = x − 2. This segment ends at
(4,2). Finally, if x is between 4 and 6,
f'(x) = −1, so y = −x + 6.

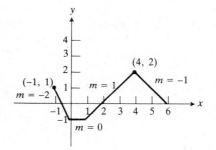

Article 1.8

1. $f'(t) = \lim_{t \to 0} \dfrac{[a(t + \Delta t)^2 + b(t + \Delta t) + c] - (at^2 + bt + c)}{\Delta t}$

$= \lim_{t \to 0} \dfrac{2at\Delta t + a(\Delta t)^2 + b\Delta t}{\Delta t} = 2at + b.$

$$S = S(2) - S(0), \quad t = 2, \quad V_{av} = \frac{t}{t}.$$

	S	V_{av}	V(t)	V(2)
2.	18m	9m/s	4t + 5	13m/s
3.	$2(V_0 + g)$	$V_0 + g$	$V_0 + gt$	$V_0 + 2g$
4.	8m	4m/s	4/ms	4m/s
5.	−2m	−1m/s	2t − 3	1m/s
6.	−8m	−4m/s	−2t − 2	−6m/s
7.	40m	20m/s	8t + 12	28m/s
8.	−4m	−2m/s	2t − 4	0m/s
9.	−8m	−4m/s	−4t	−8m/s
10.	64m	32m/s	64 − 32t	0m/s

11. a) Equation 3.

    b) It took $\frac{4}{7}$ seconds, and $V_{av} = 280$cm/sec.

    c) 17 flashes, $\frac{4}{7}$ seconds: $\frac{4/7}{17} \approx 0.034$ seconds/flash

13. The given data lie on the parabola $S = 74 - 16(t - 2)^2$. So:

    a) V(1)   = 32ft/sec

    b) V(2.5) = −16ft/sec

    c) V(2)   = 0ft/sec.

15. The engine stopped at that moment when the velocity begins to decrease, i.e., 2 seconds after launch. V(2) = 190ft/sec.

17. The rocket reaches its highest point when its velocity changes from positive to negative, which is at 8 seconds. At t = 8 seconds, the rocket's velocity is zero.

19. The rocket fell 2.8 seconds, from t = 8 seconds to t = 10.8 seconds (time when V < 0).

21. $\frac{db}{dt} = 10^4 - 2 \times 10^3 t = 10,000 = 2,000t.$

    a) b'(0)   = 10,000

b) $b'(5) = 0$

c) $b'(10) = -10,000$.

23. a) Average rate of change of area is

$$\frac{\pi(r + \Delta r)^2 - \pi r^2}{\Delta r} = 2\pi r + \pi(\Delta r).$$

b) Let $r \to 0$ in (a): $\frac{dA}{dr} = 2\pi r$.

25. Since $r = h$, $V = \frac{1}{3}\pi h^3$. Then

$$\frac{dV}{dh} = \lim_{a \to h} \frac{V(a) - V(h)}{a - h}$$

$$= \lim_{a \to h} \frac{(1/3)\pi(a^3 - h^3)}{a - h}$$

$$= \lim_{a \to h} \frac{1}{3}\pi(a^2 + ah + h^2) = \pi^2.$$

Article 1.9

1. $\lim_{x \to 2} 2x = 4 = 2 \cdot 2$

3. $\lim_{x \to 4} 4 = 4$

5. $\lim_{x \to 1} (3x - 1) = 2 = 3 \cdot 1 - 2$

7. $\lim_{x \to 5} x^2 = 25 = 5^2$

9. $\lim_{x \to 0} (x^2 - 2x + 1) = 1$

11. $\lim_{x \to 0} (2x + x) = 2x + 0 = 2x$

13. For x close to 1 (i.e., $x > 0$), $|x - 1| = x - 1$. Hence

$$\lim_{x \to 1} |x - 1| = 0.$$

15. If x is close to zero and positive, $[x] = 0$ and $x[x] = 0$. If x is close to

zero and negative, $[x] = -1$, $x[x] = -x$, which stays close to 0. Thus

$$\lim_{x \to 0} x[x] = 0.$$

17. $\lim_{x \to 2} 3x(2x - 1) = 18$

19. $\lim_{x \to 2} 3x^2(2x - 1) = 36$

21. $\lim_{x \to -1} (x + 3) = 2$

23. $\lim\limits_{x \to -2} (x + 3)^{171} = 1^{171} = 1$

25. $\lim\limits_{x \to -4} (x + 3)^{1984} = (-1)^{1984} = 1$

27. a) $\lim\limits_{x \to c} f(x)g(x) = [\lim\limits_{x \to c} f(x)][\lim\limits_{x \to c} g(x)] = (5)(-2) = -10.$

   b) $\lim\limits_{x \to c} 2f(x)g(x) = 2[\lim\limits_{x \to c} f(x)g(x)] = -20$, by part (a)

29. a) $\lim\limits_{x \to b} [f(x) + g(x)] = [\lim\limits_{x \to b} f(x)] + [\lim\limits_{x \to b} g(x)] = 4.$

   b) $\lim\limits_{x \to b} [f(x) \cdot g(x)] = [\lim\limits_{x \to b} f(x)] \cdot [\lim\limits_{x \to b} g(x)] = -21.$

   c) $\lim\limits_{x \to b} 4g(x) = 4 \cdot \lim\limits_{x \to b} g(x) = -12.$

   d) $\lim\limits_{x \to b} [f(x)/g(x)] = [\lim\limits_{x \to b} f(x)]/[\lim\limits_{x \to b} g(x)] = -\dfrac{7}{3}.$

31. $\lim\limits_{t \to 2} \dfrac{t + 3}{t + 2} = \dfrac{5}{4}$

33. $\lim\limits_{x \to 5} \dfrac{x^2 - 25}{x + 5} = \dfrac{5^2 - 25}{5 + 5} = \dfrac{0}{10} = 0$

35. If $x \neq 5$, $\dfrac{x + 5}{x^2 - 25} = \dfrac{1}{x - 5}$. But $\lim\limits_{x \to 5} \dfrac{1}{x - 5}$ does not exist.

37. If $x \neq 0$, $\dfrac{5x^3 + 8x^2}{3x^4 - 16x^2} = \dfrac{5x + 8}{3x^2 - 16}$. So,

   $\lim\limits_{x \to 0} \dfrac{5x^3 + 8x^2}{3x^4 - 16x^2} = $

   $\lim\limits_{x \to 0} \dfrac{5x + 8}{3x^2 - 16} = -\dfrac{1}{2}.$

39. If $y \neq 2$, $\dfrac{y^2 - 5y + 6}{y - 2} = \dfrac{(y - 2)(y - 3)}{y - 2} = y - 3$, so

   $\lim\limits_{y \to 2} \dfrac{y^2 - 5y + 6}{y - 2} = \lim\limits_{y \to 2} y - 3 = -1.$

41. If $x \neq 4$, $\dfrac{x - 4}{x^2 - 5x + 4} = \dfrac{x - 4}{(x - 4)(x - 1)} = \dfrac{1}{x - 1}$,

   so $\lim\limits_{x \to 4} \dfrac{x - 4}{x^2 - 5x + 4} = \lim\limits_{x \to 4} \dfrac{1}{x - 1} = \dfrac{1}{3}.$

43. If $t \neq 1$, $\dfrac{t^3 - 3t + 2}{t^2 - 1} = \dfrac{(t - 1)(t - 2)}{(t - 1)(t + 1)} = \dfrac{t - 2}{t + 1}$, so

$$\lim_{t \to 1} \frac{t^2 - 3t + 2}{t^2 - 1} = \lim_{t \to 1} \frac{t - 2}{t + 1} = -\frac{1}{2}.$$

45. $\displaystyle \lim_{x \to -3} \frac{x^2 + 4x - 3}{x - 3} = \frac{(-3)^2 + 4(-3) - 3}{-3 - 3} = \frac{-6}{-6} = 1.$

47. If $x \neq -2$, $\dfrac{x^2 + x - 2}{x^2 - 4} = \dfrac{(x + 2)(x - 1)}{(x + 2)(x - 2)} = \dfrac{x - 1}{x - 2}$, so

$$\lim_{x \to -2} \frac{x^2 + x - 2}{x^2 - 4} = \lim_{x \to -2} \frac{x - 1}{x - 2} = \frac{3}{4}.$$

49. $\displaystyle \lim_{x \to 2} \frac{x^2 - 7x + 10}{x - 2} = \lim_{x \to 2} (x - 5) = -3.$

51. If $x \neq a$, $\dfrac{x^3 - a^3}{x^4 - a^4} = \dfrac{(x - a)(x^2 + ax + a^2)}{(x^2 + a^2)(x + a)(x - a)}$

$$= \frac{x^2 + ax + a^2}{(x^2 + a^2)(x + a)} \text{ so } \lim_{x \to a} \frac{x^3 - a^3}{x^4 - a^4}$$

$$= \lim_{x \to a} \frac{x^2 + ax + a^2}{(x^2 + a^2)(x + a)} = \frac{3}{a}.$$

53. Choose, for example, $f(x) = \begin{cases} 1/x & x \neq 0 \\ 0 & x = 0 \end{cases}$, $g(x) = \begin{cases} -1/x & x \neq 0 \\ 0 & x = 0 \end{cases}$

55. Consider, for example, $f(x) = x$, $g(x) = \begin{cases} 1 & x \geq 0 \\ -1 & x < 0 \end{cases}$, in which

case $f(x)/g(x) = |x|$.

57. $f(x) = |x|$, $f'(-1) = \displaystyle \lim_{h \to 0} \frac{|-1 + h| - |-1|}{h}$

59. a) $\lim_{x\to c} f(x)$ exists for all c except 0,

  1, 2; i.e., $0 < c < 1$ or $1 < c < 2$.

  b) Only the left-hand limit exists at c = 0

  because the domain of f does not extend to

  the left of 0.

  c) Only the right-hand limit exists at c = 2

  because the domain of f does not extend to

  the right of 2.

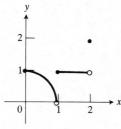

61. For $x > 0$, x close to 0, $[x] = 0 \Rightarrow \lim_{x\to 0}{}^+ [x] = 0$.

63. For x close to 0.5, $[x] = 0 \Rightarrow \lim_{x\to 0.5} [x] = 0$.

65. For $x > 0$, $|x| = x$, so $\lim_{x\to 0}{}^+ \frac{x}{|x|} = 1$.

67. For $x > 3$, $|x - 3| = x - 3$, so $\lim_{x\to 3}{}^+ \frac{x^2 - 9}{|x - 3|} = 6$.

69. For $x < 2$, $4x < 8$ and $|4x - 8| = 8 - 4x$, so

  $\lim_{x\to 2}{}^- \frac{x^2|4x - 8|}{x - 2} = \lim_{x\to 2}{}^- (-x^2) = -4$.

71. $f(x) = x/|x|$ approaches a limit as $x\to c$ for all $c \neq 0$:

  $f(x) = \begin{cases} 1 & x > 0 \\ -1 & x < 0 \end{cases}$

73. $\lim_{x\to 0}{}^+ \cos x = \cos 0 = 1$.

75. We do not immediately recognize $\lim_{h\to 0} \frac{\sin^2 h}{h}$, but we do know

  $\lim_{h\to 0} \frac{\sin h}{h} = 1$. Can we use this here?

  Yes: $\lim_{h\to 0} \frac{\sin^2 h}{h} = \lim_{h\to 0} [h \cdot \frac{\sin^2 h}{h^2}]$

  $= [\lim_{h\to 0} h] \cdot [\lim_{h\to 0} \frac{\sin h}{h}]^2 = 0 . 1^2 = 0$.

77. $\displaystyle\lim_{h\to 0}\frac{2\sin t\cos t}{t} = 2[\lim_{t\to 0}\cos t][\lim_{t\to 0}\frac{\sin t}{t}] = 2.$

79. $\displaystyle\lim_{\theta\to 0}\frac{\tan\theta}{\theta} = \lim_{\theta\to 0}\frac{\sin\theta}{\theta\cos\theta}$

$\displaystyle = [\lim_{\theta\to 0}\frac{\sin\theta}{\theta}][\lim_{\theta\to 0}\frac{1}{\cos\theta}] = 1.$

81. If $x > 0$, $|x| = x$, so $\displaystyle\lim_{x\to 0^+}\frac{\sin x}{|x|} = \lim_{x\to 0^+}\frac{\sin x}{x} = 1.$

83. $\displaystyle\lim_{x\to 0} x\cos x = 0\cos 0 = 0.$

85. $\displaystyle\lim_{x\to 0}\frac{\sin 5x}{\sin 3x} = \frac{5}{3}\left(\lim_{x\to 0}\frac{\sin 5x}{5x}\right)\left(\lim_{x\to 0}\frac{3x}{\sin 3x}\right) = \frac{5}{3}.$

87. $\displaystyle\lim_{x\to 0}\frac{\sin 2x}{2x^2 + x} = 2[\lim_{x\to 0}\frac{1}{2x + 1}][\lim_{x\to 0}\cdot\frac{\sin 2x}{2x}] = 2.$

89. $\displaystyle\lim_{x\to 0} 1 - \frac{x^2}{6} = 1$, so $\displaystyle\lim_{x\to 0}\frac{\sin x}{x} = 1$ by the Sandwich Theorem.

91. a) $L = -1$; we want $|-1 - (5 - 3t)| < \varepsilon$ if $0 < |t - 2| < \delta$.

   $|3t - 6| = 3|t - 2| < \varepsilon$, so choose $\delta = \dfrac{\varepsilon}{3}$.

   b) $L = 7$. Here, $|L - F(t)| = |7 - 7| = 0$, so any δ will suffice.

   c) Since $0 < |t - 2|$, $t \neq 2$. Hence, we may simplify:

   $F(t) = \dfrac{(t + 2)(t - 2)}{t - 2} = t + 2.$ So, $L = 4.$ $|4 - (t +$

   $2)| = |2 - t| = |t - 2| < \varepsilon$; choose $\delta = \varepsilon$.

   d) As in part (c), we may simplify: $F(t) = \dfrac{(t + 1)(t + 5)}{t + 5}$

   $= t + 1 \Rightarrow L = 6.$ $|L - F(t)| = |6 - (t + 1)| = |5 - t| = |t - 5| < \varepsilon$;

   choose $\delta = \varepsilon$.

   e) Again, $F = \dfrac{(t + 3)(3t - 1)}{2(t + 3)} = \dfrac{3t - 1}{2}$

   $\Rightarrow L = -5.$ $|-5 - \dfrac{3t - 1}{2}| = |\dfrac{9 - 3t}{2}| = \dfrac{3}{2}|t - 3| < \varepsilon$;

   choose $\delta = \dfrac{2\varepsilon}{3}$.

   f) $L = 2$: $|L - F(t)| = |2 - \dfrac{4}{t}| = \dfrac{2|t - 2|}{|t|}.$

   If $0 < |t - 2| < \delta$, then the largest possible value of the numerator is 2δ,

   while the least possible value of the denominator of $|L - F(t)|$ is $2 - \delta$

   (assuming $\delta < 2$). Thus, if $0 < |t - 2| < \delta$, $|L - F(t)|$ is bounded above by

   $\dfrac{2\delta}{2 - \delta}.$

   So, for $|L - F(t)| < \varepsilon$: $\dfrac{2\delta}{2 - \delta} < \varepsilon$ or $\delta < \dfrac{2\varepsilon}{2 + \varepsilon}.$

Hence, any value of δ less than both 2 and $\frac{2\varepsilon}{2+\varepsilon}$ will do.

g) If $t \neq 3$, $F(t) = -\frac{1}{t}$. Hence, $L = -\frac{1}{3}$.

$\left| -\frac{1}{3} - (-\frac{1}{t}) \right| = \left| \frac{1}{t} - \frac{1}{3} \right| = \left| \frac{t-3}{3t} \right|$. If $0 < |t - 3|$

$< \delta$, then (as above, in part (f)) $|L - F(t)|$ is bounded above by

$\frac{\delta}{3(3-\delta)}$. Then $\frac{\delta}{3(3-\delta)} < \varepsilon$ or $\delta < \frac{9\varepsilon}{1+3\varepsilon}$.

Hence, any value of δ less than both 3 and $\frac{9\varepsilon}{1+3\varepsilon}$ will suffice.

93. a) We want δ small enough that whenever $0 < |t - 3| < \delta$, then

$|t^2 + t - 12| < 1/10$. $|t^2 + t - 12| = |t + 4||t - 3| < \delta|t + 4|$. Now,

$-\delta < t - 3 < \delta$, $t \neq 3$ or $3 - \delta < t < 3 + \delta$, $t \neq 3$. So,

$6 - \delta < t + 3 < 6 + \delta$, $t \neq 3$. Let's take $\delta < 6$, so that $|t + 3| < 6 + \delta$

for all t under consideration. (This is all right to do because at worst we

shall merely end up taking too conservative a δ. But all we need is to find

some δ that works.) Now, since $\delta < 6$, $6 + \delta < 12$.

So, $|t^2 + t - 12| = |t + 4||t - 3| < 12\delta$. Therefore, we want

$12\delta \leq \frac{1}{10}$, or $\delta \leq \frac{1}{120}$. This δ is certainly less than 6, so we

can take $\delta = \frac{1}{120}$.

b) All the calculations from part (a) hold right up to the end. However, this

time we want $12\delta \leq \frac{1}{100}$ or $\delta \leq \frac{1}{1,200}$. So, we may take

$\delta = \frac{1}{1,200}$.

c) In general, we want $12\delta \leq \varepsilon$, or $\delta \leq \frac{\varepsilon}{12}$. Thus, any δ less than both 1

and $\frac{\varepsilon}{12}$ will work.

95. $f'(0) = \lim_{h \to 0} \frac{\sqrt{9 - h^2} - 3}{h} = \lim_{h \to 0} \frac{\sqrt{9 - h^2} - 3}{h} \cdot \frac{\sqrt{9 - h^2} + 3}{\sqrt{9 - h^2} + 3}$

$= \lim_{h \to 0} \frac{-h}{\sqrt{9 - h^2} + 3} = 0$.

Article 1.10

1. $\lim\limits_{x\to\infty} \dfrac{2x + 3}{5x + 7} = \lim\limits_{x\to\infty} \dfrac{2 + (3/x)}{5 + (7/x)} = \dfrac{2}{5}.$ Or,

$\lim\limits_{x\to\infty} \dfrac{2x + 3}{5x + 7} = \lim\limits_{h\to 0^+} \dfrac{(2/h) + 3}{(5/h) + 7} = \lim\limits_{h\to 0^+} \dfrac{2 + 3h}{5 + 7h} = \dfrac{2}{5}.$

3. $\lim\limits_{x\to\infty} \dfrac{x + 1}{x^2 + 3} = \lim\limits_{x\to\infty} \dfrac{(1/x) + (1/x^2)}{1 + 3/x^2} = 0.$ Or:

$\lim\limits_{x\to 0} \dfrac{x + 1}{x^2 + 3} = \lim\limits_{h\to 0^+} \dfrac{(1/h) + 1}{(1/h)^2 + 3} = \lim\limits_{h\to 0^+} \dfrac{h + h^2}{1 + 3h^2} = 0.$

5. $\lim\limits_{y\to\infty} \dfrac{3y + 7}{y^2 - 2} = \lim\limits_{y\to\infty} \dfrac{(3/y) + (7/y^2)}{1 - (2/y^2)} = 0.$

7. $\lim\limits_{t\to\infty} \dfrac{t^2 - 2t + 3}{2t^2 + 5t - 3} = \lim\limits_{t\to\infty} \dfrac{1 - (2/t) + (3/t^2)}{2 + (5/t) - (3/t^2)} = \dfrac{1}{2}.$

9. $\lim\limits_{x\to\infty} \dfrac{x}{x - 1} = \lim\limits_{x\to\infty} \dfrac{1}{1 - (1/x)} = 1.$ Or:

$\lim\limits_{x\to\infty} \dfrac{x}{x - 1} = \lim\limits_{h\to 0^+} \dfrac{1/h}{1/h - 1} = \lim\limits_{h\to 0^+} \dfrac{1}{1 - h} = 1.$

11. As $x\to -\infty$, $|x|$ grows without bound. Hence $\lim\limits_{x\to -\infty} |x| = \infty.$

13. $\lim\limits_{a\to\infty} \dfrac{|a|}{|a| + 1} = \lim\limits_{a\to\infty} \dfrac{1}{1 + (1/|a|)} = 1.$

15. $\lim\limits_{x\to\infty} \dfrac{3x^3 + 5x^2 - 7}{10x^3 - 11x + 5} = \lim\limits_{x\to\infty} \dfrac{3 + (5/x) - (7/x^3)}{10 - (11/x^2) + (5/x^3)} = \dfrac{3}{10}.$

17. $\lim\limits_{s\to\infty} \dfrac{s}{s + 1} = \lim\limits_{s\to\infty} \dfrac{1}{1 + 1/s} = 1; \lim\limits_{s\to\infty} \dfrac{s^2}{5 + s^2} = 1; \lim\limits_{s\to\infty} (\dfrac{s}{s + 1})(\dfrac{s^2}{5 + s^2}) = 1.$

19. $\lim\limits_{r\to -\infty} \dfrac{8r^2 + 7r}{4r^2} = \lim\limits_{r\to -\infty} \dfrac{8 + (7/4)}{4} = 2.$

21. $\lim\limits_{y\to\infty} \dfrac{y^4}{y^4 - 7y^3 + 7y^2 + 9} = \lim\limits_{y\to\infty} \dfrac{1}{1 - 7/y + 7/y^2 + 9/y^4} = 1.$

Or: $\lim\limits_{y\to\infty} \dfrac{y^4}{y^4 - 7y^3 + 7y^2 + 9} = \lim\limits_{h\to 0} \dfrac{1/h^4}{1/h^4 - 7/h^3 + 7/h^2 + 9}$

$$= \lim_{h \to 0} \frac{1}{1 - 7h + 7h^2 + 9h^4} = 1$$

23. $\lim_{x \to \infty} \dfrac{x - 3}{x^2 - 5x + 4} = \lim_{h \to 0}^+ \dfrac{1/h - 3}{1/h^2 - 5/h + 4} = \lim_{h \to 0}^+ \dfrac{h - 3h^2}{1 - 5h + 4h^2} = 0.$

25. $\lim_{x \to \infty} \dfrac{-2x^3 - 2x + 3}{3x^3 + 3x^2 - 5x} = \lim_{x \to \infty} \dfrac{-2 - 2/x^2 + 3/x^3}{3 + 3/x - 5/x^2} = -\dfrac{2}{3}.$

27. $\lim_{x \to \infty} \dfrac{x + \sin x}{x + \cos x} = \lim_{x \to \infty} \dfrac{1 + \dfrac{\sin x}{x}}{1 + \dfrac{\cos x}{x}}.$

Now, because $\dfrac{-1}{x} \le \dfrac{\sin x}{x} \le \dfrac{1}{x}$ (if $x > 0$), and because $\lim_{x \to \infty} -\dfrac{1}{x} = \lim_{x \to \infty} \dfrac{1}{x} = 0$, the Sandwich Theorem says $\lim_{x \to \infty} \dfrac{\sin x}{x} = 0$. Likewise, $\lim_{x \to \infty} \dfrac{\cos x}{x} = 0$. So, $\lim_{x \to \infty} \dfrac{x + \sin x}{x + \cos x} = 1.$

29. $\lim_{x \to \infty} \dfrac{1}{x^4} + \dfrac{1}{x} = 0 + 0 = 0.$

31. As $x \to \infty$, $\dfrac{1}{x} \to 0$, so $\cos\dfrac{1}{x} + 1 \to \cos 0 + 1 = 2.$

33. If $x > 0$, $\dfrac{1}{3x} > 0$. Hence, $\lim_{x \to 0}^+ \dfrac{1}{3x} = \infty.$

35. If $x > 0$, $\dfrac{5}{2x} > 0$. So, $\lim_{x \to 0}^+ \dfrac{5}{2x} = \infty.$

37. Since $\dfrac{t^2 - 4}{t - 2} = t + 2$ if $t = /2$, $\lim_{t \to 2} \dfrac{t^2 - 4}{t - 2} = 4.$

39. For $x > 1$, $\dfrac{x}{x - 1} > 0$. So, $\lim_{x \to 1}^+ \dfrac{x}{x - 1} = \infty.$

41. If $x < -1$, $\dfrac{1}{x + 1} < 0$. So, $\lim_{x \to -1}^- \dfrac{1}{x + 1} = -\infty.$

43. For $x > -2$, $\dfrac{1}{x + 2} > 0$. So, $\lim_{x \to -2}^+ \dfrac{1}{x + 2} = \infty.$

45. $\lim_{x \to 3} \dfrac{x - 3}{x^2} = \dfrac{3 - 3}{3^2} = \dfrac{0}{9} = 0.$

47. For $x < 2$, $\dfrac{x^2 + 5}{x - 2} < 0$. So, $\lim_{x \to 2}^- \dfrac{x^2 + 5}{x - 2} = -\infty.$

49. For $x \ne -5$, $\dfrac{x^2 + 3x - 10}{x + 5} = \dfrac{(x + 5)(x - 2)}{x + 5} = x - 2$;

so, $\lim_{x \to -5} \dfrac{x^2 + 3x - 10}{x + 5} = -7.$

51. a) $\displaystyle\lim_{x\to 0} \frac{x-1}{2x^2 - 7x + 5} = \frac{0-1}{2 \times 0^2 - 7 \times 0 + 5} = -\frac{1}{5}$.

b) $\displaystyle\lim_{x\to 0} \frac{x-1}{2x^2 - 7x + 5} = \lim_{x\to\infty} \frac{(1/x) - (1/x^2)}{2 - (7/x) + (5/x^2)} = 0$.

c) For $x \neq 1$, $\displaystyle\frac{x-1}{2x^2 - 7x + 5} = \frac{x-1}{(2x-5)(x-1)}$
$= \displaystyle\frac{1}{2x-5}$. So, $\displaystyle\lim_{x\to 1} \frac{x-1}{2x^2 - 7x + 5} = -\frac{1}{3}$.

53.

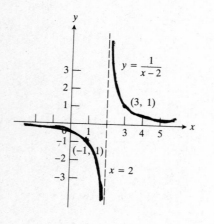

55.

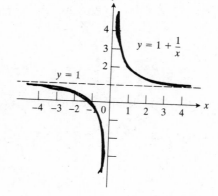

57.

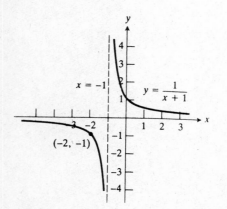

59.

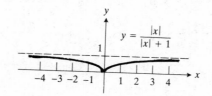

61. Let $f(x) = \frac{1}{x}$, $g(x) = \frac{1}{x^2}$, and $h(x) = \frac{1}{\sqrt{x}}$. Then, $f(x)/g(x) = x$,

$g(x)/h(x) = \frac{1}{x^{3/2}}$, and $h(x)/f(x) = \sqrt{x}$.

a) As $x \to 0$, $f(x)/g(x) \to 0$, $h(x)/f(x) \to 0$, and $g(x)/h(x) \to \infty$. Therefore, $f(x)$

grows more rapidly than $h(x)$, and $g(x)$ grows more rapidly than $f(x)$. So, as

$x \to 0$, $g(x) \gg f(x) \gg h(x)$.

b) As $x \to \infty$, $f(x)/g(x) \to \infty$ and $h(x)/f(x) \to \infty$. Therefore, $f(x)$ grows small faster

than $h(x)$, and $g(x)$ grows small faster than $f(x)$. Hence, as $x \to \infty$,

$g(x) \ll f(x) \ll h(x)$.

Article 1.11

1. a) Yes; $f(-1) = 0$ b) Yes; $\lim\limits_{x \to -1^+} f(x) = 0$.

    c) Yes   d) Yes

3. a) No, f is not defined for $x = 2$

    b) No, f cannot be continuous at $x = 2$ because $f(2)$ does not exist.

5. a) $\lim\limits_{x \to 2} f(x) = 0$

    b) If $f(2)$ is set equal to 0, then $f(x)$ will be continuous at $x = 2$.

7. f is continuous for all x except $x = 0, 1$.

9. f is continuous at all points in its domain except $x = 1$ and $x = 2$; i.e., $f(x)$

is continuous for $-1 \le x < 1$, $1 < x < 2$, and $2 < x \le 3$.

11. f is continuous for all x except $x = -1$, $x = 0$, and $x = 1$.

13. Since y is not defined at $x = -2$, y is not continuous at $x = -2$. y is

continuous at all other values of x.

15. Since y is not defined at $x = 1, 3$, y is not continuous there. y is continuous

at all other values of x.

17. Since y is undefined at $x = 5, -2$, y is not continuous there. y is continuous

for all other values of x.

19. y is continuous for all x.

21. If x > 0, y(x) = 1; if x < 0, y(x) = -1. y is undefined at x = 0, so y is

discontinuous at x = 0.

23. $\lim_{x \to 3} \frac{x^2 - 9}{x - 3} = \lim_{x \to 3} x + 3 = 6$. Hence, set f(3) = 6.

25. a)

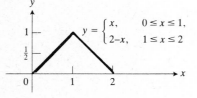

b) yes; the right- and left-hand limits at x = 1 agree.

c) No; from the graph it is clear that secants through (1,1) do not approach

any limiting position.

27. $\lim_{x \to 3^-} f(x) = 8$; $\lim_{x \to 3^+} f(x) = 6a$. To make these two equal, take

a = $\frac{4}{3}$. Then f is in fact continuous at x = 3.

29. $\lim_{x \to 4^+} \frac{x^2 - 16}{|x - 4|} = \lim_{x \to 4^+} \frac{x^2 - 16}{x - 4} = \lim_{x \to 4^+} (x + 4) = 8$.

$\lim_{x \to 4^-} \frac{x^2 - 16}{x - 4} = \lim_{x \to 4^-} \frac{x^2 - 16}{4 - x} = -8$.

The discontinuity of f at x = 4 is, therefore, not removable.

31. y is undefined if 0 ≤ x < 1. Furthermore, y is discontinuous at x = n, where n

is any integer, because $\lim_{x \to n^-} y \neq \lim_{x \to n^+} y$.

33. $\lim_{x \to 0} 1 - \frac{\sin x}{x} = 1 - 1 = 0$, so $\lim_{x \to 0} \cos(1 - \frac{\sin x}{x}) = \cos 0 = 1$.

35. From a graph, we see that the maximum value is 1 and the minimum value is 0.

37. The function has a minimum value of 0 at $x = 0$. However, the function does not take on a maximum value for $-1 < x < 1$. Again, Theorem 9 does not apply because the interval $-1 < x < 1$ is not closed.

39. The intermediate value theorem can be used here. $f(x)$ is continuous on $0 \le x \le 1$; $f(0) < 0 < f(1)$. Hence, for some c with $0 < c < 1$, $f(c) = 0$.

41. Suppose $p(x) = a_n x^n + \ldots + a_0$ (n odd), and suppose $a_n > 0$. (The argument works the same way if $a_n < 0$.) Since n is odd, $\lim_{x \to \infty} p(x) = \infty$, $\lim_{x \to -\infty} p(x) = -\infty$. So, for some values a, b we can conclude $p(a) < 0$, $p(b) > 0$. Therefore, $p(x)$ has a real root c, $a < c < b$, by the intermediate value theorem.

43. a)

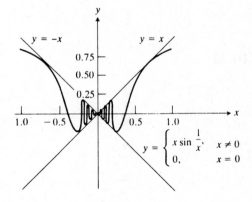

b) We must show that $\lim_{x \to 0} f(x) = f(0) = 0$. Since $0 \le |\sin\frac{1}{x}| \le 1$ for any x, $0 \le |x \sin\frac{1}{x}| \le |x|$. So, if $|x - 0| = |x| < \varepsilon$, $|x \sin\frac{1}{x} - 0|$ $= |x \sin\frac{1}{x}| \le |x| < \varepsilon$. This says $\lim_{x \to 0} x \sin\frac{1}{x} = 0$. Hence, f is continuous at $x = 0$.

Miscellaneous Problems

1. b) $M_{AB} = -\frac{3}{2}$, $M_{BC} = \frac{2}{3}$, $M_{CD} = -\frac{3}{2}$, $M_{DA} = \frac{2}{3}$, $M_{CE} = 0$, M_{BD} is

    undefined.

    c) ABCD is a parallelogram, since $M_{AB} = M_{CD}$ and $M_{BC} = M_{DA}$, AB $||$ CD and

    BC $||$ DA.

    d) No; the only pairs of lines with the same slopes are (AB,CD) and (BC,DA).

    Clearly, both pairs are composed of distinct lines, so no three points are

    collinear.

    e) Yes; the equation of line CD is $y = -\frac{3}{2}x$, and (0,0) lies on this line.

    f) AB: $y = -\frac{3}{2}x + 13$

    CD: $y = -\frac{3}{2}x$

    AD: $y = \frac{2}{3}x - \frac{13}{3}$

    CE: $y = 6$

    BD: $x = 2$

    g) AB: x-intercept is $(\frac{26}{3},0)$, y-intercept is (0,13)

    CD: x-intercept, y-intercept are (0,0)

    AD: x-intercept is $(\frac{13}{2},0)$, y-intercept is $(0,-\frac{13}{3})$

    CE: No x-intercept, y-intercept is (0,6)

    BD: x-intercept is (2,0), no y-intercept.

3. a) $M_{AB} = \frac{7}{2}$, $M_{BC} = -1$, $M_{CA} = \frac{1}{8}$. Since two

    slopes are negative reciprocals of each other, ABC

    is not a right triangle.

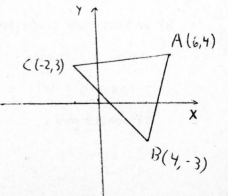

b) $AB = \sqrt{2^2 + 7^2} = \sqrt{53}$, $BC = \sqrt{6^2 + 6^2} = 6\sqrt{2}$, $CA = \sqrt{8^2 + 1^2} = \sqrt{65}$. Since

no two lengths are equal, ABC is not isosceles.

c) From a <u>carefully</u> drawn graph, we can see that (0,0) is outside the triangle.

d) $M_{C'B} = -\dfrac{1}{M_{BA}}$: $\dfrac{y + 3}{-2 - 4} = -\dfrac{2}{7} \Rightarrow y + 3 = \dfrac{12}{7} \Rightarrow y = -\dfrac{9}{7}$.

5. Let (a,b) be the coordinates of the midpoint. From the

diagram, we see that (a,b) are determined by the relations

$a - x_1, = x_2 - a$, $b - y_1, = y_2 - b$. Hence,

$a = \dfrac{x_1 + x_2}{2}$, $b = \dfrac{y_1 + y_2}{2}$.

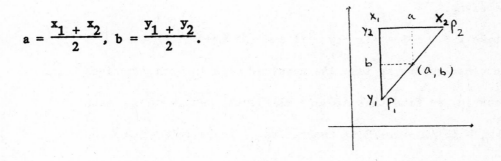

7. a) Rearranging, $y = -\dfrac{A}{B}x + \dfrac{C}{B}$. So, $m = -\dfrac{A}{B}$.

b) From (a), the y-intercept is $(0, \dfrac{C}{B})$

c) Set $y = 0$: $Ax = C$, $x = \dfrac{C}{A} \Rightarrow$ x-intercepts is $(\dfrac{C}{A}, 0)$

d) Slope of perpendicular is $\dfrac{-1}{-A/B} = \dfrac{B}{A} \Rightarrow$ required line is

$y = \dfrac{B}{A}x$.

9. Let P(x,y) denote any point in the xy-plane.

Then the distance from P to the line L_1 is

$d_1 = \dfrac{|x + y - 1|}{\sqrt{2}}$

Similarly, $d_2 = \dfrac{|x - y + 1|}{\sqrt{2}}$ and

$d_3 = \dfrac{|x - 3y - 1|}{\sqrt{10}}$

are the distances from P to L_2 and L_3,

CHAPTER 1 - Miscellaneous Problems

respectively.

Now, since the radius of a circle is perpendicular to a tangent line at the point of tangency, the conditions for the point P to be the center of a circle tangent to L_1, L_2, and L_3 is $d_1 = d_2$ and $d_2 = d_3$. Thus, we wish to solve simultaneously

$$\frac{x + y - 1}{\sqrt{2}} = \pm \frac{x - y + 1}{\sqrt{2}}$$

$$\frac{x - y + 1}{\sqrt{2}} = \pm \frac{x - 3y - 1}{\sqrt{10}}$$

There are four possible cases, depending on how we choose the signs.

Case 1: Let us take $x + y - 1 = -(x - y + 1)$ and $\sqrt{5}(x - y + 1) = -(x - 3y - 1)$ [in other words, we take the negative sign in both equations]. Solving simultaneously, we find that $A(0, \sqrt{5} - 2)$ is the center of one such circle, and that $r_A = (3\sqrt{2} - 10)/2$ is the radius of the circle (being the distance from A to L_1. The remaining cases are handled in the same manner. Summary of results:

$A(0, \sqrt{5} - 2)$	$r_A = (3\sqrt{2} - 10)/2$
$B(0, -\sqrt{5} - 2)$	$r_B = (3\sqrt{2} + 10)/2$
$C(-\sqrt{5} - 1, 1)$	$r_C = (\sqrt{10} + \sqrt{2})/2$
$D(\sqrt{5} - 1, 1)$	$r_D = (\sqrt{10} - \sqrt{2})/2.$

So, there are four such circles.

11. If L_1 is not parallel to L_2, then L_3:

$$(a_1 + ka_2)x + (b_1 + kb_2)y + (c_1 + kc_2) = 0$$

is a line passing through the point of intersection of L_1 and L_2, since all ordered pairs satisfying L_1 and L_2 simultaneously must also satisfy L_3. If L_1 and L_2 are parallel, then L_3 is a straight line parallel to L_1 and L_2.

CHAPTER 1 - Miscellaneous Problems

13. Since the slope of the given line is 5, the slope of the perpendicular must be $-\frac{1}{5}x + b$, where $(0,b)$ is the y-intercept. Now, the x-intercept of the line is $(5b,0)$. Then, since the area of the triangle is to be 5, $\frac{1}{2}(b)(5b) = 1$, or $b^2 = 2$. So, $b = \pm\sqrt{2}$. (Both answers work.)

15. $A = \pi r^2$, $C = 2\pi r$. $A = \pi r^2 = \frac{4\pi^2 r^2}{4\pi} = \frac{(2\pi r)^2}{4\pi} = \frac{C^2}{4\pi}$. Alternatively,

$$C = 2\pi r \Rightarrow r = \frac{C}{2\pi} \Rightarrow A = \pi(\frac{C}{2\pi})^2 = \frac{\pi C^2}{4\pi^2} = \frac{C^2}{4\pi}.$$

17. $|x + 1| < 4 \Rightarrow -4 < x + 1 < 4 \Rightarrow -5 < x < 3$

19. If $x \geq 2$, $2 - x \leq 0$, $|2 - x| = x - 2$, $y = 2x - 2$, $x = \frac{y + 2}{2}$, $y \geq 2$. If $x \leq 2$, $2 - x > 0$, $|2 - x| = 2 - x$, $y = 2$. So, y determines a single value of x if $y > 2$, viz. $x = \frac{y + 2}{2}$.

21. a) $f(-2) = -3$, $f(-1) = -4$, $f(x_1) = x_1^2 + 2x_1 - 3$, $f(x_1 + x) =$
$(x_1 + x)^2 + 2(x_1 + x) - 3 = (x)^2 + (x)(2x_1 + 2) + (x_1^2 + 2x_1 - 3)$.

b) $f(\frac{1}{x}) = \frac{1}{x} - \frac{1}{1/x} = \frac{1}{x} - x$; $-f(x) = -x + \frac{1}{x}$;
$f(-x) = (-x) - \frac{1}{(-x)} = \frac{1}{x} - x$. Clearly these are equal.

23. Range: $-2 \leq y \leq 6$.

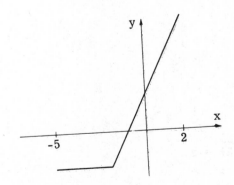

CHAPTER 1 - Miscellaneous Problems

25. a)

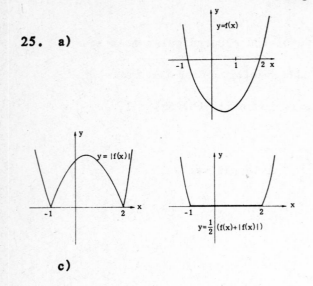

b) All three are the same.

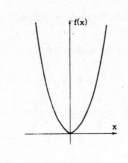

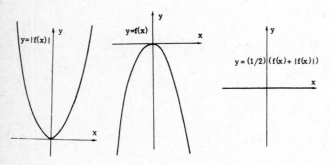

c)

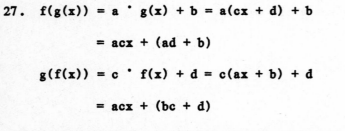

d)

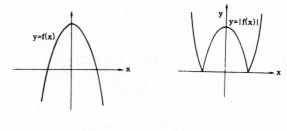

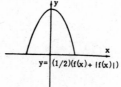

27. $f(g(x)) = a \cdot g(x) + b = a(cx + d) + b$

$$= acx + (ad + b)$$

$g(f(x)) = c \cdot f(x) + d = c(ax + b) + d$

$$= acx + (bc + d)$$

So, if $f(g(x)) = g(f(x))$, $ad + b = bc + d$.

29. a) $f(\frac{1}{x}) = \dfrac{1/x}{(1/x) - 1} = \dfrac{1}{1 - x}$

  b) $f(-x) = \dfrac{-x}{-x - 1} = \dfrac{x}{x + 1}$

  c) $f(f(x)) = \dfrac{f(x)}{f(x) - 1} = \dfrac{x/(x - 1)}{x/(x - 1) - 1} = \dfrac{1}{x - (x - 1)} = x.$

d) $f(\frac{1}{f(x)}) = \frac{1/f(x)}{1/f(x) - 1} = \frac{1}{1 - f(x)} = \frac{1}{1 - x/(x - 1)}$

$= \frac{x - 1}{x - 1 - x} = 1 - x.$

31. a) $f'(x) = \lim_{h \to 0} \frac{(x + h)^2 - 3(x + h) - 4 - (x^2 - 3x - 4)}{h}$

$= \lim_{h \to 0} 2x - 3 + h^2 = 2x - 3.$

b) $f'(x) = \lim_{h \to 0} \frac{1/(3(x + h)) + 2(x + h) - (1/3x + 2x)}{h}$

$= 2 + \lim_{h \to 0} \frac{1/(3x + 3h) - 1/3x}{h}$

$= 2 - \lim_{h \to 0} \frac{3}{3x(3x + 3h)} = 2 - \frac{1}{3x^2}$

c) $f'(t) = \lim_{h \to 0} \frac{\sqrt{t + h - 4} - \sqrt{t - 4}}{h}$

$= \lim_{h \to 0} \frac{\sqrt{t + h - 4} - \sqrt{t - 4}}{h} \cdot \frac{\sqrt{t + h - 4} + \sqrt{t - 4}}{\sqrt{t + h - 4} + \sqrt{t - 4}}$

$= \lim_{h \to 0} \frac{1}{\sqrt{t + h - 4} + \sqrt{t - 4}} = \frac{1}{2\sqrt{t - 4}}$

33. a) $f(0) = 0$, $f(-1) = 1$, $f(\frac{1}{x}) = \frac{2/x}{1/x - 1} = \frac{2}{1 - x}.$

b) $\Delta f(x) = f(x + \Delta x) - f(x) = \frac{2(x + \Delta x)}{x + \Delta x - 1} - \frac{2x}{x - 1}$

$= \frac{-2\Delta x}{(x - 1)(x + \Delta x - 1)}.$

So, $\frac{\Delta f(x)}{\Delta x} = \frac{-2}{(x - 1)(x + \Delta x - 1)}$

c) $f'(x) = \lim_{\Delta x \to 0} \frac{\Delta f(x)}{\Delta x} = \frac{-2}{(x - 1)^2}.$

35. From Problem 34, $V(t) = \frac{ds}{dt} = 180 - 32t$; the velocity vanishes at $t = \frac{45}{8}$.

37. a) $\Delta P = \frac{1}{V + \Delta V} - \frac{1}{V} = \frac{-\Delta V}{V(V + \Delta V)}$. So $\frac{\Delta P}{\Delta V} = -\frac{1}{V(V + V)}$

b) $P'(V) = \lim_{\Delta V \to 0} \frac{\Delta P}{\Delta V} = -\frac{1}{V^2}$. At $V = 2$, $P'(2) = -\frac{1}{4}$.

39. $\lim_{x \to \infty} \frac{x}{2x + 5} = \lim_{x \to \infty} \frac{1}{2 + 5/x} = \frac{1}{2}$; since $0 \le |\frac{\sin x}{2x + 5}| \le$

$|\frac{1}{2x + 5}|$ and $|\frac{1}{2x + 5}| \to 0$ as $x \to \infty$, $\lim_{x \to \infty} \frac{\sin x}{2x + 5} = 0$. Therefore,

$\lim_{x \to \infty} \frac{x + \sin x}{2x + 5} = \frac{1}{2}.$

41. $\lim_{x \to 1} \frac{x^2 - 4}{x^3 - 8} = \frac{1 - 4}{1 - 8} = \frac{3}{7}.$

CHAPTER 1 - Miscellaneous Problems

43. There is no limit: For $x = n\pi$, $\dfrac{x \sin x}{x + \sin x} = 0$;

for $x = (2n + \frac{1}{2})\pi$, $\dfrac{x \sin x}{x + \sin x} = \dfrac{\pi(2n + 1/2)}{1 + \pi(2n + 1/2)}$

which stays arbitrarily close to 1 for large n.

45. $\lim\limits_{x \to a} \dfrac{x^2 - a^2}{x + a} = \dfrac{a^2 - a^2}{a + a} = 0.$

47. $\lim\limits_{h \to 0} \dfrac{\sqrt{x + h} - \sqrt{x}}{h} = \lim\limits_{h \to 0} \dfrac{\sqrt{x + h} - \sqrt{x}}{h} \cdot \dfrac{\sqrt{x + h} + \sqrt{x}}{\sqrt{x + h} + \sqrt{x}}$

$= \lim\limits_{h \to 0} \dfrac{1}{\sqrt{x + h} + \sqrt{x}} \quad \dfrac{1}{2\sqrt{x}}.$

49. If $x > 0$, $\dfrac{1}{x} > 0$, so $\lim\limits_{x \to 0^+} \dfrac{1}{x} = \infty.$

51. If $x \neq 1$, $\dfrac{(2x - 3)(\sqrt{x} - 1)}{2x^2 + x - 3} = \dfrac{(2x - 3)(\sqrt{x} - 1)}{(2x + 3)(x - 1)}$

$\qquad = \dfrac{(2x - 3)(\sqrt{x} - 1)}{(2x + 3)(\sqrt{x} + 1)(\sqrt{x} - 1)} = \dfrac{2x - 3}{(2x + 3)(\sqrt{x} + 1)}$

So, $\lim\limits_{x \to 1} \dfrac{(2x - 3)(\sqrt{x} - 1)}{2x^2 + x - 3} = \lim\limits_{x \to 1} \dfrac{2x - 3}{(2x + 3)(\sqrt{x} + 1)} = -\dfrac{1}{10}.$

53. $\lim\limits_{x \to 1} \dfrac{\sqrt{x + 1} - \sqrt{2x}}{x^2 - x} = \lim\limits_{x \to 1} \dfrac{\sqrt{x + 1} - \sqrt{2x}}{x^2 - x} \cdot \dfrac{\sqrt{x + 1} + \sqrt{2x}}{\sqrt{x + 1} + \sqrt{2x}}$

$= \lim\limits_{x \to 1} \dfrac{1 - x}{(x^2 - x)(\sqrt{x + 1} + \sqrt{2x})} = \lim\limits_{x \to 1} \dfrac{-1}{x(\sqrt{x + 1} + \sqrt{2x})} = -\dfrac{1}{2\sqrt{2}}$

55. If $x < 0$, $|x| = -x$, $\dfrac{|x|}{x} = -1$, and $\lim\limits_{x \to 0^-} \dfrac{|x|}{x} = -1.$

57. If $x > 4$ but x is close to 4, $[x] = 4$. So, $\lim\limits_{x \to 4^+} ([x] - x) = 4 - 4 = 0.$

59. If $x < 3$ but x is close to 3, $[x] = 2$. So,

$\dfrac{[x]^2 - 9}{x^2 - 9} = \dfrac{4 - 9}{x^2 - 9} = \dfrac{-5}{x^2 - 9} \to +\infty$ as $x \to 3^-$

(since if $x < 3$, $x^2 - 9 < 0$). Thus, $\lim\limits_{x \to 3^-} \dfrac{[x]^2 - 9}{x^2 - 9} = \infty.$

61. $\lim\limits_{x \to 0^+} \dfrac{\sqrt{x}}{\sqrt{4 + \sqrt{x}} - 2} = \lim\limits_{x \to 0^+} \dfrac{\sqrt{x}}{\sqrt{4 + \sqrt{x}} - 2} \cdot \dfrac{\sqrt{4 + \sqrt{x}} + 2}{\sqrt{4 + \sqrt{x}} + 2}$

$= \lim\limits_{x \to 0^+} \sqrt{4 + \sqrt{x}} + 2 = 4.$

63. $\left. \begin{array}{l} 3x + 5y = 1 \\ (2 + c)x + 5c^2 y = 1 \end{array} \right\} \Rightarrow \left\{ \begin{array}{l} 3c^2 x + 5c^2 y = c^2 \\ (2 + cx) + 5c^2 y = 1 \end{array} \right.$

$\Rightarrow x = \dfrac{c^2 - 1}{3c^2 - \dfrac{}{}} = \dfrac{(c + 1)(c - 1)}{(3c + 2)(c - 1)} = \dfrac{c + 1}{3c + 2}$, since $3x + 5y = 1$,

$y = \dfrac{1 - 3(3(c + 1)/(3c + 2))}{5} = -\dfrac{1}{5(3c + 2)}.$

Thus, the point of intersection is $\left(\dfrac{c + 1}{3c + 2}, -\dfrac{1/5}{3c + 2}\right)$ and the limits as $c \to 1$ is

$\left(\dfrac{2}{5}, -\dfrac{1}{25}\right)$. Note that we cannot determine the limiting point simply by setting

CHAPTER 1 - Miscellaneous Problems

C = 1 in the second equation. Why not?

65. Let us require $\delta < 1$, so that $2 < t + 1 < 3$. Then, if $0 < t - 1 < \delta$,

$\sqrt{t^2 - 1} = \sqrt{t + 1}\ \sqrt{t - 1} < \sqrt{3}\ \sqrt{\delta}$. So, if $\sqrt{3}\ \sqrt{\delta} < \varepsilon$ or $\delta < \dfrac{\varepsilon^2}{3}$,

$0 < t - 1 < \delta$ implies $\sqrt{t^2 - 1} < \varepsilon$.

67. a) One zero of $f(x)$ is $x = 2$. Then, $f(x) = (x - 2)(x^2 - x + 6) =$

   $(x - 3)(x^2 - 4)$, etc.

   b) If $x \neq 3$, $h(x) = \dfrac{f(x)}{x - 3} = x^2 - 4$, and $\lim\limits_{x \to 3} h(x) = 5$. So,

   if $k = 5$, $h(x)$ is continuous at $x = 3$.

   c) If $k = 5$, $h(x)$ coincides everywhere with the function $r(x) = x^2 - 4$, so h is

   an even function.

69. a) Since $a < b$, $b - a$ is positive. But then, $b - a = (b + c) - (a + c)$ is

   positive, so $a + c < b + c$. Likewise, $(b - c) - (a - c)$ is positive, so

   $a - c < b - c$.

   b) Since $a < b$ and $c < d$, $b - a$ and $d - c$ are both positive. But the sum of

   two positive real numbers is again positive, so

   $(b - a) + (d - c) = (b + d) - (a + c)$ is positive. Hence, $b + d > a + c$.

   However, $a - c$ need not be less than $b - d$; for example, $2 < 7$ and $-5 < 4$,

   but $2 - (-5) = 7 > 3 = 7 - 4$.

   c) Suppose $a < b$; then $b - a$ is positive. Now, $\dfrac{1}{a} - \dfrac{1}{b}$

   $= \dfrac{b - a}{ab}$. If a and b are both positive or both negative, ab is

   positive, so $\dfrac{b - a}{ab}$ is positive. Hence, $\dfrac{1}{a} > \dfrac{1}{b}$, or $\dfrac{1}{b} < \dfrac{1}{a}$.

   d) $a < 0 \Rightarrow a$ is negative $\Rightarrow \dfrac{1}{a}$ is negative $\Rightarrow \dfrac{1}{a} < 0$. $b > 0 \Rightarrow b$ is

   positive $\Rightarrow \dfrac{1}{b}$ is positive $\Rightarrow \dfrac{1}{b} > 0$. Hence, $\dfrac{1}{a} < 0 < \dfrac{1}{b}$.

   e) $a < b \Rightarrow b - a$ is positive. Since c is positive, $(b - a) \cdot c = bc - ac$ is

   positive; hence, $ac < bc$.

   f) $a < b \Rightarrow b - a$ is positive. Since c is negative, $-c$ is positive, so

   $(-c)(b - a) = ac - bc$ is positive. Then $bc < ac$.

CHAPTER 1 - Miscellaneous Problems

CHAPTER 2

Article 2.2

1. $s = t^2 - 4t + 3$

$v = \dfrac{ds}{dt} = 2t - 4$

$a = \dfrac{dv}{dt} = 2$

3. $s = gt^2/2 + v_0 t + s_0$

$v = \dfrac{ds}{dt} = gt + v_0$

$a = \dfrac{dv}{dt} = g$

5. $s = (2t + 3)^2 = 4t^2 + 12t + 9$

$v = \dfrac{ds}{dt} = 8t + 12$

$a = \dfrac{dv}{dt} = 8$

7. $y = 5x^3 - 3x^5$

$y' = 15x^2 - 15x^4$

$y'' = 30x - 60x^3$

9. $y = \dfrac{x^4}{4} - \dfrac{x^3}{3} + \dfrac{x^2}{2} - x + 3$

$y' = x^3 - x^2 + x - 1$

$y'' = 3x^2 - 2x + 1$

11. $12y = 6x^4 - 18x^2 - 12x$

$y = \dfrac{1}{2}x^4 - 36x - 12$

$y' = 2x^3 - 36$

13. $y = x^2(x^3 - 1) = x^5 - x^2$

$y' = 5x^4 - 2x$

$y'' = 20x^3 - 2$

15. $y = (3x - 1)(2x + 15)$

$\quad = 6x^2 + 13x - 5$

$y' = 12x + 13$

$y'' = 12$

17. $y = x^2 + 5x$

$y' = 2x + 5$

$y'(3) = 2(3) + 5 = 11$

(c)

19. $y = x^3 - 3x + 1$

$y' = 3x^2 - 3$

$y'(2) = 3(2)^2 - 3 = 9$

slope of the curve is $m = 9$.

Slope of the perpendicular is

$-\dfrac{1}{m} = \dfrac{-1}{9}$. $y = -\dfrac{1}{9}x + 6$. At $(2,3)$

$3 = -\dfrac{1}{9}(2) + b \Rightarrow b = \dfrac{29}{9}$

perpendicular: $y = -\dfrac{1}{9}x + \dfrac{29}{9}$

$\quad\quad\quad\quad$ or $x + 9y = 29$

21. $S = 24t - 4.9t2m$. Rock reaches maximum height when it stops rising, that is when $\frac{ds}{dt} = 0$. $\frac{ds}{dt} = 24 - 9.8t = 0 \Rightarrow t = 2.449$ sec. $S(2.449) = 24(2.449) - 4.9(2.449)^2 = 29.39$ maximum height = 29.39 meters.

23. $s = t^3 - 4t^2 - 3t \Rightarrow \frac{ds}{dt} = 3t^2 - 8t - 3$. $\frac{ds}{dt} = 0 \Rightarrow 3t^2 = 8t - 3 = 0 \Rightarrow$ $t = (8 \pm (\sqrt{64 - 36}))/6 = (4 \pm \sqrt{7})/3$. Acceleration $= \frac{d^2s}{dt^2} = 6t - 8$. Acceleration at $(4 + \sqrt{3})/3$ is $2\sqrt{7}$. Acceleration at $(4 - \sqrt{3})/3$ is $-2\sqrt{7}$.

25. The tangent is parallel to the x-axis when the slope is zero.

$y = 2x^3 - 3x^2 - 12x + 20 \Rightarrow y' = 6x^2 - 6x - 12 = 0 \Rightarrow x^2 - x - 2 = 0 \Rightarrow$

$(x + 1)(x - 2) = 0$.

$x = -1$, $x = 2$. $y(-1) = 2(-1)^3 - 3(-1)^2 - 12(-1) + 20 = 27$.

$y(2) = 2(2)^3 - 3(2)^2 - 12(2) + 20 = 0$. Points are $(-1,27)$, $(2,0)$.

27. The slope of the curve must equal the slope of the tangent line. Since $y = x^n$, $y' = nx^{n-1}$ is the slope of the curve. Since $(x_0, y_0) = (t, 0)$ and $(x_1, y_1) = (x, x^n)$ are two points on the tangent line, its slope is

$\frac{y_1 - y_0}{x_1 - x_0} = \frac{x^n - 0}{x - t}$. Equating the two

slopes, $y' = \frac{y_1 - y_0}{x_1 - x_0} \Rightarrow nx^{n-1} = \frac{x^n}{x - t}$

$\Rightarrow n = \frac{x}{x - t} \Rightarrow \frac{1}{n} = \frac{x - t}{x} = 1 - \frac{t}{x} \Rightarrow \frac{t}{x} = 1 - \frac{1}{n}$

$\Rightarrow t = x(1 - \frac{1}{n})$. To construct the tangent line through (x, x^n) on the curve

$y = x^n$, draw a line through the two points (x, x^n) and $(t, 0)$ where

$t = x(1 - \frac{1}{n})$.

29. $y = x^3 - x \Rightarrow y' = 3x^2 - 1$. $y(-1) = 3(-1)^2 - 1 = 2$. At $(-1,0)$, tangent line

is $y = 2x + b$. $0 = 2(-1) + b \Rightarrow b = 2 \Rightarrow y = 2x + 2$. This line intersects $y = x^3 - x$ when $x^3 - x = 2x + 2$ or $x^3 - 3x - 2 = 0$. To solve this cubic equation,

we note that $x = -1$ is one solution since both the curve and the line pass

through the point $(-1,0)$. Therefore, $(x + 1)$ is one factor. Dividing $(x^3 - 3x - 2)$ by $(x + 1)$ gives $(x^2 - x - 2)$. Therefore $(x^3 - 3x - 2) = (x + 1)(x^2 - x -$

2) $(x + 1)(x - 2)(x + 1) = 0.$ $x = 2$ is the other root. Note that $y = x^3 - x$ = $(2)^3 - (2) = 6$ and $y = 2x + 2 = 2(2) + 2 = 6.$ The curve intersects the line at $(2,6)$.

31. The information of the problem can be summarized by:

    i) curve passes through the origin; $y(0) = 0$

    ii) curve is tangent to line $y = x$ at origin

    iii) curve passes through $(1,2)$; $y(1) = 2$

    i) $y(0) = 0 \Rightarrow a(0)^2 + b(0) + c = 0 \Rightarrow c = 0$

    ii) $y = x \Rightarrow y' = 1.$ Slope must be 1 at $(0,0)$

    $y = ax^2 + bx + c \Rightarrow$ slope $= y' = 2ax + b$

    $y'(0) = 1 \Rightarrow 1 = 2a(0) + b \Rightarrow b = 1$

    iii) $y(1) = 2 \Rightarrow 2 = a(1)^2 + b(1) + c = a + 1 + c \Rightarrow a = 1$

33. Since $y_1 = x$ has slope $y_1^1 = 1$ everywhere any tangent to this line must have slope 1. The slope of the curve $y_2 = x^2 + c$ must have slope 1 where y_1 meets y_2. $y_2^1 = 2x = 1 \Rightarrow x = \frac{1}{2}.$ Since y_1 meets y_2 at $x = \frac{1}{2}$, $y_1(\frac{1}{2}) = y_2(\frac{1}{2})$ or $y_1 = \frac{1}{2} = y_2 = (\frac{1}{2})^2 + c \Rightarrow c = \frac{1}{4}.$

Article 2.3

1. $y = \frac{x^3}{3} - \frac{x^2}{2} + x - 1 \Rightarrow y' = x^2 - x + 1$

3. $y = (x^2 + 1)^5 \Rightarrow y' = 5(x^2 + 1)^4 \cdot \frac{d(x^2 + 1)}{dx} = 10x(x^2 + 1)^4$

5. $y = (x + 1)^2(x^2 + 1)^{-3} \Rightarrow y' = 2(x + 1)^1(x^2 + 1)^{-3} - 3(x^2 + 1)^{-4}(2x) =$ $(2x + 1)(x^2 + 1)^{-3} - 6x(x + 1)^2(x^2 + 1)^{-4}$

7. $y = \frac{2x + 5}{3x - 2} = (2x + 5)(3x = 2)^{-1} \Rightarrow y' = 2(3x - 2)^{-1} -$ $(2x + 5)(3x - 2)^{-2}(3) = 2(3x - 2)^{-1} - (6x + 15)(3x - 2)^{-2}$

9. $y = (1 - x)(1 + x^2)^{-1} \Rightarrow y' = \frac{d(1 - x)}{dx} \cdot (1 + x^2)^{-1} +$ $(2x^2 - 2x)(1 + x^2)^{-2}$

11. $y = \dfrac{5}{(2x - 3)^4} = 5(2x - 3)^{-4} \Rightarrow y' = -20(2x - 3)^{-5}(2)$

    $= -40(2x - 3)^{-5}$

13. $y = (5 - x)(4 - 2x) \Rightarrow y' = (-1)(4 - 2x) - 2(5 - x) = 4x - 14$

15. $y = (2x - 1)^3(x + y)^{-3} \Rightarrow y' = 3(2x - 1)^2 \dfrac{d(2x - 1)}{dx} \cdot (x + 7)^{-3} -$

    $3(2x - 1)^3(x + 7)^{-4} \dfrac{d(x + 7)}{dx} = 6(2x - 1)^2(x + 7)^{-3}(2x - 1)^3(x + 7)^{-4}$

    $= (2x - 1)^2(x + 7)^{-4}[6(x + 7) - 3(2x - 1)] = 45(2x - 1)^2(x + 7)^{-4}$

17. $y = (2x^3 - 3x^2 + 6x)^{-5} \Rightarrow y' = -5(2x^3 - 3x^2 + 6x)^{-6}(6x^2 - 6x + 6)$

    $= -30(2x^3 - 3x^2 + 6x)^{-6}(x^2 - x + 1)$

19. $y = \dfrac{(x - 1)^2}{x^2} = \dfrac{x^2 - 2x + 1}{x^2} = 1 - 2x^{-1} + x^{-2}$

    $\Rightarrow y' = 2x^{-2} - 2x^{-3}$

21. $y = \dfrac{12}{x} - \dfrac{4}{x^3} + \dfrac{3}{x^4} = 12x^{-1} - 4x^{-3} + 3x^{-4}$

    $\Rightarrow y' = -12x^{-2} + 12x^{-4} - 12x^{-5}$

    $= -23x^{-5}(x^3 - x + 1)$, alternatively

23. $y = \dfrac{x^2 - 1}{x^2 + x - 2} = \dfrac{(x + 1)(x - 1)}{(x + 2)(x - 1)} = \dfrac{x + 1}{x + 2} - \dfrac{1}{x + 2}$

    $= 1 - (x + 2)^{-1} \Rightarrow y' = (x + 2)^{-2}$

25. $s = \dfrac{t}{t^2 + 1} = t(t^2 + 1)^{-1} \Rightarrow \dfrac{ds}{dt} = (t^2 + 1)^{-1} - t(t^2 + 1)^{-2}(2t)$

    $= (t^2 + 1)^{-1} - 2t^2(t^2 + 1)^{-2} = (t^2 + 1)^{-2}[(t^2 + 1) - 2t^2] = \dfrac{1 - t^2}{(1 + t^2)^2}$

27. $s = (t^2 - t)^{-2} \Rightarrow \dfrac{ds}{dt} = -2(t^2 - t)^{-3} \dfrac{d(t^2 - t)}{dt} = -2(t^2 - t)^{-3}(2t - 1) = \dfrac{2 - 4t}{(t^2 - t)^3}$

29. $s = \dfrac{2t}{3t^2 + 1} \Rightarrow \dfrac{ds}{dt} = \dfrac{2(3t^2 + 1) - 2t(6t)}{(3t^2 + 1)^2} = \dfrac{2 - 6t^2}{3t^2 + 1)^3}$

31. $s = (t^2 + 3t)^3 \Rightarrow \dfrac{ds}{dt} = 3(t^2 + 3t)^2(2t + 3) = (6t + 9)(t^2 + 3t)^2$

33. $u(0) = 5$, $u'(0) = -3$, $v(0) = -1$, $v'(0) = 2$

CHAPTER 2 - Article 2.3

a) $\frac{d}{dx}(uv) = u'v + uv'$ at $x = 0$ is $(-3)(-1) + (5)(2) = 13$

b) $\frac{d}{dx}(\frac{u}{v}) = \frac{u'v - uv'}{v^2}$ at $x = 0$ is $\frac{(3)(-1) - (5)(2)}{(-1)^2} = -7$

c) $\frac{d}{dx}(\frac{v}{u}) = \frac{v'u - vu'}{u^2}$ at $x = 0$ is $\frac{(2)(5) - (-1)(-3)}{(5)^2} = \frac{7}{25}$

d) $\frac{d}{dx}(7v - 28) = 7v' - 2u'$ at $x = 0$ is $7(2) - 2(-3) = 20$

e) $\frac{d}{dx}(u^3) = 3u^2u'$ at $x = 0$ is $3(5)^2(-3) = -225$

f) $\frac{d}{dx}(5v^{-3}) = -15v^{-4}v'$ at $x = 0$ is $(-15)(-1)^{-4}(2) = -30$

35. $y = x + (\frac{1}{x}) \Rightarrow y' = 1 - (\frac{1}{x^2})$;

$y'(2) = 1 - \frac{1}{4} = \frac{3}{4}$. $y(2) = 2 + \frac{1}{2} = \frac{5}{2}$.

The target line has slope $\frac{3}{4}$ and passes through $(2,\frac{5}{2})$.

$y = \frac{3}{4}x + b$, $\frac{5}{2} = \frac{3}{4}(2) + b \Rightarrow b = 1$

$y = \frac{3}{4}x + 1$, or $3x - 4y = -4$

37. $y = (3 - 2x)^{-1} \Rightarrow \frac{dy}{dx} = (-1)(3 - 2x)^{-2}(-2) = 2(3 - 2x)^{-2}$

$\frac{d^2y}{dx^2} = (-2)(2)(3 - 2x)^{-3}(-2) = 8(3 - 2x)^{-3}$

39. Note: $(a + b)^3 = a^3 + 3a^2b + 3ab + b^3$

$f(x) = 7 + 10(x - 2) + 6(x - 2)^2 + (x - 2)^3$

$= 7 + 10(x - 2) + 6(x^2 - 4x + 4) + (x^3 - 6x^2 + 12x - 8)$

$= 7 + (10x - 20) + (6x^2 - 24x + 24) + (x^3 - 6x^2 + 12x - 8)$

$= (7 - 20 + 24 - 8) + (10 - 24 + 12)x + (6 - 6)x^2 + x^3$

$= 3 - 2x + x^3$

41. Rate of growth of total production is $\frac{dy}{dt} = \frac{d}{dt}(uv) = v\frac{du}{dt} + u\frac{dv}{dt}$

$= v(0.04u) + u(0.05v) = 0.09uv = 0.09y$.

43. a) $P(t) = \frac{a^2kt}{akt + 1} = a[\frac{akt + 1}{akt + 1} - \frac{1}{akt + 1}] = a - a(akt + 1)^{-1}$

$\frac{dP}{dt} = a(akt + 1)^{-2}(ak) = \frac{a^2k}{(akt + 1)^2}$

b) Since $(akt + 1)^2$ increases with time, $\frac{dP}{dt} = \frac{a^2k}{(akt + 1)^2}$ decreases with time, and is greatest when $t = 0$ and $\frac{dP}{dt} = a^2 k$.

45. The finite product rule says

$$\frac{d}{dx}(u_1 u_2 \ldots u_n) = \frac{du_1}{dx} u_2 \ldots u_n + u_1 \frac{du_2}{dx} \ldots u_n + \ldots + u_1 u_2 \ldots u_{n-1} \frac{du_n}{dx}.$$

Setting $u_1 = u_2 = \ldots = u_1 = u$, the equation becomes

$$\frac{d}{dx}(u^n) = \frac{d}{dx}(u \cdot u \ldots u) = \frac{du}{dx} \cdot u \ldots u + u \frac{du}{dx} \ldots u + u \cdot u \ldots u \frac{dy}{dx}$$

$$= u^{n-1} \frac{du}{dx} + u^{n-1} \frac{du}{dx} + \ldots + u^{n-1} \frac{du}{dx}$$

$$\frac{d}{dx}(u^n) = n u^{n-1} \frac{du}{dx}$$

The final line is rule 6.

47. See proof in text.

Article 2.4

1. $x^2 + y^2 = 1 \Rightarrow 2x + 2y\frac{dy}{dx} = 0 \Rightarrow \frac{dy}{dx} = -\frac{2x}{2y} = -\frac{x}{y}$

3. $x^2 + xy = 2 \Rightarrow 2x + y + x\frac{dy}{dx} = 0 \Rightarrow x\frac{dy}{dx} = -(2x + y)$

$\frac{dy}{dx} = -\frac{(2x + y)}{x} = -(2 + \frac{y}{x})$

An alternative form that does not include y uses

$x^2 + xy = 2 \Rightarrow \frac{x^2}{x^2} + \frac{xy}{x^2} = \frac{2}{x^2} \Rightarrow 1 + \frac{y}{x} = \frac{2}{x^2}$

$\Rightarrow 2 + \frac{y}{x} = 1 + \frac{2}{x^2}$,

$\frac{dy}{dx} = -(2 + \frac{y}{x}) = -(1 + \frac{2}{x^2})$

5. $y^2 = x^3 \Rightarrow 2y\frac{dy}{dx} = 3x^2 \Rightarrow \frac{dy}{dx} = \frac{3x^2}{2y}$

7. $x^{1/2} + y^{1/2} = 1 \Rightarrow \frac{1}{2}x^{-1/2} + \frac{1}{2}y^{-1/2}\frac{dy}{dx} = 0$

$\Rightarrow \frac{dy}{dx} = -\frac{x^{-1/2}}{y^{-1/2}} = -(\frac{y}{x})^{1/2}$

9. $x^2 = \dfrac{x - y}{x + y} \Rightarrow x - y = x^2(x + y) = x^3 + x^2 y$

$\Rightarrow 1 - \dfrac{dy}{dx} = 3x^2 + 2xy + x^2 \dfrac{dy}{dx}$

$(x^2 + 1)\dfrac{dy}{dx} = 1 - 3x^2 + 2xy \Rightarrow \dfrac{dy}{dx} = \dfrac{1 - 3x^2 + 2xy}{x^2 + 1}$

11. $y = x \sqrt{x^2 + 1} = x(x^2 + 1)^{1/2} \Rightarrow \dfrac{dy}{dx} = (x^2 + 1)^{1/2} + \dfrac{1}{2}x(x^2 + 1)^{-1/2}(2x)$

$= \dfrac{x^2 + 1}{(x^2 + 1)^{1/2}} + \dfrac{x^2}{(x^2 + 1)^{1/2}} = \dfrac{2x^2 + 1}{(x^2 + 1)^{1/2}}$

13. $2xy + y^2 = x + y \Rightarrow 2x + 2x\dfrac{dy}{dx} + 2y\dfrac{dy}{dx} = 1 + \dfrac{dy}{dx}$

$(2x + 2y - 1)\dfrac{dy}{dx} = 1 - 2y \Rightarrow \dfrac{dy}{dx} = \dfrac{1 - 2y}{2x + 2y - 1}$

15. $y^2 = \dfrac{x^2 - 1}{x^2 + 1} = \dfrac{x^2 + 1}{x^2 + 1}\ \dfrac{-2}{x^2 + 1} = 1 - 2(x^2 + 1)^{-1}$

$2y\dfrac{dy}{dx} = 2(x^2 + 1)^{-2}(2x) \Rightarrow \dfrac{dy}{dx} = \dfrac{2x}{y(x^2 + 1)^2}$

17. $2y^3 = (3x + 7)^5 \Rightarrow 6y^2\dfrac{dy}{dx} = 5(3x + 7)^4(3)$

$\Rightarrow \dfrac{dy}{dx} = \dfrac{5(3x + 7)^4}{2y^2}$

Alternatively, $\dfrac{dy}{dx} = \dfrac{5}{3\sqrt[3]{2}}(3x + 7)^{2/3}$

19. $y^{-1} + x^- = 1 \Rightarrow -y^{-2}\dfrac{dy}{dx} - x^{-2} = 0 \Rightarrow \dfrac{dy}{dx} = -\dfrac{x^{-2}}{y^{-2}} = -\left(\dfrac{y}{x}\right)^2$

Alternatively, $\dfrac{dy}{dx} = \dfrac{1 - y}{x - 1} = -\left(\dfrac{1}{x - 1}\right)^2 = -(y - 1)^2$.

21. $y^2 = x^2 - x \Rightarrow 2y\dfrac{dy}{dx} = 2x - 1 \Rightarrow \dfrac{dy}{dx} = \dfrac{2x - 1}{2y} = \dfrac{2x - 1}{2(x^2 - x)^{1/2}}$

23. $y = \dfrac{(x^2 + 3)^{1/3}}{x} \Rightarrow \dfrac{dy}{dx} = \dfrac{(1/3)(x^2 + 3)^{-2/3}(2x)(x) - (x^2 - (x^2 + 3)^{1/3}}{x^2}$

$= \dfrac{(2/3)x^2 - (x^2 + 3)}{x^2(x^2 + 3)^{2/3}} = -\dfrac{(1/3)x^2 + 3}{x^2(x^2 + 3)^{2/3}}$

25. $x^3 + y^3 = 18xy \Rightarrow 3x^2 + 3y^2\dfrac{dy}{dx} = 18y + 18x\dfrac{dy}{dx}$

$$\Rightarrow (y^2 - 6x)\frac{dy}{dx} = 6y - x^2 \Rightarrow \frac{dy}{dx} = \frac{6y - x^2}{y^2 - 6x}$$

27. $y = (2x + 5)^{-1/5} \Rightarrow \frac{dy}{dx} = -\frac{1}{5}(2x + 5)^{-6/5}(2) = -\frac{2}{5}(2x + 5)^{-6/5}$

29. $y = (1 - x^{1/2})^{1/2} \Rightarrow \frac{dy}{dx} = \frac{1}{2}(1 - x^{1/2})^{-1/2}(-\frac{1}{2}x^{-1/2}) = -\frac{1}{4}(x - x^{3/2})^{-1/2}$

31. $T^2 = 4\pi^2 L/g \Rightarrow 2T\frac{dT}{dL} = \frac{4\pi^2}{g} \Rightarrow \frac{dT}{dL} = \frac{2\pi^2}{Tg}$

    Alternatively, $\frac{dT}{dL} = \frac{\pi}{\sqrt{Lg}}$

33. It is given that $f''(x) = x^{-1/3}$. If $f'(x) = \frac{3}{2} + c_1$, where c_1 is any

    constant, then $f''(x) = (\frac{3}{2})(\frac{2}{3})x^{-1/3} = x^{-1/3}$. Also, if

    $f(x) = \frac{9}{10}x^{5/3} + c_1 x + c_2$, where c_2 is any constant, then

    $f'(x) = (\frac{9}{10})(\frac{5}{3})x^{2/3} + c_1 = \frac{3}{2}x^{2/3} + c_1$. We conclude that $f(x)$ and

    $f'(x)$ must be of the form given.

    a) Not true.   b) May be true. Take $c_1 = 0$, $c_2 = -7$.

    c) True.   d) May be true. Take $c_1 = 6$.

35. a) $x^2 = y^2 = 1 \Rightarrow 2x - 2y\frac{dy}{dx} = 0 \Rightarrow \frac{dy}{dx} = \frac{-2x}{-2y} = \frac{x}{y}$

    b) $\frac{dy}{dx} = \frac{x}{y} = (\text{quotient rule}) = \frac{y(dx/dx) - x(dy/dx)}{y^2}$

    $= \frac{y - x(x/y)}{y^2} = \frac{y^2 - x^2}{y^3} = \frac{-1}{y^3}$

37. $x^3 + y^3 = 1 \Rightarrow 3x^2 + 3y^2\frac{dy}{dx} = 0 \Rightarrow \frac{dy}{dx} = -\frac{x^2}{y^2}$

    $\frac{dy}{dx} - \frac{x^2}{y^2} \Rightarrow \frac{d^2y}{dx^2} = -\frac{2xy^2 - 2x^2y(dy/dx)}{y^4}$

    $= -2x\frac{(y^2 - xy(-(x^2/y)))}{y^4} = -\frac{2x(y^3 + x^3)}{y^5} = -\frac{2x}{y^5}$ since $x^3 + y^3 = 1$.

39. $xy + y^2 = 1 \Rightarrow y + x\dfrac{dy}{dx} + 2y\dfrac{dy}{dx} = 0$

$(x + 2y)\dfrac{dy}{dx} = -y \Rightarrow \dfrac{dy}{dx} = -\dfrac{y}{(x + 2y)}$

$\dfrac{d^2y}{dx^2} = -\dfrac{(x + 2y)(dy/dx) - y(1 + 2(dy/dx))}{(x + 2y)^2}$

$\quad = -\dfrac{(x + 2y)(-y/(x + 2y)) - y(1 - 2y/(x + 2y))}{(x + 2y)^2} = \dfrac{y + xy/(x + 2y)}{(x + 2y)^2}$

$\quad = \dfrac{y(x + 2y) + xy}{(x + 2y)^3} = \dfrac{2xy + 2y}{(x + 2y)^3}$

41. Note $\dfrac{dx}{dt} = v$, $\dfrac{dx_0}{dt} = 0$, $\dfrac{dv_0}{dt} = 0$.

$m(v^2 - v_0^2) = k(x_0^2 - x^2) \Rightarrow m\left(2v\dfrac{dv}{dt}\right) = k\left(-2x\dfrac{dx}{dt}\right)$

$mv\dfrac{dv}{dt} = -kxv \Rightarrow \dfrac{mdv}{dt} = -kx.$

43. $x^2 + y^2 = 25 \Rightarrow 2x + 2y\dfrac{dy}{dx} = 0 \Rightarrow \dfrac{dy}{dx} = -\dfrac{x}{y}$

When $(x,y) = (3,-4)$, $\dfrac{dy}{dx} = -\left(\dfrac{3}{-4}\right) = \dfrac{3}{4}$

Tangent: $y = \dfrac{3}{4}x + b$; $-4 = \dfrac{3}{4}(3) + b \Rightarrow b = -\dfrac{25}{4}$

$\qquad y = \dfrac{3}{4}x - \dfrac{25}{4}$ or $3x - 4y = 25$

Normal: $y = -\dfrac{4}{3}x + b$; $-4 = -\dfrac{4}{3}(3) + b \Rightarrow b = 0$

$\qquad y = -\dfrac{4}{3}$ or $4x + 3y = 0$

45. $\dfrac{x - y}{x - 2y} = 2 \Rightarrow x - y = 2(x - 2y) = 2x - 4y$ or $x = 3y$. Hence the

curve $\dfrac{x - y}{x - 2y} = 2$ is simply the straight line $x = 3y$. The tangent

is the curve itself.

Tangent: $y = \dfrac{1}{3}x$ or $x - 3y = 0$ or $x = 3y$.

$\qquad$ Slope of the tangent is $m = \dfrac{1}{3}$, so the

$\qquad$ slope of the normal is $-\dfrac{1}{m} = -3$.

Normal: At $(3,1)$, $y = -3x + b$; $1 = -3(3) + b \Rightarrow b = 10$

$\qquad y = -3x + 10$ or $3x + y = 10$

47. $y^2 - 2x - 4y - 1 = 0 \Rightarrow 2y\dfrac{dy}{dx} - 2 - 4\dfrac{dy}{dx} = 0$

$\Rightarrow (y - 2)\dfrac{dy}{dx} = 1 \Rightarrow \dfrac{dy}{dx} = \dfrac{1}{y - 2}$

when $y = 1$, $\dfrac{dy}{dx} = \dfrac{1}{1 - 2} = -1$. At $(x,y) = (-2,1)$,

Tangent: $y = -x + b; \quad 1 = -(-2) + b \Rightarrow b = -1$

$$y = -x - 1 \quad \text{or} \quad x + y = -1$$

Normal: $y = x + b; \quad 1 = (-2) + b \Rightarrow b = 3$

$$y = x + 3 \quad \text{or} \quad x - y = -3$$

49. $2x + y = 0 \Rightarrow y = -2x \Rightarrow \dfrac{dy}{dx} = -2$. The line has slope -2, so the normals parallel to it must also have slope -2. That is, (slope of normal)

$= -\left(\dfrac{1}{dy/dx}\right) = -2$ or $\dfrac{dy}{dx} = \dfrac{1}{2}$ must hold. $xy + 2x - y = 0$

$\Rightarrow x\dfrac{dy}{dx} + y + 2 - \dfrac{dy}{dx} = 0.$ $(x - 1)\dfrac{dy}{dx} = -(y + 2) \Rightarrow \dfrac{dy}{dx} = \dfrac{y + 2}{1 - x}.$

Since $\dfrac{dy}{dx} = \dfrac{1}{2}, \dfrac{y + 2}{1 - x} = \dfrac{1}{2} \Rightarrow 2y + 4 = 1 - x \Rightarrow x = -(2y + 3).$

Substituting into the equation for the curve,

$xy + 2x - y = 0 \Rightarrow -(2y + 3)y - 2(2y + 3) - y = 0$

$2y^2 + 8y + 6 = 0 \Rightarrow y^2 + 4y + 3 = (y + 1)(y + 3) = 0$

$y = -1, \quad x = -(2y + 3) = -(2(-1) + 3) = -1$

$y = -3, \quad x = -(2y + 3) = -(2(-3) + 3) = 3$

There are two normals with slope -2; one passes through $(-1,-1)$, the other through $(3,-3)$.

1) $y = -2x + b; \quad -1 = -2(-1) + b \Rightarrow b = -3$

$\quad y = -2x - 3 \quad \text{or} \quad 2x + y = -3$

2) $y = -2x + b; \quad -3 = -2(3) + b \Rightarrow b = 3$

$\quad y = -2x + 3 \quad \text{or} \quad 2x + y = 3$

51. The curve $y = x^2 + 2x - 3$ has slope $y' = 2x + 2 = 2(1) + 2 = 4$ when $x = 1$. The normal has slope $-\dfrac{1}{y'}, = -\dfrac{1}{4}$ and passes through $(1,0)$. Its equation is therefore $y = -\dfrac{1}{4}x + b; \quad 0 = -\dfrac{1}{4}(1) + b \Rightarrow b = \dfrac{1}{4}.$

$y = -\dfrac{1}{4}x + \dfrac{1}{4}.$ Equating the normal line and the curve,

$-\dfrac{1}{4}x + \dfrac{1}{4} = x^2 + 2x - 3 \Rightarrow -x + 1 = 4x^2 + 8x - 12 \Rightarrow 4x^2 + 9x -$

$13 = 0. \quad (1)$

Since $x = 1$ is one solution to equation (1), $x - 1$ is a factor. Long division yields (4x + 13) as the other factor. Equation (1) is $0 = 4x^2 + 9x - 13$

$= (x - 1)(4x + 13) \Rightarrow x = -\frac{13}{4}$.

$y = -\frac{1}{4}x + \frac{1}{4} = -\frac{1}{4}(-\frac{13}{4}) + \frac{1}{4} = \frac{13}{16} + \frac{4}{16} = \frac{17}{16}$.

The other point of intersection $(-\frac{13}{4}, \frac{17}{16})$.

53. The curve $x^2 + xy + y^2 = 7$ crosses the x-axis whenever y = 0.

$x^2 + x(0) + (0)^2 = 7 \Rightarrow x = \pm\sqrt{7}$. The two points are $(\sqrt{7}, 0), (-\sqrt{7}, 0)$.

$x^2 + xy + y^2 = 7 \Rightarrow 2x + y + xy' + 2yy' = 0 \Rightarrow y' = -\frac{2x + y}{x + 2y}$.

At $(\sqrt{7}, 0)$ $y' = -\frac{2\sqrt{7} + 0}{\sqrt{7} + 0} = -2$.

At 6At $(-\sqrt{7}, 0)$ $y' = -\frac{2(-\sqrt{7}) + 0}{(-\sqrt{7}) + 0} = -2$.

55. $y^2 + (x)y + (x^2 - 7) = 0 \Rightarrow y = \frac{-x \pm \sqrt{x^2 - 4(x^2 - 7)}}{2}$

$y = -\frac{x}{2} \pm \frac{1}{2}\sqrt{28 - 3x^2}$. Since y(1) = 2, the curve in question is

$y = -\frac{x}{2} \pm \frac{1}{2}\sqrt{28 - 3x^2}$. Hence $\frac{dy}{dx} = -\frac{1}{2} +$

$(\frac{1}{2})(\frac{1}{2})(28 - 3x^2)^{-1/2}(-6x) = -\frac{1}{2} - \frac{3}{2}x(28 - 3x^2)^{-1/2}$.

When x = 1, $\frac{dy}{dx} = -\frac{1}{2} - \frac{3}{2}(1)(28 - 3(1))^{-1/2} = -\frac{4}{5}$.

Article 2.5

1. $f(x) = x^2 + 2x \Rightarrow f'(x) = 2x + 2 \Rightarrow f'(0) = 2$. Also $f(0) = (0)^2 + 2(0) = 0$.

a) L(x) = f(0) + f'(0)(x - 0) = 2x

b) $\Delta y_{tan} = f'(0)\Delta x = 2(0.1) = 0.20$

c) $\Delta y = f(.1) - f(0) = [(0.1)^2 + 2(0.1)] - [0] = 0.21$

d) $\Delta y - \Delta y_{tan} = 0.21 - 0.20 = 0.01$

3. $f(x) = x^3 - x \Rightarrow f(1) = (1)^3 - (1) = 0$

$f'(x) = 3x^2 - 1 \Rightarrow f'(1) = 3(1)^2 - 1 = 2$

a) L(x) = f(1) + f'(1)(x - 1) = 0 + 2(x - 1) = 2x - 2

b) $\Delta y_{tan} = f'(1)\Delta x = 2(0.1) = 0.2$

c) $\Delta y = f(1.1) - f(1) = [(1.1)^3 - (1.1)] - [(1)^3 - 1] = 0.231$

$f(x) = x^3 - x$

$f(x) \approx 2x - 2$

d) $\triangle y - \triangle y_{tan}$ = 0.231 = 0.200 = 0.031

5. $f(x) = x^{-1} \Rightarrow f(0.5) = (0.5)^{-1} = 2$

$f'(x) = -x^2 \Rightarrow f'(0.5) = -(0.5)^{-2} = -4$

a) $L(x) = f(0.5) + f'(0.5)(x - 0.5) = 2 - 2(x - 0.5) = -2x + 3$

b) $\triangle y_{tan} = f'(0.5) x = -4(0.1) = -0.4$

c) $\triangle y = f(0.6) - f(0.5) = (0.6)^{-1} - (0.5)^{-1} = 1.667 - 2 = -0.3333$

d) $\triangle y - \triangle y_{tan} = -0.3333 - (-0.4) = 0.0667$

7. $v = \frac{4}{3}\pi r^3; \frac{dv}{dr} = 4\pi r^2 \Rightarrow \triangle v \approx \frac{dv}{dr} r = 4\pi r^2 \triangle r$

9. $v = x^3; \frac{dv}{dx} = 3x^2 \Rightarrow \triangle v \approx \frac{dv}{dx} \triangle x = 3x^2 \triangle x$

v = volume of cube, x = edge length.

11. $v = \pi r^2 h; \frac{dv}{dr} = 2\pi rh \Rightarrow \triangle v \approx \frac{dv}{dr} \triangle r = 2\pi rh \triangle r$

r = radius of base, h = height.

13. $v = \frac{\pi r^2 h}{3}; \frac{dv}{dr} = \frac{2}{3}\pi rh \Rightarrow \triangle v \approx \frac{dv}{dr} \triangle r = \frac{2}{3}\pi rh \triangle r$

15. $f(x) = [1 + x]^{1/2} \Rightarrow f(8) = [1 + 8]^{1/2} = 3$

$f'(x) = \frac{1}{2}[1 + x]^{-1/2} \Rightarrow f'(8) = \frac{1}{2}[1 + 8]^{-1/2} = \frac{1}{2}(\frac{1}{3}) = \frac{1}{6}$

$L(x) = f(8) + f'(8)(x - 8) = 3 + \frac{1}{6}(x - 8) = \frac{5}{3} + \frac{1}{6}x$

$L(9.1) = \frac{5}{3} + \frac{1}{6}(9.1) = 3.183$

$f(9.1) = [1 + 9.1]^{1/2} = 3.178$

17. $f(x) = x^{1/3} \Rightarrow f(8) = (8)^{1/3} = 2$

$f'(x) = \frac{1}{3}x^{-2/3} \Rightarrow f'(8) = \frac{1}{3}(8)^{-2/3} = \frac{1}{3}(\frac{1}{4}) = \frac{1}{12}$

$L(x) = f(8) + f'(8)(x - 8) = 2 + \frac{1}{12}(x - 8) = \frac{4}{3} + \frac{x}{12}$

$L(8.5) = \frac{4}{3} + \frac{1}{12}(8.5) = 2.0416667$

$f(8.5) = (8.5)^{1/3} = 2.0408276$

19. $f(x) = [x^2 + 9]^{1/2} \Rightarrow f(-4) = [(-4)^2 + 9]^{1/2} = 5$

$f'(x) = \frac{1}{2}[x^2 + 9]^{-1/2}2x = x[x^2 + 9]^{-1/2} \Rightarrow f'(-4) = -4[(-4)^2 + 9]^{-1/2} = -\frac{4}{5}$

$L(x) = f(-4) + f'(-4)(x - (-4)) = 5 - \frac{4}{5}(x + 4) = \frac{9}{5} - \frac{4}{5}x$

$L(-4.2) = \frac{9}{5} - \frac{4}{5}(-4.2) = 5.1600$

$$f(-4.2) = [(-4.2)^2 + 9]^{1/2} = 5.1614$$

21. $A = \pi r^2$, $r = 2.00$, $\Delta r = 0.02$.

   a) Estimated change:

$$\frac{dA}{dr} = 2\pi r \Rightarrow \Delta A \approx \frac{dA}{dr}\Delta r = 2\pi r\Delta r = 2\pi(2)(0.002) = 0.08\pi$$

   b) Actual change:

$$\Delta A = A(2.02) - A(2) = \pi(2.02)^2 - \pi(2)^2 = (4.0804 - 9)\pi = 0.0204\pi$$

Error in estimate = Actual change − Estimated change

$$= (0.0804)\pi - (0.08)\pi = (0.0004)\pi$$

Error in estimate as percentage of origin area is

$$100 \cdot \frac{\text{error}}{\text{Area}} = 100 \cdot \frac{(0.0004)\pi}{4\pi} = 0.01 \text{ percent}$$

23. $v = \frac{4}{3}\pi r^3$; $\frac{dV}{dr} = 4\pi r^3 \Rightarrow \Delta v \propto \frac{dv}{dr}\Delta r - 4\pi(100)^2(1) = 40,000\pi cm^3$.

Estimated error is $40,000\pi cm^3$ or 3 percent of measured volume.

25. a) $s = 6x^2$; $\frac{ds}{dx} = 12x \Rightarrow \Delta s \approx \frac{ds}{dx}\Delta x = 12x\Delta x$. Percent error ≤ 2 percent $\Rightarrow$ percent error $= 100 \left|\frac{\Delta s}{s}\right| = 100 \left|\frac{12x\Delta x}{6x^2}\right| \le 2$. Therefore $100 \left|\frac{\Delta x}{x}\right| \le 1$ percent. The side length must be measured within 1 percent.

   b) $v = x^3$; $\frac{dv}{dx} = 3x^2 \Rightarrow \Delta v \approx \frac{dv}{dx}\Delta x = 3x^2\Delta x$. Percent error $= 100 \left|\frac{\Delta v}{v}\right| = 100 \left|\frac{3x^2\Delta x}{x^3}\right| = 100 \cdot 3\left|\frac{\Delta x}{x}\right| \le 3$ percent, since $100 \left|\frac{\Delta x}{x}\right| \le 1$. The volume is measured within 3 percent of its true value. In sum, when the edge is measured within 1 percent error, the surface area is calculated with 2 percent error, and the volume is calculated with 3 percent error.

27. c = circumference, r = radius, s = surface area, v = volume.

c = 10cm, Δc = 0.4cm.

   a) $c = 2\pi r \Rightarrow r = \frac{c}{2\pi} \Rightarrow \Delta r = \frac{\Delta c}{2\pi} = \frac{0.4}{2\pi} = \frac{0.2}{\pi}cm$.

   b) $s = A\pi r^2 = 4\pi(\frac{c}{2\pi})^2 = \frac{c^2}{\pi}$; $\frac{ds}{dc} = \frac{2c}{\pi} \Rightarrow \Delta s \approx \frac{ds}{dc}\Delta c = \frac{2c}{\pi}\Delta c = \frac{2(10)}{\pi}(0.4) = \frac{8}{\pi}cm^2$

c) $v = \frac{4}{3}\pi r^3 = \frac{4}{3}\pi(\frac{c}{2\pi})^3 = \frac{c^3}{6\pi^2}, \frac{dv}{dc} = \frac{c^2}{2\pi^2} \Rightarrow$

$\Delta v \approx \frac{dv}{dc}\Delta c = \frac{c^2}{2\pi^2}\Delta c = \frac{(10)^2}{2\pi^2}(0.4) = \frac{20}{\pi^2}cm^3$. Possible

errors in radius, surface area, and volume are

$\frac{0.2}{\pi}cm, \frac{8}{\pi}cm^2, \frac{20}{\pi^2}cm^3$.

29. a) (D is the interior diameter)

$V = 10\pi r^2 = 10\pi(\frac{D}{2})^2 = \frac{5}{2}\pi D^2; \frac{dV}{dD} = 5\pi D. \Delta V \approx \frac{dv}{dD} D = 5\pi D\Delta D.$

One percent error in volume calculation implies $1 \geq 100 \cdot |\frac{\Delta V}{V}| = 100 \cdot$

$|\frac{5\pi D\Delta D}{(5/2)\pi D^2}| = 100 \cdot 2|\frac{\Delta D}{D}|.$ $100 \cdot |\frac{\Delta D}{D}| \leq \frac{1}{2}$, which

indicates that the measurement error of the interior diameter must be less

than $\frac{1}{2}$ percent.

b) (D is the exterior diameter in part (b), A = area, C = circumference,

h = height).

$A = C \cdot h = (D \cdot \pi)h = 10\pi D \Rightarrow \Delta A = 10\pi\Delta D.$ Five percent error in area

calculation implies $5 \geq 100 |\frac{\Delta A}{A}|$

$= 100 |\frac{10\pi\Delta D}{\pi A D}| = 100 |\frac{\Delta D}{D}|.$ Thus the diameter should be measured

to within 5 percent of the true value.

31. a) $T^2 = 4\pi^2 L/g \Rightarrow L = \frac{T^2 g}{4\pi^2}; T = 1 \Rightarrow L = \frac{(1)^2(32.2)}{4\pi^2}$

$= \frac{8.05}{\pi^2} \approx 0.8156ft \approx 9.78in.$

b) $T = 2\pi L^{1/2} g^{-1/2}; \frac{dT}{dL} = \frac{1}{2}(2\pi)L^{-1/2}g^{-1/2} = \pi(Lg)^{-1/2}$

$\Delta T = \frac{dT}{dL}\Delta L = \pi(Lg)^{-1/2}\Delta L = \pi[(0.8156)(32.2)]^{-1/2}(0.01)$

$= 0.00195\pi \approx 0.00613s.$

c) The number of seconds in a day is $60 \cdot 60 \cdot 24 = 86,400\frac{sec}{day}$. The clock

with a long pendulum arm will record a second (it will tick) every

$T + \Delta T = 1.00613$ sec. The total number of record seconds (ticks) each day

is $\frac{86,400(sec/day)}{1.00613(sec/tick)} = 85,873.6\frac{tick}{day}$.

Therefore the clock loses $86,400 - 85.873.6 = 526.4$ seconds each day.

(526.4 seconds is 8 minutes, 46.4 seconds).

33. $(1 + x)^2 \approx 1 + 2x$ $\left(\text{Using } (1 + x)^k \approx 1 + kx\right)$

35. $(1 + x)^{-5} \approx 1 - 5x$

37. $2(1 - x)^{-4} \approx 2(1 - 4(-x)) = 2 + 8x$

39. $2(1 + x)^{1/2} \approx 2(1 + \frac{x}{2}) = 2 + x$

41. $\dfrac{1}{1 + x} = \dfrac{1}{1 - (-x)} \approx 1 + (-x) = 1 - x$

43. a) $(1.0002)^{100} = (1 + 0.0002)^{100} \approx 1 + 100(0.0002) = 1.02$

    b) $(1.009)^{1/3} = (1 + 0.009)^{1/3} \approx 1 + \frac{1}{3}(0.009) = 1.003$

    c) $(0.999)^{-1} = (1 - 0.001)^{-1} \approx 1 + (-1)(-0.001) = 1.001$

45. $g(x) = \sqrt{x} + \sqrt{1 + x} - 4$

    a) $g(3) = \sqrt{3} + \sqrt{1 + 3} - 4 = \sqrt{3} - 2 = \sqrt{3} - \sqrt{4} < 0.$

    $g(4) = \sqrt{4} + \sqrt{1 + 4} - 4 = \sqrt{5} - 2 = \sqrt{5} - \sqrt{4} > 0.$

    The intermediate value theorem says that if a continuous function f has f(a) > 0 and f(b) < 0, then f(c) = 0 for some c between a and b. The function g satisfies these conditions, so g(c) = 0 for some c between 3 and 4.

    b) If $x = 3 + \Delta x$, then $\Delta x \approx 0$ when x is near 3.

    $g(x) = \sqrt{x} + \sqrt{1 + x} - 4 = \sqrt{3 + \Delta x} + \sqrt{1 + (3 + \Delta x)} - 4$

    $= \sqrt{3}(1 + \frac{\Delta x}{3})^{1/2} + 2(1 + \frac{\Delta x}{4})^{1/2} - 4.$

    No approximation has been done yet. The $\sqrt{3}$ in the first term and the $\sqrt{4}$ in the second term were divided out in order that each term be of the form $a(1 + b\Delta x)^{1/2}$, which can be approximated by

    $a(1 + b\Delta x)^{1/2} \approx a(1 + \frac{1}{2}b\Delta x).$

    Since $\Delta x \approx 0$, equation (1) can be approximated by:

    $\sqrt{3}(1 + \frac{\Delta x}{6}) + 2(1 + \frac{\Delta x}{8}) - 4 = (\sqrt{3} - 2) + (\frac{\sqrt{3}}{6} + \frac{1}{4})\Delta x.$

    Setting this to zero and solving for Δx,

    $(\sqrt{3} - 2) + (\frac{\sqrt{3}}{6} + \frac{1}{4})\Delta x = 0 \Rightarrow \Delta x = \dfrac{2 - \sqrt{3}}{\sqrt{3}/6 + 1/4} \approx 0.49742.$

Hence, $x = 3 + \Delta x = 3.49742$ solves $g(x) = 0$, approximately.

c) $g(3.49742) = \sqrt{3.49742} + \sqrt{4.49742} - 4 = -0.00915$, which is within 0.001 of

zero.

47. The third formula says $(1 + x)^k \approx 1 + kx$, where k is any number.

1. If $k = \frac{1}{2}$, then the third formula says $(1 + x)^{1/2} \approx 1 + \frac{1}{2}x$, which is

the second formula.

2. If $k = -1$ and $x = -z$, then the third formula says $(1 + x)^{-1} \approx 1 - x$ or

replacing x by $-z$, $(1 + x)^{-1} = (1 - z)^{-1} = \frac{1}{1 - z} \approx 1 - x =$

$1 - (-z) = 1 + z$ or $\frac{1}{1 - z} \approx 1 + z$, which is the second formula,

with z used in place of x.

49. Equation (17) gives the mass m approximation $m \approx m_0 + \frac{1}{2}m_0 v^2 (\frac{1}{c^2})$ or

$\Delta m \approx \frac{1}{2}m_0 v^2 (\frac{1}{c^2})$. If the mass m is 1 percent greater, then

$1 = 100 \frac{\Delta m}{m_0} = 100 \frac{(1/2)m_0 v^2 (1/(c^2))}{m_0} = (\frac{50}{c^2})v^2$,

$v^2 = \frac{c^2}{50} \Rightarrow v = \frac{c}{\sqrt{50}} = \frac{\sqrt{2}}{10}c \approx 0.1414c$. The mass must travel

at 14 percent of the speed of light in order to have its mass increase by

1 percent.

Article 2.6

1. $y^2 = x^2 \Rightarrow \frac{dy}{dx} = 2x$; $x = 2t - 5 \Rightarrow \frac{dy}{dx} = 2$.

$\frac{dy}{dx} = \frac{dy}{dx}\frac{dx}{dt} = (2x)(2) = 4x = 4(2t - 5) = 8t - 20$.

3. $y = 8 - \frac{x}{3} \Rightarrow \frac{dy}{dx} = -\frac{1}{3}$; $x = t^3 \Rightarrow \frac{dx}{dt} = 3t^2$

$\frac{dy}{dt} = \frac{dy}{dx}\frac{dx}{dt} = (-\frac{1}{3})(3t^2 = -t^2)$

5. $2x - 3y = 9 \Rightarrow \frac{dy}{dx} = \frac{2}{3}$; $2x + \frac{t}{3} = 1 \Rightarrow x = \frac{1}{2} - \frac{t}{6} \Rightarrow \frac{dx}{dt} = -\frac{1}{6}$

$\frac{dy}{dt} = \frac{dy}{dx}\frac{dx}{dt} = (\frac{2}{3})(-\frac{1}{6}) = -\frac{1}{9}$

7. $y = (x + 2)^{1/2} \Rightarrow \frac{dy}{dx} = \frac{1}{2}(x + 2)^{-1/2}$; $x = \frac{2}{t} = 2t^{-1} \Rightarrow \frac{dx}{dt} = -2t^{-2}$

$\frac{dy}{dt} = \frac{dy}{dx}\frac{dx}{dt} = (\frac{1}{2}(x + 2)^{-1/2})(-2t^{-2}) = -(2t^{-1} + 2)^{-1/2}t^{-2}$

9. $y = x^2 + 3x - 7 \Rightarrow \dfrac{dy}{dx} = 2x + 3$; $x = 2t + 1 \Rightarrow \dfrac{dx}{dt} = 2$

$\dfrac{dy}{dt} = \dfrac{dy}{dx}\dfrac{dx}{dt} = (2x + 3)(2) = 4x + 6 = 4(2t + 1) + 6 = 8t + 10$

11. $z = w^2 - w^{-1} \Rightarrow \dfrac{dz}{dw} = 2w + w^{-2}$; $w = 3x \Rightarrow \dfrac{dw}{dx} = 3$

$\dfrac{dz}{dx} = \dfrac{dz}{dw}\dfrac{dw}{dx} = (2x + w^{-2})(3) = 6w + 3w^{-2}$. In terms of x this is

$\dfrac{dz}{dx} = 18x + \dfrac{1}{3}x^{-2}$.

13. $r = (s + 1)^{1/2} \Rightarrow \dfrac{dr}{ds} = \dfrac{1}{2}(s + 1)^{-1/2}$; $s = 16t^2 - 20t \Rightarrow \dfrac{ds}{dt} = 32t - 20$

$\dfrac{dr}{dt} = \dfrac{dr}{ds}\dfrac{ds}{dt} = (\dfrac{1}{2}(s + 1)^{-1/2})(32t - 20)$

$\quad = (s + 1)^{1/2}(16t - 10) = (16t^2 - 20t + 1)^{-1/2}(16t - 10)$

15. $u = t + t^{-1} \Rightarrow \dfrac{du}{dt} = 1 - t^{-2}$; $t = 1 - v^{-1} \Rightarrow \dfrac{dt}{dv} = v^{-2}$

$\dfrac{du}{dv} = \dfrac{du}{dt}\dfrac{dt}{dv} = (1 - t^{-2})v^{-2}$

$(1 - t^2)$ is expressed in terms of v as follows:

$t = 1 - v^{-1} = \dfrac{v - 1}{v} \Rightarrow t^{-2} = \dfrac{v^2}{(v - 1)^2} \Rightarrow 1 - t^{-2} = \dfrac{v^2 - 2v + 1}{(v - 1)^2} - \dfrac{v^2}{(v - 1)^2}$

$\quad = \dfrac{1 - 2v}{(v - 1)^2}$

$\dfrac{du}{dv} = (1 - t^{-2})v^{-2} = \dfrac{(1 - 2v)}{(v - 1)^2}v^{-2} = \dfrac{1 - 2v}{(v^2 - v)^2}$

17. $x^2 = t^2 \Rightarrow t = x^{1/2}$; $y = t^3 = (x^{1/2})^3 = x^{3/2}$

$\dfrac{dy}{dx} = \dfrac{3}{2}x^{1/2}$; $\dfrac{dx}{dt} = 2t = 2x^{1/2}$, $\dfrac{dy}{dt} = 3t^2 = 3x$

Chain Rule — $\dfrac{dy}{dx}\dfrac{dx}{dt} = \dfrac{dy}{dt}$: $(\dfrac{3}{2}x^{1/2})(2x^{1/2}) = 3x$.

19. $x = \dfrac{t}{1 + t} \Rightarrow \dfrac{1}{x} = \dfrac{1 + t}{t} = \dfrac{1}{t} + 1 = \dfrac{1}{x} - 1 = \dfrac{1 - x}{x} \Rightarrow$

$t = \dfrac{x}{1 - x}; \ y = \dfrac{t^2}{1 + t} = t(\dfrac{t}{1 + t}) = (\dfrac{x}{1 - x})x = \dfrac{x^2}{1 - x}$

$\dfrac{dy}{dx} = \dfrac{d}{dx}(\dfrac{x^2}{1 - x}) = 2x\dfrac{(1 - x) - x^2(-1)}{(1 - x)^2} = \dfrac{2x - x^2}{(1 - x)^2}$

In terms of t, this is

$\dfrac{dy}{dx} = \dfrac{2x - x^2}{(1 - x)^2} = (\dfrac{x}{1 - x})(\dfrac{2 - x}{1 - x}) = t\left[\dfrac{2 - t/(1 + t)}{1 - t/(1 + t)}\right] = t\left[\dfrac{2(1 + t) - t}{(1 + t) - t}\right]$

$\dfrac{dy}{dx} = t[\dfrac{2 + t}{1}] = 2t + t^2$

The other derivatives are computed directly

$\dfrac{dx}{dt} = \dfrac{d}{dt}(\dfrac{t}{1 + t}) = \dfrac{1(1 + t) - t(1)}{(1 + t)^2} = \dfrac{1}{(1 + t)^2};$

$\dfrac{dy}{dt} = \dfrac{d}{dt}(\dfrac{t^2}{t + 1}) = \dfrac{2t(t + 1) - t^2(1)}{(t + 1)^2} = \dfrac{2t + t^2}{(t + 1)^2};$

Chain Rule $\dfrac{dy}{dx}\dfrac{dx}{dt} = \dfrac{dy}{dt}:$ $(2t + t^2)(\dfrac{1}{(1 + t)^2}) = \dfrac{2t + t^2}{(t + 1)^2}$

Since both sides are equal, the chain rule holds.

21. $x = 1 - t \Rightarrow t = 1 - x; \ y = 1 + t = 1 + (1 - x) = 2 - x.$

The curve is the straight line $y = 2 - x$.

23. $x = t; \ y = \dfrac{1}{t} = \dfrac{1}{x};$ Curve is two halves of hyperbola $y = \dfrac{1}{x}$.

25. $x = t; \ y = \sqrt{t} \Rightarrow y^2 = t = x;$ Curve is right facing parabola $x = y^2$.

27. $y = t^2 \Rightarrow \dfrac{dy}{dt} = 2t = 4$ when $t = 2$. $x = 4t - 5 \Rightarrow \dfrac{dx}{dt} = 4$. By the chain

rule, $(\dfrac{dy}{dx})(\dfrac{dx}{dt}) = \dfrac{dy}{dt} \Rightarrow (\dfrac{dy}{dx})(4) = 4$. Therefore $\dfrac{dy}{dx} = 1$,

which is choice (c).

29. $x = t + \dfrac{1}{t} \quad \Rightarrow x(2) = 2 + \dfrac{1}{2} = \dfrac{5}{2}$

$\dfrac{dx}{dt} = 1 - \dfrac{1}{t^2} \Rightarrow \dfrac{dx}{ft}(2) = 1 - \dfrac{1}{4} = \dfrac{3}{4}$

$y = t - \dfrac{1}{t} \quad \Rightarrow y(2) = 2 - \dfrac{1}{2} = \dfrac{3}{2}$

$\dfrac{dy}{dt} = 1 + \dfrac{1}{t^2} \Rightarrow \dfrac{dy}{dx}(2) = 1 + \dfrac{1}{4} = \dfrac{5}{4}$

$\dfrac{dy}{dx} = \dfrac{dy/dt}{dx/dt} = \dfrac{5/4}{3/4} = \dfrac{5}{3}.$

The tangent has slope $\dfrac{5}{3}$ and passes through $(\dfrac{5}{2}, \dfrac{3}{2})$.

$y = \dfrac{5}{3}x + b; \ \dfrac{3}{2} = \dfrac{5}{3}(\dfrac{5}{2}) + b \Rightarrow b = -\dfrac{8}{3}$

$y = \frac{5}{3}x - \frac{8}{3}$ or $5x - 3y = 8$

31. $x = t(2t + 5)^{1/2} \Rightarrow x(2) = 2(4 + 5)^{1/2} = 6$

$\frac{dx}{dt} = (2t + 5)^{1/2} + \frac{1}{2}t(2t + 5)^{-1/2}(2) = (2t + 5)^{1/2} + t(2t + 5)^{-1/2}$

$\frac{dx}{dt}(2) = (4 + 5)^{1/2} + 2(4 + 5)^{-1/2} = 3 + \frac{2}{3} = \frac{11}{3}$

$y = (4t)^{1/3} = 4^{1/3}t^{1/3} \Rightarrow y(2) = 4 \cdot 2)^{1/3} = 2$

$\frac{dy}{dt} = \frac{4^{1/3}}{3}t^{-2/3} \Rightarrow \frac{dy}{dt}(2) = \frac{1}{3}(4)^{1/3}(2)^{-2/3} = \frac{1}{3}$

$\frac{dy}{dx} = \frac{dy/dt}{dx/dt} = \frac{1/3}{11/3} = \frac{1}{11}.$

The tangent has slope $\frac{1}{11}$ and passes through (6,2)

$y = \frac{x}{11} + b; \quad 2 = \frac{1}{11}(6) + b \Rightarrow b = \frac{16}{11}$

$y = \frac{x}{11} + \frac{16}{11}$ or $x - 11y = -16.$

33. $x = t^2, \quad y = (t^2 + 12)^{1/2} = (x + 12)^{1/2}$

$\Rightarrow \frac{dy}{dx} = \frac{1}{2}(x + 12)^{-1/2} = \frac{1}{2}(t^2 + 12)^{-1/2}$

$\frac{dx}{dt} = 2t; \quad \frac{dy}{dt} = \frac{1}{2}(t^2 + 12)^{-1/2}2t = t(t^2 + 12)^{-1/2}$

Chain Rule $\frac{dy}{dx}\frac{dx}{dt} = \frac{dy}{dt}: \quad \frac{1}{2}(t^2 + 12)^{-1/2}(2t) = t(t^2 + 12)^{-1/2}$

or $t(t^2 + 12)^{-1/2} = t(t^2 + 12)^{-1/2}$, so the chain rule holds in this case.

35. $\frac{dy}{dx} = \frac{dy/dt}{dx/dt} = 0 \Rightarrow \frac{dy}{dt} = 0.$ Since the numerator of a fraction

that equals zero must be zero.

$y = 64t - 16t^2 \Rightarrow \frac{dy}{dt} = 64 - 32t = 0 \Rightarrow t = 2.$

Also, $x = 80t$ implies $\frac{dx}{dt} = 80 \neq 0$, so $\frac{dy}{dx}$ does exist.

37. $x^2 + y^2 = 25 \Rightarrow 2x + 2y\frac{dy}{dx} = 0 \Rightarrow \frac{dy}{dx} = -\frac{2x}{2y} = -\frac{x}{y}.$

At (3,4), $\frac{dy}{dx} = -(\frac{x}{y}) = -(\frac{3}{4}).$ It is known that $\frac{dx}{dt} = 4.$ By the

Chain Rule, $(\frac{dy}{dx})(\frac{dx}{dt}) = \frac{dy}{dt}.$ Hence $(-\frac{3}{4})(4) = -3 \Rightarrow \frac{dy}{dt} = -3.$

39. $f(x) = x^2$ is differentiable everywhere including the origin. $g(x) = |x|$ is not differentiable at the origin (although it is differentiable everywhere else). $f(g(x)) = (|x|)^2 = x^2$ since $(-x)^2 = (x)^2.$ $g(f(x)) = |x^2| = x^2$ since $x^2 \geq 0$ always. Hence $f(g(x)) = g(f(x)) = x^2$, which is differentiable everywhere including zero. The Chain Rule says that if $h = f(g(x))$ then $h' = f' \cdot g'$. At zero, however, g' does not exist. It seems paradoxical that a real number, (f'), when multiplied by a non-existant number, (g'), yields a real number (h')! The paradox is resolved by noting that the Chain Rule only applies to differentiable functions. That is, f' and g' must both exist at the point in question. Since g' does not exist at zero, the Chain Rule does not apply. The

Chain Rule is not contradicted here (no valid theorem is ever contradicted), it simply does not apply.

41. By problem 21, $y = 2 - x \Rightarrow \frac{dy}{dx} = y' = -1$.

$\frac{d^2y}{dx^2} = \frac{dy'/dt}{dy/dt} = 0$ since $\frac{d}{dt}(-1) = 0$.

43. $x = t \Rightarrow \frac{dx}{dt} = 1$

$y = \frac{1}{t} = t^{-1} \Rightarrow \frac{dy}{dt} = -t^{-2}$

$y' = \frac{dy}{dx} = \frac{dy/dt}{dx/dt} = \frac{-t^{-2}}{1} = -t^{-2}$, $\frac{d^2y}{dx^2} = \frac{dy'/dt}{dx/dt} = \frac{2t^{-3}}{1} = 2t^{-3} = 2x^{-3}$.

45. a) $x = t + \frac{1}{t} \Rightarrow \frac{dx}{dt} = 1 - \frac{1}{t^2}$

$y = t - \frac{1}{t} = \frac{t^2 - 1}{1} \Rightarrow \frac{dy}{dt} = 1 + \frac{1}{t^2}$

$y' = \frac{dy}{dx} = \frac{dy/dt}{dx/dt} = \frac{1 + 1/(t^2)}{1 - 1/(t^2)} = \frac{t^2 + 1}{t^2 - 1}$

$\frac{dy'}{dt} = \frac{2t(t^2 - 1) - (t^2 + 1)}{(t + 1)^2} = \frac{-4t}{(t^2 - 1)^2}$

$\frac{d^2y}{dx^2} = \frac{dy'/dt}{dx/dt} = \frac{-4t/((t^2 - 1)^2)}{1 - 1/(t^2)} = \frac{-4t^3}{(t^2 - 1)^3} = -4y^{-3}$

Note: The equation of this curve is $x^2 - y^2 = 4$.

b) $x = \frac{t - 1}{t + 1}$; $y = \frac{t + 1}{t - 1} = \frac{1}{x} = x^{-1}$

$\frac{dy}{dx} = -x^{-2}$; $\frac{d^2y}{dx^2} = 2x^{-3}$

47. a) $T = 2\pi(L/g)^{1/2} \Rightarrow \frac{dT}{dg} = 2\pi L^{1/2}(-\frac{1}{2}g^{-3/2}) = -\pi L^{1/2}g^{-3/2}$

$\Delta T \approx \Delta T_{tan} = \frac{dT}{dg}\Delta g = -\pi \sqrt{L}g^{-3/2}\Delta g$.

b) If g increases, Δg is positive and ΔT is negative. The period decreases, which means the pendulum speed increases, and the clock speed increases. This makes sense intuitively, for if g were to decrease to zero, the speed would also go to zero and the period would become infinite.

c) $\Delta T = 0.001s = -\pi \sqrt{L}g^{-3/2}\Delta g = -\pi \sqrt{100}(980)^{-3/2}\Delta g$

$\Rightarrow \Delta g \approx -\frac{0.001}{10\pi}(980)^{3/2} \approx -\frac{3.067}{\pi} \approx -0.976 \text{ m/s}^2$

Article 2.7

1.  $y = 2 \sin x$

3. $y = \sin(-x)$

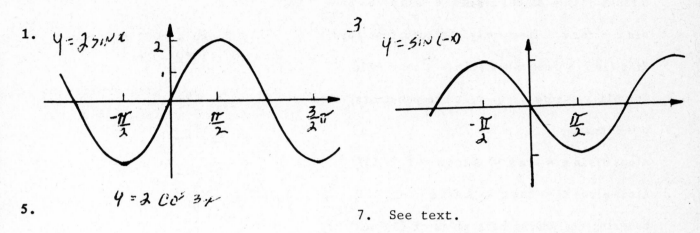

5. $y = 2 \cos 3x$

7. See text.

9. See text.

11. See text.

13. See text. Similar to sine curve.
    More flat at base, more steep
    near peaks.

15. $y = \sin x - \cos x$. Assume the form $y =$

$A \sin(x + a) = A[(\cos a)\sin x + (\sin a)\cos x] =$

$\sin x - \cos x$. Then apply the rule that says:

If $p \sin x + q \cos x = r \sin x + t \cos x$ holds

for all x, where p, q, r, t are constants, then

$p = r$ and $q = t$.

$A(\cos a)\sin x = \sin x \Rightarrow A \cos a = 1$ (1)

$A(\sin a)\cos x = -\cos x \Rightarrow A \sin a = -1$ (2)

Squaring and adding both sides of (1) and (2),

$A^2 \cos^2 a + A^2 \sin^2 a = (1)^2 + (-1)^2$ or

$A^2(\cos^2 a + \sin^2 a) = 2$

$$A^2 = 2 \Rightarrow A = \sqrt{2}.\qquad (3)$$

Equations (1), (2), and (3) are satisfied if

$a = -\dfrac{\pi}{4}$.

Hence, $y = \sin x - \cos x = \sqrt{2} \sin(x - \dfrac{\pi}{4})$.

$y = \sin x - \cos x =$
$\sqrt{2} \sin (x - \pi/4)$

Sine curve with amplitude $\sqrt{2}$, and shifted $\pi/4$ units to the right.

17.   19.   See text.

$y = 2\cos (4x - 2\pi) =$
$2 \cos 4x$

Full cycle every $\pi/2$ units.
Amplitude is 2.

21.   See text.   **23. Same as problem 21.**

25. Same as problem 21 with radius 3.

27. $x = \cos^2 t$, $y = \sin^2 t$. Hence,

$x + y = \cos^2 t + \sin^2 t = 1 \Rightarrow x + y = 1$. Since

$x \geq 0$, $y \geq 0$ the curve (which is a straight

line) is restricted to the first quadrant.

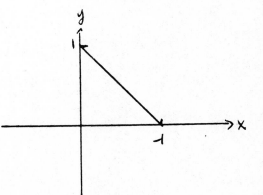

29. Referring to problem 28, with a = 2, the cartesian equation $x^2 + y^2 = 4 = (2)^2$ corresponds to the parametric equations x = 2 cos t, y = 2 sin t. As t increases, the point (x,y) moves counterclockwise. If y = -2 sin t, the motion is clockwise and the equation $x^2 + y^2 = 4$ is still satisfied. Hence

a) x = 2 cos t, y = -2 sin t

b) x = 2 cos t, y = 2 sin t

31. When y = A sin(kx - ϕ) + v, the amplitude is A, the period is $\frac{2\pi}{k}$, the horizontal shift is ϕ, and the vertical shift is v.

$f(x) = 37 \sin[\frac{2\pi}{365}(x - 101)] + 25 = 37 \sin[\frac{2\pi}{365}x - \frac{202}{365}\pi] + 25.$

a) The amplitude is 37

b) The period is $\frac{2\pi}{k} = \frac{2\pi}{2\pi/365} = 365$ days or 1 year

c) The horizontal shift is $\frac{202}{365}\pi$ days

d) The vertical shift is 25.

33. $\lim_{h\to 0} \frac{\sin h}{h} = 1$.

   $\lim_{h\to 0} \frac{1 - \cos h}{h} = \lim_{h\to 0} -\left[\frac{\cos(h + 0) - \cos(0)}{h}\right] = -\frac{d}{dx}\cos x\Big|_0 = \sin(0) = 0$

   $\lim_{h\to 0} \frac{(\sin h)(1 - \cos h)}{h^2} = (\lim_{h\to 0} \frac{\sin h}{h})(\lim_{h\to 0} \frac{1 - \cos h}{h}) = (1)(0) = 0$

35. $\frac{1 - \cos x}{\sin x} = (\frac{1 - \cos x}{x})(\frac{x}{\sin x})$

   $\lim_{x\to 0} \frac{1 - \cos x}{x} = \lim_{x\to 0} -\left[\frac{\cos(x + 0) - \cos(0)}{x}\right] = \frac{-d}{dx}\sin x\Big|_0 = \sin(0) = 0$

   $\lim_{x\to 0} \frac{\sin x}{x} = 1 \Rightarrow \lim_{x\to 0} \frac{x}{\sin x} = 1$

   $\lim_{x\to 0} \frac{1 - \cos x}{\sin x} = \lim_{x\to 0}(\frac{1 - \cos x}{x})(\frac{x}{\sin x}) = (0)(1) = 0$

37. When A = B, equation (11e) says $\cos(0) = \cos A \cos A + \sin A \sin A$ or

   $\cos^2 A + \sin^2 A = 1$.

39. In finding the angle between two curves, the formula used is

   $\tan \beta = \left|\frac{m_2 - m_1}{1 + m_1 m_2}\right|$, where m_1 and m_2 are the slopes

   of the two curves at the point of intersection and β is the angle between them.

   $3x + y = 5 \Rightarrow y' = -3 \Rightarrow m_1 = -3$

   $2x - y = 4 \Rightarrow y' = 2 \Rightarrow m_2 = 2$

   $\tan \beta = \left|\frac{2 - (-3)}{1 + (2)(-3)}\right| = \left|\frac{5}{-5}\right| = 1 \Rightarrow \beta = \frac{\pi}{4}$.

41. $y_1 = (\frac{3}{4} + x)^{1/2}$, $y_2 = (\frac{3}{4} - x)^{1/2}$. Intersection occurs when $y_1 = y_2$ or

   $(\frac{3}{4} + x)^{1/2} = (\frac{3}{4} - x)^{1/2} \Rightarrow \frac{3}{4} + x = \frac{3}{4} - x \Rightarrow 2x = 0$

   $x = 0$, $y_1 = y_2 = (\frac{3}{4})^{1/2}$

43. $y_1 = x^2$ $y_2 = x^{1/3}$.

   By inspection or algebra, the two points of intersection can be found to be

   a) (0,0)   and   b) (1,1).

   a) $y_1' = 2x = 2(0) = 0 \Rightarrow m_1 = 0$

   $y_2' = \frac{1}{3}x^{-2/3} = \frac{1}{3}(0)^{-2/3} = \infty = m_2 = \infty$.

   Since y_1 is horizontal and y_2 is vertical, the angle between them is $\frac{1}{2}$.

b) $y_1' = 2x = 2(1) = 2$ at $x = 1 \Rightarrow m_1 = 2$

$y_2' = \frac{1}{3}x^{-2/3} = \frac{1}{3}(1)^{-2/3} = \frac{1}{3}$ at $x = 1 \Rightarrow m_2 = \frac{1}{3}$

$\tan \beta = \left| \frac{1/3 - 2}{1 + (1/3)(2)} \right| = \left| \frac{-(5/3)}{5/3} \right| = 1 \Rightarrow$

$\beta = \frac{\pi}{4}$

45. A curve crosses a horizontal line at a 45° angle whenever the curve has a slope

of 45°, that is when $\frac{dy}{dx} = 1$.

$y = x^{1/2} = \frac{dy}{dx} = \frac{1}{2}x^{-1/2} = 1 \Rightarrow x^{1/2} = \frac{1}{2} \Rightarrow x = \frac{1}{4}$.

$y(\frac{1}{4}) = (\frac{1}{4})^{1/2} = \frac{1}{2}$. So the horizontal line has equation $y = \frac{1}{2}$.

47. m_1 is the slope from A to B: $m_1 = \frac{y_1 - y_0}{x_1 - x_0} = \frac{-1 - 1}{3 - 1} = -1$

m_2 is the slope from A to C: $m_2 = \frac{2 - 1}{5 - 1} = \frac{1}{4}$

m_3 is the slope from B to C: $m_3 = \frac{2 - (-1)}{5 - 3} = \frac{3}{2}$

The formula used is $\tan \theta = \left| \frac{m_2 - m_1}{1 + m_1 m_2} \right|$,

where m_1 and m_2 are the slopes of the two curves meeting at angle θ.

$\tan A = \left| \frac{m_2 - m_1}{1 + m_1 m_2} \right| = \left| \frac{1/4 - (-1)}{1 + (1/4)(-1)} \right| = \frac{5}{3}$

$\tan B = \left| \frac{m_3 - m_1}{1 + m_1 m_3} \right| = \left| \frac{3/2 - (-1)}{1 + (3/2)(-1)} \right| = 5$

$\tan C = \left| \frac{m_3 - m_2}{1 + m_2 m_3} \right| = \left| \frac{3/2 - 1/4}{1 + (3/2)(1/4)} \right| = \frac{10}{11}$

The angles are $\tan^{-1}(5/3)$, $\tan^{-1}5$, $\tan^{-1}(\frac{10}{11})$.

REMARK: Referring to problem 49, we check that

$\tan A + \tan B + \tan C = \tan A \tan B \tan C$:

$\frac{5}{3} + 5 + \frac{10}{11} = (\frac{5}{3})(5)(\frac{10}{11})$ or $\frac{250}{33} = \frac{250}{33}$.

49. $\tan(A + B) = \dfrac{\tan A + \tan B}{1 - \tan A \tan B}$. Replacing A with

A + B and B with C this becomes

$$\tan(A + B + C) = \tan((A + B) + C) = \frac{\tan(A + B) + \tan C}{1 - \tan(A + B)\tan C}$$

The numerator is $\tan(A + B) = \tan C = \dfrac{\tan A + \tan B}{1 - \tan A \tan B} + \left(\dfrac{1 - \tan A \tan B}{1 - \tan A \tan B}\right)\tan C$

$$= \frac{\tan A + \tan B + \tan C - \tan A \tan B \tan C}{1 - \tan A \tan B}$$

The denominator is $1 - \tan(A + B)\tan C = \dfrac{1 - \tan A \tan B}{1 - \tan A \tan B} - \dfrac{\tan A + \tan B}{1 - \tan A \tan B}\tan C$

$$= \frac{1 - \tan A \tan B - \tan A \tan C - \tan B \tan C}{1 - \tan A \tan B}$$

Hence, $\tan(A + B + C) = \dfrac{\tan A + \tan B + \tan C - \tan A \tan B \tan C}{1 - \tan A \tan B - \tan A \tan C - \tan B \tan C}$

If A, B, and C are the interior angles of a triangle, then $A + B + C = 180^{\circ}$ and

$\tan(A + B + C) = \tan(180^{\circ}) = 0$. The numerator in the expression for

$\tan(A + B + C)$ is therefore zero. $\tan A + \tan B + \tan C$

$- \tan A \tan B \tan C = 0$. Hence $\tan A + \tan B + \tan C = \tan A \tan B \tan C$.

51. Equation (11f) reads $\tan(A - B) = \dfrac{\tan A - \tan B}{1 + \tan A \tan B}$.

Replacing A with π and noting that $\tan \pi = 0$, this becomes

$$\tan(\pi - B) = \frac{\tan \pi - \tan B}{1 + \tan \pi \tan B} = \frac{0 - \tan B}{1 + (0)\tan B} = -\tan B.$$

53. Eq.(1) $\sin(A + B) = \sin A \cos B + \sin B \cos A$

Eq.(2) $\sin(A - B) = \sin A \cos B - \sin B \cos A$

Eq.(3) $\cos(A + B) = \cos A \cos B - \sin A \sin B$

Eq.(4) $\cos(A - b) = \cos A \cos B + \sin A \sin B$

a) Eq.(2), with $A = x$, $B = \frac{\pi}{2}$. $\sin(\frac{\pi}{2}) = 1$, $\cos(\frac{\pi}{2}) = 0$

$\sin(x - \frac{\pi}{2}) = \sin x \cos(\frac{\pi}{2}) - \sin(\frac{\pi}{2})\cos x = \cos x$

b) Eq.(1), with $A = x$, $B = \frac{\pi}{2}$

$\sin(x + \frac{\pi}{2}) = \sin x \cos(\frac{\pi}{2}) + \sin(\frac{\pi}{2})\cos x = \cos x$

c) Eq.(4), with $A = x$, $B = \frac{\pi}{2}$

$$\cos(x - \tfrac{\pi}{2}) = \cos x \cos(\tfrac{\pi}{2}) + \sin x \sin(\tfrac{\pi}{2}) = \sin x$$

d) Eq.(3), with $A = x$, $B = \tfrac{\pi}{2}$

$$\cos(x + \tfrac{\pi}{2}) = \cos x \cos(\tfrac{\pi}{2}) - \sin x \sin(\tfrac{\pi}{2}) = -\sin x$$

e) Eq.(2), with $A = \pi$, $B = x$ $(\sin \pi = 0,\ \cos \pi = -1)$

$$\sin(\pi - x) = \sin \pi \cos x - \sin x \cos \pi = \sin x$$

f) Eq.(4) with $A = \pi$, $B = x$

$$\cos(\pi - x) = \cos \pi \cos x + \sin \pi \sin x = -\cos x$$

Article 2.8

1. $y = \sin(x + 1) \Rightarrow \dfrac{dy}{dx} = \cos(x + 1)\dfrac{d}{dx}(x + 1) = \cos(x + 1)$

3. $y = \sin(\tfrac{x}{2}) \Rightarrow \dfrac{dy}{dx} = \cos(\tfrac{x}{2})\dfrac{d}{dx}(\tfrac{x}{2}) = \tfrac{1}{2}\cos(\tfrac{x}{2})$

5. $y = \cos 5x \Rightarrow \dfrac{dy}{dx} = -5 \sin 5x$

7. $y = \cos(-2x) \Rightarrow y = \cos(2x) \Rightarrow \dfrac{dy}{dx} = -2 \sin 2x$

9. $y = \sin(3x + 4) \Rightarrow \dfrac{dy}{dx} = \cos(3x + 4)\dfrac{d}{dx}(3x + 4) = 3 \cos(3x + 4)$

11. $y = x \sin x \Rightarrow \dfrac{dy}{dx} = \sin x \cdot \dfrac{dx}{dx} + x\dfrac{d}{dx}\sin x = \sin x + x \cos x$

13. By problem 11, $\dfrac{d}{dx}(x \sin x) = \sin x + x \cos x$, so $y = x \sin x + \cos x$

   $\Rightarrow \dfrac{dy}{dx} = (\sin x + x \cos x) - \sin x = x \cos x$

15. $y = \dfrac{1}{\cos x} = (\cos x)^{-1} \Rightarrow \dfrac{dy}{dx} = -(\cos x)^{-2}(-\sin x) = \dfrac{\sin x}{\cos^2 x}$

   Also, $\dfrac{dy}{dx} = \sec x \tan x$

17. $y = \sec x = (\cos x)^{-1} \Rightarrow \dfrac{dy}{dx} = -(\cos x)^{-2}(-\sin x) = \dfrac{\sin x}{(\cos x)^2}$

   Also, $\dfrac{dy}{dx} = \sec x \tan x$

19. $y = \sec(1 - x) = \sec(x - 1) \Rightarrow \dfrac{dy}{dx} = \sec(x - 1)\tan(x - 1)\dfrac{d}{dx}(x - 1)$

$= \sec(x - 1)\tan(x - 1).$ (See Problem 17)

21. $y = \tan 2x = \dfrac{\sin 2x}{\cos 2x} \Rightarrow \dfrac{dy}{dx} = \dfrac{(2\cos 2x)\cos 2x - (-2\sin 2x)\sin 2x}{(\cos 2x)^2}$

$= \dfrac{2}{(\cos 2x)^2} = 2(\sec 2x)^2$

23. $y = \sin^2 x \Rightarrow \dfrac{dy}{dx} = 2\sin x\dfrac{d}{dx}\sin x = 2\sin x\cos x$

Also, $\dfrac{dy}{dx} = \sin 2x$

25. $y = \cos^2 5x \Rightarrow \dfrac{dy}{dx} = 2\cos 5x(-\sin 5x) \cdot 5$

$= -10\sin 5x\cos 5x.$ Also, $\dfrac{dy}{dx} = -5\sin 10x$

27. $y = \tan(5x - 1) = \dfrac{\sin(5x - 1)}{\cos(5x - 1)} \Rightarrow \dfrac{dy}{dx}$

$= \dfrac{\cos(5x - 1)(5\cos(5x - 1)) - \sin(5x - 1)(-5\sin(5x - 1)}{\cos^2(5x - 1)} = \dfrac{5}{\cos^2(5x - 1)}$

$= 5\sec^2(5x - 1)$

29. $y = 2\sin x\cos x = \sin 2x \Rightarrow \dfrac{dy}{dx} = 2\cos 2x.$

Alternatively, $\dfrac{dy}{dx} = 2(\sin x\dfrac{d}{dx}\cos x + \cos x\dfrac{d}{dx}\sin x)$

$= 2(-\sin x\sin x + \cos x\cos x) = 2(\cos^2 x - \sin^2 x) = 2\cos 2x$

31. $y = (2 + \cos 2x)^{1/2} \Rightarrow \dfrac{dy}{dx} = \dfrac{1}{2}(2 + \cos 2x)^{-1/2}\dfrac{d}{dx}(2 + \cos 2x)$

$= \dfrac{1}{2}(2 + \cos 2x)^{-1/2}(-2\sin 2x) = -(2 + \cos 2x)^{-1/2}\sin 2x$

33. $y = \cos(x^{1/2}) \Rightarrow \dfrac{dy}{dx} = -\sin(x^{1/2})\dfrac{d}{dx}x^{1/2}$

$= -\dfrac{1}{2}x^{-1/2}\sin(x^{1/2})$

35. $y = \left(\dfrac{1 + \cos 2x}{2}\right)^{1/2} = \cos x \Rightarrow \dfrac{dy}{dx} = -\sin x.$ Alternatively,

$\dfrac{dy}{dx} = \dfrac{1}{2}\left(\dfrac{1 + \cos 2x}{2}\right)^{-1/2}\dfrac{d}{dx}\left(\dfrac{1 + \cos 2x}{2}\right) = \dfrac{1}{2}\left(\dfrac{1 + \cos 2x}{2}\right)^{-1/2}(-\sin 2x)$

$= \dfrac{1}{2}\left(\dfrac{1}{\cos x}\right)(-2\sin x\cos x) = -\sin x$

37. $x = \tan y \Rightarrow \dfrac{dx}{dx} = \dfrac{d}{dx}\tan y \Rightarrow 1 = \sec^2 y\dfrac{dy}{dx}$

$\Rightarrow \dfrac{dy}{dx} = \dfrac{1}{\sec^2 y} = \cos^2 y.$

(See problem 16 for second step.)

CHAPTER 2 - Article 2.8

Also, $x = \tan y \Rightarrow x^2 + 1 = \tan^2 y + 1 = \sec^2 y \Rightarrow \dfrac{dy}{dx} = \dfrac{1}{1 + x^2}$

39. $y^2 = \sin^4 2x + \cos^4 2x \Rightarrow 2y\dfrac{dy}{dx}$

$= 4\sin^3 2x \cdot \dfrac{d}{dx}\sin 2x + 4\cos^3 2x \dfrac{d}{dx}\cos 2x$

$= 8\sin^3 2x \cos 2x - 8\cos^3 2x \sin 2x$

$= 8\sin 2x \cos 2x(\sin^2 2x - \cos^2 2x)$

$= 4\sin 4x(-\cos 4x) = -2\sin 8x$

$2y\dfrac{dy}{dx} = -2\sin 8x \Rightarrow \dfrac{dy}{dx} = -\dfrac{\sin 8x}{y}$

41. $x + \tan(xy) = 0 \Rightarrow 1 + \sec^2(xy)\dfrac{d}{dx}(xy) = 0$

$\Rightarrow 1 + \sec^2(xy)(y + x\dfrac{dy}{dx}) = 0 \Rightarrow 1 + y\sec^2(xy) = -x\sec^2(xy)\dfrac{dy}{dx}$

$\Rightarrow \dfrac{dy}{dx} = -\dfrac{1 + y\sec^2(xy)}{x\sec^2(xy)}$

43. $x\sin 2y = y\cos 2x \Rightarrow \sin 2y + 2x\cos 2y\dfrac{dy}{dx} = \cos 2x\dfrac{dy}{dx} - 2y\sin 2x$

When $x = \dfrac{\pi}{4}$, $y = \dfrac{\pi}{2}$ this becomes

$\sin \pi + 2(\dfrac{\pi}{4})\cos(\pi)\dfrac{dy}{dx} = \cos 2(\dfrac{\pi}{4})\dfrac{dy}{dx} - 2(\dfrac{\pi}{2})\sin \pi$

$\Rightarrow -\dfrac{\pi}{2}\dfrac{dy}{dx} = 0 \Rightarrow \dfrac{dy}{dx} = 0.$

Therefore the tangent line has equation $y = \dfrac{\pi}{2}$.

45. $\lim\limits_{x\to\pi/4}\sin x = \dfrac{\sqrt{2}}{2}$; $\lim\limits_{x\to\pi/4}\cos x = \dfrac{\sqrt{2}}{2} \neq 0.$

Therefore, $\lim\limits_{x\to\pi/2}\dfrac{\sin x}{\cos x} = \dfrac{(\sqrt{2})/2}{(\sqrt{2})/2} = 1$

47. $\lim\limits_{x\to\pi}1 + \cos x = 1 - 1 = 0 \Rightarrow \lim\limits_{x\to\pi}\sec(1 + \cos x) = 1$

49. $\lim\limits_{x\to 0}x\csc x = \lim\limits_{x\to 0}\dfrac{x}{\sin x} = 1$

51. $\lim\limits_{h\to 0}\dfrac{\cos(a + h) - \cos a}{h} = \dfrac{d}{dx}\cos x\Big|_a = -\sin a$

53. a) $x = \cos^2 t \Rightarrow \dfrac{dx}{dt} = 2\cos t(-\sin t) = -\sin 2t$

$\Rightarrow \dfrac{dx}{dt}\Big|_{t=\pi/4} = -\sin(\dfrac{\pi}{2}) = -1$

$y = \sin^2 t \Rightarrow \dfrac{dy}{dt} = 2\sin t\cos t = \sin 2t$

$\Rightarrow \dfrac{dy}{dt}\Big|_{t=\pi/4} = \sin(\dfrac{\pi}{2}) = 1$

$\dfrac{dy}{dx} = \dfrac{dy/dt}{dx/dt} = \dfrac{1}{-1} = -1$

b) $x = \cos^3 t \Rightarrow \dfrac{dx}{dt} = 3\cos^2 t(-\sin t)$

$\Rightarrow \dfrac{dx}{dt}\Big|_{t=\pi/4} = \dfrac{-3\sqrt{2}}{4}$

$$y = \sin^3 t \Rightarrow \frac{dy}{dt} = 3 \sin^2 t \cos t$$

$$\Rightarrow \frac{dy}{dt}\Big|_{t = \pi/4} = \frac{3\sqrt{2}}{4}$$

$$\frac{dy}{dx} = \frac{dy/dt}{dx/dt} = \frac{(3\sqrt{2})/4}{(-3\sqrt{2})/4} = -1$$

c) $x = \tan^2 t \Rightarrow \frac{dx}{dt} = 2 \tan t \sec^2 t$

$$\Rightarrow \frac{dx}{dt}\Big|_{t = \pi/4} = 2(1)(\sqrt{2})^2 = 4$$

$$y = \sin 2t \Rightarrow \frac{dy}{dt} = 2 \cos 2t \Rightarrow \frac{dy}{dt}\Big|_{t = \pi/4} = 2(0) = 0$$

$$\frac{dy}{dx} = \frac{dy/dt}{dx/dt} = \frac{0}{4} = 0$$

55.

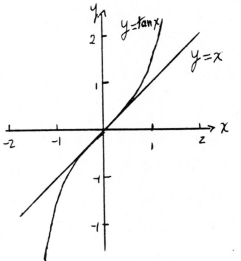

57. If $b = 1$ then $f(x) = \begin{cases} x + 1, & x < 0 \\ \cos x, & x \geq 0 \end{cases}$

$f(0) = \cos(0) = 1. \quad \lim_{x \to 0^-} f(x) = \lim_{x \to 0} x + 1 = 1 = f(0)$

$$\lim_{x \to 0^+} f(x) = \lim_{x \to 0} \cos x = 1 = f(0)$$

Hence $\lim_{x \to 0} f(x) = f(0)$. That is, $f(x)$ is continuous at $x = 0$.

Remark: $f(x)$ is not differentiable at $x = 0$.

59. a) $f(x) = x - 1 - \frac{1}{2} \sin x \Rightarrow$

$f(0) = 0 - 1 - \frac{1}{2} \sin(0) = -1 < 0$

$f(2) = 2 - 1 - \frac{1}{2} \sin(2) = 1 - \frac{1}{2} \sin(2) > 0$ since $\sin x \leq 1$. Hence

$f(x)$ has a solution between 0 and 2.

b) $\sin x \approx x$ when $x \approx 0$, so

$$f(x) \quad x - 1 - \frac{1}{2}x = \frac{1}{2}x - 1 = 0 \Rightarrow x = 2$$

c) $f(2) = 2 - 1 - \frac{1}{2}\sin(2) = 0.545$

Remark: Approximating $f(x)$ for $x \approx \frac{\pi}{4}$ gives better results:

$$f(x) \quad x - 1 - \frac{1}{2}(\frac{1}{\sqrt{2}} + \frac{1}{\sqrt{2}}(x - \frac{\pi}{4})) = 0; \ x \approx 1.664$$

solves this linear approximation. And $f(1.664) \approx 0.1665$, which is

much closer to zero.

61. $f(x) = (1 + x)^{1/2} + \sin x, \ f(0) = 1$

$f'(x) = \frac{1}{2}(1 + x)^{-1/2} + \cos x, \ f'(0) = \frac{1}{2} + 1 = \frac{3}{2}$

$f(x) \quad f(0) + f'(0)x = 1 + \frac{3}{2}x$

This is the sum of the linear approximations for $\sqrt{1 + x}$ and $\sin x$.

63. a) $f(x) = 2\cos x - \sqrt{1 + x}$

$f(0) = 2\cos(0) - \sqrt{1 + 0} = 2 - 1 = 1 > 0$

$f(\frac{\pi}{2}) = 2\cos(\frac{\pi}{2}) - \sqrt{1 + \pi/2} = -\sqrt{1 + \pi/2} < 0$

By the intermediate value theorem, $f(x)$ has a root between 0 and $\frac{\pi}{2}$.

b) Linear Approx: $\cos x = \cos(\pi/4) - \sin(\pi/4)(x - \pi/4)$

$= \sqrt{2}/2 - (\sqrt{2}/2)(x - \pi/4)$

$(1 + x)^{1/2} = (1.69)^{1/2} + \frac{1}{2}(1.69)^{-1/2}(x - 0.69) = 1.3 + (1/26)(x - 0.69)$

c) $2\cos x = (1 + x)^{1/2} \Rightarrow \sqrt{2} - \sqrt{2}(x - \pi/4) = 1.3 + (1/26)(x - 0.69)$

$x = 0.86$

$2\cos x = 1.30, \ \sqrt{1 + x} = 1.36$

65. $\frac{d}{dx}\csc u = \frac{d}{dx}(\sin u)^{-1} = -(\sin u)^{-2}\cos u\frac{dy}{dx}$

$= -\csc u \cot u\frac{du}{dx}$

Article 2.9

1. $f(x) = x^2 + x - 1$. $f(0) = 0 + 0 - 1 = -1$; $f(1) = 1 + 1 - 1 = 1$.

   Let $x_0 = \frac{1}{2}$ (Best first guess based on $f(0) = -1$, and $f(1) = 1$.)

   $$x_{n+1} = x_n - \frac{f(x_n)}{f'(x_n)} = x_n - \frac{x_n^2 + x_n - 1}{2x_n + 1}$$

   $x_0 = 0.5$, $x_1 = 0.625$, $x_2 = 0.618055$, $x_3 = 0.618025$,

   $f(x_3) = -0.000048$

3. $f(x) = x^4 + x - 3$; $f(1) = 1 + 1 - 3 = -1$; $f(2) = 16 + 2 - 3 = 15$. The root

   must be much nearer to $x = 1$ than $x = 2$, so $x_0 = 1$.

   $$x_{n+1} = x_n - \frac{f(x_n)}{f'(x_n)} = x_n - \frac{x_n^4 + x_n - 3}{4x_n^3 + 1}$$

   $x_0 = 1$, $x_1 = 1.2$, $x_2 = 1.165420$, $x_3 = 1.164037$, $x_4 = 1.164035$,

   $f(x_4) = 0.000000$

5. $f(x) = 2 - x^4 = -(x^4 - 2)$. The function for problem 4 is $f(x) = x^4 - 2$.

   Because $f(c) = 0$ implies $-f(c) = 0$, the roots in problem 5 will be the same as

   the roots to the function in problem 4. Since these functions are even

   $(f(x) = f(-x))$, $f(c) = 0$ implies $f(-c) = 0$. The root in problem 5 must be

   between -1 and -2, so it is the negative of the root in problem 4.

   $$\text{root} = -1.189207.$$

7. If $f(x_n) = 0$ for any n and $f'(x_n) \neq 0$, then

   $$x_{n+1} = x_n - \frac{f(x_n)}{f'(x_n)} = x_n - \frac{0}{f'(x_n)} = x_n.$$

   The value of x_n does not change for later iterations.

9. $f(x) = x^4 - x^3 - 75$; $f(3) = 81 - 27 - 75 = -21$;

   $f(4) = 256 - 64 - 75 = 117$. $f(3)$ is closer to zero, so $x_0 = 3$.

   $$x_{n+1} = x_n - \frac{f(x_n)}{f'(x_n)} = x_n - \frac{x_n^4 - x_n^3 - 75}{4x_n^3 - 2x_n^2}$$

   $x_0 = 3$, $x_1 = 3.2333$, $x_2 = 3.229023$, $x_3 = 3.2285773$.

To five places, the root is 3.22858.

11. $f(x) = \tan x$; $f'(x) = \sec^2 x$

$$x_{n+1} = x_n - \frac{f(x_n)}{f'(x_n)} = x_n - \frac{\tan(x_n)}{\sec^2(x_n)} = x_n - \sin(x_n)\cos(x_n)$$

To five decimal places, $\pi = 3.14159$.

13. The intersection of $y = \cos x$ and $y = x$ occurs when $f(x) = \cos x - x = 0$. This equation is solved by Newton's method.

$$x_{n+1} = x_n - \frac{f(x_n)}{f'(x_n)} = x_n - \frac{\cos(x_n) - x_n}{(-\sin(x_n) - 1)} = x_n + \frac{\cos(x_n) - x_n}{\sin(x_n) + 1}$$

A sketch helps in finding the initial guess of $x_0 = 1$.

$x_0 = 1$, $x_1 = 0.75036$, $x_2 = 0.7391128$, $x_3 = 0.7390851334$,

$x_4 = 0.7390851332$. To five decimal places, this is 0.73909.

15. $f(x) = x^4 - 2x^3 - x^2 - 2x + 2$.
Referring to the table, there is a root between 0 and 1,
and a root between 2 and 3. First take $x_0 = 0.5$, then
take $x_0 = 2.5$.

x	f(x)
-2	34
-1	6
0	2
1	-2
2	-6
3	14
4	106

$$x_{n+1} = x_n - \frac{f(x_n)}{f'(x_n)} = x_n - \frac{x_n^4 - 2x_n^3 - x_n^2 - 2x_n + 2}{4x_n^3 - 6x_n^2 - 2x_n - 2}$$

$x_0 = 0.5$, $x_1 = 0.64025$, $x_2 = 0.6301716$, $x_3 = 0.6301153978$

$x_4 = 0.6301153962$. To six decimal places, the first root is 0.630115.

$x_0 = 2.5$, $x_1 = 2.57986$, $x_2 = 2.5733190$, $x_3 = 2.573271966$,

$x_4 = 2.573271964$. To six decimal places, the second root is 2.573272.

17. For $x \geq r$, $f(x) = (x - 4)^{1/2}$, $f'(x) = \frac{1}{2}(x - 4)^{-1/2}$

$$x_{n+1} = x_n - \frac{f(x_n)}{f'(x_n)} = x_n - \frac{(x_n - r)^{1/2}}{(1/2)(x_n - 4)^{-1/2}} = x_n - 2(x_n - r) = 2r - x_n$$

For $x \leq r$, $f(x) = -(r - x)^{1/2}$, $f'(x) = \frac{1}{2}(r - x)^{-1/2}$

$$x_{n+1} = x_n - \frac{f(x_n)}{f'(x_n)} = x_n - \frac{-(r - x_n)^{1/2}}{(1/2)(r - x)^{-1/2}} = x_n + 2(r - x_n) = 2r - x_n$$

If $x_0 = r - h$, then $x_1 = 2r - x_n = 2r - (r - h) = r + h$

If $x_0 = r + h$, then $x_1 = 2r - x_n = 2r - (r + h) = r - h.$

Hence, if $x_0 \neq r$, $x_0 = x_2 = x_4 = x_6 = \ldots$ The root estimate will flip-flop

between $r + h$ and $r - h$, never improving and never worsening.

Article 2.10

1. a) $y = 2x + 3 \Rightarrow x = \frac{y}{2} - \frac{3}{2}$ b)

   $g(x) = \frac{x}{2} - \frac{3}{2}$

   c) $f(x) = 2x + 3 \Rightarrow f' = 2$ everywhere

   $g(x) = \frac{1}{2}x - \frac{3}{2} \Rightarrow g' = \frac{1}{2}$ everywhere

   At c, and f(c), $f' \cdot g' = 2 \cdot \frac{1}{2} = 1$

3. a) $y = f(x) = \frac{1}{5}x + 7 \Rightarrow x = 5y - 35.$ b)

   Interchanging x and y, $g(x) = 5x - 35$

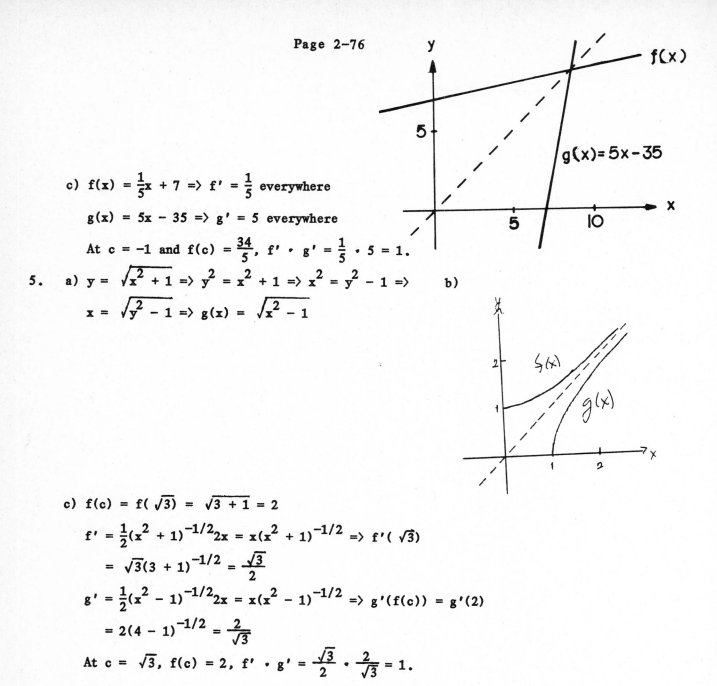

c) $f(x) = \frac{1}{5}x + 7 \Rightarrow f' = \frac{1}{5}$ everywhere

$g(x) = 5x - 35 \Rightarrow g' = 5$ everywhere

At $c = -1$ and $f(c) = \frac{34}{5}$, $f' \cdot g' = \frac{1}{5} \cdot 5 = 1$.

5. a) $y = \sqrt{x^2 + 1} \Rightarrow y^2 = x^2 + 1 \Rightarrow x^2 = y^2 - 1 \Rightarrow$ b)

$x = \sqrt{y^2 - 1} \Rightarrow g(x) = \sqrt{x^2 - 1}$

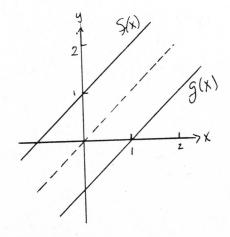

c) $f(c) = f(\sqrt{3}) = \sqrt{3 + 1} = 2$

$f' = \frac{1}{2}(x^2 + 1)^{-1/2}2x = x(x^2 + 1)^{-1/2} \Rightarrow f'(\sqrt{3})$

$= \sqrt{3}(3 + 1)^{-1/2} = \frac{\sqrt{3}}{2}$

$g' = \frac{1}{2}(x^2 - 1)^{-1/2}2x = x(x^2 - 1)^{-1/2} \Rightarrow g'(f(c)) = g'(2)$

$= 2(4 - 1)^{-1/2} = \frac{2}{\sqrt{3}}$

At $c = \sqrt{3}$, $f(c) = 2$, $f' \cdot g' = \frac{\sqrt{3}}{2} \cdot \frac{2}{\sqrt{3}} = 1$.

7. a) $y = x + 1 \Rightarrow x = y - 1 \Rightarrow g(x) = x - 1$ b)

c) $f(x) = x + 1 \Rightarrow f' = 1$ everywhere

$g(x) = x - 1 \Rightarrow g' = 1$ everywhere

At $c = 2$, and $f(c) = 3$, $f' \cdot g' = 1 \cdot 1 = 1$

9. $y = x^5 \Rightarrow x = y^{1/5} \Rightarrow f^{-1}(x) = x^{1/5}$

   $f(f^{-1}(x)) = (x^{1/5})^5 = x$; $f^{-1}(f(x)) = (x^5)^{1/5} = x$

11. $y = x^{2/3} \Rightarrow x = y^{3/2} \Rightarrow f^{-1}(x) = x^{3/2}$

   $f(f^{-1}(x)) = (x^{3/2})^{2/3} = x$; $f^{-1}(f(x)) = (x^{2/3})^{3/2} = x$

13. $y = (x - 1)^2 \Rightarrow x - 1 = y^{1/2} \Rightarrow x = y^{1/2} + 1 \Rightarrow f^{-1}(x) = x^{1/2} + 1$

   $f(f^{-1}(x)) = ((x^{1/2} + 1) - 1)^2 = (x^{1/2})^2 = x$

   $f^{-1}(f(x)) = ((x - 1)^2)^{1/2} + 1 = (x - 1) + 1 = x$

15. $y = x^{-2} \Rightarrow x = y^{-1/2} \Rightarrow f^{-1}(x) = x^{-1/2}$

   $f(f^{-1}(x)) = (x^{-1/2})^{-2} = x^{(-1/2)(-2)} = x$

   $f^{-1}(f(x)) = (x^{-2})^{-1/2} = x^{(-2)(-1/2)} = x$

17. $x = y^{1/2} + y^2 - 1 \Rightarrow \frac{dy}{dx} = \frac{1}{2}y^{-1/2}\frac{dy}{dx} + 2y\frac{dy}{dx}$

   $\Rightarrow (\frac{1}{2}y^{-1/2} + 2y)\frac{dy}{dx} = 1 \Rightarrow \frac{dy}{dx} = (\frac{1}{2}y^{-1/2} + 2y)^{-1}$

19. $x = 5(1 - y^2)^{1/2} \Rightarrow \frac{dy}{dx} = \frac{5}{2}(1 - y^2)^{-1/2}(-2y)\frac{dy}{dx}$

   $\Rightarrow 1 = -5y(1 - y^2)^{-1/2}\frac{dy}{dx} \Rightarrow \frac{dy}{dx} = -\frac{1}{5y}(1 - y^2)^{1/2}$

21. $x = y^{1/2} - y^{1/3} \Rightarrow \frac{dx}{dx} = \frac{1}{2}y^{-1/2}\frac{dy}{dx} - \frac{1}{3}y^{-2/3}\frac{dy}{dx}$

   $\Rightarrow 1 = (\frac{1}{2}y^{-1/2} - \frac{1}{3}y^{-2/3})\frac{dy}{dx} \Rightarrow \frac{dy}{dx} = (\frac{1}{2}y^{-1/2} - \frac{1}{3}y^{-2/3})^{-1}$.

   Also, $\frac{dy}{dx} = y/(\frac{1}{2}y^{1/2} - \frac{1}{3}y^{1/3}) = \frac{y}{x}$

23. b) $y = x^{-1}$, $y' = -x^{-2}$.

   At $(a,\frac{1}{a})$ $y' = -a^{-2}$

   At $(\frac{1}{a},a)$ $y' = -a^2$

   $y'(a) \cdot y'(\frac{1}{a}) = (-a^{-2}) \cdot (-a^2) = 1$

   c) $y = \frac{1}{x} \Rightarrow x = \frac{1}{y} \Rightarrow f^{-1}(x) = \frac{1}{x}$.

$$f^{-1}(x) = f(x)$$

25. The tangent is perpendicular to the x-axis when it is parallel to the y-axis and when $\frac{dx}{dy} = 0$.

$$x = y^2 + y + 1 \Rightarrow \frac{dx}{dy} = 2y + 1 = 0 \Rightarrow y = -\frac{1}{2}, \ x = \frac{3}{4}$$

The point is $(\frac{3}{4}, -\frac{1}{2})$.

27. $f(x) = y = 1 + \frac{1}{x} \Rightarrow \frac{1}{x} = y - 1 \Rightarrow x = \frac{1}{y - 1}$

$\Rightarrow f^{-1}(x) = g(x) = \frac{1}{x - 1}$

$f(g(x)) = 1 + \frac{1}{g(x)} = 1 + \frac{1}{(1/(x - 1))} = 1 + (x - 1) = x$

$g(f(x)) = \frac{1}{f(x) - 1} = \frac{1}{(1 + 1/x) - 1} = \frac{1}{1/x} = x$

$g(x) = \frac{1}{x - 1} \Rightarrow g' = \frac{-1}{(x - 1)^2}$

$g'(f(x)) = -\frac{1}{(f(x) - 1)^2} = \frac{-1}{((1 + 1/x) - 1)^2} = -\frac{1}{(1/x)^2} = -x^2.$

$f(x) = 1 + \frac{1}{x} \Rightarrow f'(x) = -\frac{1}{x^2} \Rightarrow \frac{1}{f'(x)} = x^2.$ Hence $g'(f(x)) = -\frac{1}{f'(x)}.$

29. $\sqrt{x} = x \Rightarrow x = 1$ ($x = 0$ also solves the equation, but Picard's method can only be used to find the solution $x = 1$.)

31. $1.2 \sin x = x$. Many iterations yield $x = \pm 1.026738291$

33. $\cos x = x + 1$ is solved by repeatedly striking the cosine key followed by minus one. The solution is $x = 0$.

35. $f(x) = x^2$ implies $f'(x) = 2x$ when the estimate is near one, the slope is near 2, which is greater than one. The Picard method only works when the slope is less than one. On the other hand, when the estimate is less than one half, the slope is less than one and Picard's method will find the nearest solution, which is zero.

37. $g(x) = 1 + \frac{1}{2} \sin x$. Take $x_0 = 1.5$, $x_1 = g(1.5)$, $x_2 = g(x_2)$, $x_3 = g(x_2)$...The process converges to $x = 1.498701134$.

39. If the calculator has the keys, strike repeatedly x^2, +1, $\frac{1}{x}$. The solution is 0.6823278038.

Miscellaneous Problems

1. $y = x(x^2 - 4)^{1/2} \Rightarrow y' = (x^2 - 4)^{-1/2} + x(-\frac{1}{2}(x^2 - 4)^{-3/2}(2x))$

   $= (x^2 - 4)(x^2 - 4)^{-3/2} - x^2(x^2 - 4)^{-3/2} = -4(x^2 - 4)^{-3/2}$

3. $xy + y^2 = 1 \Rightarrow y + xy' + 2yy' = 0 \Rightarrow (x + 2y)y' = -y \Rightarrow y' = \dfrac{-y}{x + 2y}$

5. $x^2y + xy^2 = 10 \Rightarrow 2xy + x^2y' + y^2 + 2xyy' = 0$

   $\Rightarrow (x^2 + 2xy)y' = -(2xy + y^2) \Rightarrow y' = \dfrac{2xy + y^2}{x^2 + 2xy}$

7. $y = \cos(1 - 2x) = \cos(2x - 1) \Rightarrow y' = -2 \sin(2x - 1)$

   Also, $y = \cos(1 - 2x) \Rightarrow y' = -\sin(1 - 2x)(-2) = 2 \sin(1 - 2x) = -2 \sin(2x - 1)$

9. $y = \dfrac{x}{x + 1} = \dfrac{x + 1 - 1}{x + 1} = 1 - \dfrac{1}{x + 1} \Rightarrow y' = \dfrac{1}{(x + 1)^2}$

   Also, $y = \dfrac{x}{x + 1} \Rightarrow y' = \dfrac{(x + 1)(dx/dx) - x(d/dx)(x + 1)}{(x + 1)^2} = \dfrac{x + 1 - x}{(x + 1)^2} = \dfrac{1}{(x + 1)^2}$

11. $y = x^2(x^2 - a^2)^{1/2} \Rightarrow y' = 2x(x^2 - a^2)^{1/2} + (1/2)x^2(x^2 - a^2)^{-1/2}(2x)$

    $= 2x(x^2 - a^2)^{1/2} + x^3(x^2 - a^2)^{-1/2}$

    Also, $y' = x(x^2 - a^2)^{-1/2}(2(x^2 - a^2) - x^2) = x(x^2 - a^2)^{-1/2}(x^2 - 2a^2)$

13. $y = \dfrac{x^2}{1 - x^2} = -\dfrac{x^2}{x^2 - 1} = -\dfrac{x^2 - 1 + 1}{x^2 - 1} = -1 - \dfrac{1}{x^2 - 1} \Rightarrow y' = \dfrac{2x}{(x^2 - 1)^2}$

    Also, $y = \dfrac{x^2}{1 - x^2} \Rightarrow y' = \dfrac{2x(1 - x^2) - x^2(-2x)}{(1 - x^2)^2} = \dfrac{2x - 2x^3 + 2x^3}{(1 - x^2)^2}$

    $= \dfrac{2x}{(1 - x^2)^2} = \dfrac{2x}{(x^2 - 1)^2}$

15. $y = \sec^2 5x \Rightarrow y' = 2 \sec 5x \dfrac{d}{dx} \sec 5x$

    $= 2 \sec 5x(5 \sec 5x \tan 5x) = 10 \sec^2 5x \tan 5x$

    Also, $y = \sec^2 5x = \cos^{-2} 5x \Rightarrow y' = -2 \cos^{-3} 5x \cdot \dfrac{d}{dx} \cos 5x$

    $= -2 \cos^{-3} 5x(-5 \sin 5x) = 10 \cos^{-3} 5x \sin 5x = 10 \sec^2 5x \tan 5x$

17. $y = \frac{1}{3}(2x^2 + 5x)^{3/2} \Rightarrow y' = (\frac{3}{2})(\frac{1}{3})(2x^2 + 5x)^{1/2}(4x + 5)$

    $= (2x + \frac{5}{2})(2x^2 + 5x)^{1/2}$

19. $xy^2 + x^{1/2}y^{1/2} = 2 \Rightarrow y^2 + 2xyy' + \frac{1}{2}x^{-1/2}y^{1/2} + \frac{1}{2}x^{1/2}y^{-1/2}y'$

$= 0 \Rightarrow (2xy + \frac{1}{2}x^{1/2}y^{-1/2})y' = -(y^2 + \frac{1}{2}x^{-1/2}y^{1/2})$

$\Rightarrow y' = -\dfrac{y^2 + (1/2)(y/x)^{1/2}}{2xy + (1/2)(x/y)^{1/2}}$.

Also, $xy^2 + (xy)^{1/2} = 2 \Rightarrow$

$\frac{1}{2}(y/x)^{1/2} = \frac{1}{x} - \frac{y^2}{2}; \frac{1}{2}(\frac{x}{y})^{1/2} = \frac{1}{y} - \frac{xy}{2}$

$\Rightarrow y' = -\dfrac{y^2 + (1/x - y^2/2)}{2xy + (1/y - xy/2)} = -\dfrac{xy^3 + 2y}{3x^2y^2 + 2x}$

21. $x^{2/3} + y^{2/3} = a^{2/3} \Rightarrow \frac{2}{3}x^{-1/3} + \frac{2}{3}y^{-1/3}y' = 0$

$\Rightarrow y' = -\dfrac{x^{-1/3}}{y^{-1/3}} = -(\frac{y}{x})^{1/3}$

23. $xy = 1 \Rightarrow y = \frac{1}{x} = x^{-1} \Rightarrow y' = -x^{-2} = \dfrac{-1}{x^2}$

25. $(x + 2y)^2 + 2xy^2 = 6 \Rightarrow 2(x + 2y)^1(1 + 2y') + 2y^2 + 4xyy' = 0$

$\Rightarrow (4(x + 2y) + 4xy)y' = -(2(x + 2y) + 2y^2)$

$\Rightarrow y' = -\dfrac{y^2 + 2y + x}{2x + 4y + 2xy}$

27. $y^2 = \dfrac{x}{x + 1} = \dfrac{x + 1 - 1}{x + 1} = 1 - \dfrac{1}{x + 1} \Rightarrow 2yy' = \dfrac{1}{(x + 1)^2} \Rightarrow y' = \dfrac{1}{2y(x + 1)^2}$

Also, $y^2 = \dfrac{x}{x + 1} \Rightarrow 2yy' = \dfrac{(x + 1)(dx/dx) - x(d/dx)(x + 1)}{(x + 1)^2} \Rightarrow 2yy'$

$= \dfrac{x + 1 - x}{(x + 1)^2} = \dfrac{1}{(x + 1)^2} \Rightarrow y' = \dfrac{1}{2y(x + 1)^2}$

29. $xy + 2x + 3y = 1 \Rightarrow y + xy' + 2 + 3y' = 0 \Rightarrow (x + 3)y' = -(y + 2) \Rightarrow y' = -\dfrac{y + 2}{x + 3}$

31. $x^3 - xy + y^3 = 1 \Rightarrow 3x^2 - y - xy' + 3y^2y' = 0 \Rightarrow (3y^2 - x)y' = y - 3x^2$

$\Rightarrow y' = \dfrac{y - 3x^2}{3y^2 - x}$

33. $\dfrac{d}{dx}(\dfrac{1 + x}{1 - x}) = \dfrac{(1 - x) \cdot d/dx(1 + x) - (1 + x)d/dx(1 - x)}{(1 - x)^2} = \dfrac{1 - x + 1 + x}{(1 - x)^2}$

$= \dfrac{2}{(1 - x)^2}$

$y = (\dfrac{1 + x}{1 - x})^{1/2} \Rightarrow y' = \frac{1}{2}(\dfrac{1 + x}{1 - x})^{-1/2}\dfrac{d}{dx}(\dfrac{1 + x}{1 - x}) = \frac{1}{2}(\dfrac{1 - x}{1 + x})^{1/2} \cdot \dfrac{2}{(1 - x)^2}$

$= (1 + x)^{-1/2}(1 - x)^{-3/2}$. Also, $y' = (1 - x)^{-1}(1 - x^2)^{-1/2}$

35. $y = (x^3 + 1)^{1/3} \Rightarrow y' = \frac{1}{3}(x^3 + 1)^{-2/3}(3x^2) = x^2(x^3 + 1)^{-2/3}$

CHAPTER 2 - Miscellaneous Problems

37. $y = \cot 2x \Rightarrow y' = -\csc^2 2x \cdot \frac{d}{dx}(2x) = -2\csc^2 2x$

Also, $y = \cot 2x = \frac{\cos 2x}{\sin 2x} \Rightarrow$

$y' = \frac{\sin 2x(-2\sin 2x) - \cos 2x(2\cos 2x)}{\sin^2 2x} = \frac{-2}{\sin^2 2x} = -2\csc^2 2x$

39. $y = \frac{\sin x}{\cos^2 x} = \tan x \sec x$

$\Rightarrow y' = \tan x \frac{d}{dx} \sec x + \sec x \frac{d}{dx} \tan x$

$= \tan x(\sec x \tan x) + \sec x(\sec^2 x) = \tan^2 x \sec x + \sec^3 x$

Also, $y = \frac{\sin x}{\cos^2 x} \Rightarrow y' = \frac{\cos^2 x(\cos x) - \sin x(2\cos x(-\sin x))}{\cos^4 x}$

$= \frac{\cos^2 x + 2\sin^2 x}{\cos^3 x} = \frac{1 + \sin^2 x}{\cos^3 x} = \sec^3 x + \tan^2 x \sec x$

41. $y = x^2 \cos 8x \Rightarrow y' = 2x \cos 8x + x^2(-8\sin 8x) = 2x\cos 8x - 8x^2\sin 8x$

43. $y = \frac{\sin x}{1 + \cos x} \Rightarrow y' = \frac{\cos x(1 + \cos x) - \sin x(-\sin x)}{(1 + \cos x)^2}$

$= \frac{(\cos^2 x + \sin^2 x) + \cos x}{(1 + \cos x)^2} = \frac{1 + \cos x}{(1 + \cos x)^2} = \frac{1}{1 + \cos x}$

Also, $y' = \frac{1 - \cos x}{\sin^2 x} = \csc^2 x - \csc x \cot x$

45. $y = \csc x = (\sin x)^{-1} \Rightarrow y' = -(\sin x)^{-2}\frac{d}{dx}\sin x$

$= -(\sin x)^{-2}\cos x = -\csc x \cot x$

47. $y = \cos(\sin^2 x) \Rightarrow y' = -\sin(\sin^2 x)\frac{d}{dx}\sin^2 x$

$= -2\sin x \cos x \sin(\sin^2 x)$

Also, $y' = -\sin 2x \sin(\sin^2 x)$

49. $y = \sec^2 x \Rightarrow y' = 2\sec x \frac{d}{dx}\sec x$

$= 2\sec x(\sec x \tan x) = 2\sec^2 x \tan x$

Also, $y = \sec^2 x = (\cos x)^{-2} \Rightarrow y' = -2(\cos x)^{-3}(-\sin x) = 2\sec^2 x \tan x$

51. $\frac{d}{dx}(\sin^2 3x) = 2\sin 3x \cos 3x(3) = 3\sin 6x$

$y = \cos(\sin^2 3x) \Rightarrow y' = -\sin(\sin^2 3x) \cdot \frac{d}{dx}\sin^2 3x$

$= 3\sin 6x \sin(\sin^2 3x)$

53. $y = (2t + t^2)^{1/2} \Rightarrow \frac{dy}{dt} = \frac{1}{2}(2t + t^2)^{-1/2}(2 + 2t)$

$= (t + 1)(2t + t^2)^{-1/2}$

CHAPTER 2 - Miscellaneous Problems

$t = 2x + 3 \Rightarrow \dfrac{dt}{dx} = 2.$

$y' = \dfrac{dy}{dx} = \dfrac{dy}{dt}\dfrac{dt}{dx} = 2(t + 1)(2t + t^2)^{-1/2}$

55. $t = \dfrac{x}{1 + x^2},\ y = x^2 + t^2 = x^2 + (\dfrac{x}{1 + x^2})^2$

$\dfrac{d}{dx}(\dfrac{x}{1 + x^2}) = \dfrac{(1 + x^2) - x(2x)}{(1 + x^2)^2} = \dfrac{1 - x^2}{(1 + x^2)^2}$

$y' = \dfrac{d}{dx}(x^2 + (\dfrac{x}{1 + x^2})^2) = 2x + 2(\dfrac{x}{1 + x^2})\dfrac{d}{dx}(\dfrac{x}{1 + x^2})$

$\quad = 2x + (\dfrac{2x}{1 + x^2})(\dfrac{1 - x^2}{(1 + x^2)^2}) = 2x + \dfrac{2x - 2x^3}{(1 + x^2)^3}$

57. $y = t^3 - 1 \Rightarrow \dfrac{dy}{dt} = 3t^2;\ x = t^2 + t \Rightarrow \dfrac{dx}{dt} = 2t + 1$

$\dfrac{dy}{dx} = \dfrac{dy/dt}{dx/dt} = \dfrac{3t^2}{2t + 1}.$

To write this a form using x and y, note that

$y = t^3 - 1 = (t - 1)(t^2 + t + 1) \Rightarrow y = (t - 1)(x + 1)$

$\Rightarrow t = \dfrac{y}{x + 1} + 1 = \dfrac{x + y + 1}{x + 1}$

$x = t^2 + t \Rightarrow t^2 = x - t = x - \dfrac{x + y + 1}{x + 1} = \dfrac{x^2 - y - 1}{x + 1}$

$\dfrac{dy}{dx} = \dfrac{3t^2}{2t + 1} = \dfrac{3((x^2 - y - 1)/(x + 1))}{2((x + y + 1)/(x + 1)) + ((x + 1)/(x + 1))} = \dfrac{3x^2 - 3y - 3}{x + 2y + 1}$

Also, $y = t^3 - 1 \Rightarrow t = (y - 1)^{1/3} \Rightarrow \dfrac{dy}{dx} = \dfrac{3(y - 1)^{2/3}}{2(y - 1)^{1/3} + 1}$

59. $y = \dfrac{x}{x^2 + 1} \Rightarrow y' = \dfrac{(x^2 + 1) - x(2x)}{(x^2 + 1)^2} = \dfrac{1 - x^2}{(x^2 + 1)^2}$

At $x = 0$, $y' = 1$. The tangent line is the line through the origin with

slope 1, or $y = x$.

61. The condition to test for in I–IV is whether $f''(x) = x^{1/3}$.

I. $f(x) = \dfrac{9}{28}x^{7/3} + 9 \Rightarrow f'(x) = \dfrac{3}{4}x^{4/3} \Rightarrow f''(x) = x^{1/3}$

I can be true.

II. $f'(x) = \dfrac{9}{28}x^{7/3} - 2 \Rightarrow f''(x) = \dfrac{3}{4}x^{4/3} \neq x^{1/3}$

II cannot be true.

III. $f'(x) = \dfrac{3}{4}x^{4/3} + 6 \Rightarrow f''(x) = x^{1/3}$

III can be true.

IV. $f(x) = \frac{3}{4}x^{4/3} - 4 \Rightarrow f'(x) = x^{1/3} \Rightarrow f''(x) = \frac{1}{3}x^{-2/3} \neq x^{1/3}$

IV cannot be true.

The correct answer is d) I and III only.

63. $x^2y + xy^2 = 6$. At $x = 1$, $y + y^2 = 6$

$\Rightarrow 0 = y^2 + y - 6 = (y + 3)(y - 2) \Rightarrow y = -3$ or $y = 2$.

$x^2y + xy^2 = 6 \Rightarrow 2xy + x^2y' + y^2 + 2xyy' = 0$

$\Rightarrow (x^2 + 2xy)y' = -(2xy + y^2) \Rightarrow y' = -\frac{2xy + y^2}{x^2 + 2xy}$.

At $(1,-3)$ $y' = -\frac{-6 + 9}{1 - 6} = \frac{3}{5}$.

One tangent has slope $\frac{3}{5}$ and passes through $(1,-3)$

$y = \frac{3}{5}x + b \Rightarrow -3 = \frac{3}{5}(1) + b \Rightarrow b = \frac{-18}{5}$

$y = \frac{3}{5}x - \frac{18}{5}$ or $3x - 5y = 18$.

At $(1,2)$ $y' = -\frac{4 + 4}{1 + 4} = -\frac{8}{5}$.

The other tangent has slope $-\frac{8}{5}$ and passes through $(1,2)$

$y = -\frac{8}{5}x + b \Rightarrow 2 = -\frac{8}{5}(1) + b \Rightarrow b = \frac{18}{5}$

$y = -\frac{8}{5}x + \frac{18}{5}$ or $8x + 5y = 18$

65. $x^2 + y^2 = 15^2 \Rightarrow 2x + 2yy' = 0 \Rightarrow y' = -\frac{x}{y}$.

At $(12,-9)$, $y' = -\frac{12}{-9} = \frac{4}{3}$. Let the point where the cable meets the

upper-right corner of the gondola be $(\frac{w}{2},-23)$, where w is the full width of

the gondola. The line from $(\frac{w}{2},-23)$ to $(12,-9)$ has slope $y' = \frac{4}{3}$ because

it is the tangent line. Hence, $y' = \frac{4}{3} = \frac{-9 - (-23)}{12 - w/2} \Rightarrow$

$48 - 2w = 42 \Rightarrow w = 3$. The width of the gondola is 3.

67. $y = 2x^2 - 6x + 3 \Rightarrow y(2) = 8 - 12 + 3 = -1$; $y' = 4x - 6 \Rightarrow y'(2) = 8 - 6 = 2$.

The curve has slope 2. The tangent has slope 2 and passes through $(2,-1)$, so

$y = 2x + b \Rightarrow -1 = 2(2) + b \Rightarrow b = -5$. $y = 2x - 5$ or $2x - y = 5$.

69. a) $y = (x^2 + 2x)^5$

$\Rightarrow y' = 5(x^2 + 2x)^4(2x + 2) = 10(x + 1)(x^2 + 2x)^4$

b) $f(t) = (3t^2 - 2t)^{1/2}$

CHAPTER 2 - Miscellaneous Problems

$$\Rightarrow f'(t) = \frac{1}{2}(3t^2 - 2t)^{1/2}(6t - 2) = (3t - 1)(3t^2 - 2t)^{-1/2}$$

c) $f(r) = (r^2 + 5)^{1/2} + (r^2 - 5)^{1/2}$

$$\Rightarrow f' = \frac{1}{2}(r^2 + 5)^{-1/2}(2r) + \frac{1}{2}(r^2 - 5)^{-1/2}(2r)$$

$$= r(r^2 + 5)^{-1/2} + r(r^2 - 5)^{-1/2}$$

d) $f(x) = \dfrac{x^2 - 1}{x^2 + 1} = \dfrac{x^2 + 1 - 2}{x^2 + 1} = 1 - \dfrac{2}{x^2 + 1} \Rightarrow f' = \dfrac{4x}{(x^2 + 1)^2}$

Also, $f(x) = \dfrac{x^2 - 1}{x^2 + 1} \Rightarrow f'(x) = \dfrac{2x(x^2 + 1) - (x^2 - 1)(2x)}{(x^2 + 1)^2}$

$$= \frac{2x^3 + 2x - 2x^3 + 2x}{(x^2 + 1)^2} = \frac{4x}{(x^2 + 1)^2}$$

71. $x^2 = 4y \Rightarrow 2x = 4y' \Rightarrow y' = \dfrac{x}{2} = \dfrac{2}{2} = 1$ when $x = 2$. The slope of the normal is $-\dfrac{1}{y'}$, $= -1$, and it passes through $(2,1)$. $y = -x + b \Rightarrow 1 = -(2) + b \Rightarrow b = 3$.

$y = -x + 3$ or $x + y = 3$

73. $\displaystyle\lim_{\Delta x \to 0} \dfrac{[2 - 3(x + \Delta x)]^2 - [2 - 3x]^2}{\Delta x} = \lim_{\Delta x \to 0} \dfrac{[(2 - 3x) - 3\Delta x]^2 - [2 - 3x]^2}{\Delta x}$

$\displaystyle\lim_{\Delta x \to 0} \dfrac{(2 - 3x)^2 - 6(2 - 3x)\Delta x + 9\Delta x^2 - (2 - 3x)^2}{\Delta x}$

$= \displaystyle\lim_{\Delta x \to 0} -6(2 - 3x) + 9\Delta x = -6(2 - 3x) = 18x - 12$.

This is the derivative of $y = (2 - 3x)^2$.

75. $V = 6\pi r^2 \Rightarrow \dfrac{dV}{dr} = 12\pi r = \Delta V_{tan} = 12\pi r \Delta r$

$\Delta V = 6\pi(r + \Delta r)^2 - 6\pi r^2 = 12\pi r \Delta r + 6\pi \Delta r^2$.

$\Delta V - \Delta V_{tan} = (12\pi r \Delta r + 6\pi \Delta r^2) - (12\pi r \Delta r) = 6\pi \Delta r^2$

ΔV represents a shell around the cylinder, and ΔV_{tan} represents this same shell, less the volume of a cylinder of radius Δr. (When Δr is negative, ΔV represents a shell inside the cylinder and ΔV_{tan} represents this interior shell plus the volume of a cylinder of radius Δr.)

77. R = Revenue = Price · Quantity = $p \cdot x = [3 - (x/40)]^2 x$

Marginal Revenue $= \dfrac{dR}{dx} = [3 - (x/40)]^2 + 2x[3 - (x/40)]'(-\dfrac{1}{40})$

$= 0 \Rightarrow 3 - (x/40) = \dfrac{2x}{40} \Rightarrow x = 40$ people.

CHAPTER 2 - Miscellaneous Problems

Price = P(40) = $[3 - (\frac{40}{40})]^2$ = 4 nickels.

79. $y = x - x^2 \Rightarrow y^2 = (x - x^2)^2 = x^2 - 2(x^2)^{3/2} + (x^2)^2$

$\Rightarrow \dfrac{dy^2}{d(x^2)} = 1 - 2(\frac{3}{2}(x^2)^{1/2} + 2(x^2)^1 = 1 - 3x + 2x^2$

81. Maximum height is achieved at time t_{max} when $\frac{ds}{dt} = 0$. $s = at - 16t^2$

$\Rightarrow \frac{ds}{dt} = a - 32t = 0 \Rightarrow t_{max} = \frac{a}{32}$. $49 = s(t_{max}) = a(\frac{a}{32})$

$-16(\frac{a}{32})^2 = \frac{a^2}{64} \Rightarrow a = ((64)(49))^{1/2} = 56$.

83. a) $y = x^2 + 1 \Rightarrow y' = 2x = 2(1) = 2$ at $x = 1$. The slope of the normal at

(1,2), on which (h,k) lies, is $-\frac{1}{y'}, = -\frac{1}{2}$. Set $\Delta x = h - x = h - 1$,

$\Delta y = k - y = k - 2$ ($\Delta y > 0$). The slope of the normal equals $\frac{\Delta y}{\Delta x}$

so $-\frac{1}{2} = \frac{\Delta y}{\Delta x} \Rightarrow x = -2\Delta y$. $(x - h)^2 + (y - k)^2 = r^2 \Rightarrow$

$(\Delta x)^2 + (\Delta y)^2 = r^2 \Rightarrow (-2\Delta y)^2 + (\Delta y)^2 = r^2 \Rightarrow y = \frac{r}{\sqrt{5}}, x = \frac{-2r}{\sqrt{5}}$.

$h = x + \Delta x = 1 - \frac{2r}{\sqrt{5}}, k = h + \Delta y = 2 + \frac{r}{\sqrt{5}}$.

b) $(x - h)^2 + (y - k)^2 = r^2 \Rightarrow 2(x - h) + 2(y - k)y' = 0 \Rightarrow y' = -\frac{x - h}{y - k}$

$y'' = -\dfrac{(y - k) - (x - h)y'}{(y - k)^2} = -\dfrac{(y - k) - (x - h)(-((x - h)/(y - k))}{(y - k)^2}$

$= -\dfrac{(y - k)^2 + (x - h)^2}{(y - k)^3} = -\dfrac{r^2}{(y - k)^3}$. From part a) $y - k = -\Delta y = \frac{-r}{\sqrt{5}}$

$y'' = -\dfrac{r^2}{(-(r/\sqrt{5}))^3} = \frac{5\sqrt{5}}{r}$.

For the parabola, $y = x^2 + 1 \Rightarrow y' = 2x \Rightarrow y'' = 2$.

Hence $2 = \frac{5\sqrt{5}}{r} \Rightarrow r = \frac{5}{2}\sqrt{5}, \Delta y = \frac{r}{\sqrt{5}} = \frac{5}{2}$,

$h = y + \Delta y = 2 + \frac{5}{2} = \frac{9}{2}; \Delta x = -2\Delta y = -5, k = x + \Delta x = 1 - 5 = -4$.

CHAPTER 2 - Miscellaneous Problems

85. $x = y^2 + y \Rightarrow \dfrac{dx}{dy} = 2y + 1; \; u = (x^2 + x)^{3/2} \Rightarrow$

$\dfrac{du}{dx} = \dfrac{3}{2}(x^2 + x)^{1/2}(2x + 1) = 3u^{1/3}(y^2 + y + \dfrac{1}{2})$.

$\dfrac{dy}{du} = \dfrac{1}{(du/dy)} = \dfrac{1}{(du/dx)(dx/dy)} = \dfrac{1}{3u^{1/3}(y^2 + y + 1/2)(2y + 1)}$

87. $u = \dfrac{2x - 1}{x + 1} \Rightarrow \dfrac{du}{dx} = \dfrac{2(x + 1) - (2x - 1)(1)}{(x + 1)^2} = \dfrac{3}{(x + 1)^2}$

$y = f(u) \Rightarrow \dfrac{dy}{dx} = f'(u)\dfrac{du}{dx} = \sin((\dfrac{2x - 1}{x + 1})^2) \cdot \dfrac{3}{(x + 1)^2}$

89. Let $f(x) = x - \sin x$, $f'(x) = 1 - \cos x > 0$ on $0 < x < \dfrac{\pi}{2}$. $f(x)$ is strictly increasing on the interval and $f(0) = 0$. Hence $f(x) = x - \sin x > 0$ on the interval. Thus $x > \sin x$. Let $g(x) = \dfrac{2}{\pi}x - \sin x$, $g'(x) = \dfrac{2}{\pi} - \cos x$, $g''(x) = \sin x > 0$. Hence $g(x)$ is strictly concave up. $g(0) = g(\dfrac{\pi}{2}) = 0$, so $g(x)$ is negative on the interval. $g(x) = \dfrac{2}{\pi}x - \sin x < 0$. Thus $\sin x > \dfrac{2}{\pi}x$.

91. $y^3 + y = x \Rightarrow (3y^2 + 1)\dfrac{dy}{dx} = 1 \Rightarrow \dfrac{dy}{dx} = (3y^2 + 1)^{-1}$

$\dfrac{d^2y}{dx^2} = -(3y^2 + 1)^{-2}6y\dfrac{dy}{dx} = -6y(3y^2 + 1)^{-3} = \dfrac{-6}{4^3}$

CHAPTER 2 — Miscellaneous Problems

$= \dfrac{3}{32}$ when $y = 1$.

93. a) $\dfrac{d^2(uv)}{dx^2} = \dfrac{d}{dx}(\dfrac{d}{dx}(uv)) = \dfrac{d}{dx}(\dfrac{du}{dx}v + u\dfrac{dv}{dx}) = \dfrac{d^2u}{dx^2}v + \dfrac{d^2u}{dx}\dfrac{dv}{dx} + \dfrac{du}{dx}\dfrac{dv}{dx} + u\dfrac{d^2v}{dx^2}$

$= \dfrac{d^2u}{dx^2}v + 2\dfrac{du}{dx}\dfrac{dv}{dx} + u\dfrac{d^2v}{dx^2}$

b) $\dfrac{d^3(uv)}{dx^3} = \dfrac{d}{dx}(\dfrac{d^2}{dx^2}(uv)) = \dfrac{d}{dx}(\dfrac{d^2u}{dx^2}v + 2\dfrac{du}{dx}\dfrac{dv}{dx} + u\dfrac{d^2v}{dx^2})$

$= \dfrac{d^3u}{dx^3}v + \dfrac{d^2u}{dx^2}\dfrac{dv}{dx} + 2\dfrac{d^2u}{dx^2}\dfrac{dv}{dx} + 2\dfrac{du}{dx}\dfrac{d^2v}{x^2} + \dfrac{du}{dx}\dfrac{d^2v}{dx^2} + u\dfrac{d^3v}{dx^3}$

$= \dfrac{d^3u}{dx^3} + 3\dfrac{d^2u}{dx^2}\dfrac{dv}{dx} + 3\dfrac{du}{dx}\dfrac{d^2v}{dx} + u\dfrac{d^3v}{dx^3}$

c) Note: $\binom{m}{k} + \binom{m}{k+1} = \dfrac{m!}{k!(m-k)!} + \dfrac{m!}{(k+1)!(m-k-1)!} =$

$\dfrac{(m+1)!}{(k+1)!(m-k)!} = \binom{m+1}{k+1}$

1) The case $n = 2$ is proven in part a).

2) The statement is assured true for $n = m$ and shown to be true for $n = m + 1$. By the Principle of Induction it will follow that the statement is true for all $n > 2$.

The statement is assumed true for $n = m$:

$$\frac{d^m(uv)}{dx^m} = \frac{d^m u}{dx^m}v + m\frac{d^{m-1}u}{dx^{m-1}}\frac{dv}{dx} + \binom{m}{2}\frac{d^{m-2}u}{dx^{m-2}}\frac{d^2 v}{dx^2} + \binom{m}{3}\frac{d^{m-3}u}{dx^{m-3}}\frac{d^3 v}{dx^3} + \ldots$$

$$+ \binom{m}{m-1}\frac{du}{dx}\frac{d^{m-1}v}{dx^{m-1}} + u\frac{d^m v}{dx^m}$$

The second equality in the following expression is the above

differentiated and regrouped.

$$\frac{d^{m+1}(uv)}{dx^{m+1}} = \frac{d}{dx}\left(\frac{d^m(uv)}{dx^m}\right) = \frac{d^{m+1}u}{dx^{m+1}}v + \left[\frac{d^m u}{dx^m}\frac{dv}{dx} + m\frac{d^m u}{dx^m}\frac{dv}{dx}\right]$$

$$+ \left[m\frac{d^{m-1}u}{dx^{m-1}}\frac{d^2 v}{dx^2} + \binom{m}{2}\frac{d^{m-1}u}{dx^{m-1}}\frac{d^2 v}{dx^2}\right] + \left[\binom{m}{2}\frac{d^{m-2}u}{dx^{m-2}}\frac{d^3 v}{dx^3} + \binom{m}{3}\frac{d^{m-2}u}{dx^{m-2}}\frac{d^3 v}{dx^3}\right]$$

$$+ \binom{m}{3}\frac{d^{m-3}u}{dx^{m-3}}\frac{d^4 v}{dx^4} + \ldots + \binom{m}{m-1}\frac{d^2 u}{dx^2}\frac{d^{m-1}v}{dx^{m-1}} + \left[\binom{m}{m-1}\frac{du}{dx}\frac{d^m v}{dx^m} + \frac{du}{dx}\frac{d^m v}{dx^m}\right] + u\frac{d^{m+1}v}{dx^{m+1}}$$

$$\frac{d^{m+1}(uv)}{dx^{m+1}} = \frac{d^{m+1}u}{dx^{m+1}}v + (m+1)\frac{d^m u}{dx^m}\frac{dv}{dx} + \binom{m+1}{2}\frac{d^{m-1}u}{dx^{m-1}}\frac{d^2 v}{dx^2}$$

$$+ \binom{m+1}{3}\frac{d^{m-2}u}{dx^{m-2}}\frac{d^3 v}{dx^3} + \ldots + (m+1)\frac{du}{dx}\frac{d^m v}{dx^m} + u\frac{d^{m+1}v}{dx^{m+1}}.$$

The last equation says that the statement holds for the case $n = m + 1$, and

that completes the proof.

95. $f(x) = (x - a)^n g(x)$; $f(a) = 0$

$f'(x) = n(x - a)^{n-1}g(x) + (x - a)^n g'(x)$; $f'(a) = 0$

By problem 93 c)

$$f^{n-1}(x) = \binom{n-1}{0}(x-a)^n g^{n-1}(x) + \ldots \binom{n-1}{n-1}n!(x-a)g(x)$$

$f^{n-1}(a) = 0$, because all terms contain a factor of $(x - a)$.

$$f^n(x) = \binom{n}{0}(x-a)^n g^n(x) + \ldots + \binom{n}{n}\frac{d^n(x-a)^n}{dx^n}g(x)$$

$\dfrac{d^n}{dx^n}(x - a)^n = n!$. Hence $f^n(a) = \binom{n}{n}(n!)g(a) = n!g(a) \neq 0$

97. a) $\theta = \frac{2\pi}{n}$ is the center

angle of each of the n

triangles. The length of each

side of the polygon is

$2r \sin(\frac{\theta}{2})$. The total

perimeter therefore is

$P_n = 2nr \sin(\frac{\theta}{2})$

$\quad\; = 2nr \sin(\frac{\pi}{n})$.

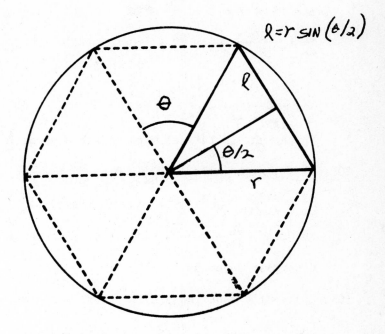

$\ell = r \sin(\theta/2)$

b) Let $\alpha = \frac{\pi}{n}$. As $n \to \infty$, $\alpha \to 0$.

$P_n = 2nr \sin(\frac{\pi}{n}) = 2\pi r \frac{\sin(\pi/n)}{\pi/n} = 2\pi r \frac{\sin(\alpha)}{\alpha} \to 2\pi r$,

as $n \to \infty$ and $\alpha \to 0$. $2\pi r$ is the exact circumference.

99. $\dfrac{a}{6} = \dfrac{a + 20}{h}$ => $h = 6 + \dfrac{120}{a}$

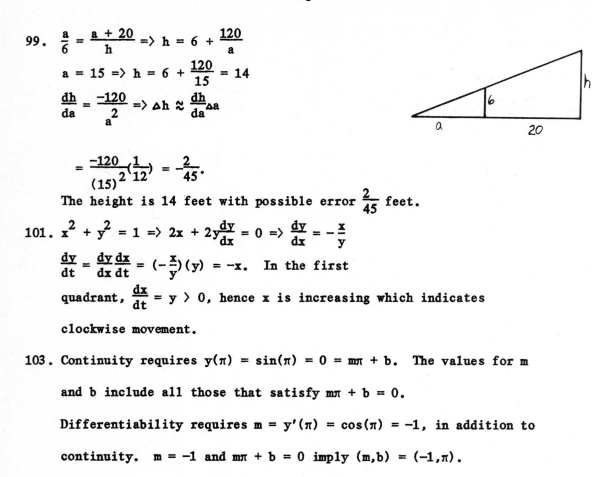

$a = 15$ => $h = 6 + \dfrac{120}{15} = 14$

$\dfrac{dh}{da} = \dfrac{-120}{a^2}$ => $\Delta h \approx \dfrac{dh}{da}\Delta a$

$= \dfrac{-120}{(15)^2}(\dfrac{1}{12}) = -\dfrac{2}{45}.$

The height is 14 feet with possible error $\dfrac{2}{45}$ feet.

101. $x^2 + y^2 = 1$ => $2x + 2y\dfrac{dy}{dx} = 0$ => $\dfrac{dy}{dx} = -\dfrac{x}{y}$

$\dfrac{dy}{dt} = \dfrac{dy}{dx}\dfrac{dx}{dt} = (-\dfrac{x}{y})(y) = -x.$ In the first

quadrant, $\dfrac{dx}{dt} = y > 0$, hence x is increasing which indicates

clockwise movement.

103. Continuity requires $y(\pi) = \sin(\pi) = 0 = m\pi + b$. The values for m

and b include all those that satisfy $m\pi + b = 0$.

Differentiability requires $m = y'(\pi) = \cos(\pi) = -1$, in addition to

continuity. $m = -1$ and $m\pi + b = 0$ imply $(m,b) = (-1,\pi)$.

105. a) $f'(x)\big|_{x=0} = \lim_{x\to 0}\dfrac{f(x) - f(0)}{x} = \lim_{x\to 0}\dfrac{x^2\sin(1/x) - 0}{x}$

$= \lim_{x\to 0}x\sin\dfrac{1}{x} = 0$ because $|\sin\dfrac{1}{x}| \leq 1.$

b) $f(x) = x^2\sin\dfrac{1}{x}$ => $f'(x) = 2x\sin\dfrac{1}{x} + x^2\cos(\dfrac{1}{x})(\dfrac{-1}{x^2})$

$= 2x\sin(\dfrac{1}{x}) - \cos(\dfrac{1}{x}) = \lim_{x\to 0}x\sin\dfrac{1}{x} - \cos(\dfrac{1}{x}).$

c) The limit of $f'(x)$ as $x\to 0$ does not exist because the limit of

$\cos(\dfrac{1}{x})$ does not exist as $x\to 0$. Hence $f'(x)$ is not

continuous at $x = 0$. ($f(x)$ is differentiable everywhere,

however.)

CHAPTER 3

Article 3.1

1. $x^2 - x - 2 = (x + 1)(x - 2) < 0 \Rightarrow x + 1 > 0$ and $x - 2 < 0$

   Hence $-1 < x < 2$.

3. $x - x^2 > 0 \Rightarrow x > x^2 \Rightarrow x > 0$ (because $x^2 \geq 0$)

   $\Rightarrow 1 > x \Rightarrow 0 < x < 1$

5. $x^3 - 4x = x(x^2 - 4) < 0 \Rightarrow (x > 0$ and $x^2 - 4 < 0)$ or

   $(x < 0$ and $x^2 - 4 > 0)$. Hence $0 < x < 2$ or $x < -2$.

7. $y = (x - 1)(x + 1)(x - 5)$. $y > 0$ when $(x + 1$, $(x - 1)$ and $(x - 5)$ are all

   positive $(x > 5)$, and when $(x + 1)$ is positive but $(x - 1)$ and $(x - 5)$ are

   negative $(-1 < x < 1)$. Hence y is positive when $x > 5$ or $-1 < x < 1$.

9. $y = x^2 - x + 1 \Rightarrow y' = 2x - 1$; $y' = 0 \Rightarrow x = \frac{1}{2}$. y is increasing for

   $x > \frac{1}{2}$ and decreasing for $x < -\frac{1}{2}$. $x = \frac{1}{2}$ is a local minimum.

11. $y = \frac{x^3}{3} - \frac{x^2}{2} - 2x + \frac{1}{3} \Rightarrow y' = x^2 - x - 2 = (x - 2)(x + 1)$.

    $y' = 0$ at $x = -1$ and $x = 2$. $y > 0$ when $x > 2$ or $x < -1$ and $y' <$ when

    $-1 < x < 1$. $x = -1$ is a local maximum and $x = 2$ is a local minimum.

13. $y = x^3 - 27x + 36 \Rightarrow y' = 3x^2 - 27 = 3(x^2 - 9)$. $y' = 0$ at $x = \pm 3$. $y' > 0$

    when $x^2 > 9$, that is when $x > 3$ or $x < -3$. $y' < 0$ when $x^2 < 9$, that is when

    $-3 < x < 3$. $x = -3$ is a local minimum and $x = 3$ is a local maximum.

15. $y = 3x^2 - 2x^3 \Rightarrow y' = 6x - 6x^2 = 6x(1 - x)$. $y' = 0$ at $x = 0$ and $x = 1$. $y' > 0$

    when x and $(1 - x)$ are both positive $(0 < x < 1)$. $y' < 0$ when x is positive

    and $1 - x$ is negative $(x > 1)$, and when x is negative and $(1 - x)$ is positive

    $(x < 0)$. The local maximum is at $x = 1$ and the local minimum is at $x = 0$.

17. $y = x^4 \Rightarrow y' = 4x^3$. $y' = 0$ at $x = 0$. $y' > 0$ when $x > 0$, and $y' < 0$ when

    $x < 0$. $x = 0$ is a local minimum.

19. $y = x^{-3} \Rightarrow y' = -3x^{-4}$. y' is never equal to zero (because $\frac{1}{x}$ is never equal

to zero). y is decreasing everywhere. There are no local maxima or minima.

21. $y = (x + 1)^{-2} \Rightarrow y' = -2(x + 1)^{-3}$. When $x < -1$, $y' < 0$, and when $x < -1$,

$y' > 0$. y' does not exist at $x = -1$, and y' is never equal to zero. Since for

no x is $y' = 0$, there are no local maxima or minima.

23. $y = \cos x \Rightarrow y' = -\sin x$. $y' = 0$ at $x = -\pi$, 0, π (when x is confined to

$-\frac{3}{2}\pi < x < \frac{3}{2}\pi$). $y' > 0$ when $\sin x < 0$, that is when $-\pi < x < 0$ or

$\pi < x < \frac{3}{2}\pi$. $y' < 0$ when $\sin x > 0$, that is when $-\frac{3}{2}\pi < x < -\pi$ or

$0 < x < \pi$. Local minima occur at $x = \pm\pi$, and a local maximum is at $x = 0$.

25. $y = x|x| = \begin{cases} x^2 & \text{for } x \geq 0 \\ -x^2 & \text{for } x < 0 \end{cases} \Rightarrow$

$y' = \begin{cases} 2x & \text{for } x \geq 0 \\ -2x & \text{for } x < 0 \end{cases} \Rightarrow y' = 2|x|$.

$y' = 0$ at $x = 0$. $y' > 0$ when $x \neq 0$, hence $x = 0$ is not a local maximum or

minimum. The function has no maximum or minimum points.

27. $y = \frac{1}{2}[|\sin x| + \sin x]$

$= \frac{1}{2}[\sin x + \sin x] = \sin x$

when $\sin x \geq 0$

$\frac{1}{2}[-\sin x + \sin x] = 0$ when $\sin x < 0$

Maximum value is 1 at $x = \frac{\pi}{2}$.

Minimum value is 0 at $x = 0$ and

$\pi \leq x \leq 2\pi$.

29. $f = \begin{cases} -x & \text{for } x \leq 0 \\ -2x & \text{for } x \geq 0 \end{cases}$

$f(x)$ is continuous for all x including $x = 0$.

$\lim\limits_{x \to 0^-} \frac{f(x) - f(0)}{x} = -1$ and

$\lim\limits_{x \to 0^+} \frac{f(x) - f(0)}{x} = -2$

Since $-1 \neq -2$, $f(x)$ has no derivative at $x = 0$.

31. $y = x - 2 \sin x \Rightarrow y' = 1 - 2 \cos x$

a) $y' = 0 \Rightarrow \cos x = \frac{1}{2}$, $x = \frac{\pi}{3}$.

$y' < 0$ if $x < \frac{\pi}{3}$; $y' > 0$ if $x > \frac{\pi}{3}$.

b) $y(0) = (0) - 2 \sin(0) = 0$; $y(\pi) = \pi - 2 \sin \pi = \pi$

$y(\frac{\pi}{3}) = \frac{\pi}{3} - 2 \sin(\frac{\pi}{3}) = \frac{\pi}{3} - \sqrt{3} \approx -0.685$

c) Use Newton's method with $f(x) = x - 2 \sin x$ and $f'(x) = 1 - 2 \cos x$.

$f(\frac{\pi}{2}) = \frac{\pi}{2} - 2 \quad -0.429$ and $f(\pi) = \pi$; there is a root between $\frac{\pi}{2}$

and π, and the root is closer to $\frac{\pi}{2}$. Take $x_0 = \frac{\pi}{2}$.

$$x_{n+1} = x_n - \frac{f(x_n)}{f'(x_n)} = x_n - \frac{x_n - 2 \sin x_n}{1 - 2 \cos x_n}.$$

$x_0 = \frac{\pi}{2} = 1.57$, $x_1 = 2$, $x_2 = 1.901$, $x_3 = 1.8955$, $x_4 = 1.895494267$,

$x_5 = 1.895494267$. The root to two places is $x = 1.90$.

Article 3.2

1. $y = x^2 - 4x + 3 \Rightarrow y' = 2x - 4 \Rightarrow y'' = 2$

a) $y' > 0$ when $x > 2$

b) $y' < 0$ when $x < 2$

c) $y'' = 2 > 0$ for all x. Concave up everywhere.

d) y'' is never less than zero. Concave down nowhere. Local minimum at $x = 2$.

No inflection points.

3. $y = x^3 - 3x + 3 \Rightarrow y' = 3x^2 - 3 \Rightarrow y'' = 6x$

   a) $y' > 0$ when $x^2 - 1 > 0$,

   that is when $x > 1$ or $x < -1$.

   b) $y' < 0$ when $x^2 - 1 < 0$, or $-1 < x < 1$.

   c) $y'' > 0$ when $x > 0$. Concave up for $x > 0$.

   d) $y'' < 0$ when $x < 0$. Concave down for

   $x < 0$. Inflection point at $x = 0$,

   local maximum at $x = -1$, local

   minimum at $x = +1$.

$$y = x^3 - 3x + 3$$

5. $y = x^3 - 6x^2 + 9x + 1 \Rightarrow y' = 3x^2 - 12x + 9 \Rightarrow y'' = 6x - 12$

   a) $y' > 0$ when $3x^2 - 12x + 9 = 3(x - 3)(x - 1) > 0$, that is when $x > 3$ or

   $x < 1$.

   b) $y' < 0$ when $3(x - 3)(x - 1) < 0$ or $1 < x < 3$.

   c) $y'' > 0$ when $6x - 12 > 0$ or $x > 2$. Concave up for $x > 2$.

   d) $y'' < 0$ when $6x - 12 < 0$ or $x < 2$. Concave down for $x < 2$. Inflection point

   at $x = 2$. Local maximum at $x = 1$, local minimum at $x = 3$.

7. $y = (x - 2)^3 + 1 \Rightarrow y' = 3(x - 2)^2 \Rightarrow y'' = 6(x - 2)$

   a) $y' > 0$ when $x \neq 2$

   b) $y' < 0$ for no x

   c) $y'' > 0$ when $6(x - 2) > 0$ or $x > 2$. Concave up for $x > 2$

   d) $y'' < 0$ when $6(x - 2) < 0$ or $x < 2$. Concave down for $x < 2$. Inflection

      point at $x = 2$. No local maxima or minima.

9. $y = \tan x \Rightarrow y' = \sec^2 x \Rightarrow y'' = 2\sec^2 x \tan x$

   a) $y' > 0$ when $\sec^2 x > 0$ which is true for all x (in the interval

      $-\frac{\pi}{2} < x < \frac{\pi}{2}$).

   b) $y' < 0$ for no x.

   c) $y'' > 0$ when $\tan x > 0$, that is when $x > 0$. Concave up for $x > 0$.

   d) $y'' < 0$ when $\tan x < 0$, that is when $x < 0$. Concave down for $x < 0$.

      Inflection point at $x = 0$. No local maxima or minima.

11. $y = -x^4 \Rightarrow y' = -4x^3 \Rightarrow y'' = -12x^2$

   a) $y' > 0$ when $x < 0$

   b) $y' < 0$ when $x > 0$

   c) $y'' > 0$ for no x. Concave up nowhere.

   d) $y'' < 0$ for all $x \neq 0$. Concave down everywhere. No inflection points.

      Local maximum at $x = 0$. Curve is parabolic, opening downward.

13. $y = (x - 1)^2$ is an example.

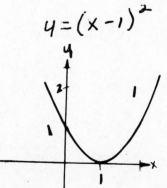

15. $y = 6 - 2x - x^2 = 7 - (1 + 2x + x^2)$

$= 7 - (x + 1)^2 \Rightarrow$ local maximum at $(-1,7)$.

Also, $y = 6 - 2x - x^2 \Rightarrow y' = -2 - 2x = 0$

$\Rightarrow x = -1 \Rightarrow y(-1) = 7$.

No inflection points

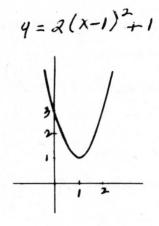
$y = 2(x-1)^2 + 1$

17. $y = x(6 - 2x)^2 = 4x(x - 3)^2$

$\Rightarrow y' = 4[3x^2 - 12x + 9] = 12(x - 1)(x - 3)$

$y'' = 12[2x - 4]$.

$y' = 0$ at $x = 1$ and $x = 3$. $y'' > 0$ for $x > 2$

and $y'' < 0$ for $x < 2$, so $x = 2$ is an inflection

point. Local maximum at $x = 1$ and local

minimum at $x = 3$.

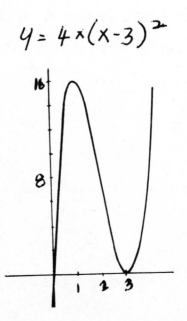
$y = 4x(x-3)^2$

19. $y = 12 - 12x + x^3$

$\Rightarrow y' = -12 + 3x^2 = 3(x^2 - 4)$

$y'' = 6x$

$y' = 0$ at $x = -2$ and $x = 2$. $y'' > 0$ for $x > 0$

and $y'' < 0$ for $x < 0$, so $x = 0$ is an inflection

point. Local maximum is at $x = -2$, and local

minimum at $x = 2$.

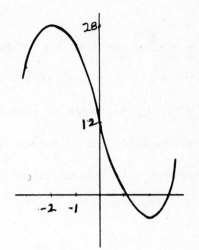

$y = 12 - 12x + x^3$

21. $y = x^3 - 6x^2 - 135x \Rightarrow y' = 3x^2 -$

$12x - 135 = 3(x - 9)(x + 5)$.

$y'' = 6x - 12$. $y' =$ at $x = -5$ and

$x = 9$. $y'' > 0$ for $x > 2$, and

$y' < 0$ for $x < 2$, so $x = 2$ is a

point of inflection. Local

maximum at $x = -5$ ($y = 400$), local

minimum at $x = 9$ ($y = -972$).

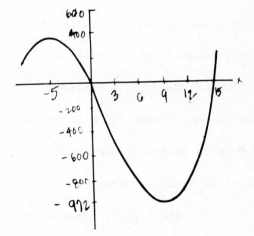

$y = x^3 - 6x^2 - 135x$

23. $y = x^4 - 32x + 48 = y' = 4x^3 - 32 = 4(x^3 - 8)$

$y'' = 12x^2$. $y' = 0$ at $x = 2$. Since

$y'' \geq 0$, the curve is concave up

everywhere, and there are no inflection

points. A minimum occurs at $(2,0)$.

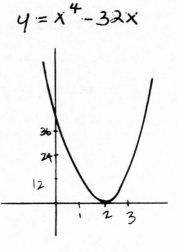

$y = x^4 - 32x$

25. $y = x + \sin 2x \Rightarrow y' = 1 + 2\cos 2x$

$y'' = -4\sin 2x$. $y' = 0$ whenever

$1 + 2\cos 2x = 0$ or $2x =$

$\pm\frac{2}{3}\pi + 2\pi k$ or $x = \pm\pi + \pi k$ (k is

any integer). $y'' > 0$ whenever

$\sin 2x < 0$ and $y'' < 0$ whenever

$\sin 2x > 0$. Hence, if m is any integer,

$x = \frac{m}{2}\pi$ is an inflection point.

There is a local maximum at

$x = \frac{\pi}{3} + k\pi$ and a local minimum at

$x = -\frac{\pi}{3} + k\pi$, for every integer k.

$y = x + \sin 2x$

27. $x = (y^3 + 3y^2 + 3y + 1) + 1 = (y + 1)^3 + 1$

$\frac{dx}{dy} = 3(y + 1)^2$. $\frac{dx}{dy} = 0$ when $y = -1$ and $x = 1$.

(Vertical tangents occur where $\frac{dx}{dy} = 0$.)

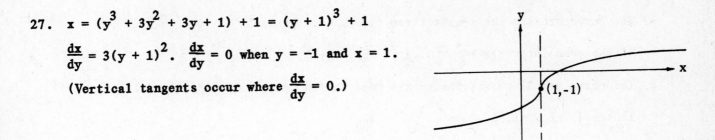

29. $x = y^4 - 2y^2 + 2 \Rightarrow \frac{dx}{dy} = 4y^3 - 4y = 4y(y^2 - 1)$.

Vertical tangents occur when $\frac{dx}{dy} = 0$ or $y = 0$ $(x = 2)$,

$y = +1$ $(x = 1)$, $y = -1$ $(x = 2)$.

31. $f' > 0$ at points P and T. $f'' < 0$ at R, S, and T. $f'' > 0$ at P and Q.

a) $f' < 0$ and $f'' < 0$ at point T.

33.   35.

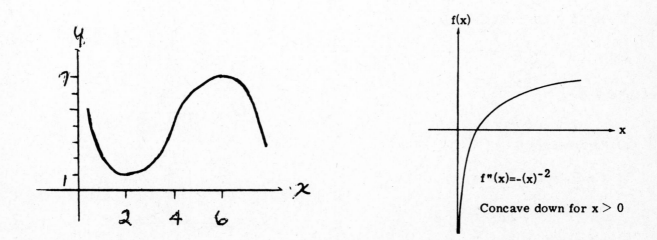

$f''(x) = -(x)^{-2}$

Concave down for $x > 0$

37. a) The curve crosses the x-axis three times, once on each

of the open intervals $(\frac{-5}{3},\frac{-4}{3})$, $(\frac{-2}{3},-\frac{1}{3})$, $(\frac{2}{3},1)$.

b) The curve y + 3 crosses the x-axis once on the open

interval (-1.9,-1.8).

c) The curve y - 3 crosses the x-axis once on the open

interval (1.2, 1.3).

39. $y = x^4 + 8x^3 - 270x^2 \Rightarrow y' = 4x^3 + 24x^2 - 540x = 4x(x + 15)(x - 9)$. $y' = 0$ at

$x = -15$, $x = 0$, $x = 9$. $y' > 0$ for $x > 9$ and $y' < 0$ for $0 < x < 9$, so $x = 9$ is

at a local minimum. $y' > 0$ for $-15 < x < 0$, so $x = 0$ is at a local maximum.

$y' < 0$ for $x < -15$, so $x = 15$ is at a local minimum. $y'' = 12x^2 + 48x - 540$

$= 12(x + 9(x - 5)$. $y'' > 0$ for $x > 5$ and $y'' < 0$ for $-9 < x < 5$, so $x = 5$ is at

an inflection point. $y'' > 0$ for $x < -9$, so $x = 9$ is also an inflection point.

41. $y' = 0$ when $x = 1, 2, 3, 4$; each of these is checked as to whether y takes on a

maximum or a minimum. Let ε be a small positive number $(0 < \varepsilon < 1)$.

$y'(1 + \varepsilon) > 0$ and $y'(1 - \varepsilon) > 0$, so at $x = 1$ there is neither a maximum nor a

minimum. $y'(2 + \varepsilon) < 0$ and $y'(2 - \varepsilon) > 0$, so $x = 2$ is at a local maximum.

$y(3 + \varepsilon) < 0$ and $y'(3 - \varepsilon) < 0$, so $x = 3$ is at neither. $y'(4 + \varepsilon) > 0$ and

$y'(4 - \varepsilon) > 0$, so $x = 4$ is at a local minimum.

a) $x = 2$

b) $x = 4$

Article 3.3

1. a) No symmetry

b) $x \neq \frac{3}{2}$, $y \neq 0$

c) $x = 0 \Rightarrow y = \frac{1}{2x - 3} = -\frac{1}{3}$

intercept at $(0,-\frac{1}{3})$

$$y = \frac{1}{2x-3}$$

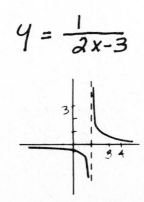

d) $y \to 0$ as $x \to \pm \alpha$. $x \to \frac{3}{2}$ as $y \to \pm \alpha$. The asymptotes

are the x-axis and the vertical line $x = \frac{3}{2}$.

e) $y' = \dfrac{-2}{(2x - 3)^2} = \dfrac{-2}{9}$ when $x = 0$.

3. a) $y(-x) = y(x) \Rightarrow$ symmetrical about

y-axis.

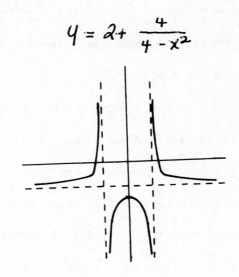

$y = 2 + \dfrac{4}{4 - x^2}$

b) If $x^2 \le 4$, then $x \ne \pm 2$.

$y \ge 3$. If $x^2 \ge 4$, then $y < 2$.

Hence $y \ge 3$ or $y < 2$.

c) If $x = 0$, then $y = 3$. $y = 0$ when

$x = \pm \sqrt{6}$. Intercepts at $(0,3)$,

$(\sqrt{6},0)$, $(-\sqrt{6},0)$

d) As $x \to \pm \infty$, $y \to 2$. As $y \to \pm \alpha$,

$x \to \pm 2$. Thus the asymptotes are the

lines $y = 2$, $x = 2$, $x = -2$.

e) $y = \dfrac{8x}{(4 - x^2)^2} = 0$
when $x = 0$. $y' = \pm 2\sqrt{6}$ when

$x = \pm \sqrt{6}$.

5. a) $y(-x) = -y(x) \Rightarrow$ symmetrical

about the origin.

b) $x \ne 0$. y can be any number.

c) $y = x - \dfrac{1}{x} = 0 \Rightarrow x = \dfrac{1}{x}$

$\Rightarrow x = \pm 1$. Intercepts at

$(1,0)$, $(-1,0)$.

d) As $x \to \pm \infty$, $y \to x$. As $x \to 0^+$,

$y \to -\alpha$. As $x \to 0^-$, $y \to +\alpha$. The

asymptotes are therefore $y = x$

See text.

and the y-axis.

e) $y' = 1 + \dfrac{1}{x^2} = 2$ at $x = \pm1$.

7. a) $y = \dfrac{x}{x+1} = \dfrac{x+1-1}{x+1} = 1 - \dfrac{1}{x+1}$.

   Symmetrical about the point $(-1,1)$.

   b) $x \neq -1$, $y \neq 1$.

   c) $x = 0 \Rightarrow y = 1 - \dfrac{1}{x+1} = 0$. Intercept at $(0,0)$.

   d) Asymptotes include lines $x = -1$ and $y = 1$.

   e) $y = 1 - \dfrac{1}{x+1} \Rightarrow y' = \dfrac{1}{(x+1)^2} = 1$ at $x = 0$.
   Graph is identical to graph of $y = -\dfrac{1}{x}$, only with "origin" at $(-1,1)$.

9. a) $y(x) = x^2 + 1 = y(-x) \Rightarrow$ symmetrical about y-axis.

   b) $x^2 \geq 0 \Rightarrow y = x^2 + 1 \geq 1$. x is any real number.

   c) $x = 0 \Rightarrow y = 1$. No x-intercept because $y \geq 1$. Intercept at $(0,1)$.

   d) No asymptotes

   e) $y' = 2x = 0$ at $x = 0$. Curve is parabola $y = x^2$ with vertex at $(0,1)$.

11. a) symmetric about y-axis.

    b) $y \geq -1$; x is any real number.

    c) $x = 0 \Rightarrow y = -1$. $y = 0 \Rightarrow x^2 - 1 = 0 \Rightarrow x = \pm1$. Intercepts are $(0,-1)$,

       $(1,0)$, $(-1,0)$.

    d) No asymptotes.

    e) $x^2 \geq 0 \Rightarrow y = x^2 + 1 \geq 1$. x is any real number.

Curve is same as parabola $y = x^2$ with vertex at $(0,-1)$.

13. $y = \dfrac{x^2}{x-1} = \dfrac{x^2-1+1}{x-1} = \dfrac{(x-1)(x+1)}{x-1} + \dfrac{1}{x-1} = x + 1 + \dfrac{1}{x-1}$

    a) $y = x + 1 + \dfrac{1}{x-1} \Rightarrow y - 2 = (x-1) + \dfrac{1}{x-1} \Rightarrow$

    $y' = x' + \dfrac{1}{x'}$, where $y' = y - 2$ and $x' = x - 1$. If (x',y') is replaced by

    $(-x',-y')$ equality still holds. The curve is therefore symmetrical about

    the origin of the new curve, that is about the point $(1,2)$ of the original system.

b) If v is any positive (negative) number then the smallest (largest) value of

$v + \frac{1}{v}$ is 2(-2). This applied to the equation $y = 2 = x - 1$

$+ \frac{1}{x - 1}$, with $v = x - 1$, reveals that $y \geq 4$ or $y \geq 0$. Hence

$x \neq 1$ and $y \geq 4$ or $y \leq 0$.

c) $x = 0 \Rightarrow y = \frac{x^2}{x - 1} = 0$. Intercept at (0,0).

d) $y = x + 1 + \frac{1}{x - 1}$. As $x \rightarrow \pm\infty$ $\frac{1}{x - 1} \rightarrow 0$. As $x \rightarrow 1$,

$y \rightarrow \pm\infty$. The asymptotes therefore are the lines $y = x + 1$ and $x = 1$.

e) $y' = 1 - \frac{1}{(x - 1)^2} = 0$ at $x = 0$. The curve is similar to

that of problem 4, with new origin at (1,2) and asymptote $y = x + 1$ instead

of $y = \frac{1}{2}x$. (See problem 15.)

15. $y = \frac{x^2 + 1}{x - 1} = \frac{x^2 - 1 + 2}{x - 1} = \frac{(x + 1)(x - 1)}{x - 1} + \frac{2}{x - 1} = x + 1 + \frac{2}{x - 1}$

a) $y = x + 1 + \frac{2}{x - 1} \Rightarrow y - 2 = x - 1 + $ See text.

$\frac{2}{x - 1} \Rightarrow y' = x' + \frac{2}{x'}$, where

$y' = y - 2$ and $x' = x - 1$. Replacing (x',y') with

$(-x',-y')$ does not effect the equality. Thus the

curve is symmetrical about the origin of the new

system, that is about the point (1,2) of the

original system.

b) If v is any positive (negative) number then the smallest (largest) value of

$v + \frac{2}{v}$ is $2\sqrt{2}$ $(-2\sqrt{2})$. This applied to the equation $y - 2 = $

$x - 1 + \frac{2}{x - 1}$ reveals that $y \geq 2 + 2\sqrt{2}$ or $y \leq 2 - 2\sqrt{2}$.

Also, $x \neq 1$.

c) $x = 0 \Rightarrow y = \frac{x^2 + 1}{x - 1} = -1$. Intercept at (0,-1)

d) $y = x + 1 + \frac{2}{x - 1}$. $\frac{2}{x - 1} \rightarrow 0$, as $x \rightarrow \pm\infty$ $y \rightarrow \infty$ as $x \rightarrow 1$.

The asymptotes are the lines $y = x + 1$ and $x = 1$.

e) $y' = 1 - \dfrac{2}{(x-1)^2} = -1$ at $x = 0$.

17. $y = \dfrac{x^2 - 4}{x - 1} = \dfrac{x^2 - 1 - 3}{x - 1} = \dfrac{(x-1)(x+1)}{x-1} - \dfrac{3}{x-1} = x\pi - \dfrac{3}{x-1}$

a) By the argument of problem 15(a) the curve is symmetrical about the point $(1,2)$.

b) $x \neq 1$, y is any real number.

c) $x = 0 \Rightarrow y = \dfrac{x^2 - 4}{x - 1} = 4$. $y = 0 \Rightarrow x^2 - 4 = 0 \Rightarrow x = 2$ or $x = -2$.

d) $y = x + 1 - \dfrac{3}{x - 1}$. $\dfrac{3}{x - 1} \to 0$ as $x \to \infty$. $y \to \infty$ as $x \to 1$.

Asymptotes are the lines $y = x + 1$ and $x = 1$.

e) $y' = 1 + \dfrac{3}{(x-1)^2}$. $y' = 4$ at $x = 0$. $y' = 4$ at $x = 2$.
$y' = \dfrac{4}{3}$ at $x = -2$. The graph is similar to that of problem 5.

19. $y = \dfrac{x^2 - 1}{2x + 4} = \dfrac{x^2 - 4 + 3}{2(x + 2)} = \dfrac{(x+2)(x-2)}{2(x+2)} + \dfrac{3}{2(x+2)} = \dfrac{x}{2} - 1 + \dfrac{3}{2x+4}$

a) Symmetrical about $(-2,-2)$

b) $y' = \dfrac{1}{2} - \dfrac{6}{(2x+4)^2} = 0 \Rightarrow x = -2 + \sqrt{3}$ or
$x = -2 - \sqrt{3}$. $y(-2 + \sqrt{3}) = \sqrt{3} - 2$ and $y(-2 - \sqrt{3}) = -2 - \sqrt{3}$.

Hence $y \geq \sqrt{3} - 2$ or $y \leq -2 - \sqrt{3}$. Also, $x \neq -2$.

c) $x = 0 \Rightarrow y = -\dfrac{1}{4}$. $y = 0 \Rightarrow x^2 - 1 = 0 \Rightarrow x = \pm 1$.

Intercepts $(0, -\dfrac{1}{4})$, $(1,0)$ $(-1,0)$.

d) $\dfrac{3}{2x + 4} \to 0$ as $x \to \pm\infty$. $y \to \infty$ as $x \to -2$.

Asymptotes are $y = \dfrac{x}{2} - 1$ and $x = -2$.

e) $y' = \dfrac{1}{2} - \dfrac{6}{(2x+4)^2}$. $y' = \dfrac{1}{8}$ at $x = 0$. $y' = \dfrac{1}{3}$
at $x = 1$. $y' = -1$ at $x = -1$. Curve is similar to that of problem 15, with
asymptote lines $y = \dfrac{x}{2} - 1$ and $x = -2$ instead of $y = x + 1$ and $x = 0$.

21. a) No symmetry

   b) y is any real number, $x \neq 0$.

See text.

   c) No y-intercept, because

   $x \neq 0$. $y = 0 \Rightarrow 6x^2 =$

   $-\dfrac{2}{x} \Rightarrow x = -\sqrt[3]{1/3}$.

   Intercept $(-\dfrac{1}{\sqrt[3]{3}}, 0)$.

   d) As $x \to \infty$, $\dfrac{2}{x} \to 0$. As $x \to 0$,

   $y \to \pm\infty$. The asymptotes are

   the x-axis and the curve

   $y = 6x^2$.

   e) $y' = -\dfrac{2}{x^2} + 12x$.

   $y' = -6\sqrt[3]{9}$ at $x = -\dfrac{1}{\sqrt[3]{3}}$.

23. $y = \dfrac{x^2 - 4}{x^2 - 1} = \dfrac{x^2 - 1 - 3}{x^2 - 1} = 1 - \dfrac{3}{x^2 - 1}$

See text.

   a) Symmetrical about y-axis

   because $y(x) = y(-x)$.

   b) $y' = \dfrac{6x}{(x^2 - 1)^2} = 0 \Rightarrow x = 0$. $y(0) = 4$. If $x^2 - 1 > 0$,

   then $x^2 - 4 < x^2 - 1$ implies $\dfrac{x^2 - 4}{x^2 - 1} < 1$. Hence $y \geq 4$ or

   $y < 1$. Also, $x \neq \pm 1$.

   c) $x = 0 \Rightarrow y = 4$. $y = 0 \Rightarrow x^2 - 4 = 0$, $x = \pm 2$.

   Intercepts are $(0,4)$, $(-2,0)$, $(2,0)$.

   d) $\dfrac{3}{x^2 - 1} \to 0$ as $x \to \pm\infty$. $y \to \pm\infty$ as $x \to \pm 1$. The asymptotes are

   the lines $y = 1$, $x = 1$, and $x = -1$.

e) $y' = \dfrac{6x}{(x^2 - 1)^2}$.

$y' = 0$ at $x = 0$. $y' = \dfrac{4}{3}$ at $x = 2$. $y' = -\dfrac{4}{3}$ at $x = -2$. The graph

is similar to that of problem 3.

25. $y = \dfrac{x^2 + 1}{x^2 - 4x + 3} = \dfrac{x^2 - 4x + 3}{x^2 - 4x + 3} + \dfrac{4x - 2}{x^2 - 4x + 3}$ See text.

   $= 1 + \dfrac{4x - 2}{x^2 - 4x + 3}$

a) No symmetry

b) $x^2 - 4x + 3 = (x - 3)(x - 1) \Rightarrow x \neq 3, x \neq 1$.

   $y' = -\dfrac{4(x^2 - x - 1)}{(x^2 - 4x + 3)^2} = 0 \Rightarrow x = \dfrac{1 + \sqrt{5}}{2}.$ $y(\dfrac{1 + \sqrt{5}}{2}) \approx -4.23$

   $y(\dfrac{1 - \sqrt{5}}{2}) \approx 0.236.$ $y \geq 0.236$ or $y \leq -4.23$.

c) $x = 0 \Rightarrow y = \dfrac{1}{3}$. Since $y = 0$ implies $x^3 + 1 = 0$, there can be no

   x-intercept. y-intercept at $(0, \dfrac{1}{3})$.

d) $y = 1 + \dfrac{4x - 2}{x^2 - 4x + 3} = 1 + \dfrac{4/x - 2/x^2}{1 - 4/x + 3/x^2} \rightarrow 1$ as

   $x \rightarrow \pm\infty$. Also, $y = 1 + \dfrac{4x - 2}{(x - 1)(x - 3)} \rightarrow \pm\infty$ as

   $x \rightarrow 1$ or $x \rightarrow 3$. The asymptotes are the lines $x = 1$, $x = 3$ and $y = 1$.

e) $y' = -\dfrac{4(x^2 - x - 1)}{(x^2 - 4x + 3)^2} = \dfrac{4}{9}$ at $x = 0$.

27. a) Replacing y with $-y$ does not affect the equality,

   so the curve is symmetrical with respect to the

   x-axis.

b) $y^2 \geq 0 \Rightarrow \dfrac{x}{x - 2} \geq 0$. If $x - 2 > 0$

   (that is $x > 2$), then $x \geq (x - 2) \cdot 0$ or $x \geq 0$;

   hence $x > 2$. If $x - 2 < 0$, then $x \leq (x - 2) \cdot 0$

   or $x \leq 0$, thus $x > 2$ or $x \leq 0$. Also,

   $x \neq x - 2$. Thus $\dfrac{x}{x - 2} \neq 1$, and

   $y \neq \pm 1$.

$y^2 = \dfrac{x}{x - 2}$

c) $x = 0 \Rightarrow y = 0$. Intercept at $(0,0)$.

d) $y^2 \to 1$ as $x \to \pm\infty$. $y \to \pm\infty$ as $x \to 2$. The asymptotes are the lines

   $y = 1$, $y = -1$, and $x = 2$.

e) $y' = -\dfrac{1}{y(x-2)^2}$. $y' = \infty$ when $y = 0$.

29. a) $F(x,y) = x^2 - \dfrac{1 + y^2}{1 - y^2}$.

   $F(x,y) = F(-x,y) = F(-x,-y) = F(x,-y)$. The curve

   is symmetrical about the x-axis and y-axis and

   origin.

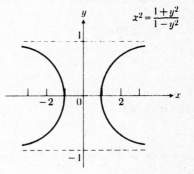

b) $x^2 \geq 0 \Rightarrow \dfrac{1 + y^2}{1 - y^2} \geq 0 \Rightarrow$

   $1 - y^2 > 0 \Rightarrow -1 < y < 1$. $1 + y^2 \geq 1 - y^2$ and 1

   $- y^2 > 0 \Rightarrow \dfrac{1 + y^2}{1 - y^2} \geq 1 \Rightarrow x \geq 1$

   or $x \leq -1$.

c) $x \neq 0 \Rightarrow$ no y-intercept. $y = 0 \Rightarrow x^2 = 1$. Intercepts at $(-1,0)$, $(1,0)$.

d) $x \to \pm\infty$ as $y \to \pm 1$. Asymptotes are the lines $y = 1$ and $y = -1$.

e) $y' = \dfrac{x}{2y}(y^2 - 1)^2$. $y' = \infty$ at both $(-1,0)$ and $(1,0)$.

31. a) No symmetry   See text.

b) $x \neq 0$, $x \neq 2$, y is any real number.

c) $y = 0 \Rightarrow x - 1 = 0$. Intercept at $(1,0)$.

d) $y = \dfrac{x-1}{x^2(x-2)} = \dfrac{x - 2 + 1}{x^2(x-2)} = \dfrac{1}{x^2} + \dfrac{1}{x^2(x-2)} \to 0$

   as $x \to \pm\infty$. $y \to \pm\infty$ as $x \to 0$ or $x \to 2$.

   Asymptotes are the lines $x = 1$, $x = 2$,

   and the y-axis.

e) $y' = \dfrac{-2}{x^3} \quad \dfrac{3x - 4}{x^3(x-2)^2}$. $y' = 1$ at $x = 1$.

33. $y = 2 + \dfrac{\sin x}{x} \Rightarrow y' = \dfrac{\cos x}{x} - \dfrac{\sin x}{x^2} \to 0$ as $x \to \infty$, because

   $|\cos x| \leq 1$ and $|\sin x| \leq 1$. The slope of the line is zero.

35. a) If $f(x)$ is even, $f(x) = f(-x)$, and $f(x)$ is symmetrical about the y-axis.

    b) If $y = f(x)$ is odd, then $-y = f(-x)$. The equation $F(x,y) = f(x) - y = 0$ is not altered when (x,y) is replaced by $(-x,-y)$. Hence there is symmetry about the origin.

    c) The functions $\sin x$ and $\tan x$ are both odd, so each is symmetrical about the origin.

Article 3.4

1. $y = x - x^2 \Rightarrow y' = 1 - 2x = 0 \Rightarrow x = \frac{1}{2}$. $y(\frac{1}{2}) = \frac{1}{4}$. $y'' = -2 < 0$. $(\frac{1}{2},\frac{1}{4})$ is a local maximum. $y(0) = 0$; $y(1) = 0$. Therefore $(\frac{1}{2},\frac{1}{4})$ is an absolute maximum and $(0,0)$ is an absolute minimum.

3. $y = x - x^3 \Rightarrow y' = 1 - 3x^2 = 0 \Rightarrow x = \frac{1}{\sqrt{3}}$. $y'' = -6x \Rightarrow y'' < 0$ when $x = \frac{1}{\sqrt{3}}$, so $(\frac{1}{\sqrt{3}}, \frac{2\sqrt{3}}{9})$ is a local maximum. $y(0) = 0$, $y(1) = 0$, and $y(\frac{1}{\sqrt{3}}) = \frac{2\sqrt{3}}{9}$. $x = \frac{1}{\sqrt{3}}$ is at an absolute maximum. An absolute minimum is at $(0,0)$.

5. $y = x^3 - 147x \Rightarrow y' = 3x^2 - 147 = 3(x^2 - 49) = 0 \Rightarrow x = \pm 7$. $y'' = 6x$. $y'' > 0$ at $x = 7$, so $(7,-686)$ is a local minimum. $y'' < 0$ at $x = -7$, so $(-7,686)$ is a local maximum. y is unbounded, so there are no absolute maxima or minima.

7. $y = |x - x^2|$. Thus $y = \pm(x - x^2) \Rightarrow y' = \pm(1 - 2x) = 0 \Rightarrow x = \frac{1}{2}$. At $x = \frac{1}{2}$, $y = x - x^2$, $y' = 1 - 2x$, $y'' = -2 < 0 \Rightarrow$ local maximum. Also, $y(-2) = 6$, $y(\frac{1}{2}) = \frac{1}{4}$, $y(2) = 2$, $y(0) = 0$. Absolute maximum is at $x = -2$. Absolute minimum at $x = 0$.

9. $y = \frac{1}{3 - x} \Rightarrow y' = \frac{+1}{(3 - x)^2}$. $y' \neq 0$, so there are no critical points. $y \rightarrow \infty$ as $x \rightarrow 3^-$ and $y \rightarrow -\infty$ as $x \rightarrow 3^+$. Thus there are no absolute maxima or minima.

11. $y = x^4 - 4x \Rightarrow y' = 4x^3 - 4 = 0 \Rightarrow x = 1$. $y'' = 12x^2 > 0$ at $x = 1$, so $(1,-3)$ is

a global minimum.

13. $y = \sec x \Rightarrow y' = \sec x \tan x = 0 \Rightarrow x = 0.$ $y'' = \sec^3 x + \sec x \tan^2 x = 1 > 0$ at $x = 0.$ $(0,1)$ is a local minimum and a global minimum. There is no absolute maximum.

15. $y = x^4 - 8x^3 - 270x^2 \Rightarrow y' = 4x^3 - 24x^2 - 540x = 4x(x - 15)(x + 9) = 0 \Rightarrow$ $x = 0, 15, -9.$ $y'' = 12x^2 - 48x - 540 = 12(x - 9)(x + 5).$ $y'' < 0$ when $x = 0$ - local maximum, $y'' > 0$ when $x = 15$ - local minimum, $y'' > 0$ when $x = -9$ - local minimum. $y(-9) = -9477,$ and $y(15) = -37125,$ so $(15, -37125)$ is the absolute minimum point. There is no absolute maximum.

17. For $x \geq 0,$ $y = \dfrac{x}{1 + x} \Rightarrow y' = \dfrac{1}{(x_1 + 1)^2} \neq 0.$
    For $x < 0,$ $y = \dfrac{x}{1 - x} \Rightarrow y' = \dfrac{1}{(x - 1)^2} \neq 0.$ There are no critical points. Also, there are no local or absolute maxima or minima.

19. (Note Some examples of greatest integer function are $[1.3] = 1,$ $[2.71] = 2,$ $[4] = 4,$ $[-2.71] = -3.$) If x is restricted to an open interval such as $(k, k + 1),$ where k is an integer, then $y' = 1.$ There are no critical points because $y' \neq 0.$ Absolute minimum is 0 and there is no absolute maximum.

21. $y = |x^3| \Rightarrow y' = 3x|x|.$ $y' = 0$ at $x = 0.$ $y \geq 0$ and $y(0) = 0,$ so $x = 0$ is at an absolute (and hence local) minimum. Absolute maximum occurs at $x = 3.$ For the domain $[-2, 3],$ there is no absolute maximum since y can always be increased by taking x closer to 3.

23. On $[0,2]$, $y = 3 - x$, $y' = -1$ and there are no critical points. On $[2,3]$, $y = \frac{1}{2}x^2 \Rightarrow y' = x \neq 0$, and there are no critical points again. Absolute minimum at $x = 2$, $y = 1$. Absolute maximum at $x = 3$, $y = \frac{9}{2}$.

25. $y = \frac{x}{x^2 + 1} \Rightarrow y' = \frac{1 - x^2}{(x^2 + 1)^2} = 0 \Rightarrow x = \pm 1$.

$y'' = \frac{2x(x^2 - 3)(x^2 + 1)}{(x^2 + 1)^4} = 0 \Rightarrow x = 0, x = \pm/3$.

Critical points at $(1, \frac{1}{2})$ and $(-1, -\frac{1}{2})$. Absolute

maximum is $\frac{1}{2}$, absolute minimum is $-\frac{1}{2}$. $y'' > 0$ for

$x > /3$, $y'' < 0$ for $0 < x < /3$, $y'' > 0$ for $-/3 < x < 0$

and $y'' < 0$ for $x < -/3$. Hence the inflection points are

$0, +/3, -/3$.

$y = \frac{x}{x^2 + 1} = \frac{1/x}{1 + 1/x^2} \rightarrow 0$ as

$x \rightarrow \pm\infty$, so the asymptote is the x-axis.

27. a) No. If $f(x) = x^2$, $f'(x) = 2x$, which is a straight line having no points of inflection.

    b) Yes. At a relative maximum or minimum (on an open interval) $f' = 0$.

    c) No. Using the example of part (a), $f'(x) = 2x$ need not have a relative maximum or minimum.

29. $y = 4\sin^2 x - 3\cos^2 x \Rightarrow y' = 8\sin x \cos x - 6\cos x(-\sin x) =$

$14 \sin x \cos x = 0 \Rightarrow \sin x = 0$ or $\cos x = 0$. If $\sin x = 0$, $y = -3$. If

$\cos x = 0$, $y = 4$. The maximum is therefore 4.

Article 3.5

1. (P = perimeter, A = area, w = width, 1 = length.) Maximize A subject to $P = 2w + 21$ or $1 = w - \frac{P}{2}$. $A = w^{.}1 = w(w - \frac{P}{2}) \Rightarrow A' = 2w - \frac{P}{2} = 0 \Rightarrow$ $w = \frac{P}{4} \Rightarrow 1 = \frac{P}{4}$. The dimensions of a square with perimeter P are $\frac{P}{4}$

by $\frac{P}{4}$. $A'' = 2 > 0$. The graph is concave up everywhere so this is an absolute maximum.

3. $A = 2xy = 2x(12 - x^2) = 24x - 2x^3 \Rightarrow A' = 24 - 6x^2 = 0 \Rightarrow x = 2$, $y = 8$. $A = 32$. $A'' = -12x < 0$, because $x > 0$. Hence the $A = 32$ is a maximum.

(0,12)

$(x, 12-x^2)$

$(-2\sqrt{3}, 0)$ $(2\sqrt{3}, 0)$

5. (1 = length parallel to river; w = width perpendicular to river; P = fixed perimeter.) $P = 1 + 2w \Rightarrow 1 = P - 2w$. $A = 1 \cdot w = (P - 2w) \cdot w \Rightarrow A' = P - 4w = 0 \Rightarrow w = \frac{P}{4}$, $1 = \frac{P}{2}$.

7. $v = 32 = w^2 h \Rightarrow h = \frac{32}{w^2}$. $m = 4wh + w^2 \Rightarrow 4w(\frac{32}{w^2}) + w^2 = \frac{128}{w} + w^2$. $m' = -\frac{128}{w^2} + 2w = 0 \Rightarrow w = 4 \Rightarrow h = 2$. $m'' = 2 + \frac{256}{w^3} > 0$,

so material is minimized. The dimensions are 4 by 4 by 2.

9. Let h and w be the height and width of the written material. $w \cdot h = 50$ and $h = \frac{50}{w}$. Total area, $A = (h + 8)(w + 4) = wh + 4h + 8w + 32 = 50 + 4(\frac{50}{w}) + 82 + 32 = 82 + \frac{200}{w} + 8w$. $A' = \frac{-200}{w^2} + 8 = 0 \Rightarrow w = 5 \Rightarrow h = 10$. $A'' = \frac{400}{w^3} > 0$, so the area is a minimum. Overall dimensions are 9 in. by 18 in.

11. (1 = length of the two long sides; s = length of the two short sides; A is the fixed area and F is the amount of fencing to be used.) $A = s1 \Rightarrow 1 = A/S$. $F = 3s + 21 = 3s + \frac{2A}{s}$. $F' = 3 - \frac{2A}{s^2} = 0 \Rightarrow s = \sqrt{\frac{2}{3} A}$, $1 = \sqrt{\frac{3}{2} A}$. $F'' = \frac{4A}{s^3} > 0$, so the area is a minimum. The total fencing is $3\sqrt{\frac{2}{3} A} + 2\sqrt{\frac{3}{2} A} = \sqrt{6A} = \sqrt{(6)(216)} = 36$; s = 12; 1 = 18.

13. (r = radius of the base; h = height.) $v = \pi r^2 h = 1000 = h = \frac{1000}{\pi r^2}$. $A = 2\pi rh + \pi r^2 = 2\pi r(\frac{1000}{\pi r^2}) + \pi r^2 = \frac{2000}{r} + \pi r^2$.

$A' = -\dfrac{2000}{r^2} + 2\pi > 0$, so the surface area and hence mass are at a minimum.

15. (ℓ = length of the base; h = height; c = cost.) $\quad 2m^3 = 1 \cdot h \cdot 1 \Rightarrow h = \dfrac{2}{\ell}$.

$c = 10(\ell \cdot 1) + 5[2(h \cdot 1) + 2(h \cdot 1)] = 10\ell + 10h + 10h\ell =$

$10(\dfrac{2}{h} + 10h + 10(\dfrac{2}{h})h = \dfrac{20}{h} + 10h + 20$.

$c' = -\dfrac{20}{h^2} + 10 = 0 \Rightarrow h = \sqrt{2} \Rightarrow c = 20 + 20\sqrt{2}$. (Note $\ell = h = \sqrt{2}$.)

17. D = Distance between (x, x^2) and $(2, -\frac{1}{2})$.

$D^2 = (\Delta x)^2 + (\Delta y)^2 = (x - 2)^2 + (x^2 + \frac{1}{2})^2$.

$\dfrac{dD^2}{dx} = 2(x - 2) + 2(x^2 + \frac{1}{2})(2x) = 4x + 4x^3 - 4 = 0$

$\Rightarrow x(1 + x^2) = 1 \Rightarrow x = \dfrac{1}{x^2 + 1}$. To check that this

is in fact the minimum,

$\dfrac{d^2 D^2}{dx^2} = 4 + 12x^2 > 0$ for all x.

19. base $= \pi r^2 = \pi(h^2 - x^2)$.

$v = \frac{1}{3}$ base. height $= \dfrac{\pi}{3}(h^2 - x^2)x$

$v' = \dfrac{\pi}{3}(h^2 - 3x^2) = 0 \Rightarrow x = \dfrac{h}{\sqrt{3}}$.

height $= \dfrac{h}{\sqrt{3}}$, radius $= (h^2 - x^2)^{1/2} = \dfrac{\sqrt{6}}{3}h$.

21. $y = \dfrac{W}{48EI}(2x^4 - 5Lx^3 + 3L^2x^2) \Rightarrow$

$\dfrac{dy}{dx} = \dfrac{W}{48EI}(8x^3 - 15Lx^2 + 6L^2x) = 0 \Rightarrow$

$x = 0$ or $x = \dfrac{L(15 \pm \sqrt{33})}{16}$.

$y'' = \dfrac{W}{8EI}(4x - L)(x - L)$. The conditions $y'' < 0$ and $0 \le X \le L$ are

both satisfied when $X = \dfrac{L(15 - \sqrt{33})}{16} \approx 0.58L$.

23. a) Need $f'(-1) = 0$ and $f'(3) = 0$. $f'(x) = 3x^2 + 2ax + b$.

$f'(3) = 27 + 6a + b = 0$. $f'(-1) = 3 - 2a + b = 0$. Solving, $a = -3$, $b = -9$.

b) Need $f''(1) = 0$ and $f'(4) = 0$. $f''(x) = 6x + 2a$; $f''(1) = 6 + 2a = 0 \Rightarrow$

$a = -3$. $f'(4) = 48 + 8a + b = 48 - 24 + b = 0 \Rightarrow b = -24$.

25. $5x^2 - 6xy + 5y^2 = 4 \Rightarrow 10x - 6y - 6x\dfrac{dy}{dx} + 10y\dfrac{dy}{dx} = 0 \Rightarrow \dfrac{dy}{dx} =$

$\dfrac{3y - 5x}{5y - 3x}$. Minimizing the distance to the origin is equivalent to minimizing the distance squared, here called f(x). $f(x) = x^2 + y^2$.

$f'(x) = 2x + 2y\frac{dy}{dx} = 0 \Rightarrow x + y(\frac{3y - 5x}{5y - 3x}) = 0 \Rightarrow 3x^2 - 5xy =$

$3y^2 - 5xy \Rightarrow x = \pm y$. If $x = y$, then $5x^2 - 6x^2 + 5x^2 = 4 \Rightarrow x = \pm 1$,

$y = \pm 1$ and $f(x) = 2$. If $x = -y$, then $5x^2 + 6xy + 5x^2 = 4 \Rightarrow x = \pm\frac{1}{2}$,

$y = \mp\frac{1}{2}$ and $f(x) = \frac{1}{2} < 2$. The closest points are therefore

$(\frac{1}{2}, -\frac{1}{2})$ and $(-\frac{1}{2}, \frac{1}{2})$.

27. Base of cone $= \pi(\sqrt{r^2 - x^2})^2$. Height of cone $= r + x$.

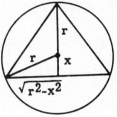

Volume is $V = \frac{1}{3}$(base)(height).

$V = \frac{\pi}{3}(r^2 - x^2)(r + x)$.

$V' = \frac{\pi}{3}(r - 3x)(r + x) = 0 \Rightarrow x = \frac{r}{3}$.

$V'' = \frac{-2\pi}{3}(r + 3x) < 0 \Rightarrow V$ is a maximum.

The maximum volume is $\frac{32}{81}\pi r^3$.

29. $\frac{y}{x} = \frac{h}{r} \Rightarrow y = (\frac{h}{r})x$. Base $= \pi x^2$.

Height $= h - y = h - (\frac{h}{r})x$.

$v = $ (base)(height) $= \pi x^2(h - \frac{h}{r}x)$.

$v' = \pi h(2x - \frac{3x^2}{r}) = 0 \Rightarrow x = \frac{2}{3}r \Rightarrow v = \frac{4\pi}{27}r^2 h$.

$v'' = \pi h(2 - \frac{6x}{r}) = -2\pi h < 0$ at $x = \frac{2}{3}r$, so the

volume is a maximum. The ratio of the volume of the

cylinder to that of the cone Base $= \pi x$

$\frac{(4/27)\pi r^2 h}{(\pi/3)r^2 h} = \frac{4}{9}$.

31. (D = fixed diameter; x = breadth; y = depth). $x^2 + y^2 = D^2 \Rightarrow$

$y = (D^2 - x^2)^{1/2}$. Strength is $S = kxy^3 = kx(D^2 - x^2)^{3/2}$.

$S' = k(D^2 - x^2)^{1/2}(D^2 - 4x^2) = 0 \Rightarrow x = \frac{D}{2}, y = \frac{\sqrt{3}}{2}D$.

$S' < 0$ for $\frac{D}{2} < x < D$ and $S' > 0$ for $0 \le x < \frac{D}{2}$, so $x = \frac{D}{2}$ is at a

maximum.

33. (P = fixed perimeter; x = width;

y = height of rectangular portion)

$P = \frac{\pi x}{2} + x + 2y \Rightarrow \frac{\pi}{2} + 1 + 2\frac{dy}{dx} = 0 \Rightarrow$

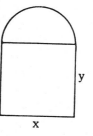

$\frac{dy}{dx} = -\frac{\pi + 2}{4}$. Area of rectangle = xy; area of

semicircle $= \frac{\pi x^2}{8}$. Total light is $L = k(2xy + \frac{\pi x^2}{8})$.

$\frac{dL}{dx} = k(2y + 2x\frac{dy}{dx} + \frac{\pi x}{4}) = 0$

$\Rightarrow 2y + 2x(-\frac{\pi + 2}{4}) + \frac{\pi 4}{4} = 0$

$\Rightarrow \frac{y}{x} = \frac{\pi + 4}{8}$.

This ratio gives maximum light, for $\frac{d^2L}{dx^2} = -k\frac{3\pi + 8}{8} < 0$.

35. <u>Area</u>

Dome $= 2\pi r^2$; Sides $= 2\pi rh$; Bottom $= \pi r^2$

$S = 2\pi r^2 + 2\pi rh + \pi r^2 = 3\pi r^2 = 2\pi rh \Rightarrow h = \frac{S - 3\pi r^2}{2\pi r}$ (1)

<u>Volume</u>

Dome $= \frac{2}{3}\pi r^3$; Cylinder $= \pi r^2 h$

$V = \frac{2}{3}\pi r^3 + \pi r^2 h = \frac{2}{3}\pi r^3 + \pi r^2(\frac{S - 3\pi r^2}{2\pi r})$

$= \frac{2}{3}\pi r^3 + \frac{Sr - 3\pi r^3}{2} = \frac{Sr}{2} - \frac{5}{6}\pi r^3$ (2)

$\frac{dV}{dr} = \frac{S}{2} - \frac{5}{2}\pi r^2 = 0 \Rightarrow S = 5\pi r^2$

(1) $\Rightarrow h = \frac{5\pi r^2 - 3\pi r^2}{2\pi r} = r$.

The ratio of r to h that maximizes volume for fixed surface area is 1.

37. $y = 3 + 4\cos\theta + \cos 2\theta \Rightarrow 0 = y' = -4\sin\theta - 2\sin 2\theta$

$= -4\sin\theta(1 + \cos\theta) \Rightarrow \theta = 0$ or $\theta = \pi$.

$y(0) = 3 + 4 + 1 = 8 > 0$. $y(\pi) = 3 - 4 + 1 = 0$. If $\frac{\pi}{2} < \theta < \pi$, then

$y' < 0$. If $\pi < \theta < \frac{3}{2}\pi$, then $y' > 0$. Hence $y(\pi) = 0$ is a minimum and

$y \geq 0$. The answer is no. Remark The second derivative test fails because

$y'' = 0$.

39. (k indicates any integer)

<u>Method 1</u>

$$y = \sin x + \cos x = A \sin(x + a) = A(\sin x \cos a + \sin a \cos x)$$

$$\Rightarrow \sin x = A \cos a \sin x \Rightarrow A \cos a = 1$$

$$\cos x = A \sin a \cos x \Rightarrow A \sin a = 1$$

$$A^2\cos^2 a + A^2\sin^2 a = 1^2 + 1^2 \Rightarrow A^2 = 2 \Rightarrow A = \sqrt{2}.$$

$$\sin a = \cos a \Rightarrow a = \frac{\pi}{4}. \quad y = \sin x + \cos x = \sqrt{2}\sin(x + \frac{\pi}{4}).$$

y attains a maximum value of $\sqrt{2}$ when $\sin(x + \frac{\pi}{4}) = 1$, or $x = \frac{\pi}{4} + 2k\pi$.

y attains the minimum value of $-\sqrt{2}$ when $\sin(x + \frac{\pi}{4}) = -1$, or

$$x = \frac{5}{4}\pi + 2k\pi.$$

<u>Method 2</u>

$$y = \sin x + \cos x \Rightarrow y' = \cos x - \sin x \Rightarrow y'' = -\sin x - \cos x.$$

$y' = 0 \Rightarrow \sin x = \cos x \Rightarrow x = \frac{\pi}{4} + k\pi$. At $x = \frac{\pi}{4} + 2k\pi$, $y = \frac{\sqrt{2}}{2} +$

$\frac{\sqrt{2}}{2} = \sqrt{2}$ and $y'' = -\sqrt{2} < 0$, so these points are maxima. At $x = \frac{5\pi}{4} +$

$2k\pi$, $y = -\sqrt{2}$ and $y'' = \sqrt{2} > 0$, so these points are minima.

41. (v = fixed volume, c = cost)

$$v = \pi r^2 h + \frac{2}{3}\pi r^3 \Rightarrow h = \frac{v}{\pi r^3} - \frac{2}{3}r.$$

$$c = k[2\pi rh + 2(2\pi r^2)] = k[2\pi r(\frac{v}{\pi r^2} - \frac{2}{3}r) + 4\pi r^2] = \frac{2v}{r} + \frac{8}{3}\pi r^2$$

$$c' = -\frac{2v}{r^2} + \frac{16}{3}\pi r = 0 \Rightarrow r = \frac{1}{2}\sqrt[3]{\frac{3v}{\pi}}, \quad h = \sqrt[3]{\frac{3v}{\pi}}. \text{ This cost}$$

is a minimum because $c'' = \frac{4v}{r^3} + \frac{16}{3}\pi > 0$.

43. A pipeline from A to B via the river requires

the same length of piping as one from B to A'

if the point of intersection with the river is

the same. The minimum distance from A to B is

therefore equal to A'B, the straight segment

between A' and B. $(b - a)^2 + h^2 = c^2$.

$(A'B)^2 = (b + a)^2 + h^2 = (b - a)^2 + h^2 + 4ab =$

$c^2 + 4ab \Rightarrow A'B = \sqrt{c^2 + 4ab}$. The minimum

distance between A and B is also $\sqrt{c^2 + 4ab}$.

45. $D(x) = f(x) - g(x)$ achieves a maximum at c, so $\frac{dD}{dx} = 0$. That is

$\frac{dD}{dx}\Big|_c = f'(c) - g'(c) = 0$. Hence $f'(c) = g'(c)$, and the tangent lines are

parallel for they both have the same slope $f'(c)$.

47. Let x be the distance from the outside ground to the wall, and y the height of

the point where the beam meets the building. Then $\frac{y}{27 + x} = \frac{8}{x}$

$\Rightarrow y = 8 + \frac{216}{x}$. Let $f(x)$ be the length of the beam squared.

$f(x) = y^2 + (27 + x^2) = (8 + \frac{216}{x})^2 + (27 + x)^2$

$f'(x) = \frac{2}{x^3}(x^4 + 27x^3 - 1728x - (216)^2)$

$\qquad = \frac{2}{x^3}(x^3 + 39x^2 + 486x + 3888)(x - 12)$

$f'(x) = 0$ at $x = 12$, $f'(x) > 0$ for $x > 12$ and $f'(x) < 0$ for $0 < x < 12$, so a

minimum occurs at $x = 12$. The shortest beam is therefore

$\sqrt{f(12)} = (13)^{3/2} = \sqrt{2197} \approx 46.87$.

49. $v = kax - kx^2 \Rightarrow v' = ka - 2kx = 0 \Rightarrow x = \frac{9}{2}$.

$v(0) = 0$, $v(a) = ka^2 - ka^2 = 0$, $v(\frac{a}{2}) = \frac{ka^2}{4}$. Since 0, a, and $(\frac{a}{2})$

include all critical points and extreme points, maximum rate is $\frac{ka^2}{4}$, and

minimum rate is 0.

Article 3.6

1. $A = \pi r^2 \Rightarrow \frac{dA}{dt} = 2\pi r \frac{dr}{dt}$

3. $V = S^3 \Rightarrow \frac{dV}{dt} = 3S^2 \frac{dS}{dt}$

5. a) $\frac{dv}{dt} = 1 \frac{\text{volt}}{\text{sec.}}$

   b) $\frac{dI}{dt} = -\frac{1}{3} \frac{\text{amp}}{\text{sec.}}$

c) $v = IR \Rightarrow \frac{dv}{dt} = R\frac{dI}{dt} + I\frac{dr}{dt}$

d) $v = 12$, $I = 2 \Rightarrow R = 6$. $\frac{dv}{dt} = 6\frac{dI}{dt} + 2\frac{dR}{dt}$

$\Rightarrow 1 = 6(-\frac{1}{3}) + 2\frac{dR}{dt} \Rightarrow \frac{dR}{dt} = \frac{3}{2} > 0$, increasing.

7. a) Let x be the runner's distance from second base, and D the distance from

   third base. $x = 60$ ft, $\frac{dx}{dt} = -16\frac{ft}{s}$, $D = 30\sqrt{13}$.

   b) $D = (x^2 + 90^2)^{1/2}$

   c) $\frac{dD}{dt} = x(x^2 + 90^2)^{-1/2}\frac{dx}{dt}$

   d) $\frac{dD}{dt} = 60(60^2 + 90^2)^{-1/2}(-16) = -\frac{32\sqrt{13}}{13} \approx -8.875\frac{ft}{sec}$.

9. $y = (26^2 - x^2)^{1/2} = A = \frac{1}{2}xy = \frac{1}{2}x(26^2 - x^2)^{1/2}$

   $\frac{dA}{dt} = \frac{1}{2}[(26^2 - x^2)^{1/2} + \frac{1}{2}x(26^2 - x^2)^{-1/2}(-2x)]\frac{dx}{dt}$

   $= \frac{1}{2}(26^2 - x^2)^{-1/2}[(26^2 - x^2) - x^2]\frac{dx}{dt}$

   $= \frac{1}{2}(26^2 - x^2)^{-1/2}[26^2 - 2x^2]\frac{dx}{dt}$

   When $x = 10$, $\frac{dx}{dt} = 4$,

   $\frac{dA}{dt} = \frac{1}{2}(26^2 - 100)^{-1/2}(26^2 - 200)4 = \frac{119}{3} \approx 39.667$.

   $\frac{dA}{dx} = 0 \Rightarrow 26^2 - 2x^2 = 0 \Rightarrow x = 13\sqrt{2}$.

   $\frac{dA}{dx} < 0$ when $13\sqrt{2} < x < 26$ and $\frac{dA}{dx} > 0$ when $0 < x < 13\sqrt{2}$, so maximum area occurs when

   $x = 13\sqrt{2}$. $\frac{dx}{dt} = 4$, so $x = 4y$, $t = \frac{1}{4}x = \frac{13\sqrt{2}}{4}$ sec.

11. $v = \frac{4}{3}\pi r^3 \Rightarrow \frac{dV}{dt} = 4\pi r^2\frac{dr}{dt}$. Accumulation is proportional to surface

   area as written. $\frac{dv}{dt} = k(\text{surface area}) = k(4\pi r^2)$.

   Hence $4\pi r^2\frac{dr}{dt} = k(4\pi r^2) \Rightarrow \frac{dr}{dt} = k$.

13. $v = \frac{4}{3}\pi r^3 \Rightarrow \frac{dv}{dt} = 4\pi r^2\frac{dr}{dt}$. When $r = 3$, $\frac{dv}{dt} = 100$,

   $100 = 4\pi(9)\frac{dr}{dt} \Rightarrow \frac{dr}{dt} = \frac{25}{9\pi}$

15. y = height of balloon; x = distance of car from directly under balloon.

   D = distance between car and balloon.

   $y = 200 + 15 = 215$; $x = 66$; $\frac{dy}{dt} = 15\frac{ft}{sec}$; $\frac{dx}{dt} = 66\frac{ft}{sec}$

   $D = (x^2 + y^2)^{1/2} \Rightarrow \frac{dD}{dt} = \frac{1}{2}(x^2 + y^2)^{-1/2}(2x\frac{dx}{dt} + 2y\frac{dy}{dt})$

   $= (x^2 + y^2)^{-1/2}(x\frac{dx}{dt} + y\frac{dy}{dt}) = (66^2 + 215^2)^{-1/2}((66)(66) + (215)(15))$

$$= \frac{7581}{\sqrt{50581}} \approx 33.708$$

17. $3x^2 - y^2 = 12 \Rightarrow 6x - 2y\frac{dy}{dx} = 0 \Rightarrow \frac{dy}{dx} = \frac{3x}{y}$

At $x = 4$, $y^2 = 3x^2 - 12 = 48 - 12 = 36 \Rightarrow y = \pm 6$.

The slope is $\frac{dy}{dx} = \frac{3x}{y} = \frac{3(4)}{\pm 6} = \pm 2$. The x-coordinate is

changing at a rate of $\frac{dx}{dt} = \frac{dx}{dy}\frac{dy}{dt} =$

$\frac{dy/dt}{dy/dt} =$

$\frac{6}{\pm 2} = \pm 3$.

19. x = distance between man and light; y = length of shadow.

By similar triangles $\frac{y}{6} = \frac{y + x}{16}$

$\Rightarrow 16y = 6y + 6x \Rightarrow y = \frac{3}{5}x$, $x + y = \frac{8}{5}x$.

Since $\frac{dx}{dt} = -5\frac{ft}{sec}$, the tip of the shadow is moving at a rate of

$-\frac{d(x + y)}{dt} = -\frac{8}{5}\frac{dx}{dt} = -\frac{8}{5}(-5) = 8\frac{ft}{sec}$. The length

of the shadow is changing at a rate of

$\frac{dy}{dt} = \frac{3}{5}\frac{dx}{dt} = \frac{3}{5}(-5) = -3$ ft/sec.

21. $\frac{50}{x + 30} = \frac{50 - 16t^2}{x}$

$\Rightarrow 50x = 50x + 1500 - 16t^2x - 480t^2$

$\Rightarrow 4t^2x + 120t^2 = 375$. At $t = \frac{1}{2}$,

$4(\frac{1}{2})^2x + 120(\frac{1}{2})^2 = 375 \Rightarrow x = 345$.

$4t^2\frac{dx}{dt} + 8tx + 240t = 0$

$\Rightarrow 4(\frac{1}{2})^2\frac{dx}{dt} + 8(\frac{1}{2})345 + 240(\frac{1}{2}) = 0$

$\frac{dx}{dt} = -1500$. Speed $= |\frac{dx}{dt}| = 1500$ ft/sec.

23. r = radius of iron and ice ball.

$v = \frac{4}{3}\pi r^3 \Rightarrow \frac{dv}{dt} = 4\pi r^2\frac{dr}{dt}$.

When $\frac{dv}{dt} = 10$ and $r = \frac{8}{2} + 2 = 6$, this is

$10 = 4\pi(36)\frac{dr}{dt} \Rightarrow \frac{dr}{dt} = \frac{5}{72\pi}$.

$S = \pi r^2 \Rightarrow \frac{ds}{dt} = 8\pi r\frac{dr}{dt} = 8\pi(6)(\frac{5}{72\pi})$

$= \frac{10}{3}\frac{in^2}{min}$

25. If θ is the angle between the light beam and the ground, then the shadow

length is $x = 80 \cot \theta$.

$\frac{d\theta}{dt} = -15^\circ/\text{hr} = -\frac{\pi}{12} \frac{\text{rad}}{\text{hr}}$.

$x = 100 \Rightarrow \cot \theta = \frac{5}{4} \Rightarrow \sin \theta = \frac{4}{\sqrt{4^2 + 5^2}} = \frac{4}{\sqrt{41}}$.

Shadow length is increasing at rate

$\frac{dx}{dt} = -80 \csc^2 \theta \frac{d\theta}{dt}$

$= -80(\frac{\sqrt{41}}{4})^2(-\frac{\pi}{12}) = (\frac{205}{12})\pi$

27. $x = 0A$; $y = 0B$; $s = AB$. $\frac{dx}{dt} = 20$, $\frac{dy}{dt} = 30$.

By the law of cosines, $s^2 = x^2 + y^2 - 2xy \cos 120^\circ = x^2 + y^2 + xy$,

$2s\frac{ds}{dt} = 2x\frac{dx}{dt} + 2y\frac{dy}{dt} + x\frac{dy}{dt} + y\frac{dx}{dt}$.

When $x = 8$, $y = 6$, $s^2 = 8^2 + 6^2 + (6)(8) = 148$, $s = 2\sqrt{37}$.

$2(2\sqrt{37})\frac{ds}{dt} - 2(8)(20) + 2(6)(30) + 8(30) + 6(20)$

$\Rightarrow \frac{ds}{dt} = \frac{1040}{4\sqrt{37}} = \frac{260}{\sqrt{37}} \frac{\text{mi}}{\text{hr}}$

Article 3.7

1. $f(x) = x^4 + 3x + 1$, $f'(x) = 4x^3 + 3$.

$f(-2) = 11$, $f(-1) = -1$, $f'(x) < -1 < 0$ if $-2 < x < -1$.

Hence exactly one root between -2 and -1.

3. $f(x) = 2x^3 - 3x^2 - 12x - 6$. $f(-1) = 1$, $f(0) = -6$.

Hence there is at least one root on $(-1,0)$.

$f'(x) = 6x^2 - 6x - 12 = 6(x + 1)(x - 2) < 0$. $f'(x) \neq 0$ on $(-1,0)$, so there

is exactly one root $(-1,0)$.

5. i) $y = x^2 - 4$, $y' = 2x$. $y = 0$ at $x = 2$ and $x = -2$.

$y' = 0$ at $x = 0$. Note that $-2 < 0 < 2$.

ii) $y = x^2 + 8x + 15 = (x + 3)(x + 5)$. $y' = 2x + 8$.

$y = 0$ at $x = -3$ and $x = -5$. $y' = 0$ at $x = -4$.

Note that $-5 < -4 < -3$.

iii) $y = x^3 - 3x^2 + 4 = (x + 1)(x - 2)^2$.

$y' = 3x^2 - 6x = 3x(x - 2)$. $y = 0$ at $x = -1$ and $x = 2$.

$y' = 0$ at $x = 0$ and at $x = 2$. Note that $y = 0$ has two roots at

$x = 2$, and $y' = 0$ also has a root at $x = 2$.

iv) $y = x^3 - 33x^2 + 216x = x(x - 9)(x - 24)$

$y' = 3x^2 - 66x + 216 = 3(x - 4)(x - 18)$.

$y = 0$ at $x = 0$, $x = 9$, and $x = 24$. $y' = 0$ at $x = 4$ and $x = 18$.

Note that $0 < 4 < 9$ and $9 < 18 < 24$.

General remark: Between every two roots of a polynomial $f(x) = 0$

is a root of $f'(x) = 0$, even when the two roots of

$f(x) = 0$ are the same.

b) Let $f(x) = x^n + a_{n-1}x^{n-1} + \ldots + a_1x + a_0$. By Rolle's theorem $f'(x)$ has a

zero between every two zeros of $f(x)$. Hence the polynomial

$nx^{n-1} + (n - 1)a_{n-1}x^{n-2} + \ldots + a_1 = f'(x)$ has a root between every two

zeros of $f(x)$.

7. Suppose $f(a) = f(b) = k$. Define $g(x) = f(x) - k$. Then $g(a) = g(b) = 0$. By

Rolle's theorem, $g'(c) = 0$ for some c on (a,b). $f'(x) = g'(x)$ everywhere on

(a,b), so $f'(c) = 0$. This contradicts $f'(x) \neq 0$ on (a,b), and proves

$f(a) \neq f(b)$.

Article 3.8

1. $f(0) = -1$; $f(1) = 2$. $\dfrac{f(1) = f(0)}{1 - 0} = 3$.

$f'(c) = 2c + 2 = 3 \Rightarrow c = \dfrac{1}{2}$.

3. $f(0) = 0$; $f(1) = 1$. $\dfrac{f(1) = f(0)}{1 - 0} = 1$.

$f'(c) = \dfrac{2}{3}c^{-1/3} = 1 \Rightarrow c = (\dfrac{2}{3})^3 = \dfrac{8}{27}$.

5. $f(1) = 0; \ f(3) = \sqrt{2}. \ \dfrac{f(3) - f(1)}{3 - 1} = \dfrac{\sqrt{2}}{2}.$

$f'(c) = \dfrac{1}{2(c - 1)^{1/2}} = \dfrac{\sqrt{2}}{2}$

$\Rightarrow \dfrac{1}{c - 1} = 2 \Rightarrow c = \dfrac{3}{2}.$

7. The cord slope equals the tangent slope: $\dfrac{y_2 - y_1}{x_2 - x_1} = f'(x_3).$

$\dfrac{y_2 - y_1}{x_2 - x_1} = \dfrac{(Ax_2{}^2 + Bx_2 + C) - (Ax_1{}^2 + Bx_1 + c)}{x_2 - x_1}$

$= \dfrac{A(x_2 + x_1)(x_2 - x_1) + B(x_2 - x_1)}{x_2 - x_1} = A(x_1 + x_2) + B.$

$f'(x_3) = 2Ax_3 + B.$ Thus, $2Ax_3 + B = A(x_1 + x_2) + B$, and $\dfrac{x_1 + x_2}{2} = x_3$

9. The function $y = x$ is continuously differentiable everywhere and hence on

$0 \le x \le 1$, so it does satisfy the hypotheses of the Mean Value Theorem.

The derivative of $y = x$ is $y' = 1$, which is independent of c, so c can be any

number in the interval.

11. There would be a contradiction if $f(x_1) = f(x_2)$ for $x_1 \ne x_2$ on $(0,1)$. Since

this never occurs, there is no contradiction.

13. Let $f(x)$ be the distance traveled at time x. $f(0) = 0$, $f(1) = 30$. By the Mean

Value Theorem there is a point c on $(0,1)$ at which

$f'(c) = \dfrac{f(1) - f(0)}{1 - 0} = \dfrac{30 - 0}{1 - 0} = 30.$ At time c the

speed is $f'(c) = 30$ miles per hour.

15. The function $f(x)$ is differentiable and hence continuous and hence has a value

at $x = 0$, to which one of the following must apply:

i) $f(0) = k > 0$

ii) $f(0) = -h < 0$

iii) $f(0) = 0.$

i)   By the M.V.T. there is a c on $(-3,0)$ such that

$f'(c) = \dfrac{f(0) - f(-3)}{0 - (-3)} = \dfrac{k + 3}{3} > 1,$

which contradicts $f'(x) \leq 1$.

ii) By the M.V.T. there is a c on $(0,3)$ such that

$$f'(c) = \frac{f(3) - f(0)}{3 - 0} = \frac{3 + h}{3} > 1,$$

which contradicts $f'(x) \leq 1$. It follows that case (iii) is correct

and $f(0) = 0$.

Remark: With slightly more work it can be shown that $f(x) = x$,

which also implies $f(0) = 0$.

17. Let $g(x) = 3x$, $g'(x) = 3$. Suppose $f'(x) = 3$. We set out to show that

$f(x) = g(x) + b = 3x + b$, where b is some constant. Define $h(x) = f(x) - g(x)$.

$h'(x) = f'(x) - g'(x) = 3 - 3 = 0$. By corrollary 2, $h(x)$ is a constant, say b.

$h(x) = b \Rightarrow f(x) - g(x) = b \Rightarrow f(x) = g(x) + b = 3x + b$.

19. Let x_0 be some point in (a,b). We show that if x is another point in (a,b),

then $f(x) = f(x_0) = c$. By the mean value theorem, there is a c between x_0 and

x with $f(x) - f(x_0) = f'(c)(x - x_0)$. Since $f'(x) = 0$ for all x, it follows

that $f'(c) = 0$. Hence $f(x) - f(x_0) = 0$, which is the conclusion.

Article 3.9

1. a) $\lim\limits_{x\to 2} \dfrac{x-2}{x^2-4} = \lim\limits_{x\to 2} \dfrac{x-2}{(x+2)(x-2)} = \lim\limits_{x\to 2} \dfrac{1}{x+2} = \dfrac{1}{4}$

    b) $\lim\limits_{x\to 2} \dfrac{x-2}{x^2-4} = \lim\limits_{x\to 2} \dfrac{1}{2x} = \dfrac{1}{4}.$

3. $\lim\limits_{x\to\infty} \dfrac{5x^2-3x}{7x^2+1} = \lim\limits_{x\to\infty} \dfrac{10x-3}{14x} = \lim\limits_{x\to\infty} \dfrac{10}{14} = \dfrac{5}{7}$

5. $\lim\limits_{t\to 0} \dfrac{\sin t^2}{t} = \lim\limits_{t\to 0} \dfrac{2t\cos t^2}{1} = 0$

7. $\lim\limits_{x\to 0} \dfrac{\sin 5x}{x} = \lim\limits_{x\to 0} \dfrac{5\cos x}{1} = 5$

9. $\lim\limits_{\theta\to\pi} \dfrac{\sin\theta}{\pi-\theta} = \lim\limits_{\theta\to\pi} \dfrac{\cos\theta}{-1} = 1$

11. $\lim\limits_{x\to\pi/4} \dfrac{\sin x - \cos x}{x - \pi/4} = \lim\limits_{x\to\pi/4} \dfrac{\cos x + \sin x}{1} = \sqrt{2}$

13. a) $\lim\limits_{x\to\pi/2} - (x - \dfrac{\pi}{2})\tan x = \lim\limits_{\theta=x-\pi/2\to 0} (\dfrac{\theta}{\sin\theta})\cos\theta = 1$

    because $\tan x = \tan(\theta + \dfrac{\pi}{2}) = -\cot\theta.$

    b) $\lim\limits_{x\to\pi/2} - (x - \dfrac{\pi}{2})\tan x = \lim\limits_{x\to\pi/2} - \dfrac{x - \pi/2}{\cot x} = \lim\limits_{x\to\pi/2} \dfrac{1}{\csc^2 x} = 1$

15. $\lim\limits_{x\to 1} \dfrac{2x^2 - (3x^{3/2} + x^{1/2}) + 2}{x - 1} = \lim\limits_{x\to 1} \dfrac{4x - ((9/2)x^{1/2} + (1/2)x^{-1/2})}{1} = 4 - 5 = -1$

17. $\lim\limits_{x\to 0} \dfrac{(a^2 + ax)^{1/2} - a}{x} = \lim\limits_{x\to 0} \dfrac{(1/2)(a^2 + ax)^{-1/2}(a)}{1} = \dfrac{1}{2}$

19. $\lim\limits_{x\to 0} \dfrac{x\cos x - x}{\sin x - x} = \lim\limits_{x\to 0} \dfrac{\cos x - x\sin x - 1}{\cos x - 1}$

    $= \lim\limits_{x\to 0} \dfrac{-\sin x - \sin x - x\cos x}{-\sin x} = \lim\limits_{x\to 0} \; 2 + \left(\dfrac{x}{\sin x}\right)\cos x = 3$

21. $\lim\limits_{r\to 1} \dfrac{a(r^n - 1)}{r - 1} = \lim\limits_{r\to 1} \dfrac{a(nr^{n-1})}{1} = an$

23. $x - \sqrt{x^2 + x} = (x - \sqrt{x^2 + x}) \dfrac{(x + \sqrt{x^2 + x})}{(x + \sqrt{x^2 + x})} = \dfrac{x^2 - (x^2 + x)}{x + \sqrt{x^2 + x}}$

$$= \dfrac{-1}{1 + \sqrt{1 + 1/x}}$$

$$\lim_{x \to \infty} x - \sqrt{x^2 + x} = \lim_{x \to \infty} \dfrac{1}{1 + \sqrt{1 + 1/x}} = -\dfrac{1}{2}$$

Remark: L'Hopital's rule was not applied.

25. $\lim\limits_{x \to \infty} \dfrac{\sqrt{10x + 1}}{\sqrt{x + 1}} = \lim\limits_{x \to \infty} \dfrac{\sqrt{10 + 1/x}}{\sqrt{1 + 1/x}} = \sqrt{10}$

27.   a) $y = r \sin \theta.$ $\lim\limits_{\theta \to \pi/2} r - y = \lim\limits_{\theta \to \pi/2} r(1 - \sin \theta) = 0$

   b) $y^2 + 1 = r^2$, hence $\lim\limits_{\theta \to \pi/2} r^2 - y^2 = 1$

   c) $r \sin \theta = y,\ r \cos \theta = 1 \Rightarrow y = \dfrac{\sin \theta}{\cos \theta} = \tan \theta.$

   $r^2 = 1 + y^2 = 1 + \tan^2 \theta = \sec^2 \theta \Rightarrow r = \sec \theta.$

   (All limits as $\theta \to \pi/2$.)

   $\lim r^3 - y^3 = \lim \sec^3 \theta - \tan^3 \theta = \lim \dfrac{1 - \sin^3 \theta}{\cos^3 \theta}$

   $= \lim \dfrac{-3 \sin^2 \theta \cos \theta}{-3 \cos^2 \theta \sin \theta} = \lim \dfrac{\sin \theta}{\cos \theta} = \infty$

29. $\dfrac{f(-2) - f(0)}{g(-2) - g(0)} = -\dfrac{1}{2}.$ $\dfrac{f'(c)}{g'(c)} = \dfrac{1}{2c} = -\dfrac{1}{2x} \Rightarrow c = -1$

31. $\dfrac{f(3) - f(0)}{g(3) - g(0)} = -\dfrac{3}{9} = -\dfrac{1}{3}.$ $\dfrac{f'(c)}{g'(c)} = \dfrac{c^2 - 4}{2c} = -\dfrac{1}{3} \Rightarrow 3c^2 + 2c - 12 = 0$

$\Rightarrow c = -\dfrac{2 + \sqrt{148}}{6} = -\dfrac{1 \pm \sqrt{37}}{3}.$ Since $0 < c < 3$, $c = -\dfrac{1 + \sqrt{37}}{3}.$

Article 3.10

1. $x \sin x \approx x(x) = x^2$

3. $\cos(1 + x)^{1/2} \approx 1 - \dfrac{[(1 + x)^{1/2}]^2}{2} = 1 - \dfrac{1 + x}{2} = \dfrac{1}{2}(1 - x)$

5. $\sqrt{x} = \sqrt{1 + (x - 1)} \approx 1 + \dfrac{x - 1}{2} - \dfrac{(x - 1)^2}{8} = \dfrac{1}{8}(3 + 6x - x^2)$

7. a) $f(x) = (1 + x)^{1/2}$; $f'(x) = \dfrac{1}{2}(1 + x)^{-1/2}$; $f''(x) = -\dfrac{1}{4}(1 + x)^{-3/2}$,

   $f'''(x) = \dfrac{3}{8}(1 + x)^{-5/2}$. $f(0) = 1$; $f'(0) = \dfrac{1}{2}$; $f''(0) = -\dfrac{1}{4}$.

   $f(x) \approx f(0) + f'(0) + \dfrac{f''(0)}{2}x^2 = 1 + \dfrac{1}{2}x - \dfrac{1}{8}x^2$.

   b) $|e_2(x)| \leq \max \left| \dfrac{f'''(x)}{6} \right| \, |\Delta x|^3 \leq \dfrac{1}{6}\left(\dfrac{3}{8}\right)(1 - 0.1)^{-5/2}(0.1)^3 \approx 0.0000813$

9. a) $f(x) = \cos x$; $f'(x) = -\sin x$; $f''(x) = -\cos x$; $f'''(x) = \sin x$; $f(0) = 1$;

   $f'(0) = 0$; $f''(0) = -1$. $f(x) \approx f(0) + f'(0)x + \dfrac{1}{2}f''(0)x^2 = 1 - \dfrac{1}{2}x^2$.

   b) $|e_2(x)| \leq \max \left| \dfrac{f'''(x)}{6} \right|(0.1)^3 \leq \left(\dfrac{1}{6}\right)(0.1)^3 \approx 0.000167$

11. a) $f(x) = \sin x$; $f'(x) = \cos x$; $f''(x) = -\sin x$; $f'''(x) = -\cos x$; $f(\pi) = 0$;

   $f'(\pi) = -1$; $f''(\pi) = 0$. $f(x) \approx f(\pi) + f'(\pi)(x - \pi) + \dfrac{1}{2}f''(\pi)(x - \pi)^2 = \pi - x$

   b) $|e_2(x)| \leq \max \left| \dfrac{f'''(x)}{6} \right|(0.1)^3 \leq \left(\dfrac{1}{6}\right)(0.1)^3 \approx 0.000167$

13. a) $f(x) = \cos x$; $f'(x) = -\sin x$; $f''(x) = -\cos x$; $f'''(x) = \sin x$; $f(\pi) = -1$;

   $f'(\pi) = 0$; $f''(\pi) = 1$. $f(x) \approx f(\pi) + f'(\pi)(x - \pi) + \dfrac{1}{2}f''(\pi)(x - \pi)^2 = -1 + \dfrac{1}{2}(x - \pi)^2$

   b) $|e_2(x)| \leq \max \left| \dfrac{f'''(x)}{6} \right|(0.1)^3 \leq \left(\dfrac{1}{6}\right)(0.1)^3 \approx 0.000167$

15. $f(x) = (1 + x)^k$; $f'(x) = k(1 + x)^{k-1}$; $f''(x) = k(k - 1)(1 + x)^{k-2}$.

   $f(0) = 1$; $f'(0) = k$; $f''(0) = k(k - 1)$

   $(1 + x)^k = f(x) \approx f(0) + f'(0)x + \dfrac{1}{2}f''(0)x^2 = 1 + kx + \dfrac{k(k - 1)}{2}x^2$, if $x \approx 0$.

17. $f(1) = 1 + 5 - 7 = -1$; $f'(1) = 3 + 5 = 8$; $f''(1) = 6(1) = 6(1) = 6$; $f'''(c_2) = 6$.

   $f(1) + f'(1)(x - 1) + \dfrac{1}{2}f''(1)(x - 1)^2 + \dfrac{1}{6}f'''(c_3)(x - 1)^3$

   $= -1 + 8(x - 1) + \dfrac{6}{2}(x - 1)^2 + \dfrac{6}{6}(x - 1)^3 = x^3 + 5x + 7 = f(x)$

19. If $f(x)$ is a polynomial of degree n, then $f^{n+1}(x) = 0$. By Taylor's Theorem,

   $f(x) = f(a) + f'(a)(x - a) + \dfrac{1}{2}f''(a)(x - a)^2 + \ldots$

   $\qquad\qquad + \dfrac{f^n(a)}{n!}(x - a)^n + \dfrac{f^{n+1}(c_{n+1})}{(n + 1)!}(x - a)^{n+1}.$

   $\qquad = f(a) + f'(a)(x - a) + \dfrac{1}{2}f''(a)(x - a)^2 + \ldots + \dfrac{f^n(a)}{n!}(x - a)^n,$

   because $f^{n+1}(c_{n+1}) = 0$.

Miscellaneous Problems

1. $y = 9x - x^2$; $y' = 9 - 2x$; $y'' = -2$

    a) $y' > 0$ when $x < \frac{9}{2}$

    b) $y' < 0$ when $x > \frac{9}{2}$

    c) For no x values is $y'' > 0$.

    d) For all x values is $y'' < 0$. Curve is downward opening parabola with vertex at $(4\frac{1}{2}, 20\frac{1}{4})$.

3. $y = 4x^3 - x^4$; $y' = 4x^2(3 - x)$; $y'' = 12x(2 - x)$

    a) $y' > 0$ when $x < 3$

    b) $y' < 0$ when $x > 3$

    c) $y'' > 0$ when $0 < x < 2$

    d) $y'' < 0$ when $x < 0$ or $x > 2$

5. $y = x^2 + 4x^{-1}$; $y' = 2x - 4x^{-2}$; $y'' = 2 + 8x^{-3}$

    a) $y' > 0$ when $x > \sqrt[3]{2}$

    b) $y' < 0$ when $x < 0$ or $0 < x < \sqrt[3]{2}$

    c) $y'' > 0$ when $x < -\sqrt[3]{4}$ or $x > 0$

    d) $y'' < 0$ when $-\sqrt[3]{4} < x < 0$

7. $y = 5 - x^{2/3}$; $y' = -\frac{2}{3}x^{-1/3}$; $y'' = \frac{2}{9}x^{-4/3}$

    a) $y' > 0$ when $x < 0$

    b) $y' < 0$ when $x > 0$

    c) $y'' > 0$ when $x \neq 0$

    d) $y'' < 0$ for no x.

CHAPTER 3 - Miscellaneous Problems

9. $y = x - 4x^{-1}$; $y' = 1 + 4x^{-2}$; $y'' = -8x^{-3}$

    a) $y' > 0$ when $x \neq 0$

    b) $y' < 0$ for no x.

    c) $y'' > 0$ when $x < 0$

    d) $y'' < 0$ when $x > 0$

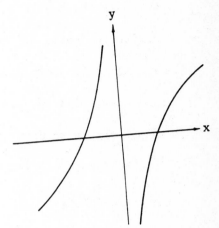

11. $y = x^2/(ax + b)$; $y' = x(ax + 2b)/(ax + b)^2$; $y'' = 2b^2/(ax + b)^3$

    a) $y' > 0$ when $x < \dfrac{-2b}{a}$ or $x > 0$

    b) $y' < 0$ when $\dfrac{-2b}{a} < x < -\dfrac{b}{a}$ or $-\dfrac{b}{a} < x < 0$.

    c) $y'' > 0$ when $x > -\dfrac{b}{a}$

    d) $y'' < 0$ when $x < -\dfrac{b}{a}$

13. $y = (x - 1)(x + 1)^2$; $y' = (3x - 1)(x + 1)$; $y'' = 2(3x + 1)$

    a) $y' > 0$ when $x < -1$ or $x > \dfrac{1}{3}$

    b) $y' < 0$ when $-1 < x < \dfrac{1}{3}$

    c) $y'' > 0$ when $x > -\dfrac{1}{3}$

    d) $y'' < 0$ when $x < -\dfrac{1}{3}$

15. $y = 2x^3 - 9x^2 + 12x - 4$; $y' = 6(x - 2)(x - 1)$; $y'' = 6(2x - 3)$

    a) $y' > 0$ when $x < 1$ or $x > 2$

CHAPTER 3 - Miscellaneous Problems

b) $y' < 0$ when $1 < x < 2$

c) $y'' > 0$ when $x > \frac{3}{2}$

d) $y'' < 0$ when $x < \frac{3}{2}$.

Curve is cubic with local maximum at $(1,1)$, local minimum at $(2,0)$ and

inflection point at $(\frac{3}{2},\frac{1}{2})$.

17. The curve is symmetrical about the x-axis, so where the curve above the x-axis

is increasing, the curve below the x-axis is decreasing and vice-versa. Also,

where the curve above the x-axis is concave up, the curve below the x-axis is

concave down. In (a) through (d) only the curve above the x-axis is referred to.

$$y = \sqrt{3}(\frac{1-x}{x})^{1/2}; \quad y' = -\frac{\sqrt{3}}{2}x^{-2}(\frac{1-x}{x})^{-1/2}; \quad y'' = \frac{\sqrt{3}}{4}x^{-5/2}(1-x)^{-3/2}(3-4x)$$

a) $y' > 0$ for no x.

b) $y' < 0$ for $0 < x < 1$

c) $y'' > 0$ for $0 < x < \frac{3}{4}$

d) $y'' < 0$ for $\frac{3}{4} < x < 1$

The curve is symmetrical about the x-axis, has no maxima or minima, is

decreasing for positive y, increasing for negative y, and is restricted to

$0 < x \leq 1$.

19. a) i) $F(x,y) = y^2 - x(4 - x) = F(x,-y)$

$\Rightarrow$ symmetrical about the x-axis

ii) $0 \leq y^2 = x(4 - x) \Rightarrow 0 \leq x \leq 4$.

Also, $-2 \leq y \leq 2$

iii) Intercepts at $(0,0),(4,0)$

iv) No asymptotes because neither x nor y goes to infinity

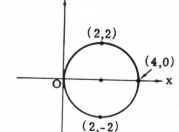

v) $y' = (2 - x)/y$. At $(0,0)$, $y' = \frac{2}{0} = \infty$.

At $(4,0)$ $y' = -\frac{2}{0} = \infty$

$F(x,y) = x^2 - 4x + y^2 = (x - 2)^2 + y^2 - 4 \Rightarrow$ circle, radius 2

CHAPTER 3 - Miscellaneous Problems

b) i) $F(x,y) = y^2 - x(x - 4) = F(x,-y)$

        => symmetrical about x-axis.

        Also symmetrical about line x = 2 and point (2,0)

ii) $0 \le y^2 = x(x - 4)$ => $x \ge 4$ or $x \le 0$; all y.

iii) Intercepts at (0,0),(4,0)

iv) $y^2 = x(x - 4)$ => $(x - 2)^2/4 - y^2/4 = 1$

        => hyperbola with asymptotes $y = \pm(x - 2)$

v) $y' = (x - 2)/y$. At (0,0) and (4,0) $y' = \infty$.

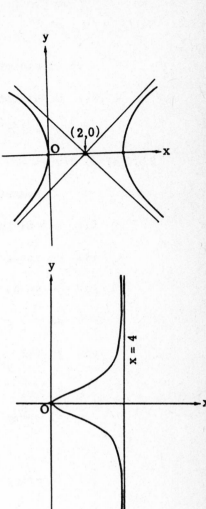

c) i) $F(x,y) = y^2 - x/(4 - x) = F(x,-y)$

        => symmetrical about x-axis

ii) $0 \le y^2 = x/(4 - x)$ => $0 \le x < 4$; all y

iii) Intercept at (0,0)

iv) $y \to \pm\infty$ as $x \to 4$. Hence asymptotes at x = 4

v) $y' = 2/(y(4 - x)^2)$. At (0,0) $y' = \infty$

21 a) i) $F(x,y) = x(x + 1)(x - 2) - y$

        => No symmetry

ii) All x, all y

iii) Intercepts at (0,0), (-1,0), (2,0)

iv) No asymptotes

v) $y' = 3x^2 - 2x - 2$. At (0,0), $y' = -2$.

        At (-1,0), $y' = 3$. At (2,0), $y' = 6$.

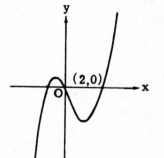

b) i) $F(x,y) = x(x + 1)(x - 2) - y^2 = F(x,-y)$

        => Symmetrical about x-axis

ii) $0 \le y^2 = x(x + 1)(x - 2)$

        => $x \ge 2$ or $-1 \le x \le 0$; all y.

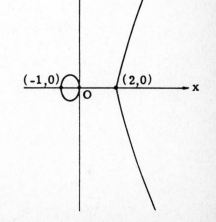

iii) Intercepts at $(0,0)$, $(-1,0)$, $(2,0)$

iv) No asymptotes

v) $y' = (3x^2 - 2x - 2)/2y$. At $(0,0)$, $(-1,0)$, $(2,0)$, $y' = \infty$.

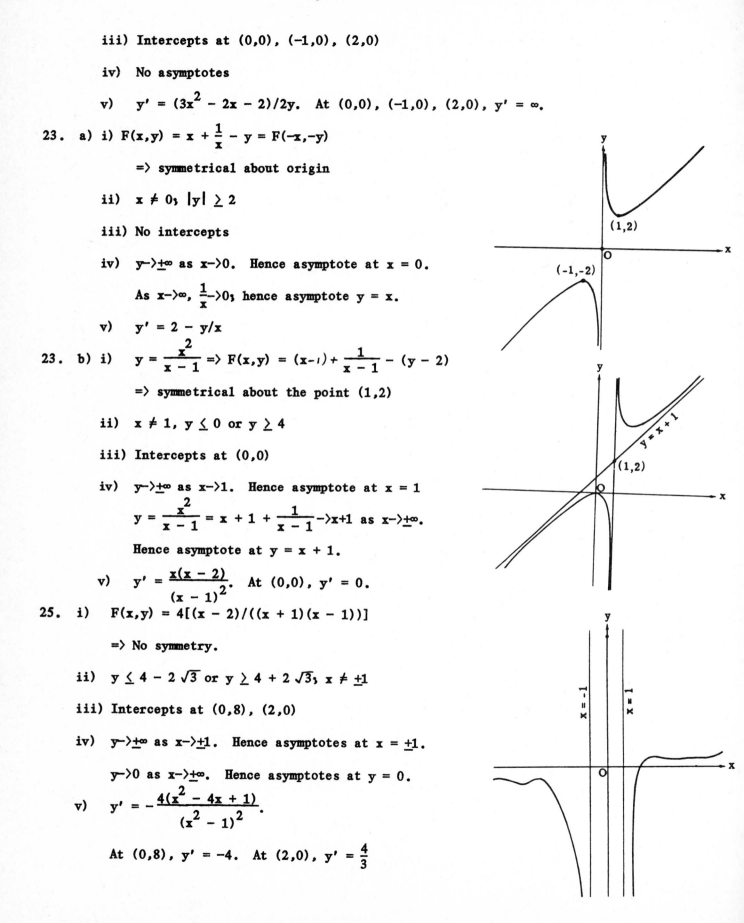

23. a) i) $F(x,y) = x + \dfrac{1}{x} - y = F(-x,-y)$

        $\Rightarrow$ symmetrical about origin

    ii) $x \neq 0$; $|y| \geq 2$

    iii) No intercepts

    iv) $y \rightarrow \pm\infty$ as $x \rightarrow 0$. Hence asymptote at $x = 0$.

        As $x \rightarrow \infty$, $\dfrac{1}{x} \rightarrow 0$; hence asymptote $y = x$.

    v) $y' = 2 - y/x$

23. b) i) $y = \dfrac{x^2}{x - 1} \Rightarrow F(x,y) = (x-1) + \dfrac{1}{x - 1} - (y - 2)$

        $\Rightarrow$ symmetrical about the point $(1,2)$

    ii) $x \neq 1$, $y \leq 0$ or $y \geq 4$

    iii) Intercepts at $(0,0)$

    iv) $y \rightarrow \pm\infty$ as $x \rightarrow 1$. Hence asymptote at $x = 1$

        $y = \dfrac{x^2}{x - 1} = x + 1 + \dfrac{1}{x - 1} \rightarrow x+1$ as $x \rightarrow \pm\infty$.

        Hence asymptote at $y = x + 1$.

    v) $y' = \dfrac{x(x - 2)}{(x - 1)^2}$. At $(0,0)$, $y' = 0$.

25. i) $F(x,y) = 4[(x - 2)/((x + 1)(x - 1))]$

       $\Rightarrow$ No symmetry.

   ii) $y \leq 4 - 2\sqrt{3}$ or $y \geq 4 + 2\sqrt{3}$; $x \neq \pm 1$

   iii) Intercepts at $(0,8)$, $(2,0)$

   iv) $y \rightarrow \pm\infty$ as $x \rightarrow \pm 1$. Hence asymptotes at $x = \pm 1$.

       $y \rightarrow 0$ as $x \rightarrow \pm\infty$. Hence asymptotes at $y = 0$.

   v) $y' = -\dfrac{4(x^2 - 4x + 1)}{(x^2 - 1)^2}$.

       At $(0,8)$, $y' = -4$. At $(2,0)$, $y' = \dfrac{4}{3}$

27. i) $F(x,y) = x^2 + xy + y^2 - 3 = F(-x,-y)$

    => Symmetrical about the origin

    ii) $x = \dfrac{-y \pm \sqrt{y^2 - 4(y^2 - 3)}}{2} \quad \dfrac{-y \pm \sqrt{12 - 3y^2}}{2}$

    => $12 - 3y^2 \geq 0$ => $|y| \leq 2$. Likewise $|x| \leq 2$

    iii) Intercepts at $(0, \pm \sqrt{3})$, $(\pm \sqrt{3}, 0)$

    iv) Since x and y do not go to infinity, there can be no asymptotes.

    v) $y' = -(2x + y)/(2y + x)$. At $(0, \pm \sqrt{3})$, $y' = -\frac{1}{2}$.

    At $(\pm \sqrt{3}, 0)$, $y' = -2$.

    Curve is ellipse with major axis along $y = -x$ and minor axis along $y = x$.

29. $y' = 0$ at $x = 1$, $x = 2$, $x = 3$, $x = 4$. A relative maximum occurs at x when $y'(x) = 0$ and $y'(x + \varepsilon) < 0$ and $y'(x - \varepsilon) > 0$, for small positive ε. A relative minimum occurs at x when $y'(x) = 0$ and $y'(x + \varepsilon) > 0$ and $y'(x - \varepsilon) < 0$. Thus there is a relative maximum at $x = 1$, and a relative minimum at $x = 3$.

31. Velocity $= \dfrac{ds}{dt} = ks^{-1/2}$. Acc $= \dfrac{d^2s}{dt^2} = -\dfrac{1}{2}ks^{-3/2}\dfrac{ds}{dt}$

    $= (-\dfrac{1}{2}ks^{-3/2})(s^{1/2}) = -\dfrac{1}{2}ks^{-2}$.

    The constant of proportionality is $-\frac{1}{2}k$.

33. $V = \dfrac{1}{3}\pi r^2 h = 4\pi h^3/3$; $dV/dt = 4\pi h^2 dh/dt$ when $dv/dt = 3 \text{ ft}^3/\text{min}$ and $h = 10$ ft, the equation is $3 = 4\pi(10)^2\dfrac{dh}{dt}$. $dh/dt = 3/400\pi$ ft/min.

35. $x^2 = a^2 + b^2 - 2ab \cos 60°$

    $2x\dfrac{dx}{dt} = (2a - b)\dfrac{da}{dt} + (2b - a)\dfrac{db}{dt}$

    a) $a = 4$ mi, $b = 2$ mi, $da/dt = 700$ mi/hr

    $db/dt = 500$ mi/hr, $x = 2\sqrt{3}$.

    $dx/dt = 350\sqrt{3} \approx 600$ mi/hr

    b) $2x\, dx/dt = (2a - b)700 + (2b - a)500$

    $x\, dx/dt = 150(3a + b)$. Also, $b/a = (2 - 500t)/(4 - 700t)$

CHAPTER 3 - Miscellaneous Problems

If $dx/dt = 0$, $b/a = -3$ and $t = 7/1300$. Therefore $b = 9/13$, $a = 3/13$,

and $x = \sqrt{117}/13 \approx 0.83$ mi. This is a minimum on $[0, 1/100]$ because at

$t = 0$, $x = 2\sqrt{3} > 0.83$ and at $t = 1/100$, $x = 3/\sqrt{3} > 0.83$

37. $A = xy/2 \Rightarrow \dfrac{dA}{dt} = \dfrac{1}{2}(x\,dy/dt + y\,dx/dt)$

$x^2 + y^2 = 26^2 \Rightarrow \dfrac{dy}{dt} = -\dfrac{x}{y}\dfrac{dx}{dt}$.

When $x = 17\sqrt{2}$, $y = 7\sqrt{2}$, and $dx/dt = 4$, $dA/dt = -480\sqrt{2}/7$ ft^2/sec.

39. $v = [v_0^2 - 2gR(1 - (R/s))]^{1/2}$

$\Rightarrow \dfrac{dv}{ds} = \dfrac{1}{2}[v_0^2 - 2gR(1 - (R/s))]^{-1/2}(-2gR^2/s^2)$

$= -v^{-1}(gR^2/s^2)$. Acceleration $= \dfrac{dv}{dt} = \dfrac{dv}{ds}\dfrac{ds}{dt} = v\dfrac{dv}{ds}$

$= -v^{-1}(gR^2/s^2)v = -gR^2/s^2$. The constant of proportionality is $-gR^2$.

41. By similar right triangles, $\dfrac{a}{r} = \dfrac{r}{x}$ and $\dfrac{b}{r} =$

$\dfrac{r}{y}$. One form of the equation of the tangent line

is $(x/a) + (y/b) = 1$. Thus $(1/a^2) + (1/b^2) = 1/r^2 \Rightarrow$

$-2a^{-3}\dfrac{da}{dt} - 2b^{-3}\dfrac{db}{dt} = 0$.

$\dfrac{3\sqrt{3}}{8r^3}\dfrac{da}{dt} + \dfrac{1}{8r^3}(0.3r) = 0$. Hence

$\dfrac{da}{dt} = \dfrac{\sqrt{3}r}{30}\dfrac{in}{sec}$, which is

increasing.

43. $x = 3h/10$, $v = 3\pi h^3/100$.

2 ft^3/min $= \dfrac{dv}{dt} = \dfrac{9\pi h^2}{100}\dfrac{dh}{dt}$. Hence $\dfrac{dh}{dt} = \dfrac{8}{9\pi}\dfrac{ft}{min}$.

45. $f(x) = 4x^3 - 8x^2 + 5x$. $f'(x) = (6x - 5)(2x - 1)$, $0 \le x \le 2$. $f'(x) = 0$ at

$x = 5/6$ and $x = 1/2$. To find the maximum, it is sufficient to check the value

of $f(x)$ at the two critical points, and the end points. The largest of these

is $f(2) = 10$.

47. $y = ax^3 + bx^2 + cx + d$; $y' = 3ax^2 + 2bx + c$; $y'' = 6ax + 2b$

1) $f(-1) = 10 \Rightarrow -a + b - c + d = 10$

2) $f'(-1) = 0 \Rightarrow 3a - 2b + c = 0$

CHAPTER 3 — Miscellaneous Problems

3) $f(1) = -6 \Rightarrow a + b + c + d = -6$

4) $f''(-1) = 0 \Rightarrow 6a + 2b = 0$

Solving the system yields $a = 1$, $b = -3$, $c = -9$, $d = 5$.

49. $2r + s = 100$; $A = 50r - r^2$. $dA/dr = 50 - 2r = 0 \Rightarrow r = 2s$. This is a maximum because $d^2A/dr^2 = -2 < 0$.

51. Let x be the radius of the inscribed cone, h its height, and v its volume. By similar triangles, $6/x = 12/(12 - h)$; or $h = 12 - 2x$.

$v = \frac{1}{3}\pi x^2 h = \frac{1}{3}\pi x^2(12 - 2x)$. $\frac{dv}{dx} = 2\pi x(4 - x) = 0 \Rightarrow x = 4$.

$h = 12 - 2(4) = 4$. The dimensions give a maximum volume because $d^2v/dx^2 < 0$.

53. $A = xy = 3x - x^3/12$. $\frac{dA}{dx} = 3 - x^2/4 = 0 \Rightarrow x = 2\sqrt{3}$, $y = 3 - x^2/12 = 2$.

Area = $4/3$. The area is a maximum because $d^2A/dx^2 = -x/2 = -\sqrt{3} < 0$.

55. Since $y^2 = 1 - x^2$, $D(x) = [(x - a)^2 + y^2]^{1/2} = [2x^2 - 2xa + a^2 - 1]^{1/2}$.

$D'(x) = \dfrac{2x - a}{[2x^2 - 2ax + a^2 - 1]^{1/2}}$; $D''(x) = \dfrac{a^2 - 2}{[2x^2 - 2ax + a^2 - 1]^{3/2}}$

a) $D'(x) = 0 \Rightarrow x = \frac{1}{2}a = 2$, $y = \pm\sqrt{3}$. $D''(2) > 0$,

   so $(2, \pm\sqrt{3})$ are minimum distance points.

b) $D'(x) = 0 \Rightarrow x = \frac{1}{2}a = 1$, $y = 0$. $D''(1) > 0$, so $(1,0)$ is the

   minimum distance point.

c) If $0 < x < 2$, $f'(x) > 0$, so the nearest point is $(1,0)$.

57. $(b - x)^2 + a^2 = y^2 \Rightarrow 2(b - x)(-1) = 2y\frac{dy}{dx} \Rightarrow \frac{dy}{dx} = \frac{x - b}{y}$.

Total cost is $C = wy + \ell x$. To minimize C, set $2C/dx = 0$.

$0 = \frac{dC}{dx} = w\frac{dy}{dx} + \ell = w(\frac{x - b}{y}) + \ell \Rightarrow \frac{b - x}{y} = \frac{\ell}{w}$.

Using the first equation,

$1 = \frac{a^2}{y^2} + \frac{(b - x)^2}{y^2} = \frac{a^2}{y^2} + \frac{\ell^2}{w^2} \Rightarrow y = \frac{wa}{\sqrt{w^2 - \ell^2}} \Rightarrow (b - x)^2 = y^2 - a^2 =$

$= \frac{w^2 a^2}{w^2 - \ell^2} - a^2 = \frac{\ell^2 a^2}{w^2 - \ell^2} \Rightarrow x = b - \frac{a\ell}{\sqrt{w^2 - \ell^2}}$

By the diagram, $x \geq 0$. Hence $x = 0$ when $b \leq \frac{a\ell}{\sqrt{w^2 - \ell^2}}$.

CHAPTER 3 - Miscellaneous Problems

To ensure that the critical values corresponds to minimum costs, we check that $\partial^2 C/\partial x^2 > 0$.

$$\frac{\partial^2 c}{\partial x^2} = w\frac{d}{dx}\left(\frac{x-b}{y}\right) = w\frac{y-(x-b)(dy/dx)}{y^2} = w\frac{y-(x-b)((x-b)/y)}{y^2}$$

$$= \frac{y^2-(x-b)^2}{y^3} = \frac{a^2}{y^3} > 0.$$

59. Let x and y be the lengths of the remaining sides, A the area, b the length of the base, and s half of the perimeter. From geometry,

$$A = [s(s-x)(s-y)(s-b)]^{1/2} = (s(s-b))^{1/2}[(s-x)(b-x+x)]^{1/2}.$$

$$A' = (s(s-b))^{1/2}\frac{2s-b-2x}{2[(s-x)(b-s-x)]^{1/2}} = 0 \Rightarrow x = y = s - \frac{b}{2}.$$

The area is a maximum because $A' > 0$ for $0 < x < s - b/2$ and $A' < 0$ for

$s - \frac{b}{2} < x < 2s - b.$

61. $f(x) = |PQ| = \left[x^2 + \frac{b^2x^2}{(x-a)^2}\right]^{1/2}$, $x > a$

$$f'(x) = \frac{x(x-a)^3 - ab^2x}{(x-a)^3(x^2+y^2)^{1/2}} = 0$$

$x = a + (ab^2)^{1/3} \Rightarrow |PQ| = (a^{2/3}+b^{2/3})^{3/2}.$

$g(x) = |OP| + |OQ| = x + (bx/(x-a))$; $g'(x) = \frac{(x-a)^2-ab}{(x-a)^2}$;

$g''(x) = \frac{2ab}{(x-a)^3} > 0.$ $g'(x) = 0 \Rightarrow x = a + a^{1/2}b^{1/2}.$

Hence min $|OP| + |OQ| = (\sqrt{a}+\sqrt{b})^2.$

$h(x) = |OP||OQ| = bx^2/(x-a)$; $h'(x) = (bx^2-2abx)/(x-a)^2$

$h'(x) = 0 \Rightarrow x = 2a.$ $h''(x) = 2a^2b/(x-a)^3 > 0.$

Hence, min $|OP||OQ| = 4ab.$

63. a) Line L, $ax + by + c = 0$ has slope $dy/dx = -a/b$. The closest point to $P_1 = (x_1,y_1)$ on L, $P_0 = (x_0,y_0)$, forms a perpendicular with P_1 to L if and only if $(y_1-y_0)/(x_1-x_0) = -(dy/dx)^{-1} = -(-a/b)^{-1} = b/a$. We find (x_0,y_0) and show that this ratio holds. (x_0,y_0) is the point that minimizes s^2, and is found by setting $\partial(s^2)/\partial x = 0$

CHAPTER 3 — Miscellaneous Problems

$$s^2 = (x - x_1)^2 + (y - y_1)^2 \Rightarrow 0 = \frac{ds^2}{dx} = 2(x_0 - x_1) + (y_0 - y_1)\frac{dy}{dx}$$

$$= 2(x_0 - x_1) + 2(y_0 - y_1)(-\frac{a}{b}) \Rightarrow \frac{y_0 - y_1}{x_0 - x_1}$$

$$= \frac{y_1 - y_0}{x_1 - x_0} = \frac{b}{a}.$$

$$\frac{d^2 s^2}{dx^2} = 2 - 2(\frac{a}{b})\frac{dy}{dx} = 2 + 2(\frac{a}{b})^2 > 0,$$

so s^2 is at a minimum.

b) $s^2 = (x_0 - x_1)^2 + (y_0 - y_1)^2 = (x_0 - x_1)^2 + (\frac{b}{a})^2(x_0 - x_1)^2$

$$= (x_0 - x_1)^2(\frac{a^2 + b^2}{a^2}).$$

Finding $x_0 - x_1$: $\frac{y_0 - y_1}{x_0 - x_1} = \frac{b}{a} \Rightarrow y_0 = y_1 + \frac{b}{a}(x_0 - x_1)$

$ax_0 + by_0 + c = 0 \Rightarrow ax_0 + b(y_1 + \frac{b}{a}(x_0 - x_1)) + c = 0$

$\Rightarrow a^2(x_0 - x_1) + a(ax_1 + by_1 + c) + b^2(x_0 - x_1) = 0$

$\Rightarrow x_0 - x_1 = -\frac{a(ax_1 + by_1 + c)}{a^2 + b^2}$. Hence,

$$s^2 = [-\frac{a(ax_1 + by_1 + c)}{a^2 + b^2}]^2(\frac{a^2 + b^2}{a^2}) \Rightarrow s = \frac{|ax_1 + by_1 + c|}{\sqrt{a^2 + b^2}}.$$

65. $f(x) = ax + (b/x) \Rightarrow f'(x) = a - (b/x^2) = 0$

$\Rightarrow x_0 = \sqrt{b/a}$. $f''(x) = 2b/x^3 > 0$, so $f(x_0)$ is the minimum value of $f(x)$.

$f(x) \geq f(x_0) \geq c \Rightarrow a\sqrt{b/a} + b\sqrt{a/b} \geq c \Rightarrow 2\sqrt{ab} \geq c \Rightarrow ab \geq c^2/4$.

67. $f(x) = ax^2 + 2bx + e \Rightarrow f'(x) = 2ax + 2b = 0 \Rightarrow x_0 = -b/a$.

$f''(x) = 2a > 0$, so $f(x_0)$ is a minimum.

$f(x) \geq f(x_0) \geq 0 \Rightarrow ax_0^2 + 2bx_0 + c \geq 0$

$\Rightarrow a(-b/a)^2 + 2b(-b/a) + c \geq 0 \Rightarrow -\frac{b^2}{a} + c \geq 0 \Rightarrow b^2 - ac \leq 0$.

69. In problem 68, $f(x) = (a_1 x + b)^2 + \ldots + (a_n x + b_n)^2 = Ax^2 + 2Bx + C$,

where $A = a_1^2 + \ldots + a_n^2$; $B = a_1 b_1 + \ldots + a_n b_n$; $C = b_1^2 + \ldots + b_n^2$.

If for all x, at least one of the n terms of $f(x)$ is positive, then $f(x)$ is

positive, or $f(x) > 0$. In this case, $Ax^2 + 2Bx + C > 0$, and by problem 67,

$B^2 < AC$, or $(a_1b_1 + \ldots + a_nb_n)^2 < (a_1^2 + \ldots + a_n^2)(b_1^2 + \ldots + b_n^2)$. If

there is an x_0 such that $b_i = -a_ix_0$ for all i, then $f(x_0) = Ax_0^2 + 2B_0 + C = 0$,

and $B^2 = AC$ or $(a_1b_1 + \ldots + a_nb_n)^2 = (a_1^2 + \ldots + a_n^2)(b_1^2 + \ldots + b_n^2)$.

71. $(12 - y)/12 = x/36$. $y = 12 - x/3$. $f(x) = A = x(12 - (x/3))$

$f'(x) = 12 - 2x/3 = 0 \Rightarrow x = 18, y = 6$.

$f''(x) = -2/3 < 0$, so area is maximum.

73. $\dfrac{h}{x - w} = \dfrac{y}{x} \Rightarrow y = \dfrac{hx}{x - w}$

Ladder length squared,

$f(x) = x^2 + y^2 = x^2 + (\dfrac{hx}{x - w})^2$

$= x^2(1 + \dfrac{h^2}{(x - w)^2})$; $f'(x) = 2x(1 + \dfrac{h^2}{(x - w)^2}) + x^2(\dfrac{-2h^2}{(x - w)^3}) = 0$

$\Rightarrow \dfrac{(x - w)^2 + h^2}{(x - w)^2} = \dfrac{xh^2}{(x - w)^3} \Rightarrow (x - w)^3 + h^2(x - w) = xh^2 = (x - w)h^2 + wh^2$

$\Rightarrow (x - w)^3 = wh^2 \Rightarrow x = w + (wh^2)^{1/3}, \quad y = \dfrac{hx}{x - w} = h + (w^2h)^{1/3}$.

$f(x,y) = x^2 + y^2 = (w^2 + 2w(wh^2)^{1/3} + (wh^2)^{2/3}) + (h^2 + 2h(hw^2)^{1/3} + (hw^2)^{2/3})$

$= w^2 + 3w^{4/3}h^{2/3} + 3w^{2/3}h^{4/3} + h^2 = (w^{2/3} + h^{2/3})^3$.

The minimum ladder length is $\sqrt{f(x)} = (w^{2/3} + h^{2/3})^{3/2}$.

75. $A = \theta r^2/2$; $P = r\theta + 2r = (2A/r) + 2r$.

$\dfrac{dP}{dr} = -\dfrac{2A}{r^2} + 2 = 0 \Rightarrow r = \sqrt{A}, \theta = 2$. This

perimeter is a minimum because $\dfrac{d^2P}{dr^2} = \dfrac{4A}{r^3} > 0$.

77. $A(t) = \pi(r_2^2 - r_1^2)$

a) $A' = 2\pi(r_2r_2' - r_1r_1') = 2\pi(0.06 - 0.08) = -0.04\pi$

b) $A' = 2\pi(r_2r_2' - r_1r_1') = 2\pi(b(bt + 5) - a(at + 3)) = 0$

$\Rightarrow t = (5b - 3a)/(a^2 - b^2)$. At this t, A is a maximum because

$A'' = b^2 - a^2 < 0$.

79. $S = \lambda t - (1 + \lambda^4)t^2$

CHAPTER 3 - Miscellaneous Problems

$$ds/dt = \lambda - 2t(1 + \lambda^4) = 0 \Rightarrow t = \lambda/(2(1 + \lambda^4)) \Rightarrow s = \lambda^2/(4(1 + \lambda^4))$$

$$ds/dx = 0 \Rightarrow 2\lambda(1 + \lambda^4) - \lambda^2(4\lambda^2) = 0 \Rightarrow \lambda = 1, \; s = (1)^2/(4(1 + 1)) = 1/8.$$

81. To minimize $f(x)$, set $f'(x) = 0$, and solve for x:

$$f(x) = (c_1 - x)^2 + (c_2 - x)^2 + \ldots + (c_n - x)^2,$$

$$f'(x) = -2[(c_1 - x) + (c_2 - x) + \ldots + (c_n - x)] = 0$$

$$\Rightarrow c_1 + c_2 + \ldots + c_n = nx \Rightarrow x = \frac{1}{n}(c_1 + c_2 + \ldots + c_n).$$

(Notice that the minimizing x is simply the average of the c's.)

83. Let $A = a_1 + a_2 + \ldots + a_{n-1}$; $B = 1/(n(a_1 a_2 \ldots a_{n-1})^{1/n})$

$$f(x) = [(a_1 + a_2 + \ldots + a_{n-1} + x)/n]/(a_1 a_2 \ldots a_{n-1} x)^{1/n} = B(A + x)x^{-1/n}$$

$$f'(x) = Bx^{-1/n}[1 - \frac{1}{nx}(A + x)] = 0 \Rightarrow nx = A + x$$

$$\Rightarrow x = \frac{A}{n - 1} = \frac{a_1 + a_2 + \ldots + a_{n-1}}{n - 1}, \text{ the arithmetic mean.}$$

85. $f(0)$ does not exist. The function is not continuous

at $x = 0$, nor can it be differentiable. The

hypotheses of the Mean Value Theorem are not

satisfied, so the theorem does not apply.

87. $f(x) = 2x^3 - 3x^2 + 6x + 6$.

$$f'(x) = 6(x^2 - x + 1) = 6(x^2 - x + \frac{1}{4} + \frac{3}{4}) = 6(x - \frac{1}{2})^2 + \frac{9}{2} > 0.$$

If $f(x)$ has two roots, say r_1 and r_2, then $f(r_1) = 0$, $f(r_2) = 0$, and by Rolle's

theorem, there is a c between r_1 and r_2 such that $f'(c) = 0$, and this

contradicts $f'(x) > 0$. Hence $f(x)$ can have at most one root. Since

$f(-1) = -5 < 0$ and $f(0) = 6 > 0$, the intermediate value theorem says there is a

root on $(-1, 0)$. Thus there is exactly one root. To find the root, use

Newton's method:

$$x_{n+1} = x_n - \frac{f(x_n)}{f'(x_n)} = x_n - \frac{2x_n^3 - 3x_n^2 + 6x_n + 6}{6(x^2 - x + 1)}$$

with $x_0 = -0.5$, four iterations give $x_4 = 0.6724958108$.

89. Profits = Revenue - Costs = Price * Quantity - y

$$= (c - ex)x - (a + bx) = -ex^2 + (c - b)x - a = F(x)$$

a) $F'(x) - 2ex + c - b = 0 \Rightarrow x = \frac{c - b}{2e}$ is the maximum

profit production level.

b) $P = c - ex = c - e(\frac{c - b}{2e}) = \frac{b + c}{2}$

c) Profit, $F = -ex^2 + (c - b)x - a$

$$= -e(\frac{c - b}{2e})^2 + (c - b)(\frac{c - b}{2e}) - a$$

$$= \frac{(c - b)^2}{4e} - a.$$

d) Price received by the firm is $P - t = (c - t) - ex$. $c - t$ is substituted

for c in the above answers. New price

$P' = \frac{b + c - t}{2} = \frac{b + c}{2} - \frac{t}{2}$, is

received by firms, and $P' + t = \frac{b + c}{2} + \frac{t}{2}$ is now paid by

consumers. Note that the price increase to consumers is $t/2$, and the price

reduction received by firms is $t/2$; the consumers and the firm split the

unit tax t evenly between them.

91. a) $\lim\limits_{x \to \infty} \frac{\sqrt{x + 5}/\sqrt{x}}{(\sqrt{x} + 5)/\sqrt{x}} = \lim\limits_{x \to \infty} \frac{\sqrt{1 + 5/x}}{1 + 5/\sqrt{x}} = \frac{1}{1} = 1.$

b) $\lim\limits_{x \to \infty} \frac{2x}{x + 7\sqrt{x}} = \lim\limits_{x \to \infty} \frac{2}{1 + 7/\sqrt{x}} = 2.$

93. We show that if there is an x_0 such that $f(x_0) < f(c)$, then x_0 cannot lie on

(a,c) or (c,b), and this proves the result. Suppose x_0 lies on (a,c). By the

Mean Value Theorem, there is a c_1 on (x_0,c) where

$$f'(c_1) = \frac{f(c) - f(x_0)}{c - x_0}.$$ $f'(c_1) > 0$ because $c - x > 0$ and $f(c) - f(x_0) > 0$.

But since c_1 is on (a,c) and $f'(c_1) \leq 0$. Hence x_0 is not on (a,c). Next

suppose that x_0 is on (c,b). By the Mean Value Theorem there is a c_2 on (c,x_0)

where $$f'(c_2) = \frac{f(c) = f(x_0)}{c - x_0}.$$ $f'(c_2) < 0$ because $c - x_0 < 0$

and $f(c) - f(x_0) > 0$. But since c_2 is on (c,b), $f'(c_2 \geq 0$. Hence x_0 cannot

be on (c,b) or (a,c), and no such x_0 exists.

CHAPTER 3 - Miscellaneous Problems

CHAPTER 4

Article 4.2

1. $y = \int (2x - 7)dx = x^2 - 7x + c$

3. $y = \int (x^2 + 1)dx = \dfrac{x^3}{3} + x + c$

5. $y = \int \dfrac{-5}{x^2}dx = \dfrac{5}{x} + c$

7. $y = \int (x - 2)^4 dx = \dfrac{(x - 2)^5}{5} + c$

9. $y = \int (\dfrac{1}{x^2} + x)dx = \int (x^{-2} + x)dx$

$\quad = -x^{-1} + \dfrac{x^2}{2} + c = \dfrac{x^2}{2} - \dfrac{1}{x} + c.$

11. $\dfrac{dy}{dx} = \dfrac{x}{y} \Rightarrow ydy = xdx \Rightarrow \int ydy = \int xdx$

$\quad \Rightarrow \dfrac{y^2}{2} = \dfrac{x^2}{2} \Rightarrow y^2 = x^2 + c$

$\quad \Rightarrow y = \sqrt{x^2 + c} \text{ since } y > 0.$

13. $\dfrac{dy}{dx} = \dfrac{x + 1}{y - 1} \Rightarrow (y - 1)dy = (x + 1)dx$

$\quad \Rightarrow \dfrac{(y - 1)^2}{2} = \dfrac{(x + 1)^2}{2} + k$

$\quad \Rightarrow (y - 1)^2 = (x + 1)^2 + c \Rightarrow y = 1 + \sqrt{(x + 1)^2 + c} \text{ since } y > 1.$

15. $\dfrac{dy}{dx} = \sqrt{xy} = \sqrt{x}\,\sqrt{y} \Rightarrow \dfrac{dy}{\sqrt{y}} = \sqrt{x}\,dx$

$\quad \Rightarrow \int y^{-1/2}dy = \int x^{1/2}dx \Rightarrow 2\sqrt{y} = \dfrac{2}{3}x^{3/2} + k$

$\quad \Rightarrow y = [\dfrac{1}{3}x^{3/2} + c]^2.$

17. $\dfrac{dy}{dx} = 2xy^2 \Rightarrow \dfrac{dy}{y^2} = 2xdx \Rightarrow -\dfrac{1}{y} = x^2 + c$

$\quad \Rightarrow y = \dfrac{-1}{x^2 + c}$

19. $x^3 \dfrac{dy}{dx} = -2 \Rightarrow \dfrac{dy}{dx} = -2x^{-3}$

$\Rightarrow y = \int -2x^{-3} dx = x^{-2} + c = \dfrac{1}{x^2} + c.$

21. $\dfrac{ds}{dt} = 3t^2 + 4t - 6$

$\Rightarrow s = \int (3t^2 + 4t - 6) dt = t^3 + 2t^2 - 6t + c.$

23. $\dfrac{du}{dv} = 2u^2(4v^3 + 4v^{-3}) \Rightarrow u^{-2} du = (8v^3 + 8v^{-3}) dv$

$\Rightarrow -\dfrac{1}{u} = 2v^4 - 4v^{-2} + c \Rightarrow u = \dfrac{-v^2}{2v^6 + cv^2 - 4}$

25. $\dfrac{dy}{dt} = (2t + t^{-1})^2 = (2t + \dfrac{1}{t})^2 = 4t^2 + 4 + \dfrac{1}{t^2}.$

So, $y = \dfrac{4t^3}{3} + 4t - \dfrac{1}{t} + c.$

27. $\int (2x + 3) dx = x^2 + 3x + c.$

29. $\int (x - 1)^{243} dx = \dfrac{(x - 1)^{244}}{244} + c.$

31. $\int (1 - x)^{1/2} dx = -\dfrac{2}{3}(1 - x)^{3/2} + c.$

33. $\int (2 - t)^{2/3}$: Guess $(2 - t)^{5/3}$. But $\dfrac{d}{dt}(2 - t)^{5/3}$

$= \dfrac{5}{3}(2 - t)^{2/3}(-1) = -\dfrac{5}{3}(2 - t)^{2/3}.$ So,

$\int (2 - t)^{2/3} = -\dfrac{3}{5}(2 - t)^{5/3} + c.$

35. $\int (1 + x^3)^2 dx = \int (1 + 2x^3 + x^6) dx = \dfrac{x^7}{7} + \dfrac{x^4}{2} + x + c.$

37. $\int 3x^2 \sqrt{1 + x^3} \, dx = \dfrac{2}{3}[1 + x^3]^{3/2} + c.$

39. $\int \sqrt{2 + 5y} \, dy$: guess $(2 + 5y)^{3/2}$. But $\dfrac{d}{dy}(2 + 5y)^{3/2}$

$= \dfrac{3}{2}(2 + 5y)^{1/2}.$ $5 = \dfrac{15}{2}(2 + 5y)^{1/2}.$

So, $\int \sqrt{2 + 5y} \, dy = \dfrac{2}{15}(2 + 5y)^{3/2} + c.$

41. $\int \dfrac{3r}{\sqrt{1 - r^2}} dr = -3 \sqrt{1 - r^2} + c$

43. $\int t^2(1 + 2t^3)^{-(2/3)} dt$: Guess $(1 + 2t^3)^{1/3}$

$\dfrac{d}{dt}(1 + 2t^3)^{1/3} = \dfrac{1}{3}(1 + 2t^3)^{-2/3} 6t^2$

So, $\int t^2(1 + 2t^3)^{-2/3} dt = \dfrac{1}{2}(1 + 2t^3)^{1/3} + c.$

45. $\int (\sqrt{x} + \dfrac{1}{\sqrt{x}}) dx = \int (x^{1/2} + x^{-1/2}) dx = \dfrac{2}{3}x^{3/2} - 2\sqrt{x} + c$

$= \dfrac{2}{3}x \sqrt{x} - 2 \sqrt{x} + c.$

47. $v(t) = \int a(t)dt = \int 1.6\,dt = 1.6\,t + c.$ Since at $t = 0$, when the rock is dropped, $v = 0$, $c = 0$. So, $v(t) = 1.6\,t$. At $t = 30$ seconds: $v(30) = 48\,m/s.$

CHAPTER 4
Article 4.3

1. $s = \int 3t^2 dt = t^3 + c.$ $s(0) = s_0$: $c = s_0$, so $s(t) = t^3 + s_0.$

3. $s(t) = \int (t + 1)^2 dt = \dfrac{(t + 1)^3}{3} + c$:

   $s(0) = s_0$: $s_0 = \dfrac{(0 + 1)^3}{3} + c \Rightarrow c = s_0 - \dfrac{1}{3}.$

   Thus, $s(t) = \dfrac{(t + 1)^3}{3} + s_0 - \dfrac{1}{3} = \dfrac{t^3}{3} + t^2 + t + s_0.$

5. $s = \int (t + 1)^{-2} dt = -\dfrac{1}{t + 1} + c.$ $s(0) = s_0$:

   $s_0 = -1 + c,$ $c = 1 + s_0.$ Hence $s(t) = 1 + s_0 - \dfrac{1}{t + 1}$

7. $v(t) = \int g\,dt = gt + c.$ $v(0) = v_0$: $c = v_0,$ $v(t) = gt + v_0.$

   $s(t) = \int (gt + v_0)dt = \dfrac{1}{2}gt^2 + v_0 t + c.$ $s(0) = s_0$

   $\Rightarrow c = s_0.$ Thus, $s(t) = s_0 + v_0 t + \dfrac{1}{2}gt^2.$

9. $v(t) = \int (2t + 1)^{1/3} dt = \dfrac{3}{8}(2t + 1)^{4/3} + c.$ $v(0) = v_0$

   $\Rightarrow v_0 = \dfrac{3}{8} + c,$ $c = v_0 - \dfrac{3}{8} \Rightarrow v = \dfrac{3}{8}(2t + 1)^{4/3} + v_0 - \dfrac{3}{8}.$

   $s(t) = \int [\dfrac{3}{8}(2t + 1)^{4/3} + (v_0 - \dfrac{3}{8})]dt$

   $= \dfrac{9}{112}(2t + 1)^{7/3} + (v_0 - \dfrac{3}{8})t + c.$ $s(0) = s_0$:

   $s_0 = \dfrac{9}{112} + c,$ $c = s_0 - \dfrac{9}{112}.$ Thus,

   $s(t) = \dfrac{9}{112}(2t + 1)^{7/3} + (v_0 - \dfrac{3}{8})t + s_0 - \dfrac{9}{112}.$

11. $v(t) = \int (t^2 + 1)^2 dt = \int (t^4 + 2t^2 + 1)dt$

    $= \dfrac{t^5}{5} + \dfrac{2t^3}{3} + t + c.$ $v(0) = v_0$

$\Rightarrow c = v_0$, and $v(t) = \frac{t^5}{5} + \frac{2t^3}{2} + t + v_0$.

$s(t) = \int (\frac{t^5}{5} + \frac{2t^3}{3} + t + v_0)dt$

$\qquad = \frac{t^6}{30} + \frac{t^4}{6} + \frac{t^2}{2} + v_0 t + c.$ $s(0) = s_0 \Rightarrow c = s_0.$

Hence, $s(t) = \frac{t^6}{30} + \frac{t^4}{6} + \frac{t^2}{2} + v_0 t + s_0.$

13. $y(x) = \int 4(x - 7)^3 dx = (x - 7)^4 + c.$ $y(8) = 10 \text{:}$ $10 = 1 + c,$

$\qquad c = 9, \ y = (x - 7)^4 + 9.$

15. $y(x) = \int \frac{x^2 + 1}{x^2} dx = \int (1 + \frac{1}{x^2})dx = x - \frac{1}{x} + c.$

$\qquad y(1) = 1 \Rightarrow c = 1.$ So, $y = x - \frac{1}{x} + 1 = \frac{x^2 + x - 1}{x}.$

17. $\frac{dy}{dx} = 2xy^2 \Rightarrow \frac{dy}{y^2} = 2x dx \Rightarrow -\frac{1}{y} = x^2 + c$

$\qquad \Rightarrow y = \frac{-1}{x^2 + c}.$ $y(1) = 1 \text{:}$ $-1 = 1 + c \Rightarrow c = -2.$

$\qquad$ So, $y = \frac{1}{2 - x^2}.$

19. $\frac{dy}{dx} = (x + x^{-1})^2 = x^2 + 2 + x^{-2} \Rightarrow y = \int (x^2 + 2 + \frac{1}{x^2})dx$

$\qquad = \frac{x^3}{3} + 2x - \frac{1}{x} + c.$ $y(1) = 1 \text{:}$ $1 = \frac{1}{3} + 2 - 1 + c$

$\qquad \Rightarrow c = -\frac{1}{3}.$ So, $y = \frac{x^3}{3} + 2x - \frac{1}{x} - \frac{1}{3} = \frac{x^4 + 6x^2 - x - 3}{3x}$

21. $y(x) = \int x \sqrt{1 + x^2} \, dx = \frac{1}{3}(1 + x^2)^{3/2} + c.$

$\qquad y(0) = -3 \Rightarrow c = -\frac{10}{3}, \ y(x) = \frac{1}{3}[(1 + x^2)^{3/2} - 10].$

23. $\frac{dy}{dx} = \int \frac{d^2 y}{dx^2} dx = \int (2 - 6x)dx = 2x - 3x^2 + c.$

$\qquad \frac{dy}{dx} = 4$ at $x = 0 = c = 4.$ So, $\frac{dy}{dx} = 2x - 3x^2 + 4.$

$\qquad$ Then, $y(x) = \int (2x - 3x^2 + 4) = x^2 - x^3 + 4x + c.$

$\qquad y(0) = 1 \text{:}$ $c = 1.$ Hence, $y(x) = x^2 - x^3 + 4x + 1.$

25. $\frac{dy}{dx} = \int \frac{3x}{8} dx = \frac{3x^2}{16} + c;$

$\qquad \frac{dy}{dx} = 3$ at $x = 4 \Rightarrow c = 0.$ So, $\frac{dy}{dx} = \frac{3x^2}{16}.$

$\qquad y(x) = \int \frac{3x^2}{16} dx = \frac{x^3}{16} + c.$

Then, $y(4) = 4$ implies $4 = \dfrac{4^3}{16} + c \Rightarrow c = 0$.

Hence, $y(x) = \dfrac{x^3}{16}$.

27. $\dfrac{df}{dx} = 3\sqrt{x} \Rightarrow y = f(x) = \int 3\sqrt{x}\,dx = \int 3x^{1/2}dx = 2x^{3/2} + c$.

$y(9) = 4 \Rightarrow 4 = 2(3)^3 + c$, $c = -50$. Hence, $f(x) = 2x^{3/2} - 50$.

29. $v(t) = \int a(t)dt = \int 9.8\,dt = 9.8t$

(because g = acceleration due to gravity = $9.8\,m/s^2$).

So, $s(t) = \int 9.8t\,dt = 4.9t^2$ (taking the diver's original position to be $s = 0$).

At $s = 30\,m$, $t = \dfrac{30}{4.9}$, $v = 9.8\sqrt{\dfrac{30}{4.9}} = 24.25\,m/s$.

31. $a = v\dfrac{dv}{ds} \Rightarrow F = m$ implies $\dfrac{-mgR^2}{s^2} = mv\dfrac{dv}{ds}$.

Then $v\,dv = \dfrac{-gR^2}{s^2}ds$, $\dfrac{v^2}{2} = \dfrac{gR^2}{s} + c$.

At $s = R$ (surface of the earth), $v = v_0 = \sqrt{2gR}$:

$\dfrac{2gR}{2} = \dfrac{gR^2}{R} + c \Rightarrow c = 0$.

So, $v = \dfrac{2gR^2}{s} = \sqrt{2gR}\sqrt{\dfrac{R}{s}} = v_0\sqrt{R/s}$.

Now, $\dfrac{ds}{dt} = v_0\sqrt{R}/\sqrt{s} \Rightarrow \sqrt{s}\,ds = v_0\sqrt{R}\,dt$

$\Rightarrow \dfrac{2}{3}s^{3/2} = v_0 t\sqrt{R} + c$. $s(0) = R$: $c = \dfrac{2}{3}R^{3/2}$.

Then, $s^{3/2} = R^{3/2} + \dfrac{3}{2}v_0 t\sqrt{R} = R^{3/2}[1 + \dfrac{3v_0 t}{2R}]^{2/3}$.

CHAPTER 4

Article 4.4

1. $\displaystyle\int \sin 3x\,dx = \dfrac{-1}{3}\cos 3x + c$

3. $\displaystyle\int \sec^2(x + 2)dx = \tan(x + 2) + c$

5. $\displaystyle\int \csc(x + \dfrac{\pi}{2})\cot(x + \dfrac{\pi}{2})dx = -\csc(x + \dfrac{\pi}{2}) + c$

7. $\displaystyle\int x\sin(2x^2)dx = -\dfrac{1}{4}\cos(2x^2) + c$

9. $\displaystyle\int \sin 2t\,dt = -\dfrac{1}{2}\cos 2t + c$

11. $\int 4 \cos 3y \, dy = \frac{4}{3} \sin 3y + c$

13. $\int \sin^2 z \cos z \, dz = \frac{\sin^3 z}{3} + c.$

15. $\int \sec^2 2\theta \, d\theta = \frac{1}{2} \tan 2\theta + c$

17. $\int \sec \frac{x}{2} \tan \frac{x}{2} \, dx = 2 \sec \frac{x}{2} + c$

19. $\int \frac{d\theta}{\sin^2 \theta} = \int \csc^2 \theta \, d\theta = -\cot \theta + c.$

21. $\cos^2 y = \frac{1 + \cos 2y}{2} \Rightarrow \int \cos^2 y \, dy$

    $= \int \frac{1 + \cos 2y}{2} dy = \frac{y}{2} + \frac{\sin 2y}{4} + c.$

23. $\int (1 - \sin^2 3t) \cos 3t = \frac{\sin 3t}{3} - \frac{\sin^3 3t}{9} + c.$

25. $\int \frac{\cos x}{\sin^2 x} dx = \int \cot x \csc x \, dx = -\csc x + c.$

27. $\int \frac{\sin 2t \, dt}{\sqrt{2 - \cos 2t}} = \sqrt{2 - \cos 2t} + c$

29. $\int \cos^2 \frac{2x}{3} \sin \frac{2x}{3} = -\frac{1}{2} \cos^3 \frac{2x}{3} + c.$

31. $\int \sec \theta (\sec \theta + \tan \theta) d\theta = \int \sec^2 \theta \, d\theta + \int \sec \theta \tan \theta \, d\theta$

    $= \sec \theta + \tan \theta + c.$

33. $\int (\sec^2 y + \csc^2 y) dy = \int \sec^2 y \, dy + \int \csc^2 y \, dy = \tan y - \cot y + c.$

35. $\int (3 \sin 2x + 4 \cos 3x) dx = -\frac{3}{2} \cos 2x + \frac{4}{3} \sin 3x + c$

37. $\int \sec x \tan x \, dx = \sec x + c$, so the only antiderivative of

    $\sec x \tan x$ is d) $\sec x - \frac{\pi}{4}$.

39. $2y \frac{dy}{dx} = 5x - 3 \sin x \Rightarrow 2y \, dy = (5x - 3 \sin x) dx$

    $\Rightarrow y^2 = \frac{5}{2} x^2 + 3 \cos x + c.$ $y(0) = 0:$ $0 = 3 + c,$ $c = -3.$

    So, $y^2 = \frac{5}{2} x^2 + 3 \cos x - 3.$

41. $\frac{d^3 y}{dx^3} = \int \frac{d^4 y}{dx^4} dx = \int \cos x \, dx = \sin x + c.$

    $y'''(0) = 0 \Rightarrow c = 0,$ $\frac{d^3 y}{dx^3} = \sin x.$

    $\frac{d^2 y}{dx^2} = \int \sin x \, dx = -\cos x + c.$ $y''(0) = 1 \Rightarrow c = 2,$

$$\frac{d^2y}{dx^2} = -\cos x + 2. \quad \frac{dy}{dx} = \int(2 - \cos x)dx = 2x - \sin x + c,$$

$$y'(0) = 2 \Rightarrow c = 2, \frac{dy}{dx} = 2 + 2x - \sin x.$$

$$y(x) = \int(2 + 2x - \sin x)dx = 2x + x^2 + \cos x + c. \quad y(0) = 3 \Rightarrow c = 2,$$

$$y(x) = 2 + 2x + x^2 + \cos x.$$

43. $v(t) = \int a(t)dt = \int \pi^2 \cos \pi t \, dt = \pi \sin \pi t + c.$

$v(0) = 8 \Rightarrow c = 8, \ v(t) = \pi \sin \pi t + 8.$

$s(t) = \int(\pi \sin \pi t + 8)dt = 8t - \cos \pi t + c.$

$s(0) = 0 \Rightarrow 0 = 1 + c \Rightarrow c = -1, \text{ and } s(t) = 8t - 1 - \cos \pi t.$

$t = 1s: \quad s(1) = 8 - 1 - \cos \pi = 8\,m.$

CHAPTER 4

Article 4.5

1. a) $A_{ins} = \frac{1}{4}(1 + 1\frac{2}{2}) + 2 + 2\frac{1}{2}) = 1\frac{3}{4}$

    b) $A_{cir} = \frac{1}{4}(1\frac{1}{2} + 2 + 2\frac{1}{2} + 3) = 2\frac{1}{4}$

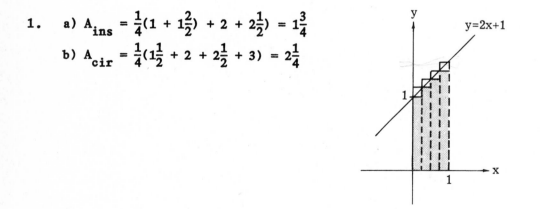

3. a) $A_{ins} = \frac{\pi}{4}[0 + \frac{\sqrt{2}}{2} + \frac{\sqrt{2}}{2} + 0] = \frac{\pi\sqrt{2}}{4}$

    b) $A_{cir} = \frac{\pi}{4}(\frac{\sqrt{2}}{2} + 1 + 1 + \frac{\sqrt{2}}{2}) = \frac{\pi}{2}(1 + \frac{\sqrt{2}}{2}).$

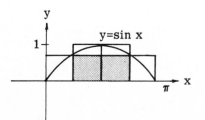

5. a) $A_{ins} = 1(0 + 1 + \sqrt{2} + \sqrt{3}) \approx 4.146$

    b) $A_{cir} = 1(1 + \sqrt{2} + \sqrt{3} + 2) \approx 6.146$

7. $\displaystyle\sum_{i=-1}^{3} 2^i = 2^{-1} + 2^0 + 2^1 + 2^2 + 2^3 = \frac{1}{2} + 1 + 2 + 4 + 8$

9. $\displaystyle\sum_{n=0}^{4} \frac{n}{4} = 0 + \frac{1}{4} + \frac{2}{4} + \frac{3}{4} + \frac{4}{4} = \frac{10}{4} = 2.5$

11. $\displaystyle\sum_{m=0}^{5} \sin\frac{m\pi}{2} = \sin 0 + \sin\frac{\pi}{2} + \sin\pi + \sin\frac{3\pi}{2} + \sin 2\pi + \sin\frac{5\pi}{2}$

    $\qquad = 0 + 1 + 0 + (-1) + 0 + 1 = 1.$

13. $\displaystyle\sum_{i=0}^{3} (i^2 + 5) = (0^2 + 5) + (1^2 + 5) + (2^2 + 5) + (3^2 + 5)$

    $\qquad = 5 + 6 + 9 + 14 = 34.$

15. All the expressions a) through d) equal $1 + 2 + 4 + 8 + 16 + 32$;

    expand each one out to see this.

17. a) $\int_1^9 f(x)dx = \int_1^7 f(x)dx + \int_7^9 f(x)dx = 4.$ Hence,

    $\int_1^9 -2f(x)dx = -2\int_1^9 f(x)dx = -8.$

    b) $\int_7^9 [2f(x) - h(x)]dx = 2\int_7^9 f(x)dx - \int_7^9 h(x)dx = 6.$

    c) $\int_9^7 f(x)dx = -\int_7^9 f(x)dx = -5.$

    d) $\int_7^7 [f(x) + h(x)]dx = \int_7^9 [f(x) + h(x)]dx + \int_9^7 [f(x) + h(x)]dx$

    $\qquad = \int_7^9 [f(x) + h(x)]dx - \int_7^9 [f(x) + h(x)]dx = 0.$

19. The hint says about all there is to be said for this problem: the base of each

    retangle in $U - L$ is x, and the height is $y_k - y_{k-1}$, $k = 1, \ldots, n$, where y_i

    is the y-coordinate of P_i. Hence, $U - L = [f(b) - f(a)] x.$

21. In both problems 19 and 20, we may write $U - L = |f(b) - f(a)| \frac{b - a}{n}.$

    Now, suppose the x_k's are unequal. Then, since $\frac{b - a}{n}$ is the _average_ of the x_k's, $x_{max} > \dfrac{b-a}{n}$. Hence, $U - L = |f(b) - f(a)|\dfrac{b-a}{n} <$

$|f(b) - f(a)|x_{max}$.

CHAPTER 4

Article 4.6

1. $k = 1$: $1^3 = (\dfrac{1 \times 2}{2})^2$. $k = 2$: $9 = 1^3 + 2^3 = (\dfrac{2 \times 3}{2})^2 = 3^2$.

    $k = 3$: $36 = 1^3 + 2^3 + 3^3 = (\dfrac{3 \times 4}{2})^2 = 6^2$.

    Then, $(1^3 + \ldots + n^3) + (h + 1)^3 = \dfrac{n^2(n + 1)^2}{4} + (n + 1)^3$

    $= \dfrac{(h + 1)^2}{4}[n^2 + 4(n + 1)] = \dfrac{(n + 1)^2}{4}(n^2 + 4n + 4) = [\dfrac{(n + 1)(n + 2)}{2}]^2$

3. The k+h circumscribed rectangle has area

    $\Delta x[ma + mk \Delta x] = \dfrac{b-a}{n}[ma + \dfrac{m}{n}(b - a)k]$

    Then the sum of these areas is

    $A_{cir} = (\dfrac{b-a}{n})[mna + \dfrac{m}{n}(b - a) \displaystyle\sum_{k=1}^{n} k]$

    $= ma(b - a) + \dfrac{m(b - a)^2}{n^2}\dfrac{n(n + 1)}{2}$.

    $\displaystyle\lim_{h \to \infty} A_{cir} = ma(b - a) + \dfrac{1}{2}m(b - a)^2$

    $= b - a[ma + \dfrac{mb - ma}{2}]$

    $= \dfrac{m}{2}(b - a)(b + a) = \dfrac{m}{2}(b^2 - a^2)$.

5. The area of the k+h circumscribed rectangle is

    $\Delta x(k \Delta x)^3 = k^3(\Delta x)^4 = \dfrac{k^3 b^4}{n^4}$.   Then

    $A_{cir} = \dfrac{b^4}{n^4}\displaystyle\sum_{k=1}^{n} k^3 = \dfrac{b^4}{n^4}\left[\dfrac{n(n + 1)}{2}\right]2 = \dfrac{b^4(n^4 + 2n^3 + n^2)}{4n^4}$

    $\displaystyle\lim_{h \to \infty} A_{cir} = \dfrac{b^4}{4}$.

7. $n = 1$: $\displaystyle\sum_{k=1}^{1} \frac{1}{k(k+1)} = \frac{1}{1\times 2} = \frac{1}{1+1}$,

so the formula holds for $n = 1$. Now suppose

$$\sum_{k=1}^{n} \frac{1}{k(k+1)} = \frac{n}{n+1}.$$

Then $\displaystyle\sum_{k=1}^{n+1} \frac{1}{k(k+1)} = \sum_{k=1}^{n} \frac{1}{k(k+1)} + \frac{1}{(n+1)(n+2)}$

$= \dfrac{n}{n+1} + \dfrac{1}{(n+1)(n+2)}$

$= \dfrac{1}{n+1}[\dfrac{n^2 + 2n + 1}{n+2}] = \dfrac{n+1}{n+2}$. Therefore the formula holds for all n.

9. $\int_0^1 (1 + x^2)dx$ is the area under the graph of $y = 1 + x^2$ between $x = 0$ and

$x = 1$. It is also the area of the square of side 1 plus the area under the

graph of $y = x^2$ from $x = 0$ to $x = 1$. Hence, $\int_0^1 (1 + x^2)dx = 1 + \frac{1}{3} = \frac{4}{3}$.

11. The area of the k+h inscribed rectangle is $x[(k-1)\ x]^2$

So, $A_{ins} = \dfrac{1}{n^3} \displaystyle\sum_{k=1}^{n} (k-1)^2 = \dfrac{1}{n^3}[1^2 + \ldots (n-1)^2]$

$= \dfrac{(n-1)(n)(2n-1)}{6n3} = \dfrac{2n3 - 3n2 + n}{6n3}$. $\displaystyle\lim_{n\to\infty} A_{ins} = \dfrac{1}{3}$, so $\lim S_n = \dfrac{1}{3}$.

13. a) The upper sum is $U = \dfrac{\pi}{2n}[\sin\dfrac{\pi}{2n} + \sin\dfrac{2n}{2n} + \ldots + \sin\dfrac{n\pi}{2n}]$

$= \dfrac{\pi}{2n}\ \dfrac{\cos \pi/4n - \cos(n + 1/2)\pi/2n}{\sin \pi/4n}$, using $h = \dfrac{\pi}{2n}$

b) As $n\to\infty$, $\dfrac{\pi/2n}{\sin(\pi/4n)} = 2\ \dfrac{\pi/4n}{\sin(\pi/4n)} \to 2$.

$\cos\dfrac{\pi}{4n} \to \cos 0 = 1$, $\cos\dfrac{(n + 1/2)\pi}{2n} \to \cos\dfrac{\pi}{2} = 0$.

Thus, $\displaystyle\lim_{n\to\infty} u = 2(1 - 0) = 2$, so $\int_0^\pi \sin x\, dx = 2$.

CHAPTER 4

Article 4.7

1. Area $= \int_0^3 (x^2 + 1)dx = \dfrac{x^3}{3} + x\Big|_0^3 = 12$.

3. We guess at an antiderivative of $\sqrt{2x + 1}$: let us try $[2x + 1]^{3/2}$.

But $\frac{d}{dx}(2x + 1)^{3/2} = \frac{3}{2}(2x + 1)^{1/2}2 = 3\sqrt{2x + 1}$.

So, an antiderivative of $\sqrt{2x + 1}$ is actually $\frac{1}{3}[2x + 1]^{3/2}$.

Then, area $= \int_0^4 \sqrt{2x + 1}\, dx = \frac{1}{3}(2x + 1)^{3/2}\Big|_0^4$

$\qquad = \frac{1}{3}[(9)^{3/2} - (1)^{3/2}] = \frac{26}{3}$.

5. Area $= \int_1^2 \frac{dx}{(2x + 1)^2} = -\frac{1}{2(2x + 1)}\Big|_1^2 = -\frac{1}{18} + \frac{1}{6} = \frac{1}{9}$.

7. Area $= \int_0^2 (x^3 + 2x + 1)dx = \frac{x^4}{4} + x^2 + x\Big|_0^2 1$

9. First we need an antiderivative of $\dfrac{x}{\sqrt{2x^2 + 1}}$. Let's suppose we

guess $\sqrt{2x^2 + 1}$; $\frac{d}{dx}(\sqrt{2x^2 + 1}) = \frac{1}{2\sqrt{2x^2 + 1}}4x$

$= \dfrac{2x}{\sqrt{2x^2 + 1}}$. Hence, the correct antiderivative is

$\frac{1}{2}\sqrt{2x^2 + 1}$. So, area $= \int_0^2 \dfrac{x\, dx}{\sqrt{2x^2 + 1}}$

$= \frac{1}{2}\sqrt{2x^2 + 1}\Big|_0^2 = 1$.

11. Area $= \int_0^1 (1 - x)dx = x - \frac{x^2}{2}\Big|_0^1 = \frac{1}{2}$.

13. Area $= \int_1^4 \frac{dx}{\sqrt{x}} = 2\sqrt{x}\Big|_1^4 = 2$.

15. The easiest way to approach this problem is to use a
different form of the area formula: area $= \int x\, dy$. This
corresponds to adding the areas of inscribed horizontal
rectangles and then making the base (in this case, y)
approach zero. Then, area $= \int_{-1}^1 (1 - y^2)dy =$
$y - \frac{y^3}{3}\Big|_1^1 = \frac{4}{3}$. Alternatively, we could take
advantage of the symmetry of the region and write the area
as $2\int_0^1 \sqrt{1 - x} = 2(\frac{2}{3}) = \frac{4}{3}$ (using the answer from
problem 14).

X = 1 - y²

4.7 /15

17. Using the arch between x = 0 and x $= \frac{\pi}{3}$, we have

area $= \int_0^{\pi/3} \sin^2 3x\, dx = \frac{1}{2}\int_0^{\pi/3}(1 - \cos 6x)dx$

$$= \frac{1}{2}(x - \frac{1}{6}\sin 6x)\Big|_0^{\pi/3} = \frac{\pi}{6}.$$

19. $\int_1^2 (2x + 5)dx = x^2 + 5x\Big|_1^2 = 8$

21. $\int_{-1}^1 (x + 1)^2 dx = \int_{-1}^1 (x^2 + 2x + 1)dx = \frac{x^3}{3} + x^2 + x\Big|_{-1}^1 = \frac{8}{3}.$

23. $\int_0^\pi \sin x\, dx = -\cos x\Big|_0^\pi = 2.$

25. $\int_{\pi/4}^{\pi/4} \frac{\cos x}{\sin^2 x}dx = \int_{\pi/2}^{\pi/2}(\frac{\cos x}{\sin x})(\frac{1}{\sin x})dx$

$$= \int_{\pi/4}^{\pi/2}\cot x \csc x\, dx = -\csc x\Big|_{\pi/4}^{\pi/4} = \sqrt{2} - 1.$$

27. $\int_0^\pi \sin^2 x\, dx = \int_0^\pi \frac{1 + \cos 2x}{2}dx$

[see hint for problem 17] $= \frac{x}{2} + \frac{\sin 2x}{4}\Big|_0^\pi = \frac{\pi}{2}.$

29. $\int_0^1 \frac{dx}{(2x + 1)^3} = -\frac{1}{4(2x + 1)^2}\Big|_0^1$

$$= -\frac{1}{36} - (-\frac{1}{4}) = \frac{2}{9}.$$

31. $\int_0^1 \sqrt{5x + 4}\, dx = \frac{2}{15}[5x + 4]^{3/2}\Big|_0^1 = \frac{2}{15}(27 - 8) = \frac{38}{15}.$

33. $\int_{-1}^0 (\frac{x^7}{2} - x^{15})dx = \frac{x^8}{16} - \frac{x^{16}}{16}\Big|_{-1}^0 = 0.$

35. $\int_1^2 \frac{2x^2 + 1}{x^2}dx = \int_1^2 (1 + \frac{1}{x^2})dx = x - \frac{1}{x}\Big|_1^2 = \frac{3}{2}.$

37. $\int_{\pi/6}^{\pi/2} \cos^2 \theta \sin \theta\, d\theta = \frac{-\cos^3 \theta}{3}\Big|_{\pi/6}^{\pi/2} = \frac{\sqrt{3}}{8}.$

39. The important observation in this problem is that

$$|x| = \begin{cases} x & \text{if } x \geq 0 \\ -x & \text{if } x \leq 0 \end{cases}$$

So we can rewrite $\int_{-4}^4 |x|dx$ as $\int_{-4}^0 -x\, dx + \int_0^4 x\, dx$

$$= \frac{-x^2}{2}\Big|_{-4}^0 + \frac{x^2}{2}\Big|_0^4 = 16.$$

41. $\int_{-1}^3 f(x)dx = \int_{-1}^2 (3 - x)dx + \int_2^3 \frac{x}{2}dx$

$$= 3x - \frac{x^2}{2}\Big|_{-1}^2 + \frac{x^2}{4}\Big|_2^3$$

$$= \frac{35}{4} = 8.75.$$

43. To find an antiderivative of $x \sin x$, guess $f = -x \cos x$. Then $\frac{df}{dx} =$

$x \sin x - \cos x$. To eliminate the unwanted $-\cos x$ term, try

$f(x) = -x \cos x + \sin x$. Then, $\dfrac{df}{dx} = (x \sin x - \cos x) + \cos x = x \sin x$,

so $f(x)$ is an antiderivative of $x \sin x$.

a) area $= \displaystyle\int_0^\pi x \sin x \, dx = (\sin x - x \cos x)\Big|_0^\pi = \pi$.

b) Since for $\pi \le x \le 2\pi$, $\sin x \le 0$,

$\quad$ area $= -\displaystyle\int_\pi^{2\pi} x \sin x \, dx = (x \cos x - \sin x)\Big|_\pi^{2\pi} = 3\pi$.

45. $\dfrac{dF(x)}{dx} = \dfrac{d}{dx}\displaystyle\int_1^x \dfrac{dt}{t} = \dfrac{1}{x}$ by the Second Fundamental Theorem.

47. $\dfrac{dF}{dx} = \dfrac{d}{dx}\displaystyle\int_0^x \dfrac{dt}{1 + t^2} = \dfrac{1}{1 + x^2}$.

49. $\dfrac{dF}{dx} = \dfrac{d}{dx}\displaystyle\int_1^{x^2} \dfrac{dt}{1 + \sqrt{1 - t}} = \dfrac{1}{1 + \sqrt{1 - x^2}} \cdot \dfrac{d(x^2)}{dx}$

$\quad = \dfrac{2x}{1 + \sqrt{1 - x^2}}$ by Leibnitz's Rule.

51. $\dfrac{dF}{dx} = \dfrac{d}{dx}\displaystyle\int_{1/x}^x \dfrac{dt}{t} = \dfrac{1}{x}\dfrac{d(x)}{dx} - \dfrac{1}{1/x}\dfrac{d}{dx}\left(\dfrac{1}{x}\right)$

$\quad = \dfrac{1}{x} - x\left(-\dfrac{1}{x^2}\right) = \dfrac{2}{x}$.

53. $\dfrac{dF}{dx} = \dfrac{d}{dx}\displaystyle\int_{\sqrt{x}}^{2\sqrt{x}} \sin(t^2)\, dt = \sin(2\sqrt{x})^2 \dfrac{d}{dx}(2\sqrt{x}) - \sin(\sqrt{x})^2 \dfrac{d}{dx}(\sqrt{x})$

$\quad = \dfrac{1}{\sqrt{x}}\sin 4x - \dfrac{1}{2\sqrt{x}}\sin x$.

55. Since $\displaystyle\lim_{x \to 0}\int_0^x \dfrac{t^2}{t^4 + 1}\, dt = \int_0^0 \dfrac{t^2}{t^4 + 1}\, dt = 0$

(the Second Fundamental Theorem assures us that $F(x) = \displaystyle\int_0^x \dfrac{t^2}{t^4 + 1}\, dt$

is continuous as a function of x, so $\displaystyle\lim_{x \to 0} F(x) = F(0) = 0$), the

expression $\displaystyle\lim_{x \to 0} \dfrac{1}{x^3}\int_0^x \dfrac{t^2}{t^4 + 1}\, dt$

is an indeterminate form. So, with $F(x) = \displaystyle\int_0^x \dfrac{t^2}{t^4 + 1}\, dt$,

we apply l'Hopital's rule:

$\displaystyle\lim_{x \to 0}\dfrac{F(x)}{x^3} = \lim_{x \to 0}\dfrac{F'(x)}{3x^2} = \lim_{x \to 0}\dfrac{1}{3x^2}\dfrac{x^2}{x^4 + 1} = \dfrac{1}{3}$.

57. We show that $y(x) = \dfrac{1}{a}\displaystyle\int_0^x f(t)\sin a(x - t)\, dt$ is a solution

Page 4-153

of $y'' + a^2 y = f(x)$ by substituting in directly for y.

$$\frac{dy}{dx} = \frac{1}{a}\int_0^x \frac{d}{dx}[f(t)\sin a(x-t)]dt + f(x)\sin a(x-x)\frac{d}{dx}(x)$$

$$= \frac{1}{a}\int_0^x af(t)\cos a(x-t)dt = \int_0^x f(t)\cos a(x-t)dt \text{ by Leibnitz's rule.}$$

Then $\dfrac{d^2y}{dx^2} = \dfrac{d}{dx}\int_0^x f(t)\cos a(x-t)dt$

$$= \int_0^x \frac{d}{dx}[f(t)\cos a(x-t)]dt + f(x)\cos a(x-x)\frac{d(x)}{dx}$$

$$= -a\int_0^x f(t)\sin a(x-t)dt + f(x)$$

So, $y'' + a^2 y = [-a\int_0^x f(t)\sin a(x-t)dt + f(x)]$

$$+ a^2[\frac{1}{a}\int_0^x f(t)\sin a(x-t)dt] = f(x).$$

59. a) $\int_0^x f(t)dt = x\cos \pi x \Rightarrow \dfrac{d}{dx}\int_0^x f(t)dt$

$$= f(x) = \frac{d}{dx}(x\cos \pi x) = \cos \pi x - \pi x \sin \pi x.$$

Then $f(4) = \cos 4\pi - 2\pi \sin 2\pi = 1.$

b) $\int_0^{x^2} f(t)dt = x\cos \pi x \Rightarrow \dfrac{d}{dx}\int_0^{x^2} f(t)dt$

$$= 2xf(x) = \frac{d}{dx}(x\cos \pi x) = \cos \pi x - \pi x \sin \pi x.$$

Then $8f(4) = 1$ or $f(4) = \frac{1}{8}.$

c) Doing the integration directly:

$$\int_0^{f(x)} t^2 dt = \frac{t^3}{3}\Big|_0^{f(x)} = \frac{[f(x)]^3}{3} = x\cos \pi x.$$

Then, $f(x) = [3x\cos \pi x]^{1/3}$. Hence $f(4) = (12\cos 4\pi)^{1/3} \approx 2.2894.$

61. a) The derivative of a constant function is zero; moreover, constant functions are the only functions with derivative identically zero. (Why?)

$$\frac{d}{da}\int_a^{a+p} f(t)dt = f(a+p)\frac{d(a+p)}{da} - f(a)\frac{d(a)}{da}.$$

$$= f(a+p) - f(a) = 0 \text{ for all a. Thus, } \int_a^{a+p} f(t)dt \text{ is constant.}$$

b) We may choose any value of a, so we take $a = 0$. Then

$$\int_a^{a+p} f(t)dt = \int_0^{2\pi}\sin x\, dx = -\cos x\Big|_0^{2\pi} = 0.$$

63. Let $S_n = \dfrac{1^5 + \ldots + n^5}{n^6} = \dfrac{1}{n}[(\frac{1}{n})^5 + (\frac{2}{n})^5 + \ldots + (\frac{n}{n})^5].$

Then, S_n is the sum of the areas of the circumscribed rectangles to $y = x^5$, $0 \le x \le 1$. Hence, $\lim_{n\to\infty} S_n = \int_0^1 x^5 dx$

CHAPTER 4 - Article 4.7

$$= \frac{x^6}{6}\Big|_0^1 = \frac{1}{6}.$$

CHAPTER 4

Article 4.8

1. $u = x^2 - a$, $du = 2x$, $\int 2x\sqrt{x^2 - 1}\, dx = \int \sqrt{u}\, du$

    $= \frac{2}{3}u^{3/2} + c = \frac{2}{3}(x^2 - 1)^{3/2} + c$

3. $u = 1 + x^3$, $du = 3x^2 dx$. When $x = 0$, $u = 1$; when $x = 2$, $u = 9$.

    So, $\int_0^2 x^2\sqrt{1 + x^3}\, dx = \int_1^9 \frac{1}{3}\sqrt{u}\, du = \frac{2}{9}u^{3/2}\Big|_1^9 = \frac{52}{9}.$

5. $u = \cos x$, $du = -\sin x\, dx$, $\int 3\cos^2 x \sin x\, dx = -3\int u^2 du = -u^3 = -\cos^3 x.$

    So, $\int_0^\pi 3\cos^2 x \sin x\, dx = -\cos^3 x\Big|_0^\pi = 2$

7. Let $u = x^2 + 1$, $du = 2x\, dx$. Then $\int x(x^2 + 1)^{10} dx = \frac{1}{2}\int u^{10} du$

    $= \frac{u^{11}}{22} + c = \frac{(x^2 + 1)^{11}}{22} + c$

9. Let $u = 1 + t$, $du = dt$. Then $\int \frac{dt}{2\sqrt{1 + t}}$

    $= \frac{1}{2}\int u^{-1/2} du = \sqrt{u} + c = \sqrt{1 + t} + c$

11. Let $u = 1 + x^{4/3}$, $du = \frac{4}{3}x^{1/3}$. Then

    $\int \frac{x^{1/3}}{(1 + x^{4/3})^2} dx = \frac{3}{4}\int \frac{du}{u^2} = -\frac{3}{4u} + c = \frac{-3}{4(1 + x^{4/3})} + c$

13. Let $u = 3 + \cos x$, $du = -\sin x$. Then $\int \frac{\sin x}{(3 + \cos x)^2} dx$

    $= -\int \frac{du}{u^2} = -\frac{1}{u} + c = \frac{1}{3 + \cos x} + c.$ Hence

    $\int_0^\pi \frac{\sin x\, dx}{(3 + \cos x)^2} = \frac{1}{3 + \cos x}\Big|_0^\pi = \frac{1}{2} - \frac{1}{4} = \frac{1}{4}.$

15. Let $u = t^5 + 2t$, $du = 5t^4 + 2$. Then $\int \sqrt{t^5 + 2t}\,(5t^4 + 2)dt$

    $= \int \sqrt{u}\, du = \frac{2}{3}u^{3/2} + c = \frac{2}{3}(t^5 + 2t)^{3/2} + c.$

    Therefore, $\int_0^1 \sqrt{t^5 + 2t}\,(5t^4 + 2)dt = \frac{2}{3}(t^5 + 2t)^{3/2}\Big|_0^1$

    $= \frac{2}{3}(\sqrt{3})^3 = 2\sqrt{3}.$

17. Let $u = \cos 2x$, $du = -2\sin 2x$. Then $\int \cos^2(2x)\sin(2x)dx = -\frac{1}{2}\int u^2 du$

$= -\frac{1}{6}u^3 + c = \frac{-\cos^3 2x}{6} + c.$

19. Let $u = \sin x$, $du = \cos x\, dx$. Then $\int \frac{\cos x}{\sin^2 x}dx = \int \frac{du}{u^2}$

$= -\frac{1}{u} + c = -\frac{1}{\sin x} + c.$ Hence

$\int_{\pi/4}^{\pi/2} \frac{\cos x}{\sin^2 x}dx = -\frac{1}{\sin x}\Big|_{\pi/4}^{\pi/2} = \sqrt{2} - 1.$

21. Let $u = 1 + x^4$, $du = 4x^3$. Then $\int \frac{x^3 dx}{\sqrt[4]{1 + x^4}}$

$= \frac{1}{4}\int u^{-1/4}du = \frac{1}{3}u^{3/4} + c = \frac{1}{3}[1 + x^4]^{3/4} +1$

24. Although the integrand looks complicated, the substitution

$u = \sqrt{3x^2 - 6}$ does the job: $du = \frac{1}{2}\frac{1}{\sqrt{3x^2 - 6}} 6x\, dx = \frac{3x\, dx}{\sqrt{3x^2 - 6}}.$

Hence, $\int \frac{x \cos\left(\sqrt{3x^2 - 6}\right)}{\sqrt{3x^2 - 6}}dx = \frac{1}{3}\int \cos u\, du = \frac{1}{3}\sin u + c = \frac{1}{3}\sin\left(\sqrt{3x^2 - 6}\right) + c$

25. $\int \frac{d2}{x^2 + 4x + 4} = \int \frac{dx}{(x + 2)^2}.$ $u = x + 2$, $dx = du$:

$\int \frac{dx}{(x + 2)^2} = \int \frac{du}{u^2} = -\frac{1}{u} + c = \frac{-1}{x + 2} + c.$

27. Let $u = a^2 - x^2$, $du = -2x\, dx$. Then $\int_0^a x\sqrt{a^2 - x^2}\, dx$

$= -\frac{1}{2}\Big|_{x=0}^{x=a} \sqrt{u}\, du = -\frac{1}{3}u^{3/2}\Big|_{x=0}^{x=a}$

$= -\frac{1}{3}(a^2 - x^2)^{3/2}\Big|_0^a = \frac{a^3}{3}.$

29. Let $u = 1 + \sqrt{x}$, $du = \frac{1}{2\sqrt{x}}dx.$ Then

$\int \frac{dx}{\sqrt{x}(1 + \sqrt{x})^2} = 2\int \frac{du}{u^2} = -\frac{2}{u} + c = \frac{-2}{1 + \sqrt{x}} + c.$

31. $dy = \frac{d}{dx}(x^3 - 3x^2 + 5x - 7)dx = (3x^2 - 6x + 5)dx$

33. $d(xy^2 + x^2 y) = d(4) = 0 \Rightarrow (y^2 dx + 2xy\, dy) + (2xy\, dx + x^2 dy) = 0$

$\Rightarrow (y^2 + 2xy)dx + (2xy + x^2)dy = 0 \Rightarrow dy = -\frac{2xy + y^2}{2xy + x^2}dx.$

Or, find $\frac{dy}{dx}$ by implicit differentiation: $\frac{d}{dx}(xy^2 + x^2 y)$

$= \frac{d}{dx}(4) = 0 \Rightarrow y^2 + 2xy\frac{dy}{dx} + 2xy + x^2\frac{dy}{dx} = 0$

$\Rightarrow y^2 + 2xy + (x^2 + 2xy)\frac{dy}{dx} = 0 \Rightarrow \frac{dy}{dx} = -\frac{y^2 + 2xy}{x^2 + 2xy}$

$\Rightarrow dy = -\frac{y^2 + 2xy}{x^2 + 2xy}dx.$

35. $\dfrac{dy}{dx} = \dfrac{d[x\sqrt{1-x^2}\,]}{dx} = \sqrt{1-x^2} + x\left(\dfrac{-2x}{2\sqrt{1-x^2}}\right) = \sqrt{1-x^2} - \dfrac{x^2}{\sqrt{1-x^2}}$

$= \dfrac{1}{\sqrt{1-x^2}}((1-x^2)-x^2) = \dfrac{1-2x^2}{\sqrt{1-x^2}}$. Therefore, $dy = \left(\dfrac{dy}{dx}\right)dx = \dfrac{1-2x^2}{\sqrt{1-x^2}}dx$

37. $\dfrac{dy}{dx} = \dfrac{d}{dx}\left[\dfrac{(1-x)^3}{2-3x}\right] = \dfrac{(2-3x)(3)(1-x)2(-1)-(1-x)3(-3)}{(2-3x)^2}$

$= \dfrac{(1-x)^2}{(2-3x)^2}(9x-6+1-x) = \dfrac{(1-x)^2(8x-5)}{(2-3x)^2}$. Then,

$dy = \dfrac{(1-x)^2(8x-5)}{(2-3x)^2}dx$.

39. $y(x) = \displaystyle\int \dfrac{dy}{dx}dx = \int \dfrac{5\cos x}{\sin x}dx$. Let $u = \sin x$, $du = \cos x$.

Then $y = \displaystyle\int \dfrac{5du}{u} = 5\ln|u| + c = 5\ln|\sin x| + c$.

To determine C: use $y(\pi/2) = 10$. $10 = 5\ln|\sin \pi/2| + c = 5\ln 1 + c = c$

$\Rightarrow c = 10$. Hence, $y = 5\ln|\sin x| + 10$.

41. $\dfrac{dy}{dx} = \dfrac{\sqrt{y^2+1}}{y}\cos x \Rightarrow \dfrac{y\,dy}{\sqrt{y^2+1}} = \cos x\,dx$.

Hence, $\displaystyle\int \dfrac{y\,dy}{\sqrt{y^2+1}} = \int \cos x\,dx = \sin x + c$.

Let $u = y^2+1$, $du = 2y\,dy$; then $\displaystyle\int \dfrac{y\,dy}{\sqrt{y^2+1}} = \dfrac{1}{2}\int \dfrac{du}{\sqrt{u}}$

$= \sqrt{u} + c = \sqrt{y^2+1} + c$. Thus, $\sqrt{y^2+1} = \sin x + c$. To determine C:

$y(\pi) = \sqrt{3} \Rightarrow 2 = \sin \pi + c = c$. Hence, $\sqrt{y^2+1} = 2 + \sin x$.

43. Both integrations are correct since $\sec^2 x = 1 + \tan^2 x$.

45. $x = 1 - t$, $dx = -dt$. At $x = 0$, $t = 1$; at $x = 1$, $t = 0$.

Then, $\displaystyle\int_0^1 f(x)dx = \int_1^0 f(1-t)(-dt) = \int_0^1 f(1-t)dt$.

47. a) Let $u = \dfrac{\pi}{x}$, $du = \dfrac{-\pi}{x^2}dx$. Then $\displaystyle\int_{0.5}^1 \dfrac{\pi}{x^2}\sin\left(\dfrac{\pi}{x}\right)dx$

$= -\displaystyle\int_{x=0.5}^{x=1}\sin u\,du = \cos u\,\Big|_{x=0.5}^{x=1} = \cos\left(\dfrac{\pi}{x}\right)\Big|_{x=1/2}^{x=1}$

$= \cos \pi - \cos 2\pi = -2$.

b) $\displaystyle\int_{x=0.01}^1 \dfrac{\pi}{x^2}\sin\left(\dfrac{\pi}{x}\right)dx = \cos\left(\dfrac{\pi}{x}\right)\Big|_{1/100}^1 = \cos \pi - \cos 100\pi = -2$.

49. Let $x = w + c$, $dx = dw$. When $x = a$, $w = a - c$; when $x = b$, $w = b - c$.

Then $\displaystyle\int_a^b f(x)dx = \int_{a-c}^{b-c} f(w+c)dw = \int_{a-c}^{b-c} f(x+c)dx$

(since we may use any dummy variable we wish).

51. a) If f is odd: let $x = -u$, $dx = -du$. At $x = 0$, $u = 0$; at $x = 1$, $u = -1$.

Then $3 = \int_0^1 f(x)dx = \int_0^{-1} f(-u)(-du) = -\int_{-1}^0 f(u)du$.

Hence, $\int_{-1}^0 f(u)du = -3$.

b) If f is even, do the same substitution as in (a). Then $3 = \int_0^{-1} f(-u)(-du)$

$= -\int_0^{-1} f(u)du = \int_{-1}^0 f(u)du$. Thus, $\int_{-1}^0 f(x) = 3$.

53. $\int_{-a}^0 h(x)dx = -\int_a^0 h(-u)du = \int_0^a h(-u)du$

$= \pm\int_0^a h(u)du$ (+ if he is even, − if h is odd).

Therefore, $\int_{-a}^a h(x)dx = \int_{-a}^0 h(x)dx + \int_0^a h(x)dx$

$$= \begin{cases} 0 & \text{if h is odd} \\ 2\int_0^a h(x)dx & \text{if h is even} \end{cases}$$

CHAPTER 4

Article 4.9

1.

i=0	1	2	3	4
x_i 0	0.5	1.0	1.5	2.0
y_i 0	0.5	1.0	1.5	2.0

a) trapezoidal rule: $\int_0^2 x\,dx \quad \frac{2-0}{2.4}(0 + 2(0.5) + 2(1.0) + 2(1.5) + 2) = 2$

b) Simpson's rule: $\int_0^2 x\,dx \quad \frac{2-0}{3.4}(0 + 4(0.5) + 2(1) + 4(1.5) + 2) = 2$

c) $\int_0^2 x\,dx = \frac{x^2}{2}\Big|_0^2 = 2$.

d,e) Since $\frac{d^2(x)}{dx^2} = 0$, $\frac{d^4(x)}{dx^4} = 0$, there is no error in either approximation.

3.

i=0	1	2	3	4
x_i 0	0.5	1.0	1.5	2.0
y_i 0	0.125	1.0	3.375	8.0

a) $\int_0^2 x^3\,dx \quad \frac{2-0}{2.4}[0 + 2(0.125) + 2(1.0) + 2(3.375) + 8.0]$

$= \frac{17}{4}$ by the trapezoidal rule.

b) $\int_0^2 x^3\,dx \quad \frac{2-0}{3.4}[0 + 4(0.125) + 2(1.0) + 4(3.375) + 8.0]$

$= \frac{24}{6} = 4$ by Simpson's rule.

c) $\int_0^2 x^3 dx = \frac{x^4}{4}\Big|_0^2 = 4.$

d) $\frac{d^2(x^2)}{dx^2} = 6x$, so $|f''(x)| \leq 12$ on $0 \leq x \leq 2$.

Hence, $|E_T| \leq \frac{b-a}{12} h^2 M = \frac{2-0}{12}(\frac{2}{n})^2.$

$12 = \frac{8}{h^2}$ with n subintervals. Now, $|E_T| < 10^{-5} \Rightarrow \frac{8}{h^2} < 10^{-5}$

$\Rightarrow n^2 > 8 \times 10^5 \Rightarrow n > 894.$ So, a minimum of 895 subintervals is required for the given accuracy.

e) Since $\frac{d^4(x^3)}{dx^4} = 0$, Simpson's rule gives the exact value of the integral.

5.

i=0	1	2	3	4
x_i 1.00	1.75	2.50	3.25	4.00
y_i 1.0000	1.3229	1.5811	1.8028	2.0000

a) $\int_1^4 \sqrt{x}\, dx = \frac{4-1}{2\cdot4}[1 + 2(1.3229) + 2(1.5811) + 2(1.8028) + 2]$

$= 4.6551$ by the trapezoidal rule.

b) $\int_1^4 \sqrt{x}\, dx = \frac{4-1}{3\cdot4}[1 + 4(1.3229) + 2(1.5811) + 4(1.8028) + 2]$

$= 4.6663$ by Simpson's rule.

c) $\int_1^4 \sqrt{x}\, dx = \frac{2}{3}x^{3/2}\Big|_1^4 = \frac{14}{3} \approx 4.6667$

d) $\frac{d^2}{dx^2}(\sqrt{x}) = -\frac{1}{4(\sqrt{x})^3}$, so $|f''(x)| \leq \frac{1}{4}$ on $1 \leq x \leq 4$.

Then, $|E_T| < \frac{4-1}{12}(\frac{3}{n})^2 \frac{1}{4} = \frac{9}{16n^2} < 10^{-5}$

$\Rightarrow n^2 > \frac{9}{16} \times 10^5$ or $n > 237.17.$ Hence, a minimum of 238 subintervals is required.

e) $\frac{d^4}{dx^4}(\sqrt{x}) = -\frac{15}{16}x^{-7/2}$, so $|f^{(4)}(x)| \leq \frac{15}{16}$ on $1 \leq x \leq 2$.

Then, $|E_s| < \frac{4-1}{180}(\frac{3}{n})^4 \frac{15}{16} = \frac{243.15}{2880n^4} < 10^{-5}.$

or $n^4 > 1.266 \times 10^5 \Rightarrow n > 18.9.$ So, we need at least 20 subdivisions to ensure the required accuracy.

7. Trapezoidal rule: $\dfrac{d^2}{dx^2}[\tfrac{1}{x}] = \dfrac{2}{x^3}$, so $|f''(x)| \le 2$ in $1 \le x \le 2$.

Then, $|E_T| < \dfrac{2-1}{12}(\tfrac{1}{10})^2 \, 2 = \dfrac{1}{300} \quad 0.0033$.

Simpson's rule: $\dfrac{d^4}{dx^4}[\tfrac{1}{x}] = \dfrac{24}{x^5}$, so $|f^{(4)}(x)| \le 24$ in $1 \le x \le 2$.

Then, $|E_s| < \dfrac{2-1}{180}(\tfrac{1}{10})^4 \, 24 \approx 1.33 \times 10^{-5}$.

9. a) If the curve is concave upward over $a \le x \le b$, then

$f''(c) > 0$ for all c between a and b. This means that the error term

$E_T = \dfrac{b-a}{12} h^2 f''(c) > 0$. Hence, the trapezoidal rule

overestimates the area under the graph of $y = f(x)$.

b) In this case, $f''(c) < 0$ for all $a < c < b$.

So, $E_T < 0$, and the trapezoidal rule underestimates the

area under the graph of $y = f(x)$.

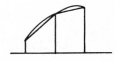

11. $\dfrac{d^2(1/t)}{dt^2} = \dfrac{2}{t^3}$, so $|f''(t)| \le 2$ on $1 \le t \le 2$.

Hence, $E_T \le (\dfrac{2-1}{12})(\tfrac{1}{4})^2 \, 2 = \dfrac{1}{48} \approx 0.0208$.

So, the maximum error in the trapezoidal rule approximation would be about 0.021.

With Simpson's rule: $\dfrac{d^4(1/t)}{dt^4} = \dfrac{24}{t^5}$,

so, $|f^{(4)}(t)| \le 24$ on $1 \le t \le 2$.

Thus, $|E_s| < \dfrac{1}{180}(\tfrac{1}{4})^4 \, 24 = \dfrac{2}{15 \cdot 4^4} = \dfrac{1}{1,920} \approx 5.2 \times 10^{-4}$.

13. For a No. 22 flashbulb, since $L(t) = 0$ if $t \ge 50$,

$A = \int_{20}^{70} L(t)\,dt = \int_{20}^{50} L(t)\,dt$

$5(\tfrac{1}{2})[4.2 + 2(3.0) + 2(.17) + 2(0.7) + 2.(0.35) + 2(0.2) + 0]$

$= 40.25$ lumen-milliseconds.

For a No. 31 flashbulb, $A = \int_{20}^{70} L(t)\,dt \approx 55.50$ lumen-milliseconds.

Hence, the No. 31 flashbulb delivers more light to the film.

15. a) $h = \dfrac{1-0}{10} = \dfrac{1}{10}$. The result of approximating

by Simpson's rule: $\operatorname{erf}(1) \approx 0.8427$

b) The given information does not help us much, since we need a bound on $|\frac{d^4(\text{erf})}{dx^4}|$ over [0,1]. We obtain such an estimate as follows.

By the Second Fundamental Theorem,

$$d(\text{erf})/dx = \frac{2}{\pi}e^{-x^2}. \text{ Thus,}$$

$$d^4(\text{erf})/dx^4 = \frac{8}{\pi}(3x - 2x^3)e^{-x^2}.$$

Now, for $0 \leq x \leq 1$, $e^{-x^2} \leq 1$, and the first derivative test together with

checking the endpoints shows that $0 \leq 3x - 2x^3 \leq \sqrt{2}$ over the interval

$0 \leq x \leq 1$. Thus,

$$|d^4(\text{erf})/dx^4| \leq 8\sqrt{2}/\pi < 4 \text{ over } [0,1].$$

So, the error in (a) is certainly less than $\frac{1}{180}(0.1)^4(4) = 2.22 \times 10^{-6}$.

CHAPTER 4

Miscellaneous Problems

1. $\frac{dy}{dx} = xy^2 \Rightarrow \frac{dy}{y^2} = x\,dx \Rightarrow -\frac{1}{y} = \frac{x^2}{2} + c$

    $\Rightarrow y = \frac{-2}{x^2 + k}(k = 2c)$

3. $\frac{dy}{dx} = \frac{x^2 - 1}{y^2 + 1} \Rightarrow (y^2 + 1)dy = (x^2 - 1)dx \Rightarrow \int(y^2 + 1)dy$

    $= \int(x^2 - 1)dx \Rightarrow \frac{y^3}{3} + y = \frac{x^3}{3} - x + c$

5. $\frac{dx}{dy} = (\frac{2 + x}{3 - y})^2 \Rightarrow \frac{dx}{(2 + x)^2} = \frac{dy}{(3 - y)^2} \Rightarrow \frac{-1}{2 + x} = \frac{1}{3 - y} + c$

7. There is such a curve. We can justify this answer by finding the equation

    of the curve $y(x)$. Since $\frac{d^2y}{dx^2} = 0$, $\frac{dy}{dx}$

    = constant = 1 ($y'(0) = 1$). Then, $y(x) = \int\frac{dy}{dx}dx = \int dx = x + c$.

    $y(0) = 0 \Rightarrow c = 0$, and $y(x) = x$ is the required curve.

9. Let the particle's velocity at $t = 0$ be v_0. Then

$$v(t) = \int a(t)dt = \int -t^2 dt = -\frac{t^3}{3} + v_0.$$

$$x(t) = \int v(t)dt = \int (-\frac{t^3}{3} + v_0)dt = -\frac{t^4}{12} + v_0 t + x_0.$$

Since $x(0) = 0$, $x_0 = 0$ and $x(t) = -\frac{t^4}{12} + v_0 t$.

Now, $x(t)$ will reach its maximum of $x = b$ at time t determined by

$$\frac{dx}{dt} = v(t) = 0; \quad -\frac{t^3}{3} + v_0 = 0, \quad t = \sqrt[3]{3v_0}.$$

$$x(\sqrt[3]{3v_0}) = -\frac{(3v_0)^{4/3}}{12} + v_0(3v_0)^{1/3} = v_0^{4/3}[3^{1/3} - \frac{3}{12}3^{1/3}]$$

$$= \frac{3}{4}\sqrt[3]{3}\, v_0^{4/3} = b. \quad \text{So, } v_0^{4/3} = \frac{4b}{3\sqrt[3]{3}},$$

$$v_0^4 = \frac{64b^3}{81}, \quad v_0 = \frac{2}{3}(4b^3)^{1/4} = \frac{2\sqrt{2}}{3}b^{3/4}.$$

11. $v(t) = \int (3 + 2t)dt = t^2 + 3t + c.$ $v(0) = 4 \Rightarrow c = 4$, $v(t) = t^2 + 3t + 4.$

$s(t) = \int (t^2 + 3t + 4)dt = \frac{t^3}{3} + \frac{3t^2}{2} + 4t + s_0.$ $s(0) = s_0,$

$s(4) = \frac{64}{3} + 24 + 16 + s_0 = 40 + \frac{64}{3} + s_0 = \frac{184}{3} + s_0.$

Thus, $s(4) = s(0) = \frac{184}{3}.$

13. $h(x)$ is a constant; $\frac{dh}{dx} = \frac{d}{dx}[f^2(x) + g^2(x)]$

$= 2f(x)\frac{df}{dx} + 2g(x)\frac{dg}{dx} = 2fg + 2g\frac{dg}{dx}.$

Since $g(x) = \frac{df}{dx}$, $\frac{dg}{dx} = \frac{d^2f}{dx^2} = -f.$ So, $\frac{dh}{dx} = 2fg + 2g(-f) = 0.$

Therefore, $h(10) = h(0) = 5.$

15. a) $a = \frac{dv}{dt} = \frac{dv}{ds}\frac{ds}{dt} = v\frac{dv}{ds} = -k \Rightarrow \int v\, dv = -\int k\, ds$

$\Rightarrow \frac{1}{2}v^2 = -ks + c.$ Suppose at $s = 0$, the brakes are applied.

Then $v(0) = 88$ ft/sec $\Rightarrow c = \frac{88^2}{2} = 3,872.$

We require $v(100) = 0; \quad 0 = -100k + 3,872 \Rightarrow k = 38.72$ ft/sec^2.

b) Here, $\frac{1}{2}v^2 = -ks + c$, $v(0) = 44$ ft/sec $\Rightarrow c = 968.$ Then, at $v = 0; \quad 0 = -38.72 s + 968,$

$s = 25.$ So, it takes 25 feet to bring the car to a stop.

17. $v(t) = \int -32\, dt = -32t + c.$ $v(0) = 96 \Rightarrow c = 96$, $v(t) = 96 - 32t.$

$s(t) = \int (96 - 32t)dt = 96t - 16t^2 + c.$ $s(0) = 0 \Rightarrow c = 0$, $s(t) = 96t - 16t^2.$

CHAPTER 4 - Miscellaneous Problems

s(t) reaches its maximum when $\frac{ds}{dt} = v(t) = 0$:

$96 - 32t = 0$, $t = 3$. At $t = 3$, $s = 144$ ft.

19. $x\,f(x) = 1$, so C is the horizontal line $y = 1$

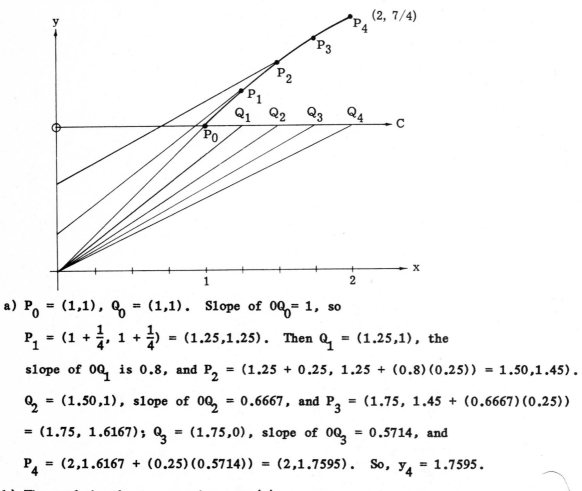

a) $P_0 = (1,1)$, $Q_0 = (1,1)$. Slope of $0Q_0 = 1$, so

$P_1 = (1 + \frac{1}{4}, 1 + \frac{1}{4}) = (1.25,1.25)$. Then $Q_1 = (1.25,1)$, the

slope of $0Q_1$ is 0.8, and $P_2 = (1.25 + 0.25, 1.25 + (0.8)(0.25)) = 1.50,1.45)$.

$Q_2 = (1.50,1)$, slope of $0Q_2 = 0.6667$, and $P_3 = (1.75, 1.45 + (0.6667)(0.25))$

$= (1.75, 1.6167)$; $Q_3 = (1.75,0)$, slope of $0Q_3 = 0.5714$, and

$P_4 = (2,1.6167 + (0.25)(0.5714)) = (2,1.7595)$. So, $y_4 = 1.7595$.

b) The work is the same as in part (a), so we present a summary.

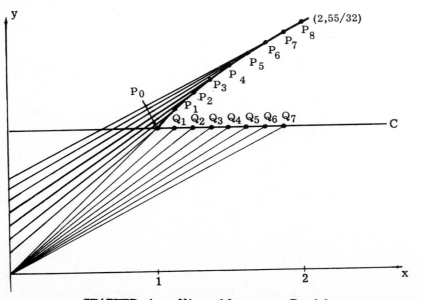

CHAPTER 4 - Miscellaneous Problems

	P_i	Q_i	Slope of OQ_i
i=0	(1,1)	(1,1)	1
i=1	(1.125,1.125)	(1.125,1)	0.8889
i=2	(1.25,1.2361)	(1.25,1)	0.8000
i=3	(1.375,1.3361)	(1.375,1)	0.7273
i=4	(1.5,1.4270)	(1.5,1)	0.6667
i=5	(1.625,1.5104)	(1.625,1)	0.6154
i=6	(1.75,1.5873)	(1.75,1)	0.5714
i=7	(1.875,1.6587)	(1.875,1)	0.5333
i=8	(2,1.7254)		

So, $y_8 = 1.7254$

21. $\int \frac{x^3 + 1}{x^2} dx = \int (x + \frac{1}{x^2}) dx = \frac{x^2}{2} - \frac{1}{x} + c$

23. Let $u = 1 + t^{4/3}$, $du = \frac{4}{3} t^{1/3}$. Then $\int_0 t^{1/3} (1 + t^{4/3})^{-7} dt$

$= \frac{3}{4} \int u^{-7} du = -\frac{1}{8} u^{-6} + c = \frac{-1}{8(1 + t^{4/3})^6} + c$

25. Let $u = 7 - 5r$, $du = -5dr$. Then $\int \frac{dr}{\sqrt[3]{(7 - 5r)^2}}$

$= -\frac{1}{5} \int u^{-2/3} du = -\frac{3}{5} u^{1/3} + c = -\frac{3}{5} (7 - 5r)^{1/3} + c$

27. Let $u = \sin 3x$, $du = 3 \cos 3x\, dx$. Then $\int \sin^2 3x \cos 3x\, dx$

$= \frac{1}{3} \int u^2 du = \frac{u^3}{9} + c = \frac{\sin^3 3x}{9} + c.$

29. Let $u = 2x - 1$, $du = 2dx$. Then $\int \cos(2x - 1) dx = \frac{1}{2} \int \cos u\, du$

$= \frac{1}{2} \sin u + c = \frac{1}{2} \sin(2x - 1) + c.$

31. $\int \frac{dt}{t \sqrt{2t}} = \frac{1}{\sqrt{2}} \int t^{-3/2} dt = -\sqrt{(2/t)} + c.$

33. Let $u = 2 - 3x$, $du = -3dx$. Then $\int \frac{dx}{(2 - 3x)^2}$

$= -\frac{1}{3} \int \frac{du}{u^2} = \frac{1}{3u} + c = \frac{1}{6 - 9x} + c.$

35. Now, $I \propto (\frac{\sin \alpha}{r})^2 (\cos \alpha)$, so $I(\alpha) = \frac{K}{r^2} \sin^2 \alpha \cos \alpha.$

$\frac{dI(\alpha)}{d\alpha} = 0: \quad 2 \sin \alpha \cos^2 \alpha - \sin^3 \alpha = 0 \text{ or}$

$\sin \alpha (2 \cos^2 \alpha - \sin^2 \alpha) = 0.$ Since $\alpha \neq 0$, $\sin \alpha \neq 0.$

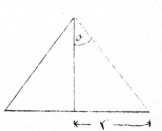

CHAPTER 4 - Miscellaneous Problems

Then, $2\cos^2\alpha - \sin^2\alpha = 0$, or $\tan\alpha = \sqrt{2}$ (since

$0 < \alpha < \pi/2$). So the height of the lamp above the table

should be $\dfrac{r}{\tan\alpha} = \dfrac{r}{\sqrt{2}}$.

37. We look at the <u>signed</u> distance $L = x_1 - x_2$

$= a[\sin bt - \sin(bt + \pi/3)]$. $\dfrac{dL}{dt} = 0$

$\Rightarrow ab[\cos bt - \cos(bt + \pi/3)] = 0$, or $\cos bt - \cos(bt + \pi/3) = 0$.

Now, $\cos(bt + \pi/3) = \cos bt \cos\pi/3 - \sin bt \sin\pi/3 = \frac{1}{2}\cos bt - \frac{\sqrt{3}}{2}\sin bt$.

Then, $\dfrac{\sqrt{3}}{2}\sin bt + \frac{1}{2}\cos bt = 0$, or $\tan(bt) = \dfrac{-1}{\sqrt{3}}$.

So, $t = \dfrac{-\pi}{3b}, \dfrac{2\pi}{3b}, \dfrac{5\pi}{3b}, \dfrac{8\pi}{3b}$.

At $t = \ldots, \dfrac{-\pi}{3b}, \dfrac{5\pi}{3b}, \dfrac{11\pi}{3b}, \ldots$,

$L = a(-\dfrac{\sqrt{3}}{2} - 0) = \dfrac{-a\sqrt{3}}{2}$; at $t = \ldots, \dfrac{2\pi}{3b}, \dfrac{8\pi}{3b}, \ldots$,

$L = \dfrac{a\sqrt{3}}{2}$. So, the greatest distance between the two particles is $\dfrac{a\sqrt{3}}{2}$.

39. From the diagram, $OP = \frac{1}{2}\tan\theta$, where θ is

the angle the beacon makes with the line OB. So, $\dfrac{d(OP)}{dt}$

$= \frac{1}{2}\sec^2\theta\dfrac{d\theta}{dt} = 2\pi\sec^2\theta$ (since the beacon makes

two revolutions per minute, $\dfrac{d\theta}{dt} = w = 4\pi/\text{min}$). When $BP = 1$,

$\sec\theta = 2$, so $\dfrac{d(OP)}{dt} = 8\pi$ miles/minute.

41. Area $A(b) = \int_1^b f(x)dx = \sqrt{b^2 + 1} - \sqrt{2}$; by the second fundamental

theorem, $f(b) = \dfrac{dA}{db} = \dfrac{b}{\sqrt{b^2 + 1}}$.

Hence, $f(x) = \dfrac{x}{\sqrt{x^2 + 1}}$.

43. a) Here, with the notation of problem 42, $f(x) = x^{15}$.

$$\lim_{n \to \infty} \frac{1}{n^{16}} [1^{15} + 2^{15} + \ldots + n^{15}] = \int_0^1 x^{15} dx = \frac{1}{16}.$$

b) Here, $f(x) = \sqrt{x}$, so $\lim_{n \to \infty} \frac{\sqrt{1} + \ldots + \sqrt{n}}{n^{3/2}} = \int_0^1 \sqrt{x}\, dx = \frac{2}{3}.$

c) $f(x) = \sin \pi x$; $\lim_{n \to \infty} \frac{1}{n}[\sin \frac{\pi}{n} + \ldots + \sin \frac{nn\pi}{n}] = \int_0^1 \sin \pi x\, dx = \frac{2}{\pi}.$

d) $\lim_{n \to \infty} \frac{1}{n^{17}} (1^{15} + \ldots + n^{15}) = [\lim_{n \to \infty} \frac{1}{n}][\lim_{n \to \infty} \frac{1^{15} + \ldots + n^{15}}{n^{16}}] = 0 \cdot \frac{1}{16} = 0,$

using the result of part (a).

e) $\lim_{n \to \infty} \frac{1}{n^{15}} (1^{15} + \ldots + n^{15}) = [\lim_{n \to \infty} n][\lim_{n \to \infty} \frac{1^{15} + \ldots + n^{15}}{n^{16}}] = \infty$

45. Differentiate both sides with respect to x:

$$1 = \frac{1}{\sqrt{1 + 4y^2}} \frac{dy}{dx}, \text{ or } \frac{dy}{dx} = \sqrt{1 + 4y^2}.$$

Then, $\frac{d^2y}{dx^2} = \frac{d}{dx}[\sqrt{1 + 4y^2}] = \frac{1}{2\sqrt{1 + 4y^2}} \cdot 8y \frac{dy}{dx}$

$= 4y.$ Thus, the constant of proportionality is 4.

47. The derivative of the left-hand side is

$$\frac{dx}{dx}[\int_0^x (\int_0^u f(t)dt)du] = \int_0^x f(t)dt.$$

The derivative of the right-hand side is

$$\frac{d}{dx}[\int_0^x f(u)(x - u)du] = \frac{d}{dx}[x\int_0^x f(u)du - \int_0^x u\, f(u)du]$$

$$= \int_0^x f(u)du + x\, f(x) - x\, f(x) = \int_0^x f(u)du = \int_0^x f(t)dt.$$

Since they have the same derivatives, the left- and right-hand sides differ

only by a constant; moreover, since both are zero for x = 0, that constant

is zero. Therefore $\int_0^x (\int_0^u f(t)dt)du = \int_0^x f(u)(x - u)du.$

CHAPTER 5

Article 5.2

1. a) $A = \lim\limits_{n \to \infty}(\frac{b - a}{n}) \sum\limits_{i=1}^{n}[f(\xi i) - g(\xi i)]$.

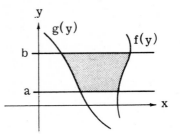

b) $A = \int_a^b (f(y) - g(y))dy$.

3. We will find the area using the method of Problem 1; that is, we will use

    horizontal rectangles bounded on the left by $x = 0$ (the y-axis) and on the

    right by $x = y^2 - y^3$. We need to determine the interval of integration, so we

    must find the points where the curve $x = y^2 - y^3$ intersects the y-axis. At

    these points $x = 0$: $y^2 - y^3 = 0 \Rightarrow y = 0,1$.   intersects the y

    $$A = \int_0^1 (y^2 - y^3)dy = \frac{y^3}{3} - \frac{y^4}{4}\Big|_0^1 = \frac{1}{12}.$$

5. First, we need the points of intersection of the two curves:

    $2x - x^2 = -3 \Rightarrow x^2 - 2x - 3 = (x - 3)(x + 1) = 0 \Rightarrow x = -1, 3$.

    The curve $y = 2x - x^2$ lies above the line $y = -3$ (to see this without graphing

    $y = 2x - x^2$, just choose some value of x, $-1 < x < 3$, say $x = 0$; then $y = 2x -$

    $x^2 = 0 > -3$), so the area is

    $$\int_{-1}^3 [(2x - x^2) - (-3)]dx = \int_{-1}^3 (3 + 2x - x^2) = \frac{32}{3}.$$

7. We will use horizontal rectangles and integrate with respect to y. The points

    of intersection of $x = 3y - y^2$ and $x + y = 3$ (or $x = 3 - y$) are given by

    $3y - y^2 = 3 - y \Rightarrow y^2 - 4y + 3 = 0 \Rightarrow (y - 3)(y - 1) = 0 \Rightarrow y = 1,3$. For $1 \leq y$

    ≤ 3, $3y - y^2 \geq 3 - y$, so

    $$A = \int_1^3 (3y - y^2 - (3 - y)dy = \int_1^3 (4y - y^2 - 3)dy = \frac{4}{3}.$$

9. We will use horizontal strips and integrate with respect to y. First, we need

    the interval over which we must integrate, so we find the points of

    intersection of the two curves:

$y^2 = y + 2 \Rightarrow y^2 - y - 2 = (y + 1)(y - 2) = 0 \Rightarrow y = -1, 2.$

So, the interval of integration is $-1 \leq y \leq 2$. On this interval, $y + 2 \geq y^2$,

so the curve $x = y + 2$ lies to the right of $x = y^2$.

Thus, the area is $\int_{-1}^{2}[(y + 2) - y^2]dy = \frac{y^2}{2} + 2y - \frac{y^3}{3}\Big|_{-1}^{2} = \frac{9}{2}.$

11. Points of intersection: $y^2 = y^3$, so $y = 0, 1$. For $0 \leq y \leq 1$,

$y^3 \leq y^2$, so $A = \int_{0}^{1}(y^2 - y^3)dy = \frac{1}{12}.$

13. Points of intersection: $\sin\frac{\pi x}{2} = x$ for $x = -1, 0, 1.$

area $= 2\int_{0}^{1}(\sin\frac{\pi x}{2} - x)dx.$

$= 2[-\frac{2}{\pi}\cos\frac{\pi x}{2} - \frac{x^2}{2}]\Big|_{0}^{1} = \frac{4}{\pi} - 1$

15. First, we find the points of intersection of the two curves:

$x^2 - 2x = x \Rightarrow x^2 - 3x = 0 \Rightarrow x = 0, 3.$ So we integrate over the interval $1 \leq x$

$\leq 3.$ In this interval, $x^2 - 2x \leq x$; hence, the area is

$\int_{1}^{3}[x - (x^2 - 2x)]dx = \int_{1}^{3}(3x - x^2)dx = \frac{3x^2}{2} - \frac{x^3}{3}\Big|_{1}^{3} = \frac{10}{3}.$

17. Integrating with respect to y, $A = \int_{-1}^{1}(3 - 3y^2)dy = 4.$

19. Points of intersection: $x = 2 - (x - 2)^2 \Rightarrow (x - 2)^2 + (x - 2) = 0$

$\Rightarrow x - 2 = 0$ or $x - 2 = -1 \Rightarrow x = 1$ or $x = 2.$ Then,

$A = \int_{1}^{2} - (x - 2)^2 - (x - 2)dx = \frac{-(x - 2)^3}{2} - \frac{(x - 2)^2}{2}\Big|_{1}^{2} = \frac{1}{6}.$

21. The curves $y = \sin x$, $y = \cos x$ intersect at the point

$x = \frac{\pi}{4}.$ From the graph, $\sin x \leq \cos x$ for

$0 \leq x \leq \pi/4$; hence, the area is

$A = \int_{0}^{\pi/4}(\cos x - \sin x)dx = (\sin x + \cos x)\Big|_{0}^{\pi/4} = \sqrt{2} - 1.$

23. The easier method is to integrate with respect to y. To

determine the limits of integration, solve $\sqrt{y} = 2 - y$, or

$y = (2 - y)^2$: $-y^2 + 5y - 4 = 0 \Rightarrow y = 1, 4.$ From the

sketch, we want the root $y = 1$; then

area $= \int_{0}^{1}[(2 - y) - \sqrt{y}]dy = 5/6.$ We could also integrate with respect to x, but this entails two

integrations:

$$A = \int_0^1 x^2 dx + \int_1^2 (2 - x) dx = \frac{1}{3} + \frac{1}{2} = 5/6.$$

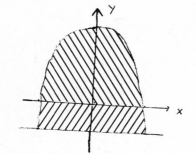

25. a) $A = \int_{-2}^2 [(3 - x^2) - (-1)] dx = \int_{-2}^2 (4 - x^2) dx = 4x - \frac{x^3}{3} \big|_{-2}^2 = \frac{32}{3}.$

b) The parabola is both the left boundary and the right

boundary of the region. However, we must make a

distinction when we solve for x in terms of y: the

right boundary is $x = \sqrt{3 - y}$, while the left boundary

is $x = -\sqrt{3 - y}$. So, the area is

$$\int_{-1}^3 [\sqrt{3 - y} - (-\sqrt{3 - y})] dy = -\frac{4}{3}(3 - y)^{3/2} \big|_1^3 = \frac{32}{3}.$$

27. We can take advantage of the symmetry of the region: the

total area is twice the area between x = 0 and x = 1. So,

the area is

$$A = 2\int_0^1 (x - x^3) dx = \frac{1}{2}.$$ [If we had done the problem

without using symmetry, we would have had to do two

integrations:

$$A = \int_{-1}^0 (x^3 - x) dx + \int_0^1 (x - x^3) dx.$$

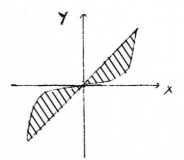

29. Area of triangle AOC $= \frac{1}{2}(2a)(a^2) = a^3.$

Area of shaded region $= \int_{-a}^a (a^2 - x^2) dx = \frac{4a^3}{2}.$

Ratio of areas $= \frac{4}{3}$, independent of a.

31. Area $= \int_{-1}^1 [(\frac{1}{1 + x^2}) - (\frac{-1}{1 + x^2})] dx = \int_{-1}^1 \frac{2\,dx}{1 + x^2} = 2\int_{-1}^1 \frac{dx}{1 + x^2}.$

Using Simpson's rule with $n = 10$, we have:

	x_i	y_i
i=0	−1.0	0.5000
i=1	−0.8	0.6098
i=2	−0.6	0.7353
i=3	−0.4	0.8261
i=4	−0.2	0.9615
i=5	0.0	1.0000
i=6	0.2	0.9615
i=7	0.4	0.8261
i=8	0.6	0.7353
i=9	0.8	0.6098
i=10	1.0	0.5000

Thus, the area is $A = 3.1033$.

Article 5.3

1. a) The velocity is positive for $0 \leq t \leq 2$.

    b) The velocity is not negative for any t, $0 \leq t \leq 2$.

    c) The displacement is $\int_0^2 (2t + 1)dt = 6$.

    d) Since $|V(t)| = V(t)$ for $0 \leq t \leq 2$,

    the total distance traveled is $\int_0^2 V(t)dt = 6$.

5. a) The velocity is positive for $0 < t < \pi/3$.

    b) The velocity is negative for $\frac{\pi}{2} < t < \frac{\pi}{2}$.

    c) The displacement is $\int_0^{\pi/2} v(t)dt = \int_0^{\pi/2} 6 \sin 3t\, dt = 2$.

    d) The total distance traveled is $\int_0^{\pi/2} |v(t)|dt$

    $= \int_0^{\pi/3} 6 \sin 3t\, dt + \int_{\pi/3}^{\pi/2} -6 \sin 3t\, dt = 6$.

7. Our work might be a little easier if we note that

    $\sin t + \cos t = \sqrt{2}(\frac{1}{\sqrt{2}} \sin t + \frac{1}{\sqrt{2}} \cos t)$

    $= \sqrt{2}(\cos \pi/4 \sin t + \sin \pi/4 \cos t) = \sqrt{2} \sin(t + \pi/4)$. Thus:

    a) The velocity is positive for $0 \leq t < \frac{3\pi}{4}$.

    b) The velocity is negative for $\frac{3\pi}{4} < t \leq \pi$.

c) The displacement is $\int_0^\pi v(t)dt = \int_0^\pi \sqrt{2} \sin(t + \pi/4)dt$

$= -\sqrt{2} \cos(t + \frac{\pi}{4})|_0^\pi = -\sqrt{2}(-\frac{1}{\sqrt{2}} - \frac{1}{\sqrt{2}}) = 2.$

d) The total distance traveled is $\int_0^\pi |v(t)|dt$

$= \int_0^{3\pi/4} \sqrt{2} \sin(t + \pi/4)dt - \int_{3\pi/r}^\pi \sqrt{2} \sin(t + \pi/4)dt = 2\sqrt{2}.$

9. We must find the velocity $v(t)$: $v(t) = \int a(t)dt = \int \sin t\, dt$

$= -\cos t + c.$ Since $v(0) = v_0 = 2$, $c = 3.$ So, $v(t) = 3 - \cos t.$

The distance traveled is $\int_0^2 |v(t)|dt$; however, since

$-1 \le \cos t \le 1$ for all t, $2 \le 3 - \cos t \le 4$ for all t.

So, the distance traveled is $\int_0^2 (3 - \cos t)dt = 6 - \sin 2.$

11. $v(t) = \int a(t)dt = \int g\, dt = gt + c.$ Since $v_0 = 0$, $g = 0$, and $v(t) = gt.$

13. $v(t) = \int a(t)dt = \int (4t + 1)^{-1/2}dt = \frac{1}{2}\sqrt{4t + 1} + c.$

$v(0) = v_0 = 1 \Rightarrow c = \frac{1}{2}$, so $v(t) = \frac{1}{2}\sqrt{4t + 1} + \frac{1}{2}.$

$v(t) \ge 0$ for $0 \le t \le 2$, so the distance traveled is

$\int_0^2 v(t)dt = \frac{1}{12}(4t + 1)^{3/2} + \frac{t}{2}|_0^2 = \frac{19}{6}.$

15. a) $s(t) = \int v(t)dt = \int -\sin t\, dt = \cos t + c.$ To make

$s(0) = 1$, we must set $c = 0$, so $s(t) = \cos t.$

b) The velocity $v(t)$ is negative for $0 < t < \pi$ and positive for

$\pi < t < 2\pi.$ So, the total distance traveled is

$-\int_0^\pi -\sin t\, dt + \int_\pi^{2\pi} -\sin t\, dt = -\cos t|_0^\pi + \cos t|_\pi^{2\pi} = 4.$

c) As an integral, displacement is $s = \int_0^{2\pi} -\sin t\, dt = \cos t|_0^{2\pi} = 1 - 1 = 0.$

From part (a): $s = s(2\pi) - s(0) = 1 - 1 = 0.$

17. We need to check two points: first, the function

$Q(t) = Q_0 + \int_0^t f(u)du$ (note the change of notation from that used

in the problem) satisfies $\frac{dQ}{dt} = f(t)$, and secondly, that

$Q(0) = Q_0.$ The relationship $\frac{dQ}{dt} = f(t)$ follows from the

Second Fundamental Theorem and the assumed continuity of $f(t)$. Also,

since $\int_0^0 f(u)du = 0$, $Q(0) = Q_0.$

Article 5.4

1. The line $x + y = 2$ intersects the x-axis at $x = 2$. Hence, the
   volume is $\pi\int_0^2 (2 - x)^2 dx = \frac{\pi(x - 2)^3}{3}\Big|_0^2 = \frac{8\pi}{3}$.

3. First, we need to find the limits of integration. To do this, we find the
   points of intersection of $y = x - x^2$ with the x-axis ($y = 0$): $x - x^2 = 0 \Rightarrow$
   $x = 0,1$. Hence, the volume is
   $V = \int_0^1 \pi(x - x^2)^2 dx = \pi\int_0^1 (x^2 - 2x^3 + x^4)dx = \pi[\frac{x^3}{3} - \frac{x^4}{2} + \frac{x^5}{5}]\Big|_0^1 = \frac{\pi}{30}$.

5. The curve $y = x^2 - 2x$ intersects the x-axis at $x = 0$ and $x = 2$.
   So, the volume is $V = \int_0^2 \pi(x^2 - 2x)^2 dx = \frac{16\pi}{15}$.

7. The volume is $V = \int_0^1 \pi x^8 dx = \frac{\pi}{9}$.

9. The volume is $V = \int_{-\pi/4}^{\pi/3} \pi y^2 dx = \pi\int_{-\pi/4}^{\pi/3} \sec^2 x dx$
   $= \pi \tan x \Big|_{-\pi/4}^{\pi/3} = \pi[\sqrt{3} - (-1)] = \pi(1 + \sqrt{3})$.

11. The volume is $V = \int_0^2 \pi(2y)^2 dy = \frac{32\pi}{3}$.

13. We shall use horizontal disks of radius $x = 1 - y^2$ and height dy. The curve
    $x = 1 - y^2$ intersects the y-axis at $y = -1$ and $y = 1$. So, the volume is
    $V = \int_{-1}^1 \pi(1 - y^2)^2 dy = \frac{16\pi}{15}$.

15. We shall use horizontal disks of radius $x = \frac{1}{y}$ and height dy.

    The volume is $V = \pi\int_1^2 x^2 dy = \pi\int_1^2 \frac{dy}{y^2} = -\frac{\pi}{y}\Big|_1^2 = \frac{\pi}{2}$.

17. The volume is $V = \pi\int_0^{\pi/3} \tan^2 x\, dx = \pi\int_0^{\pi/3} (\sec^2 x - 1)dx$
    $= \pi(\tan x - x)\Big|_0^{\pi/3} = \pi(\sqrt{3} - \frac{\pi}{3})$.

19. The curve $y = 3 - x^2$ intersects the line $y = -1$ when $3 - x^2 = -1$ or $x = \pm 2$.
    $V = \pi\int_{-2}^2 [(3 - x^2) - (-1)]^2 dx = \pi\int_{-2}^2 (4 - x^2)dx = \frac{16\pi}{3}$.

21. a) Volume $= \pi \int_{-a}^{-a+h} [\sqrt{a^2 - y^2}]^2 dy = \dfrac{\pi h^2}{3}(3a - h)$

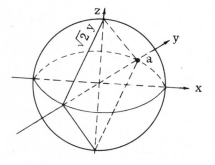

   b) From part a), we know V as a function of h:

   $V(h) = \dfrac{\pi h^2}{3}(3a - h)$. Here, $a = 5$ ft,

   $\dfrac{dV}{dt} = 0.2$ ft^3/sec. So:

   $\dfrac{dV}{dt} = \dfrac{dV}{dh} \cdot \dfrac{dh}{dt} = (2\pi a h - \pi h^2)\dfrac{dh}{dt}$;

   Substituting $a = 5$, $\dfrac{dV}{dt} = 0.2 = \dfrac{1}{5}$, $h = 4$ gives

   $\dfrac{dh}{dt} = \dfrac{1}{120\pi}$ ft/sec.

23. The curves $y = x^2$ and $y = 8 - x^2$ intersect at $x = \pm 2$. At a fixed value of x,

   the radius of the circular cross-section is $\dfrac{(8 - x^2) - x^2}{2} =$

   $4 - x^2$. So, the area of the cross section is $\pi(4 - x^2)$, whence the total

   volume of the solid is

   $V = \int_{-2}^{2} A(x)dx = \pi \int_{-2}^{2} (4 - x^2)^2 dx = \dfrac{512}{15}$.

25. At a given height h, the distance between the

   points P and Q in the figure is, by the Pythagorean

   Theorem, $\sqrt{2}\sqrt{a^2 - h^2}$; this is also the length of

   one side of the square cross-section. Hence, the

   cross-sectional area as a function of h is

   $A(h) = 2(a^2 - h^2)$, so the total volume is

   $V = \int_{-a}^{a} A(h)dh = 2\int_{-a}^{a}(a^2 - h^2)dh = \dfrac{8a^3}{3}$.

27. At a fixed value of x, the length of the base of the equilateral triangle is

   $\sin x$. So, the area of the triangle is

   $A(x) = \dfrac{(\sin x)^2 \sqrt{3}}{4}$. The total volume is therefore

   $V = \int_0^{\pi/2} A(x)dx = \int_0^{\pi/2} \dfrac{\sqrt{3} \sin^2 x}{4} dx$.

   But $\sin^2\theta = \dfrac{1 - \cos 2\theta}{2}$, so $V = \int_0^{\pi/2} \dfrac{\sqrt{3}}{4} \dfrac{1 - \cos 2\theta}{2} = \dfrac{\pi\sqrt{3}}{16}$.

Article 5.5

1. $V = \pi \int_0^2 (2 - x)^2 dx = \frac{8\pi}{3}$.

3. $V = \pi \int_0^2 [(3x - x^2) - x^2] dx$

$\quad = \pi \int_0^2 (8x^2 - 6x^3 + x^4) dx = \frac{56\pi}{15}$.

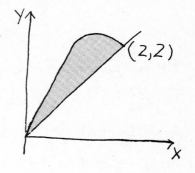

5. By washers (integrating with respect to x):

$V = \pi \int_{-2}^2 [4^2 - (x^2)^2] dx = \frac{256\pi}{5}$

By cylindrical shells (integrating with respect to y):

$V = 2\pi \int_0^4 (y)(2\ /y) dy = 4\pi \int_0^4 y^{3/2} dy = \frac{256\pi}{5}$.

7. The curves intersect when $x^2 + 1 = x + 3$ or

$x^2 - x - 2 = (x - 2)(x + 1) = 0$, whence $x = -1$ or $x = 2$.

$V = \pi \int_{-1}^2 [(x + 3)^2 - (x^2 + 1)^2] dx$

$\quad \pi \int_{-1}^2 (8 + 6x - x^2 - x^4) dx = \frac{117\pi}{5}$.

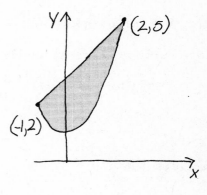

9. $V = 2\pi \int_0^1 xy\ dy = 2\pi \int_0^1 x^5 dx = \frac{\pi}{3}$.

11. a) By washers (integrating with:

$\quad V = \pi \int_1^2 [2^2 - x^2] = \frac{5\pi}{3}$.

b) By cylindrical shells (integrating with respect to x)

$\quad V = 2\pi \int_1^2 x(2 - x) dx = \frac{4\pi}{3}$.

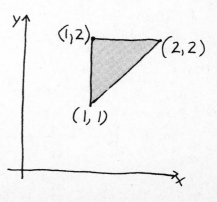

13. By cylindrical shells (integrating with respect to y):

$$V = 2\pi\int_0^1 y[1 - (y - y^3)]dy = \frac{11\pi}{15}.$$

15. The two curves intersect when $x^3 = 4x$, or $x = 0,2$. Then, using washers,

$$V = \pi\int_0^2[(4x)^2 - (x^3)^2]dx = \frac{512\pi}{21}.$$

17. By cylindrical shells: $V = 2\pi\int_0^1 x[(2x - x^2) - x]dx = \frac{\pi}{6}.$

19. Using washers: $V = \pi\int_0^{\pi/4}(\cos^2 x - \sin^2 x)dx = \pi\int_0^{\pi/4}\cos 2x\,dx = \frac{\pi}{2}.$

21. The curves are symmetric about the y-axis; we only need to consider the region

for which $0 \le x \le 2$. Using cylindrical shells:

$$V = \int_0^2 2\pi x[2x^2 - (x^4 - 2x^2)]dx$$

$$= 2\pi\int_0^2(4x^3 - x^5)dx = \frac{32\pi}{3}.$$

23. $V = 2\pi\int_0^{\pi}x \sin x\,dx = 2\pi(-x \cos x + \sin x)\Big|^{\pi} = 2\pi^2$

25. Using cylindrical shells: $V = 2\pi\int_{-a}^{a}(b - x)(2\sqrt{a^2 - x^2})dx$

$$= 2\pi^2 a^2 b + [a^2 - x^2]^{3/2}\Big|_{-a}^{a} = 2\pi^2 a^2 b.$$

Article 5.6

1. $\frac{dy}{dx} = x(x^2 + 2)^{1/2}$, $1 + (y')^2 = 1 + x^2(x^2 + 2) = (1 + x^2)^2$.

   So, the length of the curve is $\int_0^3 \sqrt{1 + (y')^2}\,dx = \int_0^3(1 + x^2)dx = 12.$

3. Use $L_{(0,0)}^{(2\sqrt{3},3)} = \int_0^3[1 + (\frac{dx}{dy})^2]^{1/2}dy$:

   $x = \frac{2}{3}y^{3/2}$, $\frac{dx}{dy} = y^{1/2} = \sqrt{y}.$ Thus,

   $L_{(0,0)}^{(2/3,3)} = \int_0^3 \sqrt{1 + (\sqrt{y})^2}\,dy = \int_0^3\sqrt{1 + y}\,dy = \frac{14}{3}.$

5. $\frac{dx}{dy} = y^3 - \frac{1}{4y^3}$, $1 + (\frac{dx}{dy})^2 = 1 + (y^6 - \frac{1}{2} + \frac{1}{16y^6})$

   $$= y^6 + \frac{1}{2} + \frac{1}{16y^6} = (y^3 + \frac{1}{4y^3})^2.$$

   So, $L = \int_1^2(y^3 + \frac{1}{4y^3})dy = \frac{123}{32}.$

7. $(\frac{dx}{d\theta})^2 + (\frac{dy}{d\theta})^2 = (-3a \cos^2\theta \sin\theta)^2 + (3a \sin^2\theta \cos\theta)^2$

$= 9a^2 \cos^2\theta \sin^2\theta(\cos^2\theta + \sin^2\theta) = 9a^2\cos^2\theta \sin^2\theta.$

$L\int_0^{2\pi}[(\frac{dx}{d\theta})^2 + (\frac{dy}{d\theta})^2]^{1/2} = 3a\int_0^{2\pi}\sqrt{\cos^2\theta \sin^2\theta}\ d\theta.$

We must, however, be careful in taking the square root; to avoid problems with

absolute values, we can use the symmetry of the hypocycloid:

$L = (4)(3\pi)\int_0^{\pi/2}\cos\theta \sin\theta\ d\theta = 6\pi.$

9. $L_0^\pi = \int_0^\pi \sqrt{(\frac{dx}{dt})^2 + (\frac{dy}{dt})^2}\ dt$

$= \int_0^\pi \sqrt{(-\sin t)^2 + (1 + \cos t)^2}\ dt = \int_0^\pi\sqrt{2 + 2\cos t}\ dt.$

But $\cos^2(\frac{t}{2}) = \frac{1 + \cos t}{2}$, so $L_0^\pi = 2\int_0^\pi\cos(\frac{t}{2}) = 2.$

11. $\frac{dx}{dt} = 1 - \cos t, \frac{dy}{dt} = \sin t$

$\sqrt{(\frac{dx}{dt})^2 + (\frac{dy}{dt})^2} = \sqrt{2 - 2\cos t} = 2\sqrt{\frac{1 - \cos t}{2}} = 2\sin(\frac{t}{2}).$

So, $L = \int_0^{2\pi}2\sin(\frac{t}{2}) = -4\cos(\frac{t}{2})\,|_0^{2\pi} = 8.$

13. $L = \int_0^3\sqrt{(\frac{dx}{dt})^2 + (\frac{dy}{dt})^2}\ dt = \int_0^3\sqrt{[(2t + 3)^{1/2}]^2 + (t + 1)^2}$

$= \int_0^3\sqrt{t^2 + 4t + 4}\ dt = \int_0^3(t + 2)dt = \frac{21}{2}.$

15. $F(x) = \sqrt{1 + (dy/dx)^2} = \sqrt{1 + [\frac{3\pi}{20}\cos(\frac{3\pi x}{20})]^2}$

i	0	1	2	3	4	5	6	7	8	9	10
xi	0	2	4	6	8	10	12	14	16	18	20
F(xi)	1.108	1.038	1.011	1.096	1.070	1.000	1.070	1.096	1.011	1.038	1.105

Simpson's rule then gives $L = \int_0^{20}F(x)dx$

$= \frac{20}{2\times10}(F(x_0) + 2F(x_1) + \ldots + 2F(x_9) + F(x_{10})) = 21.072.$

Article 5.7

1. $S = 2\pi\int_0^1 y\sqrt{1 + (dy/dx)^2}\ dx = 2\pi\int_0^1 x^3\sqrt{1 + (3x^2)^2}\ dx = 2\pi\int_0^1 x^3\sqrt{1 + 9x^4}\ dx$

$= \frac{\pi}{27}(1 + 9x^4)^{3/2}|_0^1 = \frac{\pi}{27}(10\sqrt{10} - 1).$

3. We use the formula $S = \int 2\pi\rho\ dx.$ ρ = distance from axis of revolution

$= y + 1 = \frac{x^3}{3} + \frac{1}{4x} + 1$

$$dx = \sqrt{1 + (\frac{dy}{dx})^2} \, dx = \sqrt{1 + (x^2 - \frac{1}{4x^2})^2} \, dx$$

$$= \sqrt{(x^4 + \frac{1}{2} + \frac{1}{16x^4})} dx = \sqrt{(x^2 + \frac{1}{4x^2})^2} \, dx$$

So, $S = 2\pi\int_1^3 (\frac{x^3}{3} + \frac{1}{4x} + 1)(x^2 + \frac{1}{4x^2}) dx$

$$= 2\pi\int_1^3 (\frac{x^5}{3} + x^2 + \frac{x}{3} + \frac{1}{4x^2} + \frac{1}{16x^3}) dx$$

$$= 2\pi(\frac{x^6}{18} + \frac{x^3}{3} + \frac{x^2}{6} - \frac{1}{4x} - \frac{1}{32x^2}) \Big|_1^3 = 1823\pi/18.$$

5. $S = \int_0^1 2\pi x [1 + x^2]^{1/2} dx = \frac{2\pi}{3}[2 /2 - 1].$

7. $S = \int_0^3 2\pi x [1 + (x /x^2 + 2)^2]^{1/2} dx = \int_0^3 2\pi x /1 + 2x^2 + x^4 \, dx$

$$= \int_0^3 2\pi x (1 + x^2) dx = \frac{99\pi}{2}.$$

9. a) $S = \int_0^{2\pi} 2\pi y [(\frac{dx}{dt})^2 + (\frac{dy}{dt})^2]^{1/2} dt = \int_0^{2\pi} 2\pi (1 + \sin t) dt = 4\pi^2.$

b)

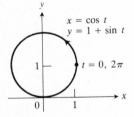

$x = \cos t$
$y = 1 + \sin t$

$t = 0, 2\pi$

11. $S = 2\int_0^{\pi/2} 2\pi (a \sin^3\theta [(-3a \cos^2\theta \sin\theta)^2 + (3a \sin^2\theta \cos\theta)^2]^{1/2} d\theta$

$$= 2\int_0^{\pi/2} 2\pi (a \sin^3\theta)(3a \cos\theta \sin\theta) d\theta = \frac{12\pi a}{5} \sin^5\theta \Big|_0^{\pi/2} = \frac{12\pi a}{5}.$$

Note: 1. We write $S = 2\int_0^{\pi/2}$ to avoid having to worry about absolute

values in taking the square root.

2. By symmetry of the hypocycloid, the surface is generated by just

that part between $\theta = 0$ and $\theta = \pi$.

13. The loop of the curve lies between $y = 0$ and

$y = 3$. We find $\frac{dx}{dy}$ by implicit

differentiation:

$18x\frac{dx}{dy} = (y - 3)^2 - 2y(3 - y)$

or $\frac{dx}{dy} = \frac{y^2 - 6y + 9 - 6y + 2y^2}{18x} = \frac{y^2 - 4y + 3}{6x}$.

$dA = 2\pi y\sqrt{1 + (dx/dy)^2} = 2\pi y\frac{\sqrt{36x^2 + (y^2 - 4y + 3)^2}}{6x}$.

Now, $36x^2 = 4y(3 - y)^2$, and $(y^2 - 4y + 3)^2 = (y - 3)^2(y - 1)^2$. So

$36x^2 + (y^2 - 4y + 3) = (3 - y)^2[4y + (y - 1)^2] = (3 - y)^2(y + 1)^2$. Therefore,

$dA = \frac{\pi y(3 - y)(1 + y)}{(3 - y)\sqrt{y}} = \pi\sqrt{y}(1 + y)$.

Surface area $= \pi\int_0^3 \sqrt{y}(1 + y)dy = \pi\int_0^3(y^{1/2} + y^{3/2})dy$

$= \pi[\frac{2}{3}y^{3/2} + \frac{2}{5}y^{5/2}]_0^3 = \frac{28\pi\sqrt{3}}{5}$.

Article 5.8

1. a) average $= (\int_0^{\pi/2}\sin x\, dx)/(\frac{\pi}{2}) = \frac{2}{\pi}$

   b) average $= (\int_0^{2\pi}\sin x\, dx)/2\pi = 0$

3. Average value $= (\int_4^{12}\sqrt{2x + 1}\, dx)/(12 - 4) = \frac{1}{8}\int_4^{12}\sqrt{2x + 1}\, dx$

   $= \frac{1}{16}\int_4^{12}[2x + 1]^{1/2}(2dx) = \frac{1}{16}\frac{2}{3}[2x + 1]^{3/2}|_4^{12} = \frac{49}{12}$.

5. $\int_a^b(\alpha x + \beta)dx = \frac{\alpha}{2}(b^2 - a^2) + \beta(b - a)$

   Average value $= \frac{(\alpha/2)(b^2 - a^2) + \beta(b - a)}{b - a} = \frac{\alpha}{2}(b + a) + \beta$

7. $V_{rms} = \sqrt{(V^2)_{avg}}$

   $(V^2)_{avg} = \frac{w}{2\pi}\int_0^{2\pi/w}V^2\sin^2 wt\, dt = (\frac{wV^2}{2\pi})(\frac{\pi}{w}) = \frac{V^2}{2}$.

   (For the calculation of the integral, please see the previous problem.)

   So, $V_{rms} = \frac{V}{\sqrt{2}}$.

9. $I_{avg} = (\int_0^{30}I(x)dx)/(30 - 0) = \frac{1}{30}\int_0^{30}(1200 - 40x)dx$

   $= \frac{1}{30}[1200x - 20x^2]|_0^{30} = 600$.

11. As in Problem 10: $I_{avg} = (\int_0^{60}(600 - 20\sqrt{15t})dt)/60 = 200$

Total holding cost $= \int_0^{60}(0.005)(600 - 20\sqrt{15t})dt = \$60$.

13. a) $\int_0^{60}L(t)dt \approx \dfrac{60 - 0}{2 \cdot 12}[0 + 2(0 \cdot 2) + 2(0 \cdot 5) + 2(2 \cdot 6)$

$+ 2(4 \cdot 2) + 2(3 \cdot 0) + 2(1 \cdot 7) + 2(0 \cdot 7) + 2(0 \cdot 35) + 2(0 \cdot 2)]$

$= 67.25$ million lumens-msec. So, average output

$= \dfrac{1}{60}\int_0^{60}L(t)dt = 1.12 \times 10^6$ lumens

b) $(765$ lumens$)$ $T = 67.25 \times 10^6$ lumens-msec.

So: $T = 87,900$ msec $= 87.9$ sec

15. Our new definition of average value says

$f'_{avg} = (\int_a^b(\dfrac{df}{dx})dx)/(b - a) = (f(x)\big|_a^b)/(b - a)$

$= \dfrac{f(b) - f(a)}{b - a}$, which agrees with our old definition.

17. Since we require that the chords intercept equal arcs, we average with respect to θ. So, the average length is

$\dfrac{1}{\pi}\int_0^\pi 2a\sin\theta\, d\theta = \dfrac{4a}{\pi}$.

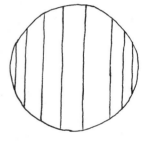

19. Averaging with respect to θ (see problem 17):

Average length $= \dfrac{4a^2}{\pi}\int_0^\pi\sin^2\theta\, d\theta$

$= \dfrac{4a^2}{\pi}\int_0^\pi\dfrac{1 - \cos 2\theta}{2}d\theta = 2a^2$.

Article 5.9

Problems 1-6: Assume $\delta = 1$, so $dm = dA$.

1. $\bar{x} = (\int_0^a x\sqrt{a^2 - x^2}\, dx)/(\pi a^2/4) = \dfrac{4a}{3\pi}$

By symmetry, $\bar{y} = \bar{x}$. So, the center of mass is $(\dfrac{4a}{3\pi}, \dfrac{4a}{3\pi})$.

3. To find $\bar{x}$, we will integrate with respect to x.

Then $x = x$, $dm = \delta(a - \sqrt{a^2 - x^2})dx$,

$\bar{x} = (\int_0^a \delta x(a - \sqrt{x^2 - x^2})dx)/(\int_0^a \delta(a - \sqrt{a^2 - x^2})dx)$

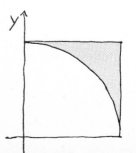

$$= \frac{\delta a^3/6}{\delta a^2(1 - \pi/4)} = \frac{2a}{3(4 - \pi)}.$$

To find $\bar{y}$, we have two choices.

I. Since the region is homogeneous and symmetric about the line $y = x$, the centroid $(\bar{x}, \bar{y})$ must lie on the line of symmetry.

So, $\bar{y} = \bar{x} = \frac{2a}{3(4 - \pi)}.$

II. We can compute $\bar{y}$ directly. We will integrate with respect to y:

$y = y$, $dm = \delta[a - \sqrt{a^2 - y^2}]dy$,

$\bar{y} = (\int_0^a \delta y[a - \sqrt{a^2 - y^2}]dy)/(\int_0^a \delta[a - \sqrt{a^2 - y^2}]dy.$

But this is exactly the same calculation as for $\bar{x}$ (the dummy variable y has merely replaced the dummy variable x).

So, $\bar{y} = \frac{2a}{3(4 - \pi)}.$

5. By symmetry, $\bar{y} = 1$. Integrate with respect to y to find $\bar{x}$: for a typical horizontal strip, $x = \frac{1}{2}x = \frac{1}{2}(2y - y^2)$, $dm = \delta x\, dy = \delta(2y - y^2)dy$.

$\bar{x} = (\int_0^2 \frac{\delta}{2}(2y - y^2)dy)/(\int_0^2 \delta(2y - y^2)dy = \frac{2}{5}.$

7. The only change to the solution in problem 6 is that now $dm = \delta dA$

$= \frac{bKy^2}{h}(h - y).$

$\bar{y} = (\int_0^h \frac{bK}{h}y^3(h - y)dy)/(\int_0^h \frac{bK}{h}y^2(h - y)dy) = \frac{3h}{5}$

9. $x = x$, $dm = \delta(x)dx = x^4 dx$

$\bar{x} = (\int_0^L x^5 dx)/(\int_0^L x^4 dx) = \frac{5L}{6}.$

11. Suppose the cone has height h and base radius a. Place the cone so that its vertex is at the origin and its axis is the z-axis. Here, $\delta(z) = K(h - z)$, K = constant or proportionality.

$\bar{z} = (\int_0^h \frac{\pi a^2}{h^2}k(h - z)z \cdot z^2 dz)/(\int_0^h \frac{\pi a^2}{h^2}k(h - z)z^2 dz = \frac{3h}{5}$

13. $dm = \delta ds = Kr \sin \theta\, d\theta$

$\bar{x} = (\int_0^\pi Kr^2 \cos \theta \sin \theta\, d\theta)/(\int_0^\pi kr \sin \theta\, d\theta)$

$$= (\frac{kr^2}{2} \sin^2\theta|_0^\pi)/(-Kr \cos\theta|_0^\pi) = 0$$

$$\bar{y} = (\int_0^\pi kr^2 \sin^2\theta \, d\theta)/(\int_0^\pi kr \sin\theta \, d\theta)$$

$$= \frac{r}{2}(\frac{1}{2} - \frac{\sin 2\theta}{4})|_0^\pi = \frac{\pi r}{4}$$

Article 5.10

1. $dA = (c^2 - x^2)dx.$ $\int_{-c}^{c} x \, dA = \int_{-c}^{c} x(c^2 - x^2)dx = 0$

$$\int_{-c}^{c} y \, dA = \int_{-c}^{c} \frac{1}{2}(c^2 - x^2)^2 dx = \frac{8c^5}{15}.$$

$$\int_{-c}^{c} dA = \int_{-c}^{c} (c^2 - x^2)dx = \frac{4c^3}{3}.$$

So, $(\bar{x}, \bar{y}) = (0, \frac{2c^2}{5})$.

3. $dA = (4 - x^2)dx,$ $\tilde{x} = x,$ $\tilde{y} = \frac{1}{2}(4 + x^2)$

$$\bar{x} = (\int_{-2}^{2} x(4 - x^2)dx)/(\int_{-2}^{2}(4 - x^2)dx) = 0$$

$$\bar{y} = (\int_{-2}^{2} \frac{1}{2}(4 + x^2)(4 - x^2)dx)/(\int_{-2}^{2}(4 - x^2)dx) = \frac{12}{5}.$$

$(\bar{x}, \bar{y}) = (0, \frac{12}{5}).$

5. Integrating with respect to y: $\tilde{x} = \frac{1}{2}(y + (y^2 - y))$

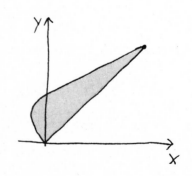

$$= \frac{y^2}{2}, \tilde{y} = y, \quad dA = [y - (y^2 - y)]dy = (2y - y^2)dy$$

$$\bar{x} = \int_0^2 \frac{y^2}{2}(2y - y^2)dy/(\int_0^2(2y - y^2)dy) = \frac{3}{5}$$

$$\bar{y} = \int_0^2 y(2y - y^2)dy/(\int_0^2(2y - y^2)dy) = 1$$

7. By symmetry, the center of gravity lies on the x-axis. To find $\bar{x}$,

we shall integrate with respect to x: $\tilde{x} = x,$ $dV = \pi y^2 dx = \pi x^4 dx.$

$$\bar{x} = (\int_0^3 \pi x^5 dx)/(\int_0^3 \pi x^4 dx) = \frac{5}{2}.$$

9. Because we are told the shell is very thin, we may make the problem simpler by

considering a right circular conical surface. By symmetry the center of

gravity will lie on the x-axis. Because we are considering a surface, we have

$\bar{y} = (\int y \, dA)/(\int dA).$ The surface is swept out by rotating the line

$y = \frac{hx}{r}$ (or $x = \frac{ry}{h}$) around the y-axis. So,

$$dA = 2\pi x \sqrt{1 + (\frac{dx}{dy})^2} = \frac{2\pi ry}{h}\sqrt{1 + \frac{r^2}{h^2}}\, dy = \frac{2\pi r\sqrt{r^2 + h^2}}{h^2} y\, dy.$$

Therefore, $\bar{y} = (\int_0^h y(\frac{2\pi r\sqrt{r^2 + h^2}}{h^2}y)ds)/(\int_0^h \frac{2\pi r\sqrt{r^2 + h^2}}{h^2}y\, ds) = \frac{2h}{3}.$

So, the center of gravity lies on the axis of the cone one-third of the way up

from the base.

11. $M_x = \int y\, ds = \int_0^4 y[1 + (\frac{dy}{dx})^2]^{1/2}dx$

$= \int_0^4 \sqrt{x}[1 + (\frac{1}{2\sqrt{x}})^2]^{1/2}dx$

$= \int_0^4 \sqrt{x + \frac{1}{4}}\, dx = \frac{17\sqrt{17} - 1}{12}.$

13. The cone frustum can be obtained by rotating the line

$y = 5 + \frac{5}{120}x = 5 + \frac{x}{24}$, $0 \le x \le 120$,

about the x-axis. $\bar{x} = (\int x\, dv)/(\int dv)$

$= (\int_0^{120}(x)(\pi)(5 + \frac{x}{24})^2 dx)/(\int_0^{120}\pi(5 + \frac{x}{24})^2 dx) = \frac{510}{7} = 72.86.$

Since the larger end is at $x = 120$, the balance point is 47.14

inches from the larger end.

Article 5.11

1. Volume = (area)(distance traveled by center of gravity)

$= (\pi)(4\pi) = 4\pi^2.$

3. By symmetry, $\bar{x} = 0$. If we rotate the arc $y = \sqrt{r^2 - x^2}$

about the x-axis, we get a sphere of radius r. The second Theorem of Pappus

says area = (length)(distance traveled by center of gravity) or

$4\pi r = (\pi r)(2\pi\bar{y})$. So, $\bar{y} = \frac{2r}{\pi}.$

5. Volume $= (\pi r^2/2)[2\pi(\bar{y} + r)] = \pi(\frac{4 + 3\pi}{3})r^3.$

7. Area $= (\pi r)(2\pi d) = (\frac{4 + 3\pi}{3})\pi r^2\sqrt{2}.$

9. If we revolve the semicircle about the line $y = -r$, the volume of the resulting

solid is

$$\pi \int_{-r}^{r} [(r + \sqrt{r^2 - x^2})^2 - r^2] dx$$

$$= \pi \int_{-r}^{r} (r^2 - x^2 + 2r\sqrt{r^2 - x^2}) dx = \frac{7\pi r^3}{3}.$$

[Observe that $\int_{-r}^{r} \sqrt{r^2 - x^2} \, dx$ = area of semicircle = $\frac{\pi r^2}{2}$.]

So, if M is the required moment, the first Theorem of Pappus gives

$\frac{7\pi r^3}{3} = (\frac{\pi r^2}{2})(2\pi M)$, or $M = \frac{7r}{3\pi}$.

Article 5.12

1. Let h be the depth of the water from the top of the trough.

   By similar triangles, $\frac{y}{3 - h} = \frac{2}{3}$. Now,

   integrating with respect to h:

   $$F = \int_0^3 (w)(2y)(h) dh = (62.6)\frac{4}{3}\int_0^3 h(3 - h) dh = 375 \text{ lbs.}$$

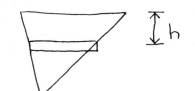

3. By similar triangles, $\frac{L}{4} = \frac{5 - (h - 1)}{5}$ or $L = \frac{4}{5}(6 - h)$

   $$F = \int p \, dA = 62.5 \int_1^6 hL \, dh$$

   $$= 62.5 \int_1^6 \frac{4}{5} h(6 - h) dh = 1,666\frac{2}{3} \text{ lbs.}$$

5. $y = \sqrt{1 - h^2}$.

   $$F = 62.5 \int_0^1 2h\sqrt{1 - h^2} \, dh = 41\frac{2}{3} \text{ lbs.}$$

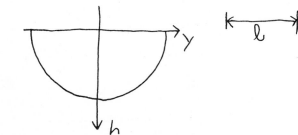

7. The distance of a horizontal strip at depth h from AB is $\frac{26}{24}(24 - h)$.

   So, the total moment about AB is

   $$M = \int_0^{24} (62.5)(\frac{13}{12})(24 - h)(h)(100)(\frac{13}{12}) dh.$$

   $$= \frac{1,056,250}{144}[12(24)^2 - \frac{(24)^3}{3}]$$

   $$= 16,900,000 \text{ ft-lbs} = 8,450 \text{ ft-tons.}$$

9. a) If the water in the container is 2 ft deep, the relationship between h and y

(as in Fig.5-60b) is $h + y = 2$, or $y = 2 - h$. For a horizontal strip at

depth h across the parabolic gate, $dA = 2x\,dh = 2\sqrt{y}\,dh = 2\sqrt{2 - h}\,dh$. Then

$F = \int_1^2 (62.5)h(2\sqrt{2 - h})dh$. To evaluate the integral, let $u = 2 - h$. At

$h = 1$, $u = 1$; at $h = 2$, $u = 0$. Then,

$F = \int_1^0 50(2 - u)(2\sqrt{u})(-du) = 100\int_0^1 (2 - u)\sqrt{u}\,du$

$= \frac{280}{3}\text{lbs} = 93\frac{1}{3}$ lbs.

b) Fill the container to a depth d. Then, as above, $h + y = d$, or $y = d - h$,

$dA = 2\sqrt{d - h}\,dh$.

$F(d) = 50\int_{d-1}^d 2h\sqrt{d - h}\,dh$. Let $u = d - h$, $h = d - u$, $dh = -du$.

$F(d) = 100\int_0^1 (d - u)\sqrt{u}\,du = [\frac{200}{3}d - 40]$lbs.

$F(d) = 160$ lbs; $d = 3$.

Article 5.13

1. $F = cx \Rightarrow 10$ lbs $= c(2$ in$) \Rightarrow c = 5$ lbs/in.

Work done by compressing the spring from 16" to 12" is

$\int_2^6 5x\,dx = 80$ lbs-in or $6\frac{2}{3}$ ft-lbs.

3. $F = cx \Rightarrow 10,000$ lbs. $= c(1$ inch$) \Rightarrow c = 10,000$ lbs/in is the spring constant.

a) Since $c = 10,000$ lbs/in and the natural length of the spring is 12 in.,

$W = \int_{12-12}^{12-11.5} 10,000x\,dx = 1,250$ in-lbs $= 104\frac{1}{6}$ ft-lbs

b) $W = \int_{12-11.5}^{12-11} 10,000x\,dx = 3,750$ in-lbs $= 312\frac{1}{2}$ ft-lbs

5. $(\frac{1}{16}$ inch$) = 150$ lbs $\Rightarrow c = 2,400$ lbs/inch.

Work to depress spring $\frac{1}{8}$ inch $= \int_0^{1/8} 2400x\,dx = 18.75$ inch-lbs

Weight of person: $w = (2400$ lbs/in$)(\frac{1}{8}$ in$) = 300$ lbs.

7. $F(x) = \dfrac{K}{(1 - x)^2}$, K is the constant of proportionality.

Work $= \int_{-1}^0 \dfrac{K\,dx}{(1 - x)^2} = \dfrac{K}{1 - x}\Big|_{-1}^0 = \dfrac{K}{2}$.

9. Work $= \int_R^0 mg(\frac{r}{R})dr = \frac{-mgR}{2}$.

The negative sign tells us that work is done on the particle as it moves from its <u>starting</u> point r = R (which must consequently be taken as the lower limit of integration) and r = 0. So, the work done on the particle is $\frac{mgr}{2}$.

11. $W = \int_{initial\ state}^{final\ state} F\,dx = \int_{(p_1,v_1)}^{(p_2,v_2)} (pA)dx$

$= \int_{(p_1,v_1)}^{(p_2,v_2)} (p)(Adx) = \int_{(p_1,v_1)}^{(p_2,v_2)} p\,dv$.

13. Integrating with respect to y:

$dV = \pi x^2 dy = \frac{25\pi}{16}y^2 dy$. The mass of water m v at a given y is moved a total distance of (14 − y) ft. So,

$work = \int_0^8 (62.5)(\frac{25\pi}{16})y^2(14 - y)dy$

$= \frac{400,000\pi}{3} ft\text{-}lbs = \frac{200\pi}{3} ft\text{-}tons$

(= 209.44 ft-tons)

15. Observe that the time taken to ascend 4,750 ft does <u>not</u> affect the work done. (It does, however, affect the <u>power</u>, which is d(work)/dt.) We need to find the weight of water remaining in the tank as a function of height h. Since the water leaks out at a steady rate and the truck climbs at a steady rate,

$\frac{dW}{dh} = (\frac{400\ gal}{4,750\ ft})(\frac{8\ lbs}{gal}) = \frac{64\ lbs}{95\ ft}$.

(Note: 1 gallon = 128 oz = 8 lbs.)

So, the weight W(h) of the water remaining at height h is

$W(h) = [(800)(8) - \frac{64}{94}h]$ lbs. The work done is therefore

$\int_0^{4,750}(6400 - \frac{64}{95}h)dh = 22,800,000$ ft-lbs = 11,400 ft-tons.

17. To fill the tower:

$work = \int_0^{25}(62.5)(300 + x)(100\pi)dx = 48,828,125\pi$ ft-lbs $\approx$ 24,400 ft-tons

To fill the pipe:

$work = \int_0^{300}(62.5)(x)(\frac{\pi}{9})dx = 312,500$ ft-lbs $\approx$ 156 ft-tons

Total work = 49,140,625 ft-lbs = 24,570 ft-tons.

Miscellaneous Problems

1. Area = $\int_{-1}^{2}(2 + x - x^2)dx = (2x + \frac{x^2}{2} - \frac{x^3}{3})\big|_{-1}^{2} = \frac{9}{2}$

3. Area = $\int_{1}^{2}(x - \frac{1}{x^2})dx = \frac{x^2}{2} + \frac{1}{x}\big|_{1}^{2} = 1$

5. Area = $\int_{-2}^{1}(2 - x - x^2)dx = 2x - \frac{x^2}{2} - \frac{x^3}{6}\big|_{-2}^{1} = \frac{9}{2}$

7. Integrating with respect to x:

    Area = $\int_{0}^{18}(3 - \sqrt{\frac{x}{2}})dx = 3x - \frac{\sqrt{2}}{3}x^{3/2}\big|_{0}^{18}$

    $= 54 - \frac{\sqrt{2}}{3}(3\sqrt{2})^3 = 18.$

CHAPTER 5 - Miscellaneous Problems

9.   Solve for y: $\sqrt{y} = \sqrt{a} - \sqrt{x}$. $y = a + x - 2\sqrt{ax}$

   Area $= \int_0^a (a + x - 2\sqrt{a}\sqrt{x})dx$

   $= ax + \dfrac{x^2}{2} - \dfrac{4}{3}\sqrt{a}(\sqrt{x})^3 \Big|_0^a = \dfrac{a^2}{6}$.

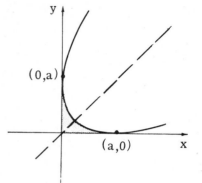

11.   Area $= \int_0^4 (3\sqrt{x} - \dfrac{3x^2}{8})dx = 2x^{3/2} - \dfrac{x^3}{8}\Big|_0^4 = 8$

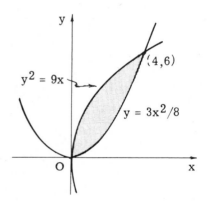

13.   Area $= \int_0^2 x\sqrt{2x^2 + 1}\,dx = (\dfrac{1}{4})(\dfrac{2}{3})(\sqrt{2x^2 + 1})^3 \Big|_0^2$

   $= \dfrac{1}{6}(27 - 1) = \dfrac{13}{3}$.

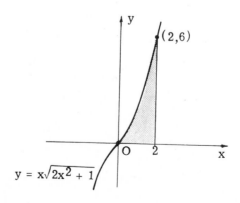

15. Area $= \int_{-2}^{2}[(2 - x^2) - (x^2 - 6)]dx = \int_{-2}^{2}(8 - 2x^2)dx$

$= (8x - \frac{2x^3}{3})\bigg|_{-2}^{2} = \frac{64}{3}.$

$y = 2 - x^2$

$(-2,-2)$ $(2,-2)$

$y = x^2 - 6$

17. a) The average rate of change is $\frac{A(8) - A(1)}{8 - 1} = \frac{4 - 1}{7} = \frac{3}{7}.$

b) $\frac{dA}{dx} = \frac{1}{2}(1 + 3x)^{-1/2}(3) = \frac{3}{2\sqrt{1 + 3x}}.$ At $x = 5$, $\frac{dA}{dx} = \frac{3}{8}.$

c) The area as a function of x is $A(x) = \int_{0}^{x} y(t)dt.$

Thus, $y(x) = \frac{dA}{dx} = \frac{3}{2\sqrt{1 + 3x}}$

(h,r)

$y = (r/h)x$

d) $y_{avg} = (\int_{1}^{8}y(x)dx)/(8 - 1) = \frac{1}{7}[A(8) - A(1)] = \frac{3}{7}.$

19. a) volume $= \int_{0}^{3} \pi(x^2)^2 dx = \frac{\pi x^5}{5}\bigg|_{0}^{3} = \frac{243\pi}{5}$

b) volume $= \int_{0}^{3} 2\pi(x + 3)x^2 dx = \frac{189\pi}{2}.$

Or, integrating with respect to y:

volume $\int_{0}^{9} \pi[6^2 - (3 + \sqrt{y})^2]dy = \frac{189\pi}{2}.$

c) Integrating with respect to x:

volume $= \int_{0}^{3}(2\pi x)(x^2)dx = \frac{81\pi}{2}.$

Integrating with respect to y:

volume $= \int_{0}^{9}\pi[3^2 - (\sqrt{y})^2]dy = \int_{0}^{9}\pi(9 - y)dy = \frac{81\pi}{2}.$

d) volume $= \int_{0}^{4}\pi(4y - y^2)^2 dy = \int_{0}^{4}\pi(16y^2 - 8y^3 + y^4)dy = \frac{512\pi}{15}.$

e) volume $= \int_{0}^{4}2\pi y(4y - y^2)dy = \frac{128\pi}{3}.$

21. The curves $y^2 = 4x$ and $y = x$ intersect at $4x = x^2$, or $x = 0,4.$

The required volume is

$V = \int_{0}^{4}\pi[(\sqrt{4x})^2 - (x)^2]dx = \int_{0}^{4}\pi(4x - x^2)dx = \frac{32\pi}{3}.$

23. Volume $= \int_{0}^{2} 2\pi x\, y\, dx = \int_{0}^{2}\frac{2\pi x^2 dx}{\sqrt{x^3 + 8}} = \int_{0}^{2}2\pi x^2(x^3 + 8)^{-1/2}dx$

CHAPTER 5 - Miscellaneous Problems

$$= \frac{4\pi}{3}(x^3 + 8)^{1/2}\Big|_0^2 = \frac{4\pi}{3}(4 - 2\sqrt{2})$$

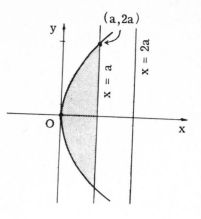

25. Integrating with respect to x:

$$\text{volume} = \int_0^a 2\pi(2a - x)2\sqrt{4ax}\,dx = 8\pi\sqrt{a}\int_0^a (2a - x)\sqrt{x}\,dx$$

$$= 8\pi\sqrt{a}\left[\frac{4}{3}ax^{3/2} - \frac{2}{5}x^{5/2}\right]\Big|_0^a = \frac{112\pi a^3}{15}.$$

27. By symmetry, $\bar{x} = \bar{y} = 0$. $dm = \pi\partial x^2 dy$, so

$$\bar{y} = \frac{\pi\partial\int_0^h (h^2 y^3 / r^2)dy}{\pi\partial\int_0^h (h^2 y^2 / r^2)dy} = \frac{3h}{4}$$

29. By cylindrical shells:

$$\text{volume bored out} = \int_0^{\sqrt{3}} (2\pi x)(2\sqrt{4 - x^2})dx = 4\pi\int_0^{\sqrt{3}} x\sqrt{4 - x^2}\,dx$$

$$= -(2\pi)(\frac{2}{3})(4 - x^2)^{3/2}\Big|_0^{\sqrt{3}} = \frac{28\pi}{3}.$$

31. Cross-sectional area $A(x) = \frac{(2y(\sqrt{3}\,y))}{2}$

$$= y^2\sqrt{3} = 4ax\sqrt{3}.$$

$$\text{Volume} = \int_0^a A(x)dx = \int_0^a 4ax\sqrt{3}\,dx = 2a^3\sqrt{3}.$$

One side of the equilateral triangle = 2y

33. $dA = 2\pi x\,dx = 2\pi x\sqrt{1 + y'^2}\,dx = 2\pi x\sqrt{1 + (\sqrt{x} - \frac{1}{4\sqrt{x}})^2}\,dx$

$$= 2\pi x\sqrt{1 + (x - \frac{1}{2} + \frac{1}{16x})}dx = 2\pi x\sqrt{(\sqrt{x} + \frac{1}{4\sqrt{x}})^2}\,dx.$$

So, the surface area generated is $\int_0^4 2\pi x(\sqrt{x} + \frac{1}{4\sqrt{x}})dx$

$$= \int_0^4 2\pi(x^{3/2} + \frac{1}{4}x^{1/2})dx = 2\pi(\frac{2}{5}x^{5/2} + \frac{1}{6}x^{3/2})\Big|_0^4 = \frac{424\pi}{15}.$$

35. $ds = 2\pi(y + 1)ds = 2\pi(y + 1)(y^{2/3} + \frac{1}{4}y^{-2/3})dy.$

$$\text{Surface area} = 2\pi\int_0^1 (y + 1)(y^{2/3} + \frac{1}{4}y^{-2/3})dy = 2\pi\int_0^1 (y^{5/3} + \frac{1}{4}y^{1/3} + y^{2/3} + \frac{1}{4}y^{-2/3})dy$$

$$= 2\pi(\frac{3}{8}y^{8/3} + \frac{3}{16}y^{4/3} + \frac{3}{5}y^{5/3} + \frac{3}{4}y^{1/3})dy = \frac{153\pi}{40}.$$

37. a) $y^2_{\text{avg}} = (\int_0^a y^2 dx)/(a - 0) = \frac{b^2}{a^3}\int_0^a (a^2 - x^2)dx = \frac{2b^2}{3}.$

b) Let $w = x^2$; then we are finding the average of y with respect to w, where now $ay = b\sqrt{a^2 - w}$.

$$y_{avg} = (\int_0^{a^2} \frac{b}{a}\sqrt{a^2 - w}\ dw)/(a^2 - 0) = \frac{2b}{3}.$$

39. a) $v(t) = \frac{ds}{dt} = 120 - 32t$.

Average value of velocity $= \frac{1}{3}\int_0^3 (120 - 32t)dt = 72$.

b) $s(3) = s(0) = 336$.

Average value of velocity with respect to $s = (\int_0^{336} v(s)ds)/336$.

It appears that we have to know v as a function of s to evaluate $\int_0^{216} v(s)ds$. But we can change variables by substituting for s:

$s = 160t - 16t^2$, $ds = 160 - 32t$. Then, $\int_0^{336} v(s)ds = \int_0^3 (v(t))(160 - 32t)dt$

$$= \int_0^3 (160 - 32t)^2 dt = -\frac{1}{96}(160 - 32t)^3 \Big|_0^3 = \frac{160^3 - 64^3}{96} = 39{,}936.$$

Therefore, the average value of the velocity with respect to distance is $\frac{39{,}936}{336} = 118.86$.

41. $dA = (\sqrt{8x} - x^2)dx$. $\bar{x} = (\int x\ dA)/(\int dA)$

$$= (\int_0^2 x(\sqrt{8x} - x^2)dx)/(\int_0^2 (\sqrt{8x} - x^2)dx = \frac{9}{10}.$$

Since, for a thin vertical strip, $y = \frac{y_1 + y_2}{2}$,

$$y = [\int_0^2 \frac{1}{2}(\sqrt{8x} + x^2)(\sqrt{8x} - x^2)dx]/(\int_0^2 (\sqrt{8x} - x^2)dx) = \frac{9}{5}.$$

43. $dA = [(4x - x^2) - 2x]dx = (2x - x^2)dx$.

$\bar{x} = (\int_0^2 x(2x - x^2)dx)/(\int_0^2 (2x - x^2)dx) = 1$

$\tilde{y} = \frac{1}{2}[(4x - x^2) + 2x] = \frac{1}{2}(6x - x^2)$

$\bar{y} = (\int_0^2 \frac{1}{2}(6x - x^2)(2x - x^2)dx)/(\int_0^2 (2x - x^2)dx) = \frac{12}{5}.$

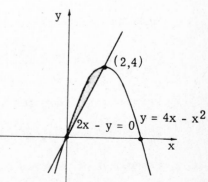

CHAPTER 5 - Miscellaneous Problems

45. The moment of the plate about the y-axis is

$\int x\, dA = M$, since x is the (signed) distance from

the y-axis. The signed distance from the line $x = b$

is $x - b$. So, the moment about the line $x = b$ is

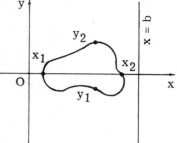

$$M_{x=b} = \int (x - b)dA = \int x\, dA - \int b\, dA$$

$$= M - b\int dA = M - bA.$$ This shows that $\bar{x}$ is

independent of the coordinate system used. In the

first coordinate system, $\bar{x} = \dfrac{M}{A}$. In the second

coordinate system, which is the first coordinate

system translated to the right by an amount b

(if $b > 0$),

$$\bar{x}_{new} = \dfrac{M_{x=b}}{A} = \dfrac{M - bA}{A} = \bar{x}_{old} - b.$$

47. By symmetry, $\bar{x} = \bar{y}$.

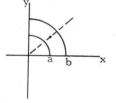

$$\int x\, dA = \int_0^a x[\sqrt{b^2 - x^2} - \sqrt{a^2 - x^2}]dx + \int_a^b x\sqrt{b^2 - x^2}\, dx$$

$$= \int_0^b x\sqrt{b^2 - x^2}\, dx - \int_0^a x\sqrt{a^2 - x^2}\, dx = \dfrac{b^3 - a^3}{3}.$$

By geometry, the area of the region is $\dfrac{\pi}{4}(b^2 - a^2)$.

So, $\bar{x} = \bar{y} = \dfrac{4(b^3 - a^3)}{3\pi(b^2 - a^2)} = \dfrac{4(b^2 + ab + a^2)}{3\pi(b + a)}$. In the limiting case $b \to a$,

$(\bar{x},\bar{y}) \to (\dfrac{2a}{3\pi},\dfrac{2a}{3\pi})$. This is the centroid of a circular arc of radius a

in the first quadrant.

49. The sides of the triangle are of length a and b; since the

area of the triangle is 36 in^2, $ab = 72$. Observe that

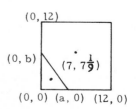

$$\bar{x}_{square} = \dfrac{\text{(area of triangle)}\bar{x}_{triangle} + \text{(area of quad.)}\bar{x}_{quad}}{\text{area of square}}$$

where "quad" denotes the region left when the triangle is cut

away. Of course, a similar relation holds for $\bar{y}_{quad}$. From

example 4 of section 5.9, we know

$(\bar{x},\bar{y})_{\text{triangle}} = (\frac{1}{3}a, \frac{1}{3}b) = (\frac{1}{3}a, \frac{24}{a})$.

Suppose $\bar{x}_{\text{quad}} = 7$. Then $6 = \frac{36(a/3) + (108)7}{144}$, or $a = 9$.

Then $b = 8$, and $\bar{y}_{\text{triangle}} = \frac{8}{3}$. Hence,

$6 = \frac{(36)(8/3) + 108\bar{y}_{\text{quad}}}{144}$, or $\bar{y}_{\text{quad}} = \frac{64}{9} = 7\frac{1}{9}$.

51. Extend the non-parallel sides of the trapezoid until

they intersect. Then, by similar triangles, we have:

$\frac{L}{200} = \frac{40 - h}{40}$, or $L = 200 - 5h$.

(Or, by a more algebraic and less geometric

approach: L and h must be linearly related. At

$L = 200$, $h = 0$; at $L = 100$, $h = 20$. The linear

relationship is then

$L = \frac{200 - 100}{0 - 20}h + 200$, or $L = 200 - 5h$).

So, $dA = Lah = (200 - 5h)dh$.

$F = \int_0^{20} 625h(200 - 5h)h = 62.5(100h^2 - \frac{5}{3}h^3)\Big|_0^{20} = 1{,}666{,}666\frac{2}{3}$ lbs $= 8{,}333\frac{1}{3}$ lbs.

53. $L = 2y - 4$, $m = 28 - 2y$.

The force on the plate is then

$F = \int_2^8 d_1 y(2y - 4)dy + \int_8^{14} d_2 y(28 - 2y)dy$

$= 2d_1(y^3/3 - y^2)\Big|_2^8 + 2d_2(7y^2 - y^3/3)\Big|_8^{14}$

$= 108d_1 + 360d_2$.

55. Note that it is irrelevant that the particle starts at rest.

Work $= \int_b^a \frac{k}{x^2}dx = \frac{k}{b} - \frac{k}{a}$.

57. Work $= \int_0^4 (w)(10 + x)(20)(2\sqrt{16 - x^2})dx = 40w\int_0^4 (10 + x)\sqrt{16 - x^2}dx$

$= (400w)\int_0^4 \sqrt{16 - x^2}\,dx - (20w)(\frac{2}{3})(16 - x^2)^{3/2}\Big|_0^4 = 400w\int_0^4 \sqrt{16 - x^2}\,dx + \frac{2{,}560w}{3}$.

But $\int_0^4 \sqrt{16 - x^2}\,dx = 4\pi$, since the integral represents the area of

the region of the circle $x^2 + y^2 = 16$ in the first quadrant. So,

work $= \frac{320w(15\pi + 8)}{3}$.

CHAPTER 6

Article 6.2

1. $\sin \alpha = \frac{1}{2}$; $\cos \alpha = \sqrt{1 - \sin^2 \alpha} = \sqrt{1 - \frac{1}{4}} = \sqrt{3}/2$

   $\sec \alpha = 2\sqrt{3}/3$; $\tan \alpha = \sqrt{3}/3$; $\csc \alpha = 2$

3. Let $\alpha = \cos^{-1}\sqrt{2}/2$. That is, $\cos \alpha = \sqrt{2}/2$.

   Hence $\sin \alpha = \sqrt{1 - \cos^2 \alpha} = \sqrt{1 - (\sqrt{2}/2)^2} = \sqrt{2}/2$.

5. $\cos(\cos^{-1}\frac{1}{2}) = \frac{1}{2}$. Hence $\sec(\cos^{-1}\frac{1}{2}) = 2$.

7. $\cos(\sec^{-1}2) = \frac{1}{2}$; $\csc(\sec^{-1}2) = (1 - (\frac{1}{2})^2)^{-1/2} = 2\sqrt{3}/3$

9. Let $\alpha = \cot^{-1}1$. Then $\cot \alpha = 1$.

   $\cot^2 \alpha + 1 = \dfrac{\cos^2 \alpha}{\sin^2 \alpha} + \dfrac{\sin^2 \alpha}{\sin^2 \alpha} = \dfrac{1}{\sin^2 \alpha} = 2$.

   $\sin^2 \alpha = \frac{1}{2} \Rightarrow \cos \alpha = \sqrt{1 - \sin^2 \alpha} = \sqrt{1 - \frac{1}{2}} = \sqrt{2}/2$.

   Hence $\cos(\cot^{-1}1) = \sqrt{2}/2$.

11. $\cos^{-1}0 = \pi/2$. Hence $\cot(\cos^{-1}0) = \dfrac{\cos \pi/2}{\sin \pi/2} = 0$.

13. $\sec^{-1}1 = 0$. Hence $\tan(\sec^{-1}1) = \dfrac{\sin 0}{\cos 0} = 0$.

15. $\sin^{-1}(1) - \sin^{-1}(-1) = (\pi/2) - (-\pi/2) = \pi$

17. $\sec^{-1}(2) - \sec^{-1}(-2) = (\pi/3) - (2\pi/3) = -\pi/3$

19. $\alpha = \sin^{-1}0.8$. $\sin \alpha = 0.8$, so $\cos \alpha = \sqrt{1 - (0.8)^2} = 0.6$

21. $\tan^{-1}(\tan \pi/3) = \pi/3$

23. $\cos x = \cos(-x)$, so $\sec^{-1}(\sec - 30^\circ) = \sec^{-1}(\sec 30^\circ) = \sec^{-1}(\sec \pi/6) = \pi/6$

25. $\lim\limits_{x \to -1+} \cos^{-1}x = \cos^{-1}(-1) = \pi$

27. Let $y = \tan^{-1}x$. Then $x = \tan y$. As $x \to -\infty$, $\dfrac{\sin y}{\cos y} \to -\infty$,

   and $\cos y \to 0$. Since $x < 0$, $-\dfrac{\pi}{2} \le y \le 0$. It follows that

   $y \to -\dfrac{\pi}{2}$, or $\lim\limits_{x \to -\infty} \tan^{-1}x = -\dfrac{\pi}{2}$.

29. $\lim\limits_{x \to -\infty} \sec^{-1}x = \lim\limits_{x \to -\infty} \cos^{-1}(\frac{1}{x}) = \cos^{-1}0 = \dfrac{\pi}{2}$

31. $\lim\limits_{x \to -\infty} \csc^{-1}x = \lim\limits_{x \to -\infty} \sin^{-1}(\frac{1}{x}) = \sin^{-1}0 = 0$

33. $\alpha = \angle \text{TOR} - \angle \text{TOS}$

$= \cot^{-1}(\dfrac{x}{a+b}) - \cot^{-1}(\dfrac{x}{b})$

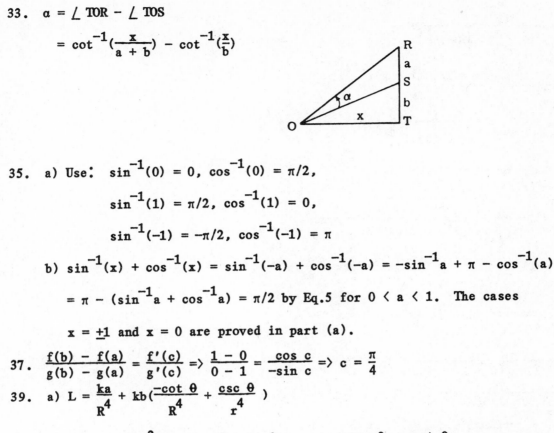

35. a) Use: $\sin^{-1}(0) = 0$, $\cos^{-1}(0) = \pi/2$,

$\sin^{-1}(1) = \pi/2$, $\cos^{-1}(1) = 0$,

$\sin^{-1}(-1) = -\pi/2$, $\cos^{-1}(-1) = \pi$

b) $\sin^{-1}(x) + \cos^{-1}(x) = \sin^{-1}(-a) + \cos^{-1}(-a) = -\sin^{-1}a + \pi - \cos^{-1}(a)$

$= \pi - (\sin^{-1}a + \cos^{-1}a) = \pi/2$ by Eq.5 for $0 < a < 1$. The cases

$x = \pm 1$ and $x = 0$ are proved in part (a).

37. $\dfrac{f(b) - f(a)}{g(b) - g(a)} = \dfrac{f'(c)}{g'(c)} \Rightarrow \dfrac{1 - 0}{0 - 1} = \dfrac{\cos c}{-\sin c} \Rightarrow c = \dfrac{\pi}{4}$

39. a) $L = \dfrac{ka}{R^4} + kb(\dfrac{-\cot \theta}{R^4} + \dfrac{\csc \theta}{r^4})$

$\dfrac{dL}{d\theta} = kb(\dfrac{\csc^2\theta_c}{R^4} - \dfrac{\csc \theta_c \cot \theta_c}{r^4}) = 0 \Rightarrow \dfrac{\csc \theta_c}{R^4} = \dfrac{\cot \theta_c}{r^4}$

$\Rightarrow \cos \theta_c = \dfrac{r^4}{R^4}$ or $\theta_c = \cos^{-1} r^4/R^4$

b) $\theta_c = \cos^{-1}(\dfrac{r^4}{R^4}) = \cos^{-1}(\dfrac{5}{6})^4 = \cos^{-1}(0.482) = 61^{\circ}\ 10'$

Article 6.3

1. $\cos^{-1}u = y$, $0 \le y \le \pi$

$u = \cos y$ and $\sin y = \sqrt{1 - u^2}$.

$\dfrac{du}{dx} = -\sin y \dfrac{dy}{dx} \Rightarrow \dfrac{dy}{dx} = -\dfrac{1}{\sin y}\dfrac{du}{dx} = -\dfrac{1}{\sqrt{1 - u^2}}\dfrac{du}{dx}$.

3. $\cot^{-1}u = y$, $-\pi/2 < y < \pi/2$. $u = \cot y$, $u^2+1 = \cot^2y + 1 = \csc^2y$

$\dfrac{du}{dx} = -\csc^2y\dfrac{dy}{dx} = -(u^2 + 1)\dfrac{dy}{dx} \Rightarrow \dfrac{dy}{dx} = \dfrac{-1}{u^2 + 1}\dfrac{du}{dx}$.

5. $\dfrac{dy}{dx} = \dfrac{-1}{\sqrt{1 - (x^2)^2}}\dfrac{d(x^2)}{dx} = -\dfrac{2x}{\sqrt{1 - x^2}}$

7. $\dfrac{dy}{dx} = \dfrac{1}{1 + (\sqrt{x})^2}\dfrac{d}{dx}(\sqrt{x}) = 1/(2(1 + x)x^{1/2})$

9. $y = 5\tan^{-1}3x \Rightarrow \dfrac{dy}{dx} = 5\left(\dfrac{1}{1 + (3x)^2}\right)\dfrac{d}{dx}(3x) = \dfrac{15}{1 + 9x^2}.$

11. $\dfrac{dy}{dx} = \dfrac{1}{2}\dfrac{1}{\sqrt{1 - (x^2/4)}} = \dfrac{1}{\sqrt{4 - x^2}}$

13. $\dfrac{d}{dx}(\sec^{-1}5x) = 5\dfrac{1}{|5x|\sqrt{25x^2 - 1}} = \dfrac{1}{|x|\sqrt{25x^2 - 1}}$

15. $\dfrac{d}{dx}(\csc^{-1}(x^2 + 1)) = \dfrac{-1}{|x^2 + 1|}\dfrac{1}{\sqrt{(x^2 + 1)^2 - 1}}\dfrac{d}{dx}(x^2 + 1)$

$= \dfrac{-2x}{(x^2 + 1)\sqrt{x^4 + 2x^2}}$, because $|x^2 + 1| = x^2 + 1$.

17. $\dfrac{dy}{dx} = -\dfrac{1}{|\sqrt{x + 1}|}\cdot\dfrac{1}{\sqrt{(x + 1) - 1}}\dfrac{d}{dx}(\sqrt{x + 1})$

$= -\dfrac{1}{\sqrt{x + 1}}\dfrac{1}{\sqrt{x}}\dfrac{1}{2}\left(\dfrac{1}{\sqrt{x + 1}}\right) = -1/(2(x + 1)x^{1/2})$

19. $\dfrac{dy}{dx} = \sqrt{1 - x^2} + \dfrac{x}{\sqrt{1 - x^2}}\left(\dfrac{1}{2}\right)(-2x) - \dfrac{-1}{\sqrt{1 - x^2}}$

$= \sqrt{1 - x^2} + \dfrac{1 - x^2}{\sqrt{1 - x^2}} = \sqrt{1 - x^2} + \sqrt{1 - x^2} = 2\sqrt{1 - x^2}$

21. $\dfrac{d}{dx}\left(\cot^{-1}\left(\dfrac{2}{x}\right) + \tan^{-1}\left(\dfrac{x}{2}\right)\right) = \dfrac{2/x^2}{1 + (4/x^2)} + \dfrac{1/2}{1 + (x^2/4)} = \dfrac{4}{4 + x^2}$

Alternatively, $y = \cot^{-1}\left(\dfrac{2}{x}\right) + \tan^{-1}\left(\dfrac{x}{2}\right) = \tan^{-1}\left(\dfrac{x}{2}\right) + \tan^{-1}\left(\dfrac{x}{2}\right) = 2\tan^{-1}\left(\dfrac{x}{2}\right)$

$\dfrac{dy}{dx} = 2\left(\dfrac{1}{2}\right)\dfrac{1}{1 + (x^2/4)} = \dfrac{4}{4 + x^2}.$

23. $\dfrac{d}{dx}\left(\tan^{-1}\dfrac{x - 1}{x + 1}\right) = \dfrac{2/(x + 1)2}{1 + [(x - 1)/(x + 1)]2} = \dfrac{1}{x^2 + 1}.$

25. $y = x(\sin^{-1}x)^2 - 2x + 2\sqrt{1 - x^2}\sin^{-1}x$

$\dfrac{dy}{dx} = \dfrac{2x\sin^{-1}x}{\sqrt{1 - x^2}} + (\sin^{-1}x)^2 - 2 + 2 - \dfrac{2x\sin^{-1}x}{\sqrt{1 - x^2}} = (\sin^{-1}x)^2$

27. $\displaystyle\int_0^{1/2}(1/\sqrt{1 - x^2}dx = \sin^{-1}x]_0^{1/2} = \dfrac{\pi}{6} - 0 = \dfrac{\pi}{6}$

29. $\displaystyle\int_{\sqrt2}^2 \dfrac{dx}{x\sqrt{x^2 - 1}} = \sec^{-1}x]_{\sqrt2}^2 = \left(\dfrac{\pi}{3}\right) - \left(\dfrac{\pi}{4}\right) = \dfrac{\pi}{12}.$

31. $\int \dfrac{dx}{\sqrt{1 - 4x^2}} = \dfrac{1}{2} \sin^{-1} 2x + C$

33. Let $u = x^2$, $du = 2x\,dx$. Then

$$\dfrac{x\,dx}{1 + x^4} = \dfrac{(1/2)\,du}{1 + u^2} \quad \text{and}$$

$$\int_{x=0}^{1} \dfrac{x}{1 + x^2}\,dx = \int_{u=0}^{1} \dfrac{(1/2)\,du}{1 + u^2} = \dfrac{1}{2} \tan^{-1} u]_0^1 = \dfrac{1}{2}(\dfrac{\pi}{4} - 0) = \dfrac{\pi}{8}.$$

35. $\int_0^1 \dfrac{4\,dx}{1 + x^2} = 4 \tan^{-1}]_0^1 = 4(\dfrac{\pi}{4} - 0) = \pi.$

37. Let $u = \sqrt{x}$, $du = \dfrac{dx}{2\sqrt{x}}$.

$$\int \dfrac{dx}{\sqrt{x}\sqrt{1 - x}} = \int \dfrac{2\,du}{\sqrt{1 - u^2}} = 2 \sin^{-1} \sqrt{x} + C$$

39. Let $u = (x - 1)^2$, $du = 2(x - 1)\,dx$.

$$\int \dfrac{2(x - 1)}{\sqrt{1 - (x - 1)^4}}\,dx = \int \dfrac{du}{\sqrt{1 - u^2}} = \sin^{-1} u + C = \sin^{-1}(x - 1)^2 + C.$$

41. $\lim\limits_{x \to 0} \dfrac{\sin^{-1} 2x}{x} = \lim\limits_{x \to 0} \dfrac{2/\sqrt{1 - 4x^2}}{1} = 2$

43. $\lim\limits_{x \to 0} \dfrac{\sin^{-1} x - x}{x^3} = \lim\limits_{x \to 0} \dfrac{(1 - x^2)^{-1/2} - 1}{3x^2}$

$= \lim\limits_{x \to 0} [\dfrac{1 - (1 - x^2)^{1/2}}{3x^2(1 - x^2)^{1/2}}] [\dfrac{1 + (1 - x^2)^{1/2}}{1 + (1 - x^2)^{1/2}}]$

$= \lim\limits_{x \to 0} \dfrac{1}{3[1 + (1 - x^2)^{1/2}](1 - x^2)^{1/2}} = \dfrac{1}{6}.$

45. $\int_0^b \dfrac{dx}{\sqrt{1 - x^2}} = \sin^{-1} x]_0^b = \sin^{-1} b - 0 = \sin^{-1} b$

$\lim\limits_{b \to 1^-} \int_0^b \dfrac{dx}{\sqrt{1 - x^2}} = \lim = 0.$

47. $\alpha = \cot^{-1}(\dfrac{x}{a + b}) - \cot^{-1}(\dfrac{x}{b})$. We find where $\dfrac{d\alpha}{dx} = 0$.

$\dfrac{d\alpha}{dx} = \dfrac{-1/(a + b)}{1 + (x/(a + b))^2} + \dfrac{1/b}{1 + (x/b)^2} = -\dfrac{a + b}{(a + b)^2 + x^2} + \dfrac{b}{b^2 + x^2} = 0$

$\dfrac{b^2 + x^2}{b} = \dfrac{(a + b)^2 + x^2}{a + b}$ $\quad \dfrac{x^2}{b} = a + \dfrac{x^2}{a + b}$ $\quad x = \sqrt{b(a + b)}$

49. $f(x) = \sin^{-1} x + \cos^{-1} x = \pi/2$ (see #35, Sec. 6-2).

a) $f'(x) = \dfrac{d}{dx}(\pi/2) = 0$

b) $f(0.32) = \dfrac{\pi}{2}.$

51. Yes. $\sin^{-1}x + \cos^{-1}x = \pi/2$. Hence

$$\sin^{-1}x = -\cos^{-1}x + \pi/2 \text{ or } \sin^{-1}x + c = -\cos^{-1}x + (c + \pi/2).$$

53. Using the washer method, let $R = 1/(1+x^2)^{1/2}$ be the outer radius and $r = x/\sqrt{2}$

be the inner radius. The curves intersect when $1/\sqrt{1 + x^2} = x/\sqrt{2}$, or $x = 1$.

Hence the volume is

$$\text{vol} = \int_{x=0}\pi(R^2 - r^2)dx = \int_0^1\pi(\frac{1}{1 + x^2} - \frac{x^2}{2})dx$$

$$= \pi(\tan^{-1}x - \frac{x^3}{6}]_0^1) = \pi(\frac{\pi}{4} - \frac{1}{6}) = \frac{\pi^2}{4} - \frac{\pi}{6}.$$

Article 6.4

1. $\frac{d}{dx}\ln 2x = \frac{1}{2x}\frac{d}{dx}(2x) = \frac{1}{x}.$

3. $\frac{d}{dx}\ln kx = \frac{1}{kx}\frac{d}{dx}(kx) = \frac{1}{kx}k = \frac{1}{x}.$

    Alternatively, $y = \ln x = \ln k + \ln x \Rightarrow \frac{dy}{dx} = 0 + \frac{1}{x}.$

5. $\frac{d}{dx}\ln(\frac{10}{x}) = \frac{1}{(10/x)}\frac{d}{dx}(\frac{10}{x}) = x(-\frac{1}{x^2}) = -\frac{1}{x}.$

    Alternatively, $y = \ln(\frac{10}{x}) = \ln 10 - \ln x \Rightarrow \frac{dy}{dx} = -\frac{1}{x}.$

7. $\frac{d}{dx}(\ln x)^3 = 3(\ln x)^2\frac{d}{dx}(\ln x) = \frac{3(\ln x)^2}{x}$

9. $\frac{d}{dx}\ln(\tan x + \sec x) = \frac{1}{\tan x + \sec x}\frac{d}{dx}(\tan x + \sec x)$

$$= \frac{\sec^2 x + \sec x \tan x}{\tan x + \sec x} = \sec x\frac{\sec x + \tan x}{\tan x + \sec x} = \sec x$$

11. $\frac{d}{dx}(x^3\ln 2x) = 3x^2\ln 2x + x^3(\frac{1}{2x})\frac{d}{dx}(2x) = 3x^2\ln 2x + x^2$

13. $\frac{d}{dx}\tan^{-1}(\ln x) = \frac{1}{1 + (\ln x)^2}\frac{d}{dx}\ln x = \frac{1}{1 + (\ln x)^2}\frac{1}{x}$

15. $\frac{d}{dx}(x^2\ln x^2) = 2x\ln x^2 + x^2(\frac{1}{x^2})dx(x^2) = 2x\ln x^2 + 2x.$

17. $\frac{d}{dx}(\ln(x^2 + 4) - x\tan^{-1}\frac{x}{2}) = \frac{2x}{x^2 + 4} - \tan^{-1}\frac{x}{2} - x\frac{1/2}{1 + (x/2)^2} = -\tan^{-1}\frac{x}{2}$

19. $\frac{d}{dx}(x(\ln x)^3) = (\ln x)^3 + 3x(\ln x)^2(\frac{1}{x}) = (\ln x)^3 + 3(\ln x)^2$

21. $y = x \sec^{-1} x - \ln(x + \sqrt{x^2 - 1})$, $x > 1$

$$\frac{dy}{dx} = x\left(\frac{1}{x\sqrt{x^2 - 1}}\right) - \frac{1 + x/\sqrt{x^2 - 1}}{x + \sqrt{x^2 - 1}} + \sec^{-1} x$$

$$\frac{1 + x/\sqrt{x^2 - 1}}{x + \sqrt{x^2 - 1}} \cdot \left(\frac{x - \sqrt{x^2 - 1}}{x - \sqrt{x^2 - 1}}\right) = \frac{x^2/\sqrt{x^2 - 1} - \sqrt{x^2 - 1}}{x^2 - (x^2 - 1)} = \frac{x^2 - (x^2 - 1)}{\sqrt{x^2 - 1}} = \frac{1}{\sqrt{x^2 - 1}}$$

Hence $\frac{dy}{dx} = \sec^{-1} x$.

23. $\int \frac{dy}{x} = \ln|x| + C$

25. $\int \frac{dx}{2x} = \frac{1}{2}\ln|x| + C$

27. Let $u = x + 1$, $du = dx$. Then,

$$\int_{x=0}^{1} \frac{dx}{x + 1} = \int_{u=1}^{2} \frac{du}{u} = \ln|u|\,\big]_1^2 = \ln 2 - \ln 1 = \ln 2.$$

29. Let $u = 2x + 3$, $du = 2dx$. Then,

$$\int \frac{dx}{2x + 3} = \int \frac{(1/2)du}{u} = \frac{1}{2}\ln|u| + C = \frac{1}{2}\ln|2x + 3| + C.$$

31. Let $u = 4x^2 + 1$, $du = 8x\,dx$. Then,

$$\int \frac{x\,dx}{4x^2 + 1} = \frac{1}{8}\int \frac{du}{u} = \frac{1}{8}\ln(4x^2 + 1) + C.$$

33. Let $u = \sin x$, $du = \cos x\,dx$. Then,

$$\int \frac{\cos x\,dx}{\sin x} = \int \frac{du}{u} = \ln|u| + C = \ln|\sin x| + C.$$

35. $\frac{x}{x + 1} = \frac{x + 1}{x + 1} - \frac{1}{x + 1} = 1 - \frac{1}{x + 1}.$

$$\int \frac{x\,dx}{x + 1} = \int 1\,dx - \int \frac{1}{x + 1}dx = x - \ln|x + 1| + C.$$

37. Let $u = 1 - x^2$, $du = -2x\,dx$. Then,

$$\int \frac{x\,dx}{1 - x^2} = -\frac{1}{2}\int \frac{du}{u} = -\frac{1}{2}\ln|u| + C = -\frac{1}{2}\ln|1 - x^2| + C.$$

39. Let $u = \ln x$, $du = dx/x$. Then,

$$\int (\ln x)^2 \frac{dx}{x} = \int u^2 du = \frac{u^3}{3} + C = \frac{(\ln x)^3}{3} + C.$$

41. $\int \tan x\,dx = \int \frac{\sin x}{\cos x}dx = \int \frac{1}{\cos x}\left(-\frac{d}{dx}\cos x\right) = -\ln|\cos x| + C.$

43. Let $u = \ln x$, $du = dx/x$. Then,

$$\int \frac{\ln x}{x}dx = \int u\,du = \frac{u^2}{2} + C = \frac{(\ln x)^2}{2} + C.$$

45. We need to find a function $f(x)$ such that $\int x \ln dx = f(x) + C$, or,

equivalently, $f'(x) = x \ln x$. Try $f(x) = x^2 \ln x$. Then, $f'(x) = 2x \ln x +$

$x^2(1/x) = 2x \ln x + x$. The factor of 2 and the x term must be eliminated.

Next try

$f(x) = \frac{1}{2}x^2 \ln x - \frac{x^2}{4} \Rightarrow f'(x) - x\ln x + \frac{x}{2} - \frac{x}{2} = x\ln x.$

Hence, $\int x\ln x\,dx = \frac{1}{2}x^2\ln x - \frac{x^2}{4} + C.$

47. $\lim\limits_{x\to\infty}\frac{\ln x}{x} = \lim\limits_{x\to\infty}\frac{1/x}{1} = 0.$

49. $\lim\limits_{\theta\to 0^+}\frac{\ln(\sin\theta)}{\cot\theta} = \lim\limits_{\theta\to 0^+}\frac{\cos\theta/\sin\theta}{-\cot\theta\csc\theta}.$

51. The space consists of disks of radius $r = 1/\sqrt{x}$ and volume $\pi r^2 dx$. The total

volume is

$\text{Vol} = \int_{x=1}^{2}\pi r^2 dx = \int_1^2 \pi\frac{dx}{x} = \pi\ln x]_1^2 = \pi(\ln 2 - \ln 1) = \pi\ln 2.$

53. $\frac{dy}{dx} = \frac{1 + (1/x)}{1 + (1/y)} = (1 + \frac{1}{y})dy = (1 + \frac{1}{x}dx.$ $y + \ln y = x + \ln x + C.$

$(x,y) = (1,1) \Rightarrow 1 + \ln 1 = 1 + \ln 1 + C \Rightarrow C = 0.$ Solution is $y + \ln y = x + \ln x.$

55. a) $f(x) = \ln(1 + x),\ f'(x) = 1/(1 + x),\ f''(x) = -1/(1 + x)^2$

$f(0) = 1, \qquad f'(0) = 1, \qquad f''(0) = -1$

linear approximation: $f(x) \approx f(0) + f'(0) = 0 + 1\cdot x = x$

quadratic approx.: $f(x) \approx f(0) + f'(0)\cdot x + \frac{f''(0)}{2!}x^2 = x - \frac{x^2}{2}.$

b) (13) $\ln(1+0.16) \approx 0.16$

(14) $\ln(1+0.16) \approx 0.16 - (0.16)^2/2 = 0.1472$

(13) $\ln(1-0.16) \approx -0.16$

(14) $\ln(1-0.16) \approx -0.16 - (0.16)^2/2 = -0.1728$

c) $\ln(1.16) = \int_1^{1.16}\frac{1}{x}dx \approx \frac{0.08}{3}(1 + 4(\frac{1}{1.08}) + (\frac{1}{1.16}) = 0.148420605$

$\ln(0.84) = \int_1^{0.84}\frac{1}{x}dx \quad \frac{-0.08}{3}(1 + 4(\frac{1}{0.92}) + (\frac{1}{0.84}) = -0.1743547274$

d) $\ln(1.16) = 0.1484200051$

$\ln(0.84) = -0.174353387.$

57. a) $(1 + u)(1 - u + u^2 - \frac{u^3}{1 + u}) = 1 - u + u^2 - u^3/(1 + u) + u - u^2 + u^3 - u^4/(1 + u)$

$= 1 + \frac{u^3(1 + u) - u^3 - u^4}{1 + u} = 1.$ Hence, $\frac{1}{1 + u} = 1 - u + u^2 - \frac{u^3}{1 + u}.$

b) Let $u = t - 1,\ du = dt.$ $t = 1 \Rightarrow u = 0,\ t = x \Rightarrow u = x - 1$

$\ln x = \int_{t=1}^{x}\frac{1}{t}dt = \int_{u=0}^{x-1}\frac{1}{u + 1}du$

c) $\ln x = \int_{u=0}^{x-1}\frac{1}{u + 1}du = \int_0^{x-1}(1 - u + u^2 - \frac{u^3}{1 + u})du$

$$= u - \frac{u^2}{2} + \frac{u^3}{3}\Big]_0^{x-1} - \int_0^{x-1} \frac{u^3}{1+u}du = (x-1) - \frac{(x-1)^2}{2} + \frac{(x-1)^3}{3} - R$$

d) $x > 1$, $0 \le u \le x - 1 \Rightarrow 1 + u \ge x > 1 \Rightarrow u^3/(1+u) \le u^3$.

$$R = \int_0^{x-1} \frac{u^3}{1+u}du \le \int_0^{x-1} u^3 du = \frac{u^4}{4}\Big]_0^{x-1} = \frac{(x-1)^4}{4}$$

e) Let $h(x) = (x-1) - (1/2)(x-1)^2 + (1/3)(x-1)^3$.

$\ln(x) = h(x) - R \ge h(x) - \frac{(x-1)^4}{4}$, because $-R \ge -\frac{(x-1)^4}{4}$.

Since $R > 0$, $\ln x < h(x)$, so $h(x)$ over estimates $\ln x$, but not by

more than $(x-1)^4/4$.

f) $\ln(1.2) \approx h(1.2) = 0.2 - \frac{(0.2)^2}{2} + \frac{(0.2)^3}{3} = 0.182667$.

This figure exceeds $\ln(1.2)$ by not more than $(0.2)^4/4 = 0.0004$.

Article 6.5

1. $\ln 16 = \ln 2^4 = 4 \ln 2 = 4a$

3. $\ln 2 \sqrt{2} = \ln 2^{3/2} = 3a/2$

5. $\ln(4/9) = \ln 4 - \ln 9 = 2 \ln 2 - 2 \ln 3 = 2a - 2b$

7. $\ln(9/8) = \ln 9 - \ln 8 = \ln 3^2 - \ln 2^3 = 2b - 3a$

9. $\ln 4.5 = \ln 9 - \ln 2 = \ln 3^2 - \ln 2 = 2b - a$

11. $y = \ln \sqrt{x^2 + 5} = \frac{1}{2}\ln(x^2 + 5) \Rightarrow \frac{dy}{dx} = \frac{x}{x^2 + 5}$

13. $y = \ln \frac{1}{x\sqrt{x+1}} = -\ln x - \frac{1}{2}\ln(x+1) \Rightarrow \frac{dy}{dx} = -\frac{1}{x} - \frac{1}{2(x+1)}$

15. $y = \ln(\sin x \sin 2x) = \ln \sin x + \ln \sin 2x$

$\Rightarrow \frac{dy}{dx} = \frac{\cos x}{\sin x} + 2\frac{\cos 2x}{\sin 2x} = \cot x + 2 \cot 2x$

17. $y = \ln(3x \sqrt{x+2}) = \ln 3 + \ln x + \frac{1}{2}\ln(x+2) \Rightarrow \frac{dy}{dx} = \frac{1}{x} + \frac{1}{2x+4}$

19. $y = \frac{1}{3}\ln\frac{x^3}{1+x^3} = \frac{1}{3}(\ln(x^3) - \ln(1+x^3)) = \ln x - \frac{1}{3}\ln(1+x^3) \Rightarrow \frac{dy}{dx} = \frac{1}{x} - \frac{x^2}{1+x^3}$

21. $y = \ln\frac{(x^2+1)^5}{\sqrt{1-x}} = 5\ln(x^2+1) - \frac{1}{2}\ln(1-x) \Rightarrow \frac{dy}{dx} = \frac{10x}{x^2+1} + \frac{1}{2-2x}.$

23. $2 \ln y = \ln x + \ln(x + 1) \Rightarrow \dfrac{dy'}{y} = \dfrac{1}{x} + \dfrac{1}{x + 1}$, $y' = \dfrac{y}{2}(\dfrac{1}{x} + \dfrac{1}{x + 1})$.

Also, $y' = \dfrac{2x + 1}{2\sqrt{x(x + 1)}}$.

25. $\ln y = \dfrac{1}{2}\ln(x + 3) + \ln \sin x + \ln \cos x \Rightarrow \dfrac{y'}{y} = \dfrac{1}{2(x + 3)} + \dfrac{\cos x}{\sin x} - \dfrac{\sin x}{\cos x}$

$= \dfrac{1}{2(x + 3)} + \cot x - \tan x$.

$y' = y(\dfrac{1}{2x + 6} + \cot x - \tan x)$. Also, $y' = y(\dfrac{1}{2x + 6} + 2 \cot 2x)$.

27. $\ln y = \dfrac{1}{3}[\ln x + \ln(x - 2) - \ln(x^2 + 1)] \Rightarrow \dfrac{y'}{y} = \dfrac{1}{3}[\dfrac{1}{x} + \dfrac{1}{x - 2} - \dfrac{2x}{x^2 + 1}]$

$\Rightarrow y' = \dfrac{y}{3}[\dfrac{1}{x} + \dfrac{1}{x - 2} - \dfrac{2x}{x^2 + 1}]$. Also, $y' = \dfrac{2y}{3}[\dfrac{x^2 + x - 1}{x(x - 2)(x^2 + 1)}]$.

29. $\ln y = \dfrac{1}{2}[\ln x + \ln(x + 1) + \ln(x - 2) - \ln(x^2 + 1) - \ln(2x + 3)]$

$\Rightarrow \dfrac{y'}{y} = \dfrac{1}{2}[\dfrac{1}{x} + \dfrac{1}{x + 1} + \dfrac{1}{x - 2} - \dfrac{2x}{x^2 + 1} - \dfrac{2}{2x + 3}]$

$\Rightarrow y' = \dfrac{1}{2}[\dfrac{1}{x} + \dfrac{1}{x + 1} + \dfrac{1}{x - 2}]y - \dfrac{xy}{x^2 + 1} - \dfrac{y}{2x + 3}$.

31. $\dfrac{1}{2}\ln y = 5\ln x + \ln \tan^{-1} x - \ln(3 - 2x) - \dfrac{1}{3}\ln x$

$\Rightarrow \dfrac{y'}{2y} = \dfrac{5}{x} + \dfrac{1}{\tan^{-1} x (1 + x^2)} + \dfrac{2}{3 - 2x} - \dfrac{1}{3x}$

$\Rightarrow y' = 2y[\dfrac{14}{3x} + \dfrac{2}{3 - 2x} + \dfrac{1}{(1 + x^2)\tan^{-1} x}]$.

33. Let $u = x + 3$, $du = dx$. Then,

$\int_{x=-1}^{1} \dfrac{dx}{x + 3} = \int_{x=-1}^{1} \dfrac{du}{u} = \ln|u|]_{x=-1}^{1} - \ln(x + 3)]_{-1}^{1} = \ln 4 - \ln 2 = 2 \ln 2 = \ln 2$.

35. $\int_{\pi/4}^{\pi/2} \dfrac{\cos x}{\sin x} dx = \ln|\sin x|]_{\pi/4}^{\pi/2} = \ln(\dfrac{1}{/2}) = \dfrac{1}{2}\ln 2$.

37. a) The graph of $y = \ln x$, $x > 0$ is the same as for $y = e^x$, reflected about the

line $y = x$. The graph of $y = \ln|x|$ is the graph of $y = \ln x$ with $x > 0$,

plus its reflection about the y-axis.

b) The graph of $y = |\ln x|$ is the same as for $y = \ln x$ with $x > 0$ except when 0

$< x < 1$, where it is the reflection of $y = \ln x$ about the x-axis.

39. $\int_{x}^{2x} \dfrac{dy}{t} = \ln t]_{x}^{2x} = \ln 2x - \ln x = (\ln x + \ln 2) - \ln x = \ln 2$.

Hence, $\lim\limits_{x \to \infty} \int_{x}^{2x} \dfrac{dt}{t} = \lim\limits_{x \to \infty} \ln 2 = \ln 2$.

41. a) $dA = (y - 0)dx = \tan x\, dx$. $0 \le x \le \dfrac{\pi}{3}$

$\Rightarrow$ Area $= \int_0^{\pi/3} \frac{\sin x}{\cos x} dx = -\ln \cos x]_0^{\pi/3} = \ln 1 - \ln(\frac{1}{2}) = \ln 2.$

b) Vol $= \int_{x=0}^{\pi/3} \pi y^2 dx = \int \pi \tan^2 x\, dx = \int \pi(\sec^2 x - 1)dx = \pi[\tan x - x]_0^{\pi/3} = \pi(\sqrt{3} - \frac{\pi}{3}).$

43. $dA = y\,ds = \ln x\, dx.$ $y_1 = \ln x,\ y_2 = 0,\ y_1 = y_2 \Rightarrow x = 1.$

Area $= \int dA = \int_{x=1}^{2} \ln x\, dx = x \ln x - x]_1^2 = 2 \ln 2 - 1$

$M_y = \int x\, dA = \int_{x=1}^{2} x \ln x\, dx = \frac{x^2}{2} \ln x - \frac{x^2}{4}]_1^2 = 2 \ln 2 - \frac{3}{4}$

$M_x = \int \frac{y}{2} dA = \int_{x=1}^{2} \frac{1}{2}(\ln x)^2 dx = \frac{1}{2}[x \ln x - 2x \ln x + 2x]_1^2 = \frac{1}{2}(2(\ln 2)^2 - 4 \ln 2 + 2)$

$\bar{x} = \frac{M_y}{\text{Area}} = \frac{2 \ln 2 - (3/4)}{2 \ln 2 - 1} \approx 1.65$

$\bar{y} = \frac{M_x}{\text{Area}} = \frac{2(\ln 2)^2 - 4 \ln 2 + 2}{2 \ln 2 - 1} \approx 0.24$

45. $y = \ln x - (x - 1) + \frac{1}{2}(x - 1)^2 \Rightarrow y' = \frac{1}{x} - 1 + (x - 1) = \frac{1}{x} + x - 2.$

$y' = 0$ when $x = 1.$ $y'' = -\frac{1}{x^2} + 1.$ $y'' > 0$ when $x > 1$, and $y'' < 0$ when

$0 < x < 1$, so there is an inflection point at $x = 1$. This implies that $y \approx 0$

near $x = 1$, or that $\ln x \approx x - 1 + \frac{1}{2}(x - 1)^2$ near $x = 1$.

47. $y = x \ln x;\ y' = 1 + \ln x;\ y'' = 1/x.$ $\ln x_0 = 1.$ Since $x > 0,\ y'' > 0$ and the

curve is concave up. At $x = 1/x_0,\ y' = 0$ and there is a relative minimum at

$(1/x_0,\ -1/x_0).$ (See Fig. 47)

49. a) $\ln 2 =$ Area under $y = 1/x \leq$ Area under $y = 1 = (2-1)(1) = 1.$ Hence, $\ln 2 \leq 1.$

b) $A_1 = 1/2,\ A_2 = 1/3,\ A_3 = 1/4$

$\ln 4 \geq A_1 + A_2 + A_3 = 13/12 > 1.$

Since $\ln x$ is continuous, the Intermediate Value Theorem can be applied.

$\ln 2 < 1$ and $\ln 4 > 1$, so there is a c on (2.4) with $\ln c = 1.$

(Fig. 47)

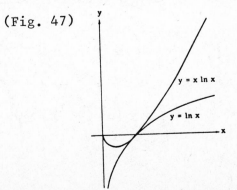

Article 6.6

1. $e^{\ln x} = x$

3. $e^{-\ln(x^2)} = e^{\ln(x^2)^{-1}} = e^{\ln(x^{-2})} = 1/x^2.$

5. $\ln(e^{1/x}) = 1/x$.

7. $e^{\ln(1/x)} = 1/x$.

9. $e^{\ln 2 + \ln x} = (e^{\ln 2})(e^{\ln x}) = 2x$.   Alternatively, $e^{\ln 2 + \ln x} = e^{\ln 2x} = 2x$.

11. $\ln(e^{x-x^2}) = x - x^2$.

13. $e^{x+\ln x} = e^x e^{\ln x} = xe^x$.

15. $\ln(e^{\sqrt{y}}) = \ln x^2 \Rightarrow \sqrt{y} = 2\ln x \Rightarrow y = 4(\ln x)^2$.

17. $e^{x^2} \cdot e^{(2x+1)} = e^y \Rightarrow x^2 + 2x + 1 = y$ or $y = (x+1)^2$.

19. $\ln(y - 2) = \ln(\sin x) - x \Rightarrow y - 2 = e^{(\ln(\sin x) - x)} = e^{-x}\sin x \Rightarrow y = e^{-x}\sin x + 2$.

21. $y = e^{3x} \Rightarrow dy/dx = e^{3x}(d/dx)(3x) = 3e^{3x}$.

23. $y = e^{5-7x} \Rightarrow dy/dx = -7e^{5-7x}$.

25. $y = x^2 e^x \Rightarrow dy/dx = 2xe^x + x^2 e^x$.

27. $y = e^{\sin x} \Rightarrow dy/dx = e^{\sin x}(d/dx)(\sin x) = \cos x\, e^{\sin x}$.

29. $y = \ln(3x\, e^{-x}) = \ln 3 + \ln x - x \Rightarrow dy/dx = 1/x - 1$.

31. $y = \ln\dfrac{e^x}{1 + e^x} = x - \ln(1 + e^x) \Rightarrow \dfrac{dy}{dx} = 1 - \dfrac{e^x}{1 + e^x} = \dfrac{1}{1 + e^x}$.

33. $y = (1/2)(e^x + e^{-x}) \Rightarrow dy/dx = (1/2)(e^x - e^{-x})$.

35. $\dfrac{dy}{dx} = e^{\sin^{-1}x}\dfrac{d}{dx}(\sin^{-1}x) = \dfrac{e^{\sin^{-1}x}}{\sqrt{1 - x^2}}$.

37. $y = (9x^2 - 6x + 2)e^{3x} \Rightarrow \dfrac{dy}{dx} = (18x - 6)e^{3x} + 3e^{3x}(9x^2 - 6x + 2) = 27\, x^2\, e^{3x}$.

39. $y = x^2 e^{-x^2} \Rightarrow \dfrac{dy}{dx} = 2xe^{-x^2} + x^2 e^{-x^2}\dfrac{d}{dx}(-x^2) = 2xe^{-x^2} - 2x^3 e^{-x^2}$.

41. $y = \tan^{-1}(e^x) \Rightarrow \dfrac{dy}{dx} = \dfrac{1}{1 + (e^x)^2}\dfrac{d}{dx}(e^x = \dfrac{e^x}{1 + e^{2x}}$.

43. $y = e^{1/x} \Rightarrow \dfrac{dy}{dx} = e^{1/x}\dfrac{d}{dx}(\dfrac{1}{x}) = -\dfrac{e^{1/x}}{x^2}$.

45. $y - \displaystyle\int_0^{\ln x} \sin e\, dt \Rightarrow \dfrac{dy}{dx} = \sin(e^{\ln x})\dfrac{d}{dx}(\ln x) = \dfrac{\sin x}{x}$.

47. $\ln xy = \ln x + \ln y = e^{x+y} \Rightarrow \dfrac{1}{x} + \dfrac{y'}{y} = e^{x+y}(1 + y')$

$\Rightarrow \dfrac{1}{x} - e^{x+y} = y'(e^{x+y} - \dfrac{1}{y}) \Rightarrow y' = \dfrac{xy\, e^{x+y} - x}{y - xy\, e^{x+y}}$.

49. $\tan y = e^x + \ln x \Rightarrow \sec^2 y \frac{dy}{dx} = e^x + \frac{1}{x} \Rightarrow \frac{dy}{dx} = \cos^2 y (e^x + \frac{1}{x}).$

51. Let $u = x^2$, $du = 2x\,dx$.

$$\int x e^{x^2}\,dx = \int e^u \frac{1}{2}\,du = \frac{1}{2}e^u + C = \frac{1}{2}e^{x^2} + C.$$

53. $\int e^{x/3}\,dx = 3e^{x/3} + C.$

55. $\int_0^2 e^{x/2}\,dx = 2e^{x/2}]_0^2 = 2(e-1) = 2e - 2$

57. $\int_0^1 \frac{dx}{e^x} = \int_0^1 e^{-x}\,dx = -e^{-x}]_0^1 = -(e^{-1}-1) = 1 - \frac{1}{e}$

59. Let $u = 1 + e^x$, $du = e^x\,dx$.

61. Let $u = \ln x$, $du = dx/x$.

$$\int_e^{e^2} \frac{dx}{x \ln x} = \int_{x=e}^{e^2} \frac{du}{u} = \ln|u|]_{x=e}^{e^2} = \ln 2.$$

63. Let $u = e^x$, $du = e^x\,dx$, $u^2 = e^{2x}$.

$$\int_{x=0}^{\ln 2} \frac{e^x\,dx}{1 + e^{2x}} = \int_{x=0}^{\ln 2} \frac{du}{1 + u^2} = \tan^{-1} u]_{x=0}^{\ln 2} = \tan^{-1} e^x]_0^{\ln 2}$$

$$= \tan^{-1} 2 - \tan^{-1} 1 = \tan^{-1} 2 - \frac{\pi}{4}.$$

65. $\lim_{h \to 0} \frac{e^h - (1+h)}{g^2} = \lim_{h \to 0} \frac{e^h - 1}{2h} = \lim_{h \to 0} \frac{e^h}{2} = \frac{1}{2}$

67. $\lim_{h \to 0} \frac{\sin x}{e^x - 1} = \lim_{h \to 0} \frac{\cos x}{e^x} = \frac{1}{1} = 1$

69. $\lim_{x \to \infty} x e^{-x} = \lim_{x \to \infty} \frac{x}{e^x} = \lim_{x \to \infty} \frac{1}{e^x} = 0$

71. $\lim_{x \to 0} \frac{e^x - 1}{x} = \lim_{x \to 0} \frac{e^x}{1} = 1.$ Let $f(x) = e^x$. Then

$$f'(0) = \lim_{x \to 0} \frac{f(x + 0) - f(0)}{x - 0} = \lim_{x \to 0} \frac{e^x - 1}{x}.$$

The limit is the derivative of $y = e^x$ at $x = 0$.

73. $y = \frac{\ln x}{x}$; $y' = \frac{1 - \ln x}{x^2}$; max at $x = 3$. $y'' = \frac{2 \ln x - 3}{x^3}$; inflection at $e = e^{3/2}$.

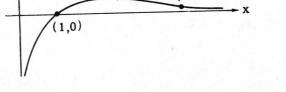

75. $y = e^{-t} \cos t$; $y' = -e^{-t}(\cos t + \sin t)$, $y'' = e^{-t}(2 \sin t)$. The curve is the same as for $y = \cos x$, except that the amplitude shrinks toward zero as x gets large, and expands toward infinity when x becomes negative.

77. $f(x) = \ln x - 1$, $f'(x) = \frac{1}{x}$. $x_{n+1} = s_n - \frac{f(x_n)}{f'(x_n)} = x_n - (\ln x_n - 1)x_n = 2x_n - x_n \ln x_n$. $x_0 = 3$, $x_1 = 2.704$, $x_2 = 2.7182451$, $x_3 = 2.718281828$.

79. a) $y = e^{\sin x}$, $-\pi \leq x \leq 2\pi$

$y' = \cos x \, e^{\sin x}$. $y' = 0 \Rightarrow \cos x = 0 \Rightarrow x = \pm \pi/2$ or $x = 3\pi/2$.

Absolute maxima and minima will occur either at an end point or at a critical point ($y' = 0$).

End points: $e^{\sin(-\pi)} = 1$; $e^{\sin(2\pi)} = 1$.

Critical points: $e^{\sin(-\pi/2)} = e^{\sin(3\pi/2)} = 1/e$; $e^{\sin(\pi/2)} = e$.

Thus local and absolute minima occur at $x = -\pi/2$ and $x = 3\pi/2$, and the local and absolute maximum occurs at $x = \pi/2$.

b) The curve is similar to the curve $y = \sin x$, except that its peaks are sharper, its trough more shallow, and its extremes are $1/e$ and e instead of -1 and 1.

81. $f(x) = x^2 \ln(1/x) = -x^2 \ln x$; $f'(x) = -[2x \ln x + x]$

$f'(x) = 0 \Rightarrow \ln x = -1/2 \Rightarrow x = e^{-1/2}$.

$f''(x) = -[2 \ln x + 2x(1/x) + 1] = -[2 \ln x + 3]$

$f''(e^{-1/2}) = -[2(-1/2) + 3] = -2 < 0$. Hence $f(e^{-1/2}) = 1/2e$ is a maximum.

83. $\text{Vol} = \int \pi y^2 dx = \int \pi (e^{-x})^2 dx = \int_{x=1}^{\ln 10} \pi e^{-2x} dx = \frac{\pi}{-2}[e^{-2x}]_1^{\ln 10} = \frac{\pi}{2}(e^{-2} - 0.01)$

85. a) $y = Ce^{ax}$, $dy/dx = Cae^{ax} = ay$. Hence $y = Ce^{ax}$ is a solution.

b) Use $a = -2$, $y = Ce^{-2t}$. $y = 3$ when $t = 0$ implies $3 = Ce^0 = C$.

The solution is $y = 3e^{-2t}$.

87. $\frac{dy}{dx} = e^{-x} \Rightarrow y = -e^{x} + C.$ $(x,y) = (4,0) \Rightarrow 0 = -e^{-4} + C \Rightarrow C = e^{-4}.$

Solution is $y = -e^{-x} + e^{-4}.$

89. $\frac{dy}{y+1} = \frac{dx}{2x} \Rightarrow \ln|y+1| = \frac{1}{2}\ln|x| + C.$ $(x,y) = (2,1) \Rightarrow \ln 2 = \frac{1}{2}\ln 2 + C$

$\Rightarrow C = \frac{1}{2}\ln 2.$ Solution is $(y+1)^2 = 2x.$

91. a) $dx = dt/(t+2) \Rightarrow x = \ln|t+2| + C_1$

$(x,t) = (\ln 2, 0) \Rightarrow \ln 2 = \ln 2 + C_1 \Rightarrow C_1 = 0,$

$x = \ln(t+2)$ or $t = e^x - 2.$

$dy = 2t\,dt \Rightarrow y = t^2 + c_2$ $(y,t) = (1,0) \Rightarrow$

$1 = 0 + c_2 \Rightarrow c_2 = 1$

$y = t^2 + 1$ or $t = \sqrt{y-1}$

b) $y = t^2 + 1 = (e^x - 2)^2 + 1 = e^{2x} - 4e^x + 5.$

c) $x = \ln(t+2) = \ln(\sqrt{y-1} + 2).$

d) $x(0) = \ln 2$ $y(0) = 1$

$x(2) = \ln 4$ $y(2) = 5.$

Average change in y is $\frac{y(2) - y(0)}{x(2) - x(0)} = \frac{5-1}{\ln 4 - \ln 2} = \frac{4}{\ln 2}.$

e) At $t = 1, \frac{dx}{dt} = \frac{1}{t+2} = \frac{1}{3}, \frac{dy}{dt} = 2t = 2, \frac{dy}{dx} = \frac{(dy/dt)}{(dx/dt)} = 6.$

93. $f(x) = \frac{2}{\sqrt{\pi}}\int_0^x e^{-t^2} dt \Rightarrow f'(x) = \frac{2}{\sqrt{\pi}}e^{-x^2}$, by the Fundamental Theorem of Calculus.

$f''(x) = -\frac{4x}{\sqrt{\pi}}e^{-x^2}.$

$f(0) = \frac{2}{\sqrt{\pi}}\int_0^0 e^{-t^2} dt = 0.$ $f'(0) = \frac{2}{\sqrt{\pi}}e^0 = \frac{2}{\sqrt{x}}.$

$f''(0) = -\frac{40)}{\sqrt{\pi}}e^{-1} = 0.$

$f(x) \approx f(0) + f'(0)(x-0) + \frac{f''(0)}{2}(x=0)^2 = \frac{2x}{\sqrt{\pi}}.$

95. $\frac{d}{dx}(\cosh x) = \frac{d}{dx}(\frac{e^x + e^{-x}}{2}) = \frac{e^x - e^{-x}}{2} = \sinh x.$

$\frac{d}{dx}(\sinh x) = \frac{d}{dx}(\frac{e^x - e^{-x}}{2}) = \frac{e^x + e^{-x}}{2} = \cosh x.$

97. $\cosh(-x) = (\frac{e^x + e^{-x}}{2}) = \frac{e^x + e^{-x}}{2} = \cosh x.$

$\sinh(-x) = (\frac{e^x - e^{-x}}{2}) = \frac{e^x - e^{-x}}{2} = \sinh x.$

Article 6.7

1. $y = 2^x \Rightarrow dy/dx = 2^x \ln 2$.

3. $y = 8^x \Rightarrow dy/dx = 8^x \ln 8 = 8^x(3 \ln 2)$.

   Alternatively, $y = 8^x = 2^{3x} \Rightarrow dy/dx = 2^{3x}(3 \ln 2) = 8^x(3 \ln 2)$.

5. $y = 9^x = 3^{2x} \Rightarrow dy/dx = 2(\ln 3)3^{2x}$.

7. $y = x^{\sin x} \Rightarrow \ln y = \sin x \cdot \ln x$

   $\dfrac{y'}{y} = \cos x \ln x + \dfrac{\sin x}{x} \Rightarrow \dfrac{dy}{dx} = x^{\sin x}(\cos x \ln x + \dfrac{\sin x}{x})$.

9. $y = 2^{\sec x} \Rightarrow \ln y = \ln(2^{\sec x}) = \sec x \cdot \ln 2$

   $\dfrac{1}{y}\dfrac{dy}{dx} = \sec x \tan x (\ln 2) \Rightarrow \dfrac{dy}{dx} = y \sec x \tan x (\ln 2) = 2^{\sec x} \sec x \tan x (\ln 2)$.

11. $\ln y = x \ln \cos x \Rightarrow \dfrac{y'}{y} = \ln \cos x + x\dfrac{(-\sin x)}{\cos x} \Rightarrow \dfrac{dy}{dx} = y(\ln \cos x - x \tan x)$.

13. $y = x \cdot 2^{(x^2)} \Rightarrow \ln y = \ln x + x^2 \ln 2$

   $\Rightarrow \dfrac{y'}{y} = \dfrac{1}{x} + 2x \ln 2 \Rightarrow \dfrac{dy}{dx} = x \cdot 2^{(x^2)}(\dfrac{1}{x} + 2x \ln 2) = 2^{(x^2)}(1 + 2x^2 \ln 2)$.

15. $y = (\cos x)^{\sqrt{x}} \Rightarrow \ln y = \sqrt{x} \ln \cos x \Rightarrow \dfrac{1}{x}\dfrac{dy}{dx} = \dfrac{1}{2\sqrt{x}}\ln \cos x + \sqrt{x}\dfrac{(-\sin x)}{\cos x}$

   $\Rightarrow \dfrac{dy}{dx} = (\cos x)^{\sqrt{x}}(\dfrac{\ln \cos x}{2\sqrt{x}} - \sqrt{x}\tan x)$.

17. $\displaystyle\int 3^{2x}dx = \dfrac{3^{2x}}{2 \ln 3} + C$.

19. $\displaystyle\int 5^{-x}dx = -\dfrac{5^{-x}}{\ln 5} + C$.

21. $\displaystyle\int_0^1 \dfrac{1}{2^x}dx = \int_0^1 2^{-x}dx = -\dfrac{2^{-x}}{\ln 2}\Big]_0^1 = -\dfrac{1}{\ln 2}(\dfrac{1}{2} - 1) = \dfrac{1}{2 \ln 2}$.

23. $\displaystyle\int_0^{\ln 2} e^{-2x}dx = \dfrac{e^{-2x}}{-2}\Big]_0^{\ln 2} = -\dfrac{1}{2}[\dfrac{1}{4} - 1] = \dfrac{3}{8}$.

25. Let $u = -x^2$; $du = -2x\, dx$.

   $\displaystyle\int_{x=1}^2 x\, 2^{(-x^2)}dx = \int_{x=1}^2 -\dfrac{1}{2}2^u du = -\dfrac{1}{2}2^{(-x^2)}\Big]_1^{\sqrt{2}} = \dfrac{1}{4}$.

27. Let $u = \ln x$; $du = dx/x$

   $\displaystyle\int_{x=1}^{e^2} \dfrac{2 \ln x}{x}dx = \int_{x=1}^{e^2} 2u\, du = u^2\Big]_{x=1}^{e^2} = (\ln x)^2\Big]_{x=1}^{e^2} = 4 - 0 = 4$.

29. Let $u = -\sin\theta$, $du = -\cos\theta\, d\theta$

$$\int_{\theta=0}^{\pi/6} (\cos\theta)4^{-\sin\theta}d\theta = \int_{\theta=0}^{\pi/6} -4^u du = -\frac{4^{-\sin\theta}}{\ln 4}\Big]_0^{\pi/6} = \frac{1}{2\ln 4}.$$

31. a) $2^x = 4^x = 2^{(2x)} \Rightarrow x = 2x \Rightarrow x = 0.$

    b) $x^x = 2^x \Rightarrow x\ln x = x\ln 2 \Rightarrow \ln x = \ln 2 \Rightarrow x = 2.$

    c) $x\ln 3 = (x+1)\ln 2 \Rightarrow \ln 2 = x(\ln 3 - \ln 2) = x\ln(3/2) \Rightarrow x = (\ln 2)/\ln(3/2).$

    d) $-x\ln 4 = (x+2)\ln 3 \Rightarrow -2\ln 3 = x(\ln 3 + \ln 4) = x\ln 12$

    $$x = -\frac{2\ln 3}{\ln 12}.$$

33. a) $3^x = (e^{\ln 3})^x = e^{x\ln 3} = f(g(x))$, where $g(x) = x\ln 3$ and

    $f(y) = e^y$ are the functions.

    b) $\lim_{x\to 3} 3^x = 3^3 = 27$ (because $y = 3^x$ is continuous)

35. $\lim_{x\to -\infty} 3^x = \lim_{x\to\infty} 3^{-x} = \lim_{x\to\infty} \frac{1}{3^x} = 0.$

37. $\lim_{x\to 1} \ln x^{1/(x-1)} = \lim_{x\to 1}\frac{\ln x}{x-1} = \lim_{x\to 1}\frac{1/x}{2} = 1$ (L'Hopital's rule).

    Hence $\lim_{x\to 1} x^{1/(x-1)} = e.$

39. $\lim_{x\to\infty} \ln(x^{1/x}) = \lim_{x\to\infty}\frac{\ln x}{x} = \lim_{x\to\infty}\frac{1/x}{1} = 0$ (1'Hopital's rule).

    Hence, $\lim_{x\to\infty} x^{1/x} = e^0 = 1.$

41. a) $\dfrac{a^{x_1}}{a^{x_2}} = a^{(x_1 - x_2)} > 1$, because $x_1 - x_2 > 0$, and $a > 1$.

    Hence $a^{x_1} > a^{x_2}$ (this uses $a^{x_2} > 0$).

    b) $y = a^x \Rightarrow y' = (\ln a)a^x \Rightarrow y'' = (\ln a)^2 a^x > 0.$

    $y'' > 0$, so $y = a^x$ is concave up everywhere.

    c) At $x = 0$, $y = 1 > 0$. Suppose $y > 0$ for some x. Then since $y = a^x$ is

    continuous the Intermediate Value Theorem says that $y_0 = a^{x_0}$ is zero for

    some $x_0 \neq 0$. That is, $0 = a^{x_0}$. But $\ln a^{x_0} = x_0 \ln a$ exists, and $\ln(0)$

    does not. The contradiction proves that $y = a^x > 0$ everywhere.

    d) $y = a^x \Rightarrow y' = (\ln a)a^x$. $y'(0) = \ln a$. At (x, a^x) the slope is $y' = (\ln a)a^x$, so the slope is proportional to the ordinate $y = a^x$ and the factor is ln a, the slope at $x = 0$.

e) $\lim\limits_{x \to -\infty} a^x = \lim\limits_{x \to -\infty} a^{-x} = \lim\limits_{x \to -\infty} 1/a^x = 0,$

because $a > 1$. Thus y approaches the x-axis as $x \to -\infty$.

43. a) $f(x) = 2^x - x^2$; $f'(x) = 2^x \ln 2 - 2x$.

$f'(x) = 0 \Rightarrow 2^x \ln 2 = 2x$.

$y = 2^x \ln 2 \Rightarrow y' = 2^x (\ln 2)^2$; $y'' = 2^x (\ln 2)^3 > 0$.

$y = 2x$ is a straight line with slope 2.

b) $f'(x)$ can have at most two zeros, so $f(x)$ can have at most three zeros.

c) The zeros of $f'(x)$ are 0.485 and 3.2124.

45. The function $y = \ln u$ is monotonically increasing. So if x_0 maximizes $f(x)$, then x_0 also maximizes $\ln f(x)$.

a) $g(x) = \ln(x^{1/x}) = (\ln x)/x$. Therefore, $g'(x) = [(1/x)x - \ln x]/x^2$

$= (1 - \ln x)/x^2$.

$g'(x) = 0 \Rightarrow 1 - \ln x = 0 \to x = e$. $g'' = [-x - 2x(1 - \ln x)]/x^4$.

Since $g''(e) = -1/e^3$ is negative, $x = e$ is a max.

b) $g(x) = \ln(x^{1/x^2}) = (\ln x)/x^2$ and $g'(x) = (1 - 2\ln x)/x^3$.

$g'(x) = 0 \Rightarrow \ln x = 1/2 \Rightarrow x = e^{1/2}$. $g''(x) = [-2 - 3(1 - 2\ln 3)]/x^4$.

Since $g''(e^{1/2}) = -2/e^2 < 0$, $x = e^{1/2}$ is a max.

c) $g(x) = \ln(x^{1/x^n}) = (\ln x/x^n)$, $g'(x) = (1 - n\ln x)/x^{n+1}$.

$g'(x) = 0 \Rightarrow \ln x = 1/n$, $x = e^{1/n}$. $g''(x) = [-n - (1 + n)(1 - n\ln x)]/x^{n+2}$

$g''(e^{1/n}) = -n/e^{(1+(2/n))} < 0$, so there is a maximum at $x = e^{1/n}$.

47. a,b > 0, so the following manipulations are valid:

$(ab)^u = e^{\ln(ab)^u} = e^{u \ln(ab)} = e^{u(\ln a + \ln b)} = e^{(u \ln a + u \ln b)}$

$= e^{(\ln a^u + \ln b^u)} = \qquad (e^{\ln a^u})(e^{\ln b^u}) = a^u b^u$.

49. $(x^x)^x = x^{(x^x)} \Rightarrow x \ln(x^x) = (x^x)\ln x \Rightarrow x^2 \ln x = x^x \ln x$. If $x = 1$ this becomes $0 = 0$. If $x \neq 1$, $x^2 = x^x \Rightarrow 2 \ln x = x \ln x \Rightarrow x = 2$. Thus, the original equation is true for $x = 1$ and $x = 2$.

Article 6.8

1. a) $\log_4 16 = \log_4 (4)^2 = 2$

   b) $\log_8 32 = \log_8 (8)^{5/3} = 5/3$

   c) $\log_5 0.04 = \log_5 5^{-2} = -2$

   d) $\log_{0.5} 4 = \log_{0.5} (0.5)^{-2} = -2$

3. $3^{(\log_3 7)} + 2^{(\log_2 5)} = 5^{(\log_5 x)} \Rightarrow 7 + 5 = x \Rightarrow x = 12.$

5. $y = \log_4 x^2 = 2 \log_4 x \Rightarrow dy/dx = 2/x \ln 4 = 1/x \ln 2.$

7. $y = \log_5 \sqrt{x} = (1/2)\log_5 x \Rightarrow dy/dx = 1/2x \ln 5.$

9. $y = \log_2 (1/x) = -\log_2 x \Rightarrow dy/dx = -1/x \ln 2.$

11. $y = \ln 10^x \quad x \ln 10 \Rightarrow dy/dx = \ln 10.$

13. $y = \log_2 (\ln x) \Rightarrow \dfrac{dy}{dx} = \dfrac{1}{\ln 2 \ \ln x} \dfrac{d}{dx}(\ln x) = \dfrac{1}{x = \ln 2 \ \ln x}.$

15. $y = e^{\log_{10} x} \Rightarrow \dfrac{dy}{dx} = e^{\log_{10} x} \dfrac{d}{dx} \log_{10} x = \dfrac{e^{\log_{10} x}}{x \ln 10}.$

    Alternatively, $y = e^{\log_{10} x} = e^{\left(\frac{\ln x}{\ln 10}\right)} = (e^{\ln x})^{\frac{1}{\ln 10}} = x^{(1/\ln 10)}$

    $\dfrac{dy}{dx} = \dfrac{1}{\ln 10} x^{((1/\ln 10)-1)} = \dfrac{e^{\log_{10} x}}{x \ln 10}.$

17. a) $\log_{10} 20 = \log_{10} 2 + \log_{10} 10 = \log_{10} 2 + 1 = 1.30103$

    $\log_{10} 200 = \log_{10} 2 + 2 \log_{10} 10 = \log_{10} 2 + 2 = 2.30103$

    $\log_{10} 0.2 = \log_{10} 2 - \log_{10} 10 = 0.30103 - 1 = -0.69897$

    $\log_{10} 0.02 = \log_{10} 2 - 2 \log_{10} 10 = 0.30103 - 2 = -1.69897$

    b) $\ln 20 = \ln 2 + \ln 10 = 2.99574$

    $\ln 200 = \ln 2 + 2 \ln 10 = 5.29833$

    $\ln 0.2 = \ln 2 - \ln 10 = -1.60944$

    $\ln 0.02 = \ln 2 - 2 \ln 10 = -3.91202$

19. $\log_2 x = (\ln x)/\ln 2, \quad \log_3 x = (\ln x)/\ln 3.$

Thus, $\lim\limits_{x \to \infty} \dfrac{\log_2 x}{\log_3 x} = \lim\limits_{x \to \infty} \dfrac{(\ln x)/(\ln 2)}{(\ln x)/(\ln 3)} = \dfrac{\ln 3}{\ln 2}.$

21. Note: "Faster" indicates that the function grows faster than $y = x^2 - 1$, likewise for "slower."

a) $\lim\limits_{x \to \infty} \dfrac{x^2 + 4x}{x^2 - 1} = \lim\limits_{x \to \infty} \dfrac{2x + 4}{2x} - \lim\limits_{x \to \infty} \dfrac{2}{2} = 1.$ Same.

b) $\lim\limits_{x \to \infty} \dfrac{x^3 + 3}{x^2 - 1} = \lim\limits_{x \to \infty} \dfrac{6x}{2} = \infty.$ Faster.

c) $\lim\limits_{x \to \infty} \dfrac{x^5}{x^2 - 1} = \lim\limits_{x \to \infty} \dfrac{20x^3}{2} = \infty.$ Faster.

d) $\lim\limits_{x \to \infty} \dfrac{15x + 3}{x^2 - 1} = \lim\limits_{x \to \infty} \dfrac{15}{2x} = 0.$ Slower.

e) $x^2 < \sqrt{x^4 + 5x} < x^2 + 5x.$

$\lim\limits_{x \to \infty} \dfrac{x^2}{x^2 - 1} = \lim\limits_{x \to \infty} \dfrac{2x}{2x} = 1$

$\lim\limits_{x \to \infty} \dfrac{x^2 + 5x}{x^2 - 1} - \lim\limits_{x \to \infty} \dfrac{2x + 5}{2x} = \lim\limits_{x \to \infty} \dfrac{2}{2} = 1.$

$y = x^2 - 1$ grows at the same rate as x^2 and $x^2 + 5x$, so it grows at the same rate as $\sqrt{x^4 + 5x}.$

f) $\lim\limits_{x \to \infty} \dfrac{(x + 1)^2}{x^2 - 1} = \lim\limits_{x \to \infty} \dfrac{2(x + 1)}{2x} = \lim\limits_{x \to \infty} \dfrac{2}{2} = 1.$ Same.

g) $\lim\limits_{x \to \infty} \dfrac{\ln x}{x^2 - 1} = \lim\limits_{x \to \infty} \dfrac{\frac{1}{x}}{2x^2} = 0.$ Slower.

h) $\lim\limits_{x \to \infty} \dfrac{\ln(x^2)}{x^2 - 1} = \lim\limits_{x \to \infty} \dfrac{2 \ln x}{x^2 - 1} - 0.$ Slower.

i) $\lim\limits_{x \to \infty} \dfrac{\ln(10^x)}{x^2 - 1} = \lim\limits_{x \to \infty} \dfrac{x \ln 10}{x^2 - 1} = \lim\limits_{x \to \infty} \dfrac{\ln 10}{2x} = 0.$ Slower.

$y = 0.1 = e^{kt} = e^{-[(\ln 2)/5700]t} \Rightarrow \ln(0.1) = -[(\ln 2)/5700]t$

$\Rightarrow t = (5700 \ln 10)/\ln 2 \approx 18{,}935$ years.

c) $y = 0.445 = e^{-[(\ln 2)/5700]t} \Rightarrow \ln(0.445) = -[(\ln 2)/5700]t$

$\Rightarrow t = -(5700 \ln(0.445))/\ln 2 \approx 6{,}658$ years.

9. Let y be the ratio of the level of radioactivity to its level when the polonium is created. $y = y_0 e^{kt}$. At $t = 0$ level is 100%, or $y = 1$, so $y_0 = 1$. $y = 1/2$ when $t = 140$, so

$1/2 = e^{140k} \Rightarrow k = [\ln(1/2)]/140$

$y = 0.90 = e^{kt} = e^{[(\ln(1/2))/140]t} \Rightarrow t = 140[\ln(0.9)]/\ln(0.5) \approx 21.28$ days

11. $y = y_0(0.9)^t$, $y_0 = 10{,}000$

$y = 1{,}000 = 10{,}000(0.9)^t \Rightarrow (0.9)^t = 0.1$

$t = \log_{(0.9)}(0.1) = [\ln(0.1)]/\ln(0.9) \approx 21.9$ years.

13. By Eq.(19) of the text, $i = \dfrac{V}{R}(1 - e^{-Rt/L})$.

Steady state is $i = V/R$, and half of this is achieved when

$1 - e^{-Rt/V} = \dfrac{1}{2}$ or $\dfrac{-Rt}{L} = \ln(\dfrac{1}{2}) = -\ln 2 \Rightarrow t = \dfrac{L \ln 2}{R}$.

15. a) $L\dfrac{di}{dt} + Ri = 0 \Rightarrow \dfrac{di}{i} = -\dfrac{R}{L}dt \Rightarrow \ln i = -\dfrac{R}{L}t + D$

$\Rightarrow i = (e^D)e^{(-Rt/L)} = i_0 e^{(-Rt/L)}$.

b) $i = \dfrac{1}{2}i_0 \Rightarrow e^{(-Rt/L)} = \dfrac{1}{2} \Rightarrow \dfrac{-Rt}{L} = \ln(\dfrac{1}{2}) = -\ln 2 \Rightarrow t = \dfrac{L \ln 2}{R}$.

c) $t = \dfrac{L}{R} \rightarrow i = i_0 e^{((-R/L)(L/R))} = \dfrac{i_0}{e}$.

Article 6-10

1. $A = A_0(1.04)^t = 2A_0 \Rightarrow \ln(1.04)^t = \ln 2 \Rightarrow t = (\ln 2)/\ln(1.04) \approx 17.67$.

3. i) If $r = 5$, then $A = A_0 e^{0.05t} = 2A_0$

$\Rightarrow 0.05t = \ln 2 \approx 0.70 \Rightarrow t \approx \dfrac{0.70}{0.05} = \dfrac{70}{5} = 14$.

ii) In general, $A = A_0 e^{(r/100)t} = 2A_0 \Rightarrow rt/100 = \ln 2 \approx 0.70 \Rightarrow t_0 \approx \dfrac{70}{r}$

is the number of years the balance takes to double. Thus in 70 years, or $t_0 r$ years, the balance doubles r times. That is, it increases by a factor of 2^r

5. a) To compound an interest rate of 100% a total of n times in t years is equivalent to multiplying it by the factor $(1 + (t/n))$ a total of n times in the period of t years. Money stock grows according to the formula $M = M_0 (1 + (t/n))^n$. To compound continuously is to let $n \to \infty$. $\lim_{n \to \infty} (1 + (t/n))^n = e^t$. Therefore, $M = M_0 e^t$.

b) $M_0 e^t = 3M_0 \Rightarrow t = \ln 3 \approx 1.1$ years ≈ 401 days.

c) $M_0 e^1 \approx 2.72 M_0$, which is 172 percent of holdings.

CHAPTER 6

Miscellaneous Problems

1. $y = \sin 2x$

   $y' = 2 \sin 2x$

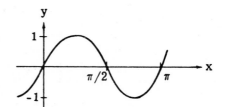

3. $y = 2 \sin x + \sin 2x$

   $y' = 2 \cos x + 2 \cos 2x$

5. $x = \cos y, \quad y = \cos^{-1} x \Rightarrow y' = \dfrac{-1}{\sqrt{1 - x^2}}$

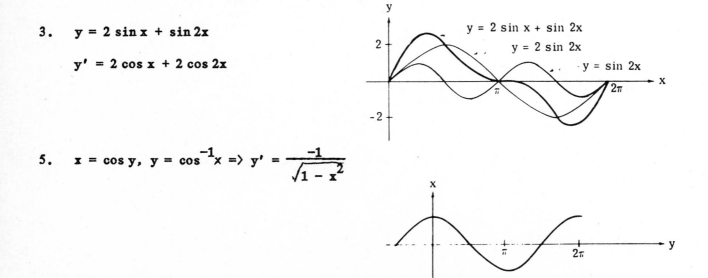

7. $y = 4 \sin((x/2) + \pi)$

   $y' = -2 \cos(x/2)$

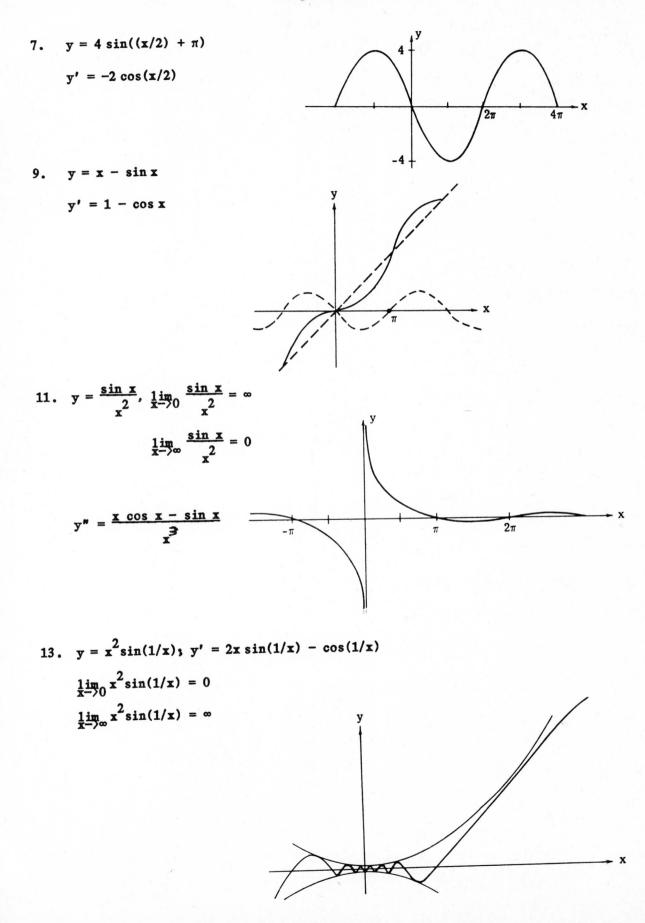

9. $y = x - \sin x$

   $y' = 1 - \cos x$

11. $y = \dfrac{\sin x}{x^2}$, $\lim\limits_{x \to 0} \dfrac{\sin x}{x^2} = \infty$

   $\lim\limits_{x \to \infty} \dfrac{\sin x}{x^2} = 0$

   $y'' = \dfrac{x \cos x - \sin x}{x^3}$

13. $y = x^2 \sin(1/x)$; $y' = 2x \sin(1/x) - \cos(1/x)$

   $\lim\limits_{x \to 0} x^2 \sin(1/x) = 0$

   $\lim\limits_{x \to \infty} x^2 \sin(1/x) = \infty$

CHAPTER 6 - Miscellaneous Problems

15. $y = \dfrac{e^x - e^{-x}}{e^x + e^{-x}}$

$y' = \dfrac{4}{(e^x + e^{-x})^2}$

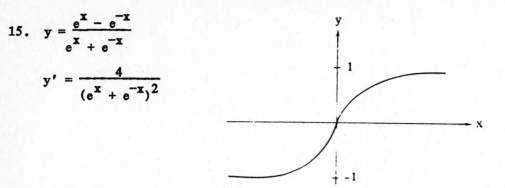

17. a) $y = x \ln x$

$y' = \ln x + 1$

$y'' = 1/x > 0$ for $x > 0$

$y' = 0$ when $x = e^{-1}$

b) $y = \sqrt{x} \ln x$

$y' = \dfrac{\ln x}{2\sqrt{x}} + \dfrac{1}{\sqrt{x}}$

$y'' = -\ln x/4x^{3/2}$

If $0 < x < 1$, $y'' > 0$, concave up.

If $x > 1$, $y'' < 0$, concave down.

If $x = e^{-2}$, $y' = 0$, relative min.

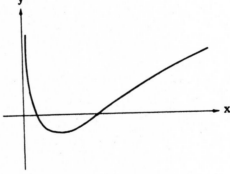

19. $y = e^{-x} \sin 2x$

$y' = e^{-x}(2 \cos 2x - \sin 2x)$

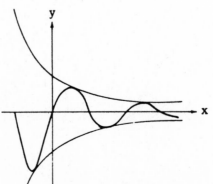

21. a) $y = x - e^x$; $y' = 1 - e^x$; $y'' = -e^x$

   Relative maximum at $(0,-1)$.

   Curve looks similar to inverted parabola.

   Curve tends toward $y = x$ for $x < 0$ and toward $y = e^x$ for $x > 0$.

   b) $y = e^{(x - e^x)}$

   $y' = (1 - e^x)e^{(x - e^x)}$

   $y'' = (1 - 3e^x + e^{2x})e^{(x - e^x)}$

   $\lim_{t \to \pm\infty} x - e^x = -\infty$.

   Hence, $\lim_{x \to \infty} e^{(x - e^x)} = \lim_{n \to -\infty} e^n = 0$

23. $y = \frac{1}{2}\ln(\frac{1 + x}{1 - x})$

   $y'' = 1/(1 - x^2)$

   $y'' = 2x/(1 - x^2)^2$

   Domain is $|x| < 1$. $y' > 0$ for all x. $y'' > 0$ if $x > 0$.

   $y'' < 0$ if $x < 0$.

   $y = \ln[(1 + x)/(1 - x)]/2$

25. $y = x\tan^{-1}(x/2)$; $y' = \tan^{-1}(\frac{x}{2}) + \frac{2x}{4 + x^2}$

   $y'' = 16/(4 + x^2)^2$. Relative min at $(0,0)$.

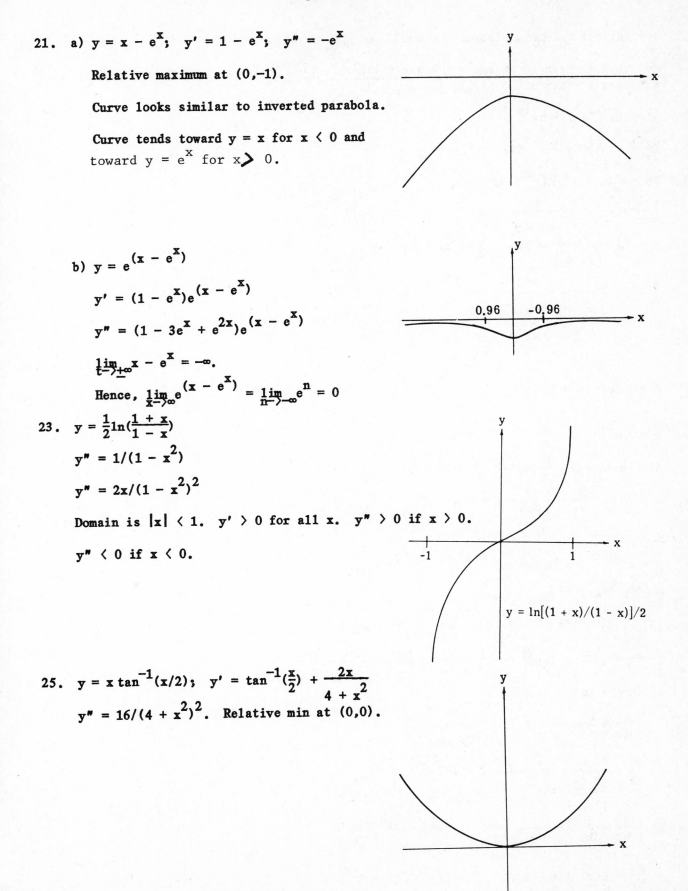

CHAPTER 6 - Miscellaneous Problems

27. $y = \ln[(\sec x + \tan x)/(\sec x - \tan x)]$

$$\frac{dy}{dx} = \frac{\sec x - \tan x}{\sec x + \tan x} \cdot \frac{2\sec^3 x - 2\sec x \tan^2 x}{(\sec x - \tan x)^2} = 2\sec x.$$

29. $y = \sin^{-1}(x^2) - xe^{(x^2)}$

$$\frac{dy}{dx} = \frac{2x}{\sqrt{1 - x^4}} - e^{x^2}(2x^2 + 1)$$

31. $\ln y = (2 - x)\ln(x^2 + 2)$

$$\frac{1}{y}\frac{dy}{dx} = (2 - x)\frac{2x}{x^2 + 2} - \ln(x^2 + 2)$$

33. $y = \ln \dfrac{x}{\sqrt{x^2 + 1}} = \ln x - \dfrac{1}{2}\ln(x^2 + 1)$

$$\frac{dy}{dx} = \frac{1}{x} - \frac{x}{x^2 + 1} = \frac{1}{x(x^2 + 1)}$$

35. $y = \ln(3x^2 + 4x)$ $\quad \dfrac{dy}{dx} = \dfrac{6x + 4}{3x^2 + 4x} = \dfrac{2(3x + 2)}{x(3x + 4)}$

37. $y = x\ln(x^3) = 3x\ln x.$ $\quad \dfrac{dy}{dx} = 3\ln x + 3$

39. $y = \ell n\, e^x = x.$ $\quad \dfrac{dy}{dx} = 1.$

Alternatively, $\dfrac{dy}{dx} = \dfrac{1}{e^x}\dfrac{d}{dx}e^x = \dfrac{e^x}{e^x} = 1.$

41. $x = \ln y.$ $\quad \dfrac{dx}{dx} = \dfrac{1}{y}\dfrac{dy}{dx} \Rightarrow \dfrac{dy}{dx} = y = e^x.$

43. $y = \ln(2xe^{2x}) = 2x + \ln 2 + \ln x.$ $\quad \dfrac{dy}{dx} = 2 + \dfrac{1}{x}$

45. $y = x^{\tan 3x} \Rightarrow \ln y = \ln x^{\tan 3x} = \tan 3x \ln x.$

47. $\dfrac{d}{dx}e^{\sec x} = e^{\sec x}\sec x \tan x$

49. $\ln(x - y) = e^{xy} \Rightarrow \dfrac{1}{x - y}(1 - \dfrac{dy}{dx}) = e^{xy}(y + x\dfrac{dy}{dx})$

$$\frac{1}{x - y} - ye^{xy} = (xe^{xy} + \frac{1}{x - y})\frac{dy}{dx} \Rightarrow \frac{dy}{dx} = \frac{1 - (x - y)ye^{xy}}{1 + (x - y)xe^{xy}}.$$

51. a) $\ln y = \ln x - \ln(x^2 + 1)$

$$\frac{1}{y}\frac{dy}{dx} = \frac{1}{x} - \frac{2x}{x^2 + 1} = \frac{1 - x^2}{x(x^2 + 1)}$$

$$\frac{dy}{dx} = \frac{x}{x^2 + 1}\frac{1 - x^2}{x(x^2 + 1)} = \frac{1 - x^2}{x(x^2 + 1)}$$

b) $\dfrac{3}{y}\dfrac{dy}{dx} = \dfrac{1}{x} + \dfrac{1}{x - 2} - \dfrac{2x}{x^2 + 1} = \dfrac{2x^2 + 2x - 2}{x(x - 2)(x^2 + 1)}$

$$\frac{dy}{dx} = 2(x^2 + x - 1)/[3(x^2 - 2x)^{2/3}(x^2 + 1)^{4/3}]$$

CHAPTER 6 - Miscellaneous Problems

c) $\ln y = \ln(2x - 5) + \frac{1}{2}\ln(8x^2 + 1) - 2\ln(x^3 + 2)$

$$\frac{1}{y}\frac{dy}{dx} = \frac{2}{2x - 5} + \frac{8x}{8x^2 + 1} - \frac{6x^2}{x^3 + 2}$$

$$\frac{dy}{dx} = y[\frac{2}{2x - 5} + \frac{8x}{8x^2 + 1} - \frac{6x^2}{x^3 + 2}]$$

53. a) $y = a^{(x^2 - x)}$. $\frac{dy}{dx} = a^{(x^2 - x)}(2x - 1)\ln a$

b) $y = \ln\frac{e^x}{1 + e^x} = x - \ln(1 + e^x)$. $\frac{dy}{dx} = 1 - \frac{e^x}{1 + e^x} = \frac{1}{1 + e^x}$.

c) $y = x \ln x - x \Rightarrow \frac{dy}{dx} = \ln x + x(\frac{1}{x}) - 1 = \ln x$

d) $\ln y = \frac{1}{x}\ln x \Rightarrow \frac{1}{y}\frac{dy}{dx} = -\frac{\ln x}{x^2} + \frac{1}{x^2}$

$$\Rightarrow \frac{dy}{dx} = x^{1/x}(\frac{1}{x^2} - \frac{\ln x}{x^2}) = x^{(1 - 2x)/x}(1 - \ln x).$$

55.

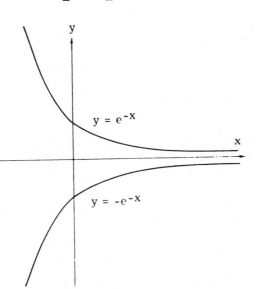

$y = e^{-x}$

$y = -e^{-x}$

57. $\frac{dy}{dx} = \frac{e^x - e^{-x}}{e^x + e^{-x}}$. $\frac{d^2y}{dx^2} = \frac{(e^x + e^{-x})^2 - (e^x - e^{-x})^2}{(e^x + e^{-x})^2}$

$$= \frac{(e^{2x} + 2 + e^{-2x}) - (e^{2x} - 2 + e^{-2x})}{(e^x + e^{-x})^2} = \frac{4}{(e^x + e^{-x})^2}$$

Finding y: Let $u = e^x + e^{-x}$, $du = (e^x - e^{-x})dx$. Then,

$$y = \int dy = \int \frac{e^x - e^{-x}}{e^x + e^{-x}}dx = \int \frac{du}{u} = \ln|u| + C = \ln(e^x + e^{-x}) + C$$

59. Assume that $x_2 > x_1$ and $y_1 > y_2$:

$$A_1 = \int_{x_1}^{x_2}(R/x)dx = R \ln x]_{x_1}^{x_2} = R \ln(\frac{x_2}{x_1})$$

CHAPTER 6 - Miscellaneous Problems

$$A_2 = \int_{y_2}^{y_1}(R/y)\,dy = R\ln y\Big]_{y_2}^{y_1} = R\ln\left(\frac{y_1}{y_2}\right)$$

Since $x_1 = R/y_1$ and $x_2 = R/y_2$, $\frac{x_2}{x_1} = \frac{y_1}{y_2}$.

It follows that $A_1 = A_2$.

61. $\frac{dE}{dt} = -\frac{E}{40}$. $\int_{E_0}^{E}\frac{dE}{E} = -\int_0^t\frac{dt}{40}$

$\ln(E/E_0) = -t/40$; $E = E_0 e^{-t/40}$; $\frac{E_0}{10} = E_0 e^{-t/40}$

$\Rightarrow \frac{t}{40} = \ln 10$, t is about 92.1 sec.

63. We use

1) Distance from $(0,0)$ to (x,y) is $s = \int_0^x[1 + (\frac{dy}{dx})^2]^{1/2}dx$

2) $\frac{d}{dx}\int_0^x f(x)\,dx = f(x)$

$\quad s = e^x + y - 1 = \int_0^x[1 + (dy/dx)^2]^{1/2}dx \Rightarrow e^x + \frac{dy}{dx} = [1 + (\frac{dy}{dx})^2]^{1/2}$

$\quad\quad \Rightarrow (e^x)^2 + 2e^x\frac{dy}{dx} + (\frac{dy}{dx})^2 = 1 + (\frac{dy}{dx})^2 \Rightarrow \frac{dy}{dx} = \frac{1 - e^{2x}}{2e^x} = \frac{1}{2e^x} - \frac{e^x}{2}$

$y = \int dy = \int\frac{1}{2}(e^{-x} - e^x)\,dx = -\frac{1}{2}(e^{-x} + e^x) + C$

$(x,y) = (0,0) \Rightarrow C = 1$. Hence $y = 1 - \frac{1}{2}(e^x + e^{-x})$

$$(\text{Also, } y = 1 - \cosh x.)$$

65. $\frac{dv}{dt} = \frac{4}{(4 - t)^2}$. $v = \int\frac{4}{(4 - t)^2}\,dt = \frac{4}{4 - t} + C$

$v_0 = 2 \Rightarrow C = 1$. $\frac{ds}{dt} = \frac{4}{4 - t} + 1$

$s = \int_1^2(\frac{4}{4 - t} + 1)\,dt = -4\ln|4 - t| + t]_1^2$

$\quad = 1 + 4\ln(3/2) \approx 1 + (4)(0.4055) \approx 2.62$

67. $\frac{dy}{y^2} = e^{-x}dx \Rightarrow -\frac{1}{y} = -e^{-x} + C$

$(x,y) = (0,2) \Rightarrow c = \frac{1}{2} \Rightarrow y = 2e^x/(2 - e^x)$

69. $y = (e^{2x} - 1)/(e^{2x} + 1)$

$\frac{dy}{dx} = \frac{2e^{2x}(e^{2x} + 1) - 2e^{2x}(e^{2x} - 1)}{(e^{2x} + 1)^2} = \frac{2e^{2x}}{e^{2x} + 1}(1 - \frac{e^{2x} - 1}{e^{2x} + 1})$

$\quad = \frac{e^{2x} + 1 + e^{2x} - 1}{e^{2x} + 1}(1 - y) = (1 + y)(1 - y) = 1 - y^2.$

71. $V = \pi\int_0^{\pi/3}\sec^2 y\,dy = \pi\tan y]_0^{\pi/3} = \sqrt{3}\,\pi$

73. $V = \pi \int_0^2 e^{2x} dx = (\frac{\pi}{2}) e^{2x}]_0^2 = \frac{\pi}{2}(e^4 - 1) = 84.19$

75. The tangent to the curve at (x,y) also goes through $(-x,0)$, because the y-axis bisects the cord from $(-x,0)$ to (x,y). The slope at (x,y), dy/dx_1, is the slope of the tangent:

$\frac{dy}{dx} = \frac{\Delta y}{\Delta x} = \frac{y - 0}{x - (-x)} = \frac{y}{2x}$. Hence $2\int \frac{dy}{y} = \int \frac{dx}{x}$ and

$2 \ln y = \ln x + C$. $(x,y) = (1,2) \Rightarrow C = 2 \ln 2 = \ln 4$.

$2 \ln y = \ln y^2 = \ln x + \ln 4 = \ln 4x \Rightarrow y^2 = 4x$.

77. $\int 5 \, dx/(x - 3) = 5 \ln|x - 3| + C$

79. $\int_0^2 x \, dx/(x^2 + 2)^2 = -(1/2)(x^2 + 2)^{-1}]_0^2 = \frac{1}{6}$

81. $\frac{x}{2x + 1} = \frac{1}{2}(\frac{2x}{2x + 1}) = \frac{1}{2}(\frac{2x + 1}{2x + 1} - \frac{1}{2x + 1}) = \frac{1}{2}(1 - \frac{1}{2x + 1})$.

$\int \frac{x}{2x + 1} dx = \int \frac{1}{2}(1 - \frac{1}{2x + 1}) dx = \frac{1}{2}x - \frac{1}{4}\ln|2x + 1| + C$.

83. Let $u = x^2 - 1$, $du = 2x \, dx$. Then,

$\int_{x=0}^3 x e^{(x^2 - 1)} dx = \frac{1}{2}\int_{x=0}^3 e^u du = \frac{1}{2}e^{(x^2 - 1)}]_{x=0}^3 = \frac{e^8 - e^{-1}}{2}$

85. $\int_0^5 \frac{x \, dx}{x^2 + 1} = \frac{1}{2}\ln|x^2 + 1|]_0^5 = \ln \sqrt{26}$

87. $\int \frac{\sec^3 x + e^{\sin x}}{\sec x} dx = \int \sec^2 x \, dx + \int e^{\sin x} \cos x \, dx$

$= \tan x + e^{\sin x} + C$

89. $\lim_{t \to \infty} \frac{t e^t}{4 - 4e^t} = \lim_{t \to \infty} \frac{t}{4e^{-t} - 4} = \infty$.

91. $\lim_{x \to \infty} \ln(e^x + x)^{1/x} = \lim_{x \to \infty} \frac{\ln(e^x + x)}{x} = \lim_{x \to \infty} \frac{((e^x + 1)/(e^x + x))}{1} = 1$.

Hence, $\lim_{x \to \infty} (e^x + x)^{1/x} = e^1 = e$.

93. We must show that for all n,

$\lim_{x \to \infty} \frac{(\ln x)^n}{x} = 0$.   (1)

Case of $n = 1$: $\lim_{x \to \infty} \frac{(\ln x)^1}{x} = \lim_{x \to \infty} \frac{(1/x)}{1} = 0$.

Assume true for case $n = k$:

$\lim_{x \to \infty} \frac{(\ln x)^k}{x} = 0$.   (2)

Show true for case $n = k + 1$:

$$\lim_{x \to \infty} \frac{(\ln x)^{k+1}}{x} = \lim_{x \to \infty} \frac{(k + 1)(\ln x)^k(1/x)}{1},$$

by l'Hopital's rule,

$$= (k + 1)(\lim_{x \to \infty} \frac{(\ln x)^k}{x}) = 0, \text{ by (2)}.$$

Thus, if (1) holds for $n = k$, then it holds for $n = k + 1$. Since it holds for $n = 1$, it holds for $n = 1,2,3,\ldots$ That is, it holds for all n.

95. $x = \ln t \Rightarrow t = e^x$. $y = t \sin(t - 1) = e^x \sin(e^x - 1)$

$A = \int y\,dx = \int_{x=0}^{\ln(\pi+1)} e^x \sin(e^x - 1)dx = -\cos(e^x - 1)]_0^{\ln(\pi+1)} = 2.$

97. $V = 100\pi y$; $\frac{dV}{dt} = ky = 100\pi(\frac{dy}{dt})$; $100\pi \ln y = kt + C.$

If $\ln y_0 = 20$, $C = 100\pi \ln 20.$

If $y_1 = 20 - (1/20)20 = 18$, $k = 100\pi \ln(9/10)$

$100\pi \ln y = 100\pi t \ln(9/10) + 100\pi \ln 20$

$y/20 = (9/10)^t$

$V = 100\pi 20(9/10)^t = 2000\pi(9/10)^t$

99. a) $\frac{\ln x}{x} = \frac{\ln 2}{2} \Rightarrow 2\ln x = x \ln 2 \Rightarrow$

$\ln x^2 = \ln 2^x$ or $x^2 = 2^x$. $x = 2$ or $x = 4.$

b) $\frac{\ln x}{x} = -2 \ln 2 \Rightarrow \ln x^{-1/2} = \ln 2x$

$x^{-1/2} = 2^x \Rightarrow x = 1/2.$

101. The definition of the derivative of $f(x)$ at x_0 is

$$f'(x_0) = \lim_{h \to 0} \frac{f(x_0 + h) - f(x_0)}{h}.$$

Hence, $\frac{d}{dx}e^x\big|_{x=0} = \lim_{h \to 0} \frac{e^{0+h} - e^0}{h} = \lim_{h \to 0} \frac{e^h - 1}{h} = e^0 = 1,$

because $\frac{d}{dx}e^x\big|_{x=0} = e^0 = 1.$

103. $0 < x < 1 \Rightarrow 1 < x+1 < 2 \Rightarrow \frac{1}{2} < \frac{1}{x + 1} < \frac{1}{1} \Rightarrow \frac{x}{2} < \frac{x}{x + 1} < x.$

Since $\frac{x}{x + 1} = 1 - \frac{1}{x + 1} < x$, $\frac{x}{2} < 1 - \frac{1}{x + 1} < x.$

Hence, $\int_0^x (\frac{x}{2})dx < \int_0^x (1 - \frac{1}{x + 1})dx < \int_0^x x\,dx.$

$$\frac{x^2}{4} < x - \ln(x + 1) < \frac{x^2}{2}.$$

CHAPTER 6 - Miscellaneous Problems

105. Let $x = \frac{1}{n}$, $x_k = k \triangle x$, $f(x) = e^x$.

$$L_n = \frac{e^{1/n} + e^{2/n} + \ldots + e^{n/n}}{n} = \sum_{k=1}^{n} f(x_k) \triangle x$$

$$\lim_{n \to \infty} L_n - \lim_{n \to \infty} \sum_{k=1}^{n} f(x_k) \triangle x = \lim_{\triangle x \to 0} \sum_{k=1}^{n} f(x_k) \triangle x$$

$$= \int_{x=0}^{1} f(x)dx = \int_{0}^{1} e^x dx = e^1 - 1.$$

107. Let $q = 0.99$, $k = -\ln q > 0$, $a = -\ln k > 0$.

$$y = N(1 - q^x + \frac{1}{x}) \Rightarrow y' = N(-q^x \ln q - \frac{1}{x^2}) = N(kq^x - \frac{1}{x^2})$$

$$y' > 0 \Longleftrightarrow kq^x > \frac{1}{x^2} \Longleftrightarrow \ln k + x \ln q > -2 \ln x$$

$$\Longleftrightarrow -a - kx > -2 \ln x$$

$$\Longleftrightarrow \ln x > \frac{k}{2}x + \frac{a}{2}$$

Also, $y' < 0 \Longleftrightarrow \ln x > \frac{k}{2}x + \frac{a}{2}$.

The straight line $y = \frac{k}{2}x + \frac{a}{2}$ intersects the curve $y = \ln x$ in exactly two places (sketching helps to see this). Between these points $y > 0$, and outside $y' < 0$. Thus the sign of y' can change only twice. Once the extreme values have been checked ($x = 1$, and $x = 1000$), it suffices to show that y' changes sign near $x = 11$ and $x = 895$ to show that these points are a minimum and a maximum, respectively.

$y(1) = N(1 - (0.99)^1 + 1) = N(1.01)$; $y(1000) = N(1.000956)$.

$y(10) = N(0.19562)$ $y(894) = N(1.000993293)$

$y(11) = N(0.19557)$ $y(895) = N(1.000993296)$

$y(12) = N(0.19694)$ $y(896) = N(1.000993290)$

We see that the sign of y' changes near $x = 11$, and $x = 895$. By inspecting the values of x and y, we see that the integer value that minimizes y is $x = 11$ and the integer value that maximizes y is $x = 895$.

CHAPTER 7

INTEGRATION METHODS

Article 7.1

1. $u = 3x^2 + 5$, $du = 6x\ dx$:

$$\int 6x\sqrt{3x^2 + 5}\ dx = \int \sqrt{u}\ du = \frac{2}{3}u^{3/2} + C = \frac{2}{3}\left(3x^2 + 5\right)^{3/2} + C$$

3. $u = x^2$, $du = 2x\ dx$; when $x = 0$, $u = 0$; when $x = \sqrt{\ln 2}$, $u = \ln 2$.   Then,

$$\int_0^{\sqrt{\ln 2}} xe^{x^2}\ dx = \int_0^{\ln 2} \frac{1}{2}e^u\ du = \frac{1}{2}e^u \Big|_0^{\ln 2} = \frac{1}{2}$$

5. $u = 8x^2 + 1$, $du = 16x\ dx$: $\displaystyle\int_0^1 \frac{x\ dx}{\sqrt{8x^2 + 1}} = \int_1^9 \frac{1}{16}\cdot\frac{du}{\sqrt{u}} = \frac{1}{8}\sqrt{u}\Big|_1^9 = \frac{1}{4}$

7. $u = e^x$, $du = e^x\ dx$:

$$\int e^x \sec^2 (e^x)\ dx = \int \sec^2 u\ du = \tan u + C = \tan (e^x) + C$$

9. $u = 3 + 4e^x$, $du = 4e^x\ dx$:

$$\int e^x\sqrt{3 + 4e^x}\ dx = \frac{1}{4}\int \sqrt{u}\ du = \frac{1}{6}u^{3/2} + C = \frac{1}{6}\left(3 + 4e^x\right)^{3/2} + C$$

11. $u = \ln x$, $du = \frac{dx}{x}$: $\displaystyle\int \frac{dx}{x\ \ln x} = \int \frac{du}{u} = \ln u + C = \ln (\ln x) + C$

13. $u = \sqrt{x}$, $du = \frac{dx}{\sqrt{x}}$ or $dx = \sqrt{x}\ du = u\ du$.   When $x = 0$, $u = 0$; when $x = 1$, $u = 1$.
    Therefore,

$$\int_0^1 e^{\sqrt{x}}\ dx = \int_0^1 ue^u\ du.$$

As an antiderivative of ue^u, guess $f(u) = ue^u$.   Then, $\frac{df}{du} = ue^u + e^u$.   To
eliminate the e^u term, try $f(u) = ue^u - e^u$:

$$\frac{df}{du} = (ue^u + e^u) - e^u = ue^u.$$

So,

$$\int_0^1 ue^u\ du = (ue^u - e^u)\Big|_0^1 = 1$$

15. $u = 2x + 3$, $du = 2\ dx$:

$$\int \sqrt{2x + 3}\ dx = \frac{1}{2}\int \sqrt{u}\ du = \frac{1}{2}\left(\frac{2}{3}u^{3/2}\right) + C = \frac{1}{3}u^{3/2} + C = \frac{1}{3}(2x + 3)^{3/2} + C$$

(formula 4)

17. $u = 2x - 7$, $du = 2\ dx$: $\quad \int \frac{dx}{(2x-7)^2} = \frac{1}{2}\int \frac{du}{u^2} = -\frac{1}{2u} + C = -\frac{1}{4x-14} + C$

(formula 4)

19. $u = 2 + \cos x$, $du = -\sin x\ dx$:

$$\int \frac{\sin x}{2 + \cos x}\ dx = -\int \frac{du}{u} = -\ln |u| + C = -\ln (2 + \cos x) + C$$

21. $u = 1 - 4x^2$, $du = -8x\ dx$:

$$\int \frac{x\ dx}{\sqrt{1 - 4x^2}} = -\frac{1}{8}\int \frac{du}{\sqrt{u}} = -\frac{1}{4}\sqrt{u} + C = -\frac{1}{4}\sqrt{1 - 4x^2} + C$$

Or: $u = \sqrt{1 - 4x^2}$, $du = \frac{-8x}{2\sqrt{1 - 4x^2}}\ dx$:

$$\int \frac{x\ dx}{\sqrt{1 - 4x^2}} = -\frac{1}{4}\int du = -\frac{1}{4}u + C = -\frac{1}{4}\sqrt{1 - 4x^2} + C$$

23. By formula 13: $\displaystyle \int_0^{\sqrt{2}/2} \frac{dx}{\sqrt{1 - x^2}} = \sin^{-1} x \Big|_0^{\sqrt{2}/2} = \frac{\pi}{4} - 0 = \frac{\pi}{4}$

25. $u = 1 + x^2$, $du = 2x\ dx$; when $x = 0$, $u = 1$, and when $x = 1$, $u = 2$. So,

$$\int_0^1 \frac{3x}{1 + x^2}\ dx = \frac{3}{2}\int_1^2 \frac{du}{u} = \frac{3}{2}(\ln 2 - \ln 1) = \frac{3}{2}\ln 2$$

27. $\displaystyle \int_0^{\pi/4} \frac{\sin^2 2x}{1 + \cos 2x}\ dx = \int_0^{\pi/4} \frac{1 - \cos^2 2x}{1 + \cos 2x}\ dx = \int_0^{\pi/4} \frac{(1 - \cos 2x)(1 + \cos 2x)}{1 + \cos 2x}\ dx$

$$= \int_0^{\pi/4} (1 - \cos 2x)\ dx = \frac{\pi}{4} - \int_0^{\pi/4} \cos 2x\ dx$$

To evaluate $\int_0^{\pi/4} \cos 2x\ dx$, set $u = 2x$, $du = 2\ dx$:

$$\int_0^{\pi/4} \cos 2x \, dx = \frac{1}{2}\int_0^{\pi/2} \cos u \, du = \frac{1}{2} \sin u \Big|_0^{\pi/2} = \frac{1}{2} \ .$$

So,

$$\int_0^{\pi/4} \frac{\sin^2 2x}{1 + \cos 2x} \, dx = \frac{\pi}{4} - \frac{1}{2} \ .$$

29. Let $u = 2x$, $du = 2 \, dx$: $\displaystyle\int \frac{2 \, dx}{\sqrt{1 - 4x^2}} = \int \frac{du}{\sqrt{1 - u^2}} = \sin^{-1} u + C = \sin^{-1} 2x + C$

31. $u = 3x^2 + 4$, $du = 6x \, dx$. Then,

$$\int \frac{x \, dx}{\left(3x^2 + 4\right)^3} = \frac{1}{6}\int \frac{du}{u^3} = \frac{1}{6}\left(-\frac{1}{2u^2}\right) + C = -\frac{1}{12} \frac{1}{\left(3x^2 + 4\right)^2} + C$$

33. $u = x^3 + 5$, $du = 3x^2 \, dx$. Then, $\displaystyle\int \frac{x^2 \, dx}{\sqrt{x^3 + 5}} = \frac{1}{3}\int \frac{du}{\sqrt{u}} = \frac{1}{3}(2\sqrt{u}) + C = \frac{2}{3}\sqrt{x^2 + 5} + C$

35. $u = e^{2x}$, $du = 2e^{2x} \, dx$. Then, $\displaystyle\int e^{2x} \, dx = \frac{1}{2}\int du = \frac{1}{2}u + C = \frac{e^{2x}}{2} + C$

37. $$\int \frac{dx}{e^{3x}} = \int e^{-3x} \, dx$$

$u = -3x$, $du = -3 \, dx$. Then,

$$\int e^{3x} \, dx = -\frac{1}{3}\int e^u \, du = -\frac{1}{3}e^u + C = -\frac{1}{3}e^{-3x} + C$$

39. $u = e^x$, $du = e^x \, dx$. Then,

$$\int \frac{e^x \, dx}{1 + e^{2x}} = \int \frac{du}{1 + u^2} = \tan^{-1} u + C = \tan^{-1} (e^x) + C$$

41. $u = \cos x$, $du = -\sin x \, dx$. Then,

$$\int \cos^2 x \sin x \, dx = -\int u^2 \, du = -\frac{u^3}{3} + C = -\frac{\cos^3 x}{3} + C$$

43. $u = \cot x$, $du = -\csc^2 x \, dx$. Then,

$$\int \cot^3 x \csc^2 x \, dx = -\int u^3 \, du = -\frac{u^4}{4} + C = -\frac{\cot^4 x}{4} + C$$

45. $u = e^{2x} - e^{-2x}$, $du = (e^{2x} + e^{-2x})\,dx$. Then,

$$\int \frac{e^{2x} + e^{-2x}}{e^{2x} - e^{-2x}}\,dx = \int \frac{du}{u} = \ln|u| + C = \ln\left|e^{2x} - e^{-2x}\right| + C$$

47. $u = 1 + \cos\theta$, $du = -\sin\theta\,d\theta$. Then,

$$\int (1 + \cos\theta)^3 \sin\theta\,d\theta = \int -u^3\,du = -\frac{u^2}{4} + C = -\frac{(1 + \cos\theta)^4}{4} + C$$

49. $u = 2t$, $du = 2\,dt$. Then,

$$\int \frac{dt}{t\sqrt{4t^2 - 1}} = \int \frac{2\,dt}{2t\sqrt{4t^2 - 1}} = \int \frac{du}{u\sqrt{u^2 - 1}} = \sec^{-1} u + C = \sec^{-1}(2t) + C$$

51. $u = \sin x$, $du = \cos x\,dx$. Then,

$$\int \frac{\cos x\,dx}{\sin x} = \int \frac{du}{u} = \ln|u| + C = \ln|\sin x| + C$$

53. $u = \sec x$, $du = \sec x \tan x\,dx$. Then,

$$\int \sec^3 x \tan x\,dx = \int u^2\,du = \frac{u^3}{3} + C = \frac{\sec^3 x}{3} + C$$

55. $u = \tan 3x$, $du = 3\sec^2 3x\,dx$, so

$$\int e^{\tan 3x} \sec^2 3x\,dx = \frac{1}{3}\int e^u\,du = \frac{1}{3}e^u + C = \frac{1}{3}e^{\tan 3x} + C$$

57. Let $u = 2x$, $du = 2\,dx$. Then,

$$\int \frac{1 + \cos 2x}{\sin^2 2x}\,dx = \frac{1}{2}\int \frac{1 + \cos u}{\sin^2 u}\,du = \frac{1}{2}\int \frac{du}{\sin^2 u} + \frac{1}{2}\int \frac{\cos u\,du}{\sin^2 u}$$

$$= \frac{1}{2}\int \csc^2 u\,du + \frac{1}{2}\int \cot u \csc u\,du$$

$$= -\frac{1}{2}(\cot u + \csc u) + C$$

59. $u = 1 + \cot 2t$, $du = -2\csc^2 2t\,dt$. Then,

$$\int \frac{\csc^2 2t\,dt}{\sqrt{1 + \cot 2t}} = -\frac{1}{2}\int u^{-1/2}\,du = -\frac{1}{2}\cdot 2\sqrt{u} + C = -\sqrt{1 + \cot 2t} + C$$

61. $u = \tan^{-1} 2t$, $du = \dfrac{2\ dt}{1 + (2t)^2} = \dfrac{2\ dt}{1 + 4t^2}$. Then,

$$\int \frac{e^{\tan^{-1} 2t}}{1 + 4t^2}\ dt = \frac{1}{2}\int e^u\ du = \frac{1}{2}e^u + C = \frac{1}{2}e^{\tan^{-1} 2t} + C$$

63. Note that $3 = e^{\ln 3}$, so $3^x = \left(e^{\ln 3}\right)^x = e^{x\ \ln 3}$. Therefore,

$$\int 3^x\ dx = \int e^{x\ \ln 3}\ dx\ .$$

Now set $u = x\ \ln 3$, $du = \ln 3\ dx$. Then,

$$\int e^{x\ \ln 3}\ dx = \frac{1}{\ln 3}\int e^u\ du = \frac{1}{\ln 3}\ e^u + C = \frac{1}{\ln 3}\ e^{x\ \ln 3} + C = \frac{3^x}{\ln 3} + C$$

65. $u = \ln x$, $du = \dfrac{dx}{x}$. Then, $\displaystyle\int \frac{\ln x}{x}\ dx = \int u\ du = \frac{1}{2}u^2 + C = \frac{1}{2}(\ln x)^2 + C$

67. $u = 1 + \sqrt{x}$, $du = \dfrac{dx}{2\sqrt{x}}$. Then,

$$\int \frac{dx}{\sqrt{x}(1 + \sqrt{x})} = \int \frac{2\ du}{u} = 2\ \ln u + C = 2\ \ln (1 + \sqrt{x}) + C$$

69. a) $u = x^2$, $du = 2x\ dx$. Then,

$$\int xe^{x^2}\ dx = \frac{1}{2}\int e^u\ du = \frac{1}{2}e^u + C = \frac{1}{2}e^{x^2} + C$$

   b) $u = e^{x^2}$, $du = \left(e^{x^2}\right)(2x)\ dx = 2xe^{x^2} + C$. Then,

$$\int xe^{x^2}\ dx = \frac{1}{2}\int du = \frac{1}{2}u + C = \frac{1}{2}e^{x^2} + C$$

71. a) Suppose $\theta = \tan^{-1} u$. Then, $\tan \theta = u$. But $\tan \theta = \cot (\pi/2 - \theta)$, so $\cot (\pi/2 - \theta) = u$. Therefore, $\cot^{-1} u = \pi/2 - \theta$. Hence,

$$\tan^{-1} u + \cot^{-1} u = \theta + \left(\frac{\pi}{2} - \theta\right),$$

so $C = \pi/2$.

   b) Suppose $\theta = \sec^{-1} |u|$. Then, $|u| = \sec \theta$, so $|u| = \csc (\pi/2 - \theta)$. Therefore, $\pi/2 - \theta = \csc^{-1} |u|$. Hence,

$$\sec^{-1} |u| + \csc^{-1} |u| = \frac{\pi}{2}\ .$$

Article 7.2

1. Let $u = x$, $dv = \sin x\, dx$. Then, $du = dx$, $v = -\cos x$, and

$$\int x \sin x\, dx = -x \cos x - \int -\cos x\, dx = -x \cos x + \int \cos x\, dx$$

$$= -x \cos x + \sin x + C$$

3. Let $u = x^2$, $dv = \sin x\, dx$. Then, $du = 2x\, dx$, $v = -\cos x$, and

$$\int x^2 \sin x\, dx = -x^2 \cos x - \int -2x \cos x\, dx = -x^2 \cos x + 2\int x \cos x\, dx$$

Now, to find $\int x \cos x\, dx$ (if you have not already done so in Problem 2), let $u = x$, $dv = \cos x\, dx$. Then, $du = dx$, $v = \sin x$, and

$$\int x \cos x\, dx = x \sin x - \int \sin x\, dx = x \sin x + \cos x + C.$$

Putting this together with our earlier work:

$$\int x^2 \sin x\, dx = -x^2 \cos x + 2x \sin x + 2 \cos x + C.$$

5. Let $u = \ln x$, $dv = x\, dx$; then, $v = x^2/2$, $du = 1/x\, dx$, and

$$\int x \ln x\, dx = \frac{x^2}{2} \ln x - \int \frac{x^2}{2} \cdot \frac{dx}{x} = \frac{x^2}{2} \ln x - \frac{x^2}{4} + C$$

7. Let $u = \ln ax$, $dv = x^n$. Then,

$$v = \frac{x^{n+1}}{n+1}, \quad du = \frac{1}{ax} \cdot a\, dx = \frac{dx}{x},$$

and $\displaystyle \int x^n \ln ax\, dx = \frac{x^{n+1}}{n+1} \ln ax - \int \frac{x^{n+1}}{n+1} \cdot \frac{dx}{x} = \frac{x^{n+1}}{n+1} \ln ax - \frac{x^{n+1}}{(n+1)^2} + C$

9. Using tabular integration:

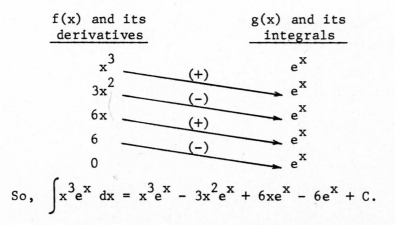

f(x) and its derivatives	g(x) and its integrals
x^3	e^x
$3x^2$	e^x
$6x$	e^x
6	e^x
0	e^x

So, $\displaystyle \int x^3 e^x\, dx = x^3 e^x - 3x^2 e^x + 6x e^x - 6e^x + C.$

11. Using tabular integration:

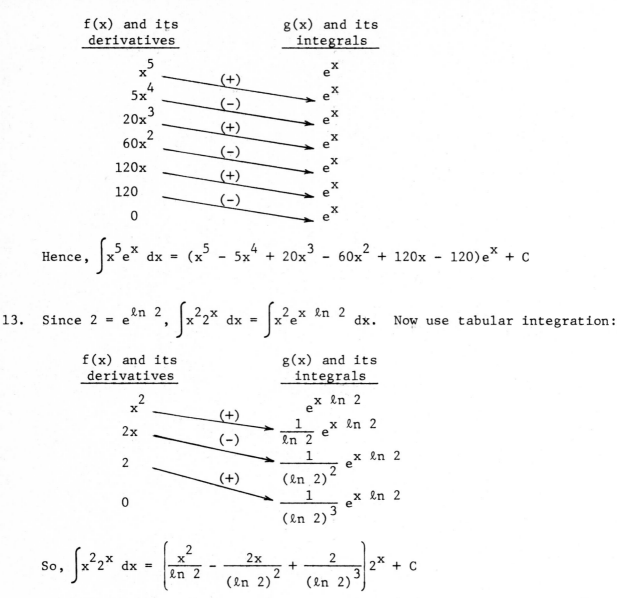

Hence, $\int x^5 e^x \, dx = (x^5 - 5x^4 + 20x^3 - 60x^2 + 120x - 120)e^x + C$

13. Since $2 = e^{\ln 2}$, $\int x^2 2^x \, dx = \int x^2 e^{x \ln 2} \, dx$. Now use tabular integration:

f(x) and its derivatives	g(x) and its integrals
x^2	$e^{x \ln 2}$
$2x$	$\dfrac{1}{\ln 2} e^{x \ln 2}$
2	$\dfrac{1}{(\ln 2)^2} e^{x \ln 2}$
0	$\dfrac{1}{(\ln 2)^3} e^{x \ln 2}$

So, $\int x^2 2^x \, dx = \left[\dfrac{x^2}{\ln 2} - \dfrac{2x}{(\ln 2)^2} + \dfrac{2}{(\ln 2)^3} \right] 2^x + C$

15. To determine an antiderivative of $x^3 \cos 2x$, use tabular integration:

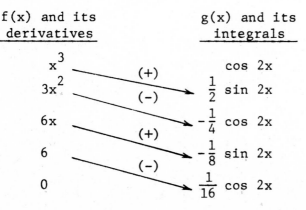

Hence,

$$\int_0^{\pi/2} x^3 \cos 2x \, dx = \frac{x^3}{2} \sin 2x + \frac{3x^2}{4} \cos 2x - \frac{3x}{4} \sin 2x - \frac{3}{8} \cos 2x \Big|_0^{\pi/2}$$

$$= \frac{3}{4} - \frac{3\pi^2}{16}$$

17. Tabular integration:

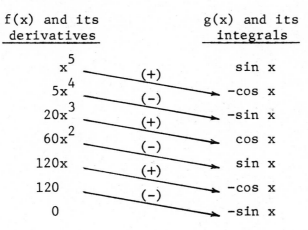

f(x) and its derivatives		g(x) and its integrals
x^5	(+)	sin x
$5x^4$	(−)	−cos x
$20x^3$	(+)	−sin x
$60x^2$	(−)	cos x
$120x$	(+)	sin x
120	(−)	−cos x
0		−sin x

Thus,

$$\int x^5 \sin x \, dx = (5x^4 - 60x^2 + 120) \sin x - (x^5 - 20x^3 + 120x) \cos x$$

19. Using Equation (6) from the text:

$$\int \sin^2 x \, dx = -\frac{\sin x \cos x}{2} + \frac{1}{2} \int dx = \frac{x - \sin x \cos x}{2} + C$$

21. Let $u = \sin^{-1} ax$, $dv = dx$. Then,

$$v = x, \quad du = \frac{a \, dx}{\sqrt{1 - (ax)^2}} \, ,$$

and

$$\int \sin^{-1} ax \, dx = x \sin^{-1} ax - \int \frac{ax \, dx}{\sqrt{1 - a^2 x^2}}$$

In the last integral, let $u = 1 - a^2 x^2$, $du = -2a^2 x \, dx$. Then,

$$\int \frac{ax \, dx}{\sqrt{1 - a^2 x^2}} = -\frac{1}{2a} \int \frac{du}{\sqrt{u}} = -\frac{1}{2a}(2\sqrt{u}) + C$$

$$= \frac{-\sqrt{u}}{a} + C = \frac{-\sqrt{1 - (ax)^2}}{a} + C$$

Hence,

$$\int \sin^{-1} ax \, dx = x \sin^{-1} ax + \frac{1}{a} \sqrt{1 - (ax)^2} + C$$

23. By tabular integration:

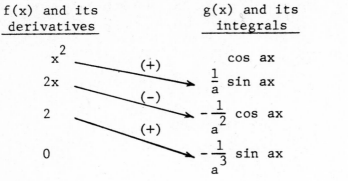

f(x) and its derivatives		g(x) and its integrals
x^2	(+)	$\cos ax$
$2x$		$\frac{1}{a} \sin ax$
	(−)	
2		$-\frac{1}{a^2} \cos ax$
	(+)	
0		$-\frac{1}{a^3} \sin ax$

Hence,

$$\int x^2 \cos ax \, dx = \left(\frac{x^2}{a} - \frac{2}{a^3}\right) \sin ax + \frac{2x}{a^2} \cos ax$$

25. Let $u = \sec^{-1} x$, $dv = x \, dx$. Then,

$$v = \frac{x^2}{2}, \quad du = \frac{dx}{x\sqrt{x^2 - 1}} \,,$$

and

$$\int x \sec^{-1} x = \frac{x^2}{2} \sec^{-1} x - \int \frac{x \, dx}{2\sqrt{x^2 - 1}}$$

In this last integral, set $w = x^2 - 1$, $dw = 2x \, dx$. Then,

$$\int \frac{x \, dx}{2\sqrt{x^2 - 1}} = \frac{1}{4} \int \frac{dw}{\sqrt{w}} = \frac{1}{2} \sqrt{w} = \frac{1}{2}\sqrt{x^2 - 1}$$

Thus, an antiderivative of $x \sec^{-1} x$ is

$$F(x) = \frac{x^2}{2} \sec^{-1} x - \frac{1}{2}\sqrt{x^2 - 1} \,,$$

and

$$\int_1^2 x \sec^{-1} x \, dx = F(2) - F(1) = \frac{2\pi}{3} - \frac{\sqrt{3}}{2}$$

27. Let $w = \ln x$, $dw = dx/x$. Then,

$$\int \frac{\ln x}{x} \, dx = \int w \, dw = \frac{w^2}{2} + C = \frac{1}{2}(\ln x)^2 + C$$

29. Let $I = \int e^{2x} \cos 3x \, dx$. Let $u = e^{2x}$, $dv = \cos 3x \, dx$. Then,

$$v = \frac{1}{3} \sin 3x, \quad du = 2e^{2x} \, dx,$$

and we have

$$I = \frac{1}{3}e^{2x} \sin 3x - \frac{2}{3}\int e^{2x} \sin 3x \, dx$$

At first glance, this might seem as though it has not helped us much. However, we again use integration by parts on the last integral.

Let $u = e^{2x}$, $dv = \sin 3x \, dx$. Then, $v = -1/3 \cos 3x$, $du = 2e^{2x} \, dx$, and

$$\int e^{2x} \sin 3x \, dx = -\frac{1}{3}e^{2x} \cos 3x + \frac{2}{3}\int e^{2x} \cos 3x \, dx$$

$$= -\frac{1}{3}e^{2x} \cos 3x + I$$

Therefore, we have the equation

$$I = \frac{1}{3}e^{2x} \sin 3x + \frac{2}{9}e^{2x} \cos 3x - \frac{4}{9}I$$

Solving for I:

$$I = \int e^{2x} \cos 3x \, dx = \frac{e^{2x}}{13}(3 \sin 3x + 2 \cos 3x).$$

31. Let $u = e^{ax}$, $dv = \sin bx \, dx$. Then, $v = -1/b \cos bx$, $du = ae^{ax} \, dx$, and

$$\int e^{ax} \sin bx \, dx = -\frac{e^{ax}}{b} \cos bx + \int \frac{ae^{ax}}{b} \cos bx \, dx.$$

Let $u = e^{ax}$, $dv = \cos bx \, dx$. Then, $du = ae^{ax}$, $v = 1/b \sin bx$, so we have

$$\int e^{ax} \sin bx = -\frac{e^{ax}}{b} \cos bx + \frac{a}{b}\left[\frac{e^{ax} \sin bx}{b} - \frac{a}{b}\int e^{ax} bx \, dx\right]$$

Hence,

$$\left(\frac{a^2 + b^2}{b^2}\right)\int e^{ax} \sin bx \, dx = e^{ax}\left(\frac{a \sin bx}{b^2} - \frac{\cos bx}{b}\right)$$

Simplifying:

$$\int e^{ax} \sin bx = \frac{e^{ax}}{a^2 + b^2}(a \sin bx - b \cos bx).$$

33. Let $u = \cos(\ln x)$, $dv = dx$. Then, $v = x$, $du = \frac{-\sin(\ln x)}{x} \, dx$, and

$$\int \cos(\ln x) \, dx = x \cos(\ln x) + \int \sin(\ln x) \, dx.$$

Now, for $\int \sin(\ln x) \, dx$, let $u = \sin(\ln x)$, $dv = dx$. Then,

$$v = x, \quad du = \frac{\cos(\ln x)}{x} \, dx,$$

and we have

$$\int \sin(\ln x) \, dx = x \cos(\ln x) - \int \cos(\ln x) \, dx.$$

Combining this with our previous equation:

$$\int \cos (\ln x)\, dx = x \cos (\ln x) + x \sin (\ln x) - \int \cos (\ln x)\, dx \,.$$

Therefore, we have

$$\int \cos (\ln x)\, dx = \frac{x \cos (\ln x) + x \sin (\ln x)}{2}$$

35. $$\text{Volume} = 2\pi \int_0^{\pi/2} x \cos x\, dx.$$

To find $\int x \cos x\, dx$: Let $u = x$, $dv = \cos x\, dx$. Then, $du = dx$, $v \sin x$, and

$$\int x \cos x\, dx = x \sin x - \int \sin x\, dx = x \sin x + \cos x + C.$$

Hence, the volume of the solid is

$$2\pi (x \sin x + \cos x) \Big|_0^{\pi/2} = \pi^2 - 2\pi.$$

37. The area of the region is

$$\int_1^e (1 - \ln x)\, dx = e - 1 - \int_1^e \ln x\, dx.$$

To find $\int \ln x\, dx$, let $u = \ln x$, $dv = dx$. Then, $v = x$, $du = dx/x$, and

$$\int \ln x\, dx = x \ln x - \int dx = x \ln x - x.$$

So,

$$\int_1^e \ln x\, dx = (x \ln x - x) \Big|_1^e = e - e + 1 = 1,$$

and the area of the region is $(e - 1) - 1 = e - 2$.

$$\tilde{x} = x, \quad \text{so} \quad \bar{x} = \frac{1}{e - 2} \int_1^e x(1 - \ln x)\, dx.$$

To find $\int x(1 - \ln x)\, dx$, let $u = 1 - \ln x$, $dv = x\, dx$. Then, $du = -dx/x$, $v = x^2/2$, and

$$\int x(1 - \ln x)\, dx = \frac{x^2}{2}(1 - \ln x) + \int \frac{x}{2}\, dx = \frac{3x^2}{4} - \frac{x^2 \ln x}{2} + C,$$

so

$$\int_1^e x(1 - \ln x)\, dx = \frac{3x^2}{4} - \frac{x^2 \ln x}{2} \Big|_1^e = \frac{e^2 - 3}{4}\,.$$

Hence,

$$\bar{x} = \frac{e^2 - 3}{4(e - 2)} \quad \text{and} \quad \tilde{y} = \frac{1}{2}(1 + \ln x),$$

so

$$\bar{y} = \frac{1}{2(e - 2)} \int_1^e [1 - (\ln x)^2] \, dx.$$

To find $\int [1 - (\ln x)^2] \, dx$, let $u = 1 - (\ln x)^2$, $dv = dx$. Then,

$$v = x, \quad du = -\frac{2 \ln x}{x} \, dx,$$

and

$$\int [1 - (\ln x)^2] \, dx = x[1 - (\ln x)]^2 + 2 \int \ln x \, dx$$

$$= x - x(\ln x)^2 + 2x \ln x - 2x$$

(we did $\int \ln x \, dx$ at the beginning of the problem). So,

$$\int_1^e [1 - (\ln x)^2] \, dx = 2x \ln x - x(\ln x)^2 - x \Big|_1^e = 1.$$

Therefore,

$$\bar{y} = \frac{1}{2(e - 2)} \, .$$

39.
$$M_y = \int_0^\pi xy\delta \, dx = \int_0^\pi x(1 + x) \sin x \, dx$$

To find $\int (x + x^2) \sin x \, dx$, use tabular integration:

$$
\begin{array}{lll}
x + x^2 & \xrightarrow{\ (+)\ } & \sin x \\
1 + 2x & \xrightarrow{\ (-)\ } & -\cos x \\
2 & \xrightarrow{\ (+)\ } & -\sin x \\
0 & \xrightarrow{} & \cos x
\end{array}
$$

Then,

$$M_y = (1 + 2x) \sin x + (2 - x - x^2) \cos x \Big|_0^\pi = \pi^2 + \pi - 4.$$

41. Let $w = \sqrt{x}$, $x = w^2$, $dx = 2w \, dw$. Then,

$$\int_0^1 e^{\sqrt{x}} \, dx = \int_0^1 2we^w \, dx.$$

Now we can use integration by parts: Let $u = w$, $dv = e^w \, dw$. Then,

$$v = e^w, \quad du = dw, \quad \text{and} \quad \int we^w \, dw = we^w - \int e^w \, dw = we^w - e^w + C.$$

Therefore,

$$\int_0^1 2we^w \, dw = 2we^w - 2e^w \Big|_0^1 = 2.$$

43. a) Let $u = (\ln x)^n$, $dv = x^m \, dx$. Then,

$$v = \frac{x^{m+1}}{m+1}, \quad du = \frac{n(\ln x)^{n-1}}{x} \, dx,$$

and

$$\int x^m (\ln x)^n \, dx = \frac{x^{m+1}(\ln x)^n}{m+1} - \int \frac{n(\ln x)^{n-1}}{x} \cdot \frac{x^{m+1}}{m+1} \, dx$$

$$= \frac{x^{m+1}(\ln x)^n}{m+1} - \frac{n}{m+1} \int x^m (\ln x)^{n-1} \, dx,$$

as desired.

b)
$$\int x^3 (\ln x)^2 = \frac{x^4 (\ln x)^2}{4} - \frac{2}{4} \int x^3 \ln x \, dx$$

$$= \frac{x^4 (\ln x)^2}{4} - \frac{x^4 \ln x}{8} + \frac{1}{8} \int x^3 \, dx$$

$$= \frac{x^4 (\ln x)^2}{4} - \frac{x^4 \ln x}{8} + \frac{x^4}{32} + C.$$

45. a) Let $u = x^n$, $dv = e^x \, dx$. Then, $v = e^x$, $du = nx^{n-1} \, dx$, and

$$\int x^n e^x \, dx = x^n e^x - n \int x^{n-1} e^x \, dx$$

b)
$$\int x^3 e^x \, dx = x^3 e^x - 3 \int x^2 e^x \, dx = x^3 e^x - 3x^2 e^x + 6 \int xe^x \, dx$$

$$= x^3 e^x - 3x^2 e^x + 6xe^x - 6 \int e^x \, dx = (x^3 - 3x^2 + 6x - 6)e^x$$

Article 7.3

1. Let $u = \sin x$, $du = \cos x \, dx$. Then, at $x = 0$, $u = 0$; at $x = \pi/4$, $u = \sqrt{2}/2$, and

$$\int_0^{\pi/4} \sin^5 x \cos x \, dx = \int_0^{\sqrt{2}/2} u^5 \, du = \frac{1}{48}.$$

3. Let $u = \cos x$, $du = -\sin x \, dx$. Then,

$$\int \sqrt{\cos x} \sin x \, dx = \int -\sqrt{u} \, du = -\int u^{1/2} \, du$$

$$= -\frac{2}{3} u^{3/2} + C = -\frac{2}{3} (\cos x)^{3/2} + C.$$

5. Let $u = 2 - \cos \theta$, $du = \sin \theta\, d\theta$. Then,

$$\int \frac{\sin \theta\, d\theta}{2 - \cos \theta} = \int \frac{du}{u} = \ln |u| + C = \ln (2 - \cos \theta) + C.$$

7. $$\int \tan 3x\, dx = \int \frac{\sin 3x}{\cos 3x}\, dx.$$

Let $u = \cos 3x$, $du = -3 \sin 3x\, dx$. Then,

$$\int \frac{\sin 3x}{\cos 3x}\, dx = -\frac{1}{3} \int \frac{du}{u} = -\frac{1}{3} \ln |u| + C = -\frac{1}{3} \ln |\cos 3x| + C.$$

9. Use the identity $\cos^2 x = 1 - \sin^2 x$:

$$\int \cos^3 x\, dx = \int \cos x\, (1 - \sin^2 x)\, dx = \int \cos x\, dx - \int \sin^2 x \cos x\, dx$$

$$= \sin x - \int \sin^2 x \cos x\, dx$$

In the last integral let $u = \sin x$, $du = \cos x\, dx$. Then,

$$\int \sin^2 x \cos x\, dx = \int u^2\, du = \frac{1}{3} u^3 + C = \frac{1}{3} \sin^3 x + C.$$

Therefore,

$$\int \cos^3 x\, dx = \sin x - \frac{1}{3} \sin^3 x + C.$$

11. $$\int \tan^2 4x\, dx = \int \frac{\sin^2 4x}{\cos^2 4x}\, dx = \int \frac{1 - \cos^2 4x}{\cos^2 4x}\, dx$$

$$= \int \sec^2 4x\, dx - \int dx = \left(\int \sec^2 4x\, dx \right) - x + C.$$

Now, let $u = 4x$, $du = 4\, dx$:

$$\int \sec^2 4x\, dx = \frac{1}{4} \int \sec^2 u\, du = \frac{1}{4} \tan u + C = \frac{1}{4} \tan 4x + C.$$

Hence,

$$\int \tan^2 4x\, dx = \frac{1}{4} \tan 4x - x + C.$$

13. Using formula (4):

$$\int \tan^5 t \, dt = \frac{\tan^4 t}{4} - \int \tan^3 t \, dt = \frac{\tan^4 t}{4} - \frac{\tan^2 t}{2} + \int \tan t \, dt$$

$$= \frac{\tan^4 t}{4} - \frac{\tan^2 t}{2} + \ln |\sec t| + C$$

using the result of Problem 10 for $\int \tan t \, dt$. Therefore,

$$\int_{\pi/6}^{\pi/3} \tan^5 t \, dt = \frac{\tan^4 t}{4} - \frac{\tan^2 t}{2} + \ln |\sec t| \Big|_{\pi/6}^{\pi/3}$$

$$= \left[\frac{(\sqrt{3})^4}{4} - \frac{(\sqrt{3})^2}{2} + \ln 2 \right] - \left[\frac{1}{4}\left(\frac{1}{\sqrt{3}}\right)^4 - \frac{1}{2}\left(\frac{1}{\sqrt{3}}\right)^2 + \ln \left| \frac{2}{\sqrt{3}} \right| \right]$$

$$= \frac{9}{4} - \frac{3}{2} - \frac{1}{36} + \frac{1}{6} + \ln \left(\frac{2}{2/\sqrt{3}}\right) = \frac{8}{9} + \ln \sqrt{3}$$

15. a)

$$\int \sin^3 x \cos^2 x \, dx = \int (\sin x)(1 - \cos^2 x)(\cos^2 x) \, dx$$

$$= \int \cos^2 x \sin x \, dx - \int \cos^4 x \sin x \, dx$$

In both integrals let $u = \cos x$, $du = -\sin x \, dx$. Then,

$$\int \sin^3 x \cos^2 x \, dx = -\int u^2 \, du + \int u^4 \, du = \frac{u^5}{5} - \frac{u^3}{3} + C$$

$$= \frac{\cos^5 x}{5} - \frac{\cos^3 x}{3} + C \ .$$

b)

$$\int \frac{\sin^3 x}{\cos^2 x} \, dx = \int \frac{\sin x (1 - \cos^2 x)}{\cos^2 x} \, dx = \int \frac{\sin x}{\cos^2 x} \, dx - \int \sin x \, dx$$

$$= \int \tan x \sec x \, dx + \cos x + C = \sec x + \cos x + C.$$

17. Let $u = \tan x$, $du = \sec^2 x \, dx$. Then,

$$\int \tan^n x \sec^2 x \, dx = \int u^n \, du = \frac{u^{n+1}}{n+1} + C = \frac{\tan^{n+1} x}{n+1} + C.$$

19. Let $u = \cos x$, $du = -\sin x \, dx$. Then,

$$\int \cos^n x \sin x \, dx = -\int u^n \, du = -\frac{u^{n+1}}{n+1} + C = -\frac{\cos^{n+1} x}{n+1} + C.$$

21. Let $u = \cos 2x$, $du = -2 \sin 2x$. Then,

$$\int \cos^3 2x \sin 2x \, dx = -\frac{1}{2}\int u^3 \, du = -\frac{u^4}{8} + C = -\frac{\cos^4 2x}{8} + C.$$

23. To find

$$\int \sin^2 y \cos^3 y \, dy = \int \sin^2 y(1 - \sin^2 y) \cos y \, dy,$$

let $u = \sin y$, $du = \cos y \, dy$. Then,

$$\int \sin^2 y \cos^3 y \, dy = \int u^2 \, du - \int u^4 \, du = \frac{u^3}{3} - \frac{u^5}{5} + C = \frac{\sin^3 y}{3} - \frac{\sin^5 y}{5} + C.$$

Hence,

$$\int_{-\pi/2}^{\pi/2} \sin^2 y \cos^3 y \, dy = \frac{\sin^3 y}{3} - \frac{\sin^5 y}{5}\bigg|_{-\pi/2}^{\pi/2} = \frac{4}{15}$$

25. To find $\int \tan^3 2x \sec^2 2x \, dx$, let $u = \tan 2x$, $du = 2 \sec^2 2x \, dx$. Then,

$$\int \tan^3 2x \sec^2 2x \, dx = \frac{1}{2}\int u^3 \, du = \frac{u^4}{8} + C = \frac{\tan^4 2x}{8} + C.$$

Hence,

$$\int_{\pi/6}^{\pi/3} \tan^3 2x \sec^2 2x \, dx = \frac{\tan^4 2x}{8}\bigg|_{\pi/6}^{\pi/3} = \frac{\tan^4 (2\pi/3) - \tan^4 (\pi/3)}{8} = 0$$

27. Let $u = \cot x$, $du = -\csc^2 x \, dx$. Then,

$$\int \cot^3 x \csc^2 x \, dx = \int -u^3 \, du = -\frac{u^4}{4} + C = -\frac{\cot^4 x}{4} + C.$$

29. Let $u = \tan 3x$, $du = 3 \sec^2 3x$. Then,

$$\int \sec^4 3x \, dx = \int (1 + \tan^2 3x) \sec^2 3x = \frac{1}{3}\int du + \frac{1}{3}\int u^2 \, du$$

$$= \frac{u}{3} + \frac{u^3}{9} + C = \frac{\tan 3x}{3} + \frac{\tan^3 3x}{9} + C.$$

31. Let $u = 2x$, $du = 2 \, dx$. Then,

$$\int \tan^3 2x \, dx = \frac{1}{2}\int \tan^3 u \, du.$$

Now apply formula (4):

$$\frac{1}{2}\int \tan^3 u \, du = \frac{\tan^2 u}{4} - \frac{1}{2}\int \tan u \, du = \frac{\tan^2 u}{4} + \frac{1}{2} \ln |\sec u| + C.$$

So,

$$\int \tan^3 2x \, dx = \frac{\tan^2 2x}{4} + \frac{1}{2} \ln |\sec 2x| + C.$$

33. Let $u = \cos x$, $du = -\sin x$:

$$\int \sin^3 x \, dx = \int (1 - \cos^2 x) \sin x \, dx = -\int (1 - u^2) \, du$$

$$= \frac{u^3}{3} - u + C = \frac{\cos^3 x}{3} - \cos x + C.$$

35. Let $u = 2 + \tan x$, $du = \sec^2 x \, dx$. Then,

$$\int \frac{\sec^2 x \, dx}{2 + \tan x} = \int \frac{du}{u} = \ln |u| + C = \ln |2 + \tan x| + C.$$

37.

$$\int \cot^3 x \, dx = \int \cot x \, (\csc^2 x - 1) \, dx = \int \cot x \, \csc^2 x \, dx - \int \cot x \, dx$$

To find $\int \cot x \, \csc^2 x \, dx$, let $u = \cot x$, $du = -\csc^2 x \, dx$:

$$\int \cot x \, \csc^2 x \, dx = -\int u \, du = -\frac{u^2}{2} + C = -\frac{\cot^2 x}{2} + C.$$

To find $\int \cot x \, dx = \int \cos x / \sin x \, dx$, let $u = \sin x$, $du = \cos x \, dx$; then,

$$\int \cot x \, dx = \int \frac{du}{u} = \ln |u| = \ln |\sin x| + C.$$

So,

$$\int \cot^3 x \, dx = -\frac{\cot^2 x}{2} + \ln |\sin x| + C.$$

39.

$$\int \csc^4 x \, dx = \int (1 + \cot^2 x) \csc^2 x \, dx.$$

Let $u = \cot x$, $du = -\csc^2 x \, dx$; then,

$$\int \csc^4 x \, dx = \int -(1 + u^2) \, du = -u - \frac{u^3}{3} + C = -\cot x - \frac{\cot^3 x}{3} + C.$$

41.

$$\int_0^{\pi/4} \frac{dx}{\sqrt{1 - \sin^2 x}} = \int_0^{\pi/4} \frac{dx}{\cos x} = \int_0^{\pi/4} \sec x \, dx$$

$$= \ln |\sec x + \tan x| \Big|_0^{\pi/4} = \ln (1 + \sqrt{2}).$$

43. $$\int_0^{\pi/3} \frac{dx}{1 + \sin x} = \int_0^{\pi/3} \frac{(1 - \sin x)}{1 - \sin^2 x} \, dx = \int_0^{\pi/3} \frac{(1 - \sin x) \, dx}{\cos^2 x}$$

$$= \int_0^{\pi/3} (\sec^2 x - \sec x \tan x) \, dx = (\tan x - \sec x) \Big|_0^{\pi/3}$$

$$= \sqrt{3} - 1.$$

45. Recall the trigonometric identity

$$\sin A \cos B = \frac{1}{2}[\sin (A + B) + \sin (A - B)].$$

Therefore,

$$\int_{-\pi}^{\pi} \sin 3x \cos 2x \, dx = \frac{1}{2}\int_{-\pi}^{\pi} \sin 5x \, dx + \frac{1}{2}\int_{-\pi}^{\pi} \sin x \, dx$$

$$= -\frac{1}{10} \cos 5x \Big|_{-\pi}^{\pi} - \frac{1}{2} \cos x \Big|_{-\pi}^{\pi} = 0.$$

47. $$\sin^2 3x = \frac{1 - \cos 6x}{2},$$

so

$$\int_{-\pi}^{\pi} \sin^2 3x \, dx = \int_{-\pi}^{\pi} \frac{1 - \cos 6x}{2} \, dx = \frac{x}{2} - \frac{\sin 6x}{12} \Big|_{-\pi}^{\pi} = \pi.$$

49. $$\int \frac{\sec 2x \csc 2x}{2} \, dx = \int \frac{dx}{2 \sin 2x \cos 2x} = \int \frac{dx}{\sin 4x}$$

$$= \int \csc 4x \, dx = \frac{1}{4}\int \csc w \, dw \quad (w = 4x)$$

$$= -\frac{1}{4} \ln |\csc w + \cot w| + C$$

$$= -\frac{1}{4} \ln |\csc 4x + \cot 4x| + C.$$

51. $$ds = \left[1 + \left(\frac{dy}{dx}\right)^2\right]^{1/2} dx; \quad \frac{dy}{dx} = \frac{1}{\cos x} \cdot (-\sin x) = -\tan x \, dx.$$

Then,

$$\left[1 + y'^2\right]^{1/2} = \left[1 + \tan^2 x\right]^{1/2} = \sec x$$

(for $0 \le x \le \pi/3$, the positive square root is correct). So, the length of the curve is

$$\int_0^{\pi/3} \sec x \, dx = \ln |\sec x + \tan x| \Big|_0^{\pi/3} = \ln (2 + \sqrt{3}).$$

53. The area of the region is

$$\int_{-\pi/4}^{\pi/4} \sec x \, dx = 2 \int_0^{\pi/4} \sec x \, dx = 2 \ln \left| \sec x + \tan x \right| \Big|_0^{\pi/4}$$

$$= 2 \ln (1 + \sqrt{2}) = \ln (1 + \sqrt{2})^2 = \ln (3 + 2\sqrt{2})$$

$\tilde{x} = x$, so

$$\bar{x} = \frac{1}{\ln (3 + 2\sqrt{2})} \int_{-\pi/4}^{\pi/4} x \sec x \, dx = 0 \quad \text{(by symmetry)}$$

$\tilde{y} = \frac{1}{2} y = \frac{1}{2} \sec x$, so

$$\bar{y} = \frac{1}{2 \ln (3 + 2\sqrt{2})} \int_{-\pi/4}^{\pi/4} \sec^2 x \, dx = \frac{\tan x}{2 \ln (3 + 2\sqrt{2})} \Big|_{-\pi/4}^{\pi/4} = \frac{1}{\ln (3 + 2\sqrt{2})} \cdot$$

Hence,

$$(\bar{x}, \bar{y}) = \left[0, \frac{1}{2 \ln (1 + \sqrt{2})} \right] = \left[0, \frac{1}{\ln (3 + 2\sqrt{2})} \right].$$

55. a) $\quad \sin mx \sin nx = \frac{1}{2} [\cos (m - n)x - \cos (m + n)x],$

so

$$\int_0^{2\pi} \sin mx \sin nx \, dx = \frac{1}{2} \int_0^{2\pi} \cos (m - n)x \, dx - \frac{1}{2} \int_0^{2\pi} \cos (m + n)x \, dx$$

$$= \frac{\sin (m - n)x}{2(m - n)} \Big|_0^{2\pi} - \frac{\sin (m + n)x}{2(m + n)} \Big|_0^{2\pi} = 0$$

since $\sin k\pi = 0$ for any integer k. The condition $m^2 \ne n^2$ is needed to insure $m + n \ne 0$, $m - n \ne 0$.

b) $\quad \sin px \cos qx = \frac{1}{2} [\sin (p + q)x + \sin (p - q)x],$

so

$$\int_0^{2\pi} \sin px \cos qx \, dx = \frac{1}{2} \int_0^{2\pi} \sin (p + q)x \, dx + \frac{1}{2} \int_0^{2\pi} \sin (p - q)x \, dx$$

$$= \frac{-\cos (p + q)x}{2(p + q)} \Big|_0^{2\pi} - \frac{\cos (p - q)x}{2(p - q)} \Big|_0^{2\pi} = 0$$

because $\cos 0 = \cos 2k\pi = 1$ if k is an integer.

57. Note that $\cot^2 ax = \csc^2 ax - 1$, so

$$\int \cot^n ax \, dx = \int \cot^{n-2} ax \, (\csc^2 ax - 1) \, dx$$

$$= \int \cot^{n-2} ax \csc^2 ax \, dx - \int \cot^{n-2} ax \, dx$$

Let $u = \cot ax$, $du = -a \csc^2 ax \, dx$; then,

$$\int \cot^{n-2} ax \csc^2 ax \, dx = -\frac{1}{a}\int u^{n-2} \, du = -\frac{u^{n-1}}{a(n-1)} = -\frac{\cot^{n-1} ax}{a(n-1)} \; .$$

Thus,

$$\int \cot^n x \, dx = -\frac{\cot^{n-1} ax}{a(n-1)} - \int \cot^{n-2} ax \, dx \; .$$

To evaluate $\int \cot^4 3x \, dx$:

$$\int \cot^4 3x \, dx = -\frac{\cot^3 3x}{9} - \int \cot^2 3x \, dx$$

$$= -\frac{\cot^3 3x}{9} - \left[-\frac{\cot 3x}{3} - \int dx \right] = -\frac{\cot^3 3x}{9} + \frac{\cot 3x}{3} + x \; .$$

Article 7.4

1. $\sin x \cos x = \frac{1}{2} \sin 2x,$

   so

$$\int \sin^2 x \cos^2 x \, dx = \frac{1}{4}\int \sin^2 2x \, dx = \frac{1}{4}\int \frac{1 - \cos 4x}{2} \, dx$$

$$= \frac{x}{8} - \frac{\sin 4x}{32} + C.$$

3. $\int \sin^2 2t \, dt = \int \left[\frac{1 - \cos 4t}{2} \right] dt = \frac{t}{2} - \frac{\sin 4t}{8} + C.$

5. $\int \sin^4 ax \, dx = \int \left[\frac{1 - \cos 2ax}{2} \right]^2 dx = \int \left[\frac{1}{4} - \cos 2ax + \frac{\cos^2 2ax}{4} \right] dx$

$$= \int \left[\frac{1}{4} - \cos 2ax + \frac{1}{8} + \frac{\cos 4ax}{8} \right] dx$$

$$= \frac{3x}{8} - \frac{\sin 2ax}{2a} + \frac{\sin 4ax}{32a} + C \; .$$

7. $\int \frac{dx}{\cos^2 x} = \int \sec^2 x \, dx = \tan x + C \; .$

9. $\int \frac{\sin^2 z}{\cos z} \, dz = \int \tan z \sec z \, dz = \sec z + C.$

11.
$$\int \frac{\sin^3 x}{\cos^2 x}\,dx = \int \frac{\sin x(1 - \cos^2 x)}{\cos^2 x}\,dx$$

$$= \int \tan x \sec x\,dx - \int \sin x\,dx = \sec x + \cos x + C.$$

13.
$$\int \frac{\cos 2t}{\sin^2 2t}\,dt = \int \csc 2t \cot 2t\,dt = -\frac{1}{2}\csc 2t + C.$$

15. Since $\sin^2\left(\frac{t}{2}\right) = \frac{1 - \cos t}{2}$ and $\sqrt{\frac{1 - \cos t}{2}} = \left|\sin\left(\frac{t}{2}\right)\right|$.

But for $0 \le t \le 2\pi$, $0 \le t/2 \le \pi$, so $\sin t/2 \ge 0$. Therefore,
$$\left|\sin\left(\frac{t}{2}\right)\right| = \sin\left(\frac{t}{2}\right),$$
and
$$\int_0^{2\pi} \sqrt{\frac{1 - \cos t}{2}}\,dt = \int_0^{2\pi} \sin\left(\frac{t}{2}\right)\,dt = -2\cos\left(\frac{t}{2}\right)\Big|_0^{2\pi} = 4.$$

17.
$$\int_0^{\pi/10} \sqrt{1 + \cos 5\theta}\,d\theta = \int_0^{\pi/10} \sqrt{2}\cos\left(\frac{5\theta}{2}\right)\,d\theta$$

$$= \frac{2\sqrt{2}}{5}\sin\left(\frac{5\theta}{2}\right)\Big|_0^{\pi/10} = \frac{2}{5}.$$

19. Using the suggested identity,
$$\int \theta\sqrt{1 - \cos\theta}\,d\theta = \sqrt{2}\int \theta\sin\frac{\theta}{2}\,d\theta.$$

Now, integrate by parts:
$$u = \theta,\ dv = \sin\theta/2,\ v = -2\cos\theta/2,\ du = d\theta,$$
and
$$\sqrt{2}\int \theta\sin\left(\frac{\theta}{2}\right)\,d\theta = -2\sqrt{2}\theta\cos\left(\frac{\theta}{2}\right) + 2\sqrt{2}\int \cos\left(\frac{\theta}{2}\right)\,d\theta$$

$$= -2\sqrt{2}\theta\cos\left(\frac{\theta}{2}\right) + 4\sqrt{2}\sin\left(\frac{\theta}{2}\right) + C.$$

21. $\sqrt{1 + \tan^2 x} = |\sec x| = \sec x$ for $-\pi/4 \le x \le \pi/4$.
So,
$$\int_{-\pi/4}^{\pi/4} \sqrt{1 + \tan^2 x}\,dx = \int_{-\pi/4}^{\pi/4} \sec x\,dx = \ln|\sec x + \tan x|\Big|_{-\pi/4}^{\pi/4}$$

$$= \ln\left(\frac{\sqrt{2} + 1}{\sqrt{2} - 1}\right).$$

23. $\displaystyle\int_0^\pi \sqrt{1 - \cos^2 \theta}\; d\theta = \int_0^\pi |\sin \theta|\; d\theta = \int_0^\pi \sin \theta\; d\theta = -\cos \theta \Big|_0^\pi = 2 \;.$

25. The element of surface area is

$$dA = 2\pi y \sqrt{\left(\frac{dx}{dt}\right)^2 + \left(\frac{dy}{dt}\right)^2} = 2\pi(1 - \cos t)\sqrt{(1 - \cos t)^2 + \sin^2 t}\; dt$$

$$= 2\pi(1 - \cos t)\sqrt{2 - 2\cos t}\; dt = 4\pi(1 - \cos t)\left|\sin \frac{t}{2}\right| dt$$

$$= 4\pi(1 - \cos t)\left(\sin \frac{t}{2}\right) dt \quad \text{for} \quad 0 \le t \le 2\pi$$

$$= \left[4\pi \sin \frac{t}{2} - 2\pi\left(\sin \frac{3t}{2} - \sin \frac{t}{2}\right)\right] dt.$$

Therefore, the surface area is

$$2\pi\int_0^{2\pi}\left(3 \sin \frac{t}{2} - \sin \frac{3t}{2}\right) dt = 2\pi\left[-6 \cos \frac{t}{2} + \frac{2}{3}\cos \frac{3t}{2}\right]\Bigg|_0^{2\pi} = \frac{64\pi}{3}\;.$$

27. $\displaystyle \text{Area} = \int_0^\pi \sqrt{1 + \cos 4x}\; dx = \sqrt{2}\int_0^\pi |\sin 2x|\; dx$

$$= \sqrt{2}\left(\int_0^{\pi/2}\sin 2x\; dx + \int_{\pi/2}^\pi -\sin 2x\; dx\right)$$

$$= \sqrt{2}\left(-\frac{\cos 2x}{2}\Bigg|_0^{\pi/2} + \frac{\cos 2x}{2}\Bigg|_\pi^{2\pi}\right) = 2\sqrt{2}\;.$$

Article 7.5

1. $\displaystyle\int_{-2}^2 \frac{dx}{4 + x^2} = \frac{1}{2}\tan^{-1}\left(\frac{x}{2}\right)\Bigg|_{-2}^2 = \frac{\pi}{4}\;.$

3. $\displaystyle\int \frac{dx}{1 + 4x^2} = \frac{1}{4}\int \frac{dx}{\left(\frac{1}{4}\right) + x^2} = \frac{1}{2}\tan^{-1}(2x) + C\;.$

5. $\displaystyle\int \frac{2\;dx}{\sqrt{1 - 4x^2}} = \int \frac{2\;dx}{2\sqrt{\left(\frac{1}{4}\right) - x^2}} = \sin^{-1}(2x) + C\;.$

7. $\displaystyle\int \frac{dy}{\sqrt{25 + y^2}} = \ln\left|\sqrt{25 + y^2} + y\right| + C\;.$

9. $$\int \frac{dy}{\sqrt{25 + 9y^2}} = \frac{1}{3} \int \frac{dy}{\sqrt{\left(\frac{25}{9}\right) + y^2}} = \frac{1}{3} \ln \left| \sqrt{\frac{25}{9} + y^2} + y \right| + C .$$

11. $$\int \frac{3\ dz}{\sqrt{9z^2 - 1}} = \int \frac{3\ dz}{3\sqrt{z^2 - \left(\frac{1}{9}\right)}} = \ln \left| z + \sqrt{z^2 - \frac{1}{9}} \right| + C .$$

13. $$\int \frac{dx}{x\sqrt{x^2 - 16}} = \frac{1}{4} \sec^{-1} \left(\frac{x}{4}\right) + C .$$

15. Let $u = 2x$, $du = 2\ dx$. Then,

$$\int \frac{dx}{2x\sqrt{4x^2 - 1}} = \frac{1}{2} \int \frac{du}{u\sqrt{u^2 - 1}} = \frac{1}{2} \sec^{-1} |u| + C = \frac{1}{2} \sec^{-1} |2x| + C .$$

17. To find $\int \sqrt{x^2 - 4} \, dx$, let $x = 2\sec u$, $dx = 2\sec u \tan u \, du$. Then,

$$\int \sqrt{x^2 - 4} \, dx = \int 4\tan^2 u \sec u \, du = 4\int (\sec^2 u - 1) \sec u \, du$$

To find $\int \sec^3 u \, du$, use integration by parts:

$$w = \sec u, \quad dv = \sec^2 u \, du, \quad v = \tan u, \quad dw = \sec u \tan u \, du.$$

So,

$$\int \sec^3 u \, du = \sec u \tan u - \int \sec u \tan^2 u \, du$$

$$= \sec u \tan u + \int \sec u \, du - \int \sec^3 u \, du$$

$$= \sec u \tan u + \ln |\sec u + \tan u| - \int \sec^3 u \, du.$$

Hence,

$$\int \sec^3 u \, du = \frac{1}{2} \sec u \tan u + \frac{1}{2} \ln |\sec u + \tan u|,$$

and

$$\int \sqrt{x^2 - 4} \, dx = 2\sec u \tan u - \ln |\sec u + \tan u| + C$$

$$= \frac{1}{2} x \sqrt{x^2 - 4} - 2\ln \left| \frac{x}{2} + \frac{\sqrt{x^2 - 4}}{2} \right| + C$$

So,

$$\int_2^4 \sqrt{x^2 - 4} = 4\sqrt{3} - 2\ln (2 + \sqrt{3}) \ .$$

19. If $x = \cos u$, $dx = -\sin u \, du$, and

$$\int \frac{x^3 \, dx}{\sqrt{1 - x^2}} = \int \frac{-\cos^3 u \, \sin u}{\sin u} \, du = \int -\cos^3 u \, du = \int \cos u \, (\sin^2 u - 1) \, du$$

$$= \frac{\sin^3 u}{3} - \sin u + C = \frac{\left(1 - x^2\right)^{3/2}}{3} - \sqrt{1 - x^2} + C .$$

21. To find

$$\int \frac{dx}{\sqrt{1 + x^2}} ,$$

let $x = \tan u$, $dx = \sec^2 u \, du$. Then,

$$\int \frac{dx}{\sqrt{1 + x^2}} = \int \frac{\sec^2 u \, du}{\sec u} = \int \sec u \, du$$

$$= \ln \left| \sec u + \tan u \right| + C = \ln \left| x + \sqrt{x^2 + 1} \right| + C .$$

Then,

$$\int_0^{1/2} \frac{dx}{\sqrt{1 + x^2}} = \ln \left| x + \sqrt{x^2 + 1} \right| \Big|_0^{1/2} = \ln \left(\frac{1 + \sqrt{5}}{2} \right) .$$

23. $x = \csc u$, $dx = -\csc u \cot u \, du$, so

$$\int \frac{dx}{x^2 \sqrt{x^2 - 1}} = \int \frac{-\csc u \cot u}{\csc^2 u \cot u} \, du = \int -\sin u \, du = \cos u + C = \frac{\sqrt{x^2 - 1}}{x} + C .$$

25. $$\int \frac{dx}{\sqrt{1 - 4x^2}} = \frac{1}{2} \int \frac{dx}{\sqrt{\left(\frac{1}{4}\right) - x^2}} = \frac{1}{2} \sin^{-1} (2x) + C .$$

27. Let $w = x - 1$, $dw = dx$. Then,

$$\int \frac{dx}{\sqrt{4 - (x - 1)^2}} = \int \frac{dw}{\sqrt{4 - w^2}} = \frac{1}{2} \sin^{-1} \left(\frac{w}{2} \right) + C = \frac{1}{2} \sin^{-1} \left(\frac{x - 1}{2} \right) + C .$$

29. $$\int_0^2 \frac{dx}{\sqrt{4 + x^2}} = \ln \left| \sqrt{4 + x^2} + x \right| \Big|_0^2 = \ln (2 + \sqrt{8}) - \ln 2 = \ln (1 + \sqrt{2}) .$$

31. Let $u = 4 + x^2$, $du = 2x \, dx$. Then,

$$\int \frac{x \, dx}{\sqrt{4 + x^2}} = \frac{1}{2} \int \frac{du}{\sqrt{u}} = \sqrt{u} + C = \sqrt{4 + x^2} + C .$$

33. Let $x = \tan u$, $dx = \sec^2 u \, du$. At $x = 0$, $u = 0$; at $x = 1$, $u = \pi/4$. Then,

$$\int_0^1 \frac{x^3 \, dx}{\sqrt{x^2 + 1}} = \int_0^{\pi/4} \frac{\tan^3 u \sec^2 u \, du}{\sec u} = \int_0^{\pi/4} \tan^3 u \sec u \, du$$

$$= \int_0^{\pi/4} (\sec^2 u - 1)(\tan u \sec u \, du)$$

$$= \frac{\sec^3 u}{3} - \sec u \Big|_0^{\pi/4} = \frac{2 - \sqrt{2}}{3}$$

35.
$$\int \frac{x + 1}{\sqrt{4 - x^2}} \, dx = \int \frac{x}{\sqrt{4 - x^2}} \, dx + \int \frac{dx}{\sqrt{4 - x^2}} = -\sqrt{4 - x^2} + \frac{1}{2} \sin^{-1} \frac{x}{2} + C ,$$

using the substitution $u = 4 - x^2$, $du = -2x \, dx$ in the first integral on the right-hand side.

37. Let $u = \cos \theta$, $du = -\sin \theta \, d\theta$. Then,

$$\int \frac{\sin \theta \, d\theta}{\sqrt{2 - \cos^2}} = \int \frac{-du}{\sqrt{2 - u^2}} = -\frac{1}{\sqrt{2}} \sin^{-1} \left(\frac{u}{\sqrt{2}} \right) = -\frac{1}{\sqrt{2}} \sin^{-1} \left(\frac{\cos \theta}{\sqrt{2}} \right)$$

39. Let $x = a \cos u$, $dx = -a \sin u \, du$. Then,

$$\int \frac{dx}{x\sqrt{a^2 - x^2}} = \int \frac{-a \sin u \, du}{a^2 \sin u \cos u} = -\frac{1}{a} \int \sec u \, du$$

$$= -\frac{1}{a} \ln |\sec u + \tan u| + C = -\frac{1}{a} \ln \left| \frac{a}{x} + \frac{\sqrt{a^2 - x^2}}{x} \right| + C .$$

41. Let $x = a \tan u$, $dx = a \sec^2 u \, du$. Then,

$$\int \frac{dx}{\left(a^2 + x^2\right)^2} = \int \frac{a \sec^2 u \, du}{a^4 \sec^4 u \, du} = \frac{1}{a^3} \int \cos^2 u \, du = \frac{1}{a^3} \int \frac{1 + \cos 2u}{2} \, du$$

$$= \frac{u}{2a^3} + \frac{\sin 2u}{4a^3} + C = \frac{1}{2a^3} \tan^{-1} \left(\frac{x}{a} \right) + \frac{x}{2a^2 (a^2 + x^2)}$$

43. Let $x = \sin u$, $dx = \cos u \, du$. Then,

$$\int x \sin^{-1} x \, dx = \int u \sin u \cos u \, du = \frac{1}{2} \int u \sin 2u \, du .$$

Now, integrate by parts: let $w = u$, $dv = \sin 2u \, du$. Then,

$$v = -\frac{1}{2} \cos 2u, \quad dw = du,$$

and

$$\frac{1}{2}\int u \sin 2u \, du = -\frac{u}{4} \cos 2u + \frac{1}{4}\int \cos 2u \, du = -\frac{u \cos 2u}{4} + \frac{\sin 2u}{8}$$

$$= \frac{-u(1 - 2\sin^2 u)}{4} + \frac{\sin u \cos u}{4}$$

$$= \frac{(2x^2 - 1) \sin^{-1} x}{4} + \frac{x\sqrt{1 - x^2}}{4} + C \ .$$

45. Let $x = \sin u$, $dx = \cos u \, du$. Then,

$$\int \frac{dx}{x^2\sqrt{1 - x^2}} = \int \frac{\cos u \, du}{\sin^2 u \cos u} = \int \csc^2 u \, du = -\cot u + C = -\frac{\sqrt{1 - x^2}}{x} + C \ .$$

47. a) Let $u = 16 - y^2$, $du = -2y \, dy$. Then,

$$\int \frac{y \, dy}{\sqrt{16 - y^2}} = -\frac{1}{2}\int \frac{du}{\sqrt{u}} = -\sqrt{u} + C = -\sqrt{16 - y^2} + C \ .$$

b) Let $y = 4\sin u$, $dy = 4\cos u \, du$. Then,

$$\int \frac{y \, dy}{\sqrt{16 - y^2}} = \int \frac{16\sin u \cos u}{4\cos u} \, du = 4\int \sin u \, du = -4\cos u + C$$

$$= -4\sqrt{1 - \frac{y^2}{16}} + C = -\sqrt{16 - y^2} + C \ .$$

49. a) Let $u = x^2 - 1$, $du = 2x \, dx$. Then,

$$\int \frac{x \, dx}{\left(x^2 - 1\right)^{3/2}} = \frac{1}{2}\int u^{-3/2} \, du = -u^{-1/2} + C = \frac{-1}{\sqrt{x^2 - 1}} + C \ .$$

b) Let $x = \sec u$, $dx = \sec u \tan u \, du$. Then,

$$\int \frac{x \, dx}{\left(x^2 - 1\right)^{3/2}} = \int \frac{\sec^2 u \tan u}{\tan^3 u} \, du = \int \csc^2 u \, du$$

$$= -\cot u + C = \frac{-1}{\sqrt{x^2 - 1}} + C \ .$$

51. Area $= \displaystyle\int_0^3 \sqrt{1 - \left(\frac{x^2}{9}\right)} \, dx$

Let $x = 3\sin u$, $dx = 3\cos u \, du$. At $x = 0$, $u = 0$; at $x = 3$, $u = \pi/2$. So,

$$\int_0^3 \sqrt{1 - \frac{x^2}{9}}\, dx = 3\int_0^{\pi/2} \cos^2 u\, du = \frac{3}{2}\int_0^{\pi/2} (1 + \cos 2u)\, du$$

$$= \left(\frac{3u}{2} + \frac{3\sin 2u}{4}\right)\Bigg|_0^{\pi/2} = \frac{3\pi}{4}\ .$$

53. We have

$$\frac{dy}{y^2} = \left(1 - \frac{1}{\sqrt{4 - x^2}}\right) dx\ ,$$

so

$$\int\frac{dy}{y^2} = \int dx - \int\frac{dx}{\sqrt{4 - x^2}}\ , \quad \text{or} \quad -\frac{1}{y} = x - \sin^{-1}\left(\frac{x}{2}\right) + C\ .$$

To determine C use the given point $(\sqrt{2},\ 4/\pi)$ to find

$$-\frac{\pi}{4} = \sqrt{2} - \sin^{-1}\left(\frac{\sqrt{2}}{2}\right) + C, \quad \text{or} \quad C = -\sqrt{2}$$

Therefore,

$$y = \frac{-1}{x - \sin^{-1}\left(\frac{x}{2}\right) - \sqrt{2}}$$

55. a)

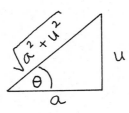

b)

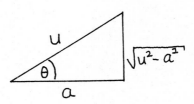

From the diagram,

$$\sin\left(\tan^{-1}\frac{u}{a}\right) = \frac{u}{\sqrt{u^2 + a^2}}$$

From the diagram,

$$\cos\left(\sec^{-1}\frac{u}{a}\right) = \frac{a}{u}$$

Article 7.6

1.

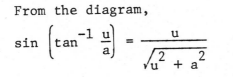
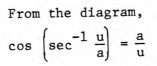

$$\int_1^3 \frac{dx}{x^2 - 2x + 5} = \int_1^3 \frac{dx}{(x^2 - 2x + 1) + 5} = \int_1^3 \frac{dx}{(x - 1)^2 + 4}$$

$$= \frac{1}{2}\tan^{-1}\left(\frac{x - 1}{2}\right)\Bigg|_1^3 = \frac{\pi}{8}\ .$$

3.
$$\int \frac{x\ dx}{x^2 - 2x + 5} = \int \frac{x\ dx}{(x^2 - 2x + 1) + 4} = \int \frac{x\ dx}{(x - 1)^2 + 4}$$

$$= \int \frac{(x - 1)\ dx}{(x - 1)^2 + 4} + \int \frac{dx}{(x - 1)^2 + 4}$$

$$= \frac{1}{2} \ln [(x - 1)^2 + 4] + \frac{1}{2} \tan^{-1} \left(\frac{x - 1}{2}\right) + C .$$

5.
$$\int_1^2 \frac{dx}{x^2 - 2x + 4} = \int_1^2 \frac{dx}{(x^2 - 2x + 1) + 3} = \int_1^2 \frac{dx}{(x - 1)^2 + 3}$$

$$= \frac{1}{\sqrt{3}} \tan^{-1} \left(\frac{x - 1}{\sqrt{3}}\right) \Big|_1^2 = \frac{\pi}{6\sqrt{3}}$$

7.
$$\int \frac{dx}{\sqrt{9x^2 - 6x + 5}} = \int \frac{dx}{\sqrt{(3x - 1)^2 + 4}} = \frac{1}{3} \ln |(3x - 1) + \sqrt{(3x - 1)^2 + 4}| + C .$$

9.
$$\int \frac{x\ dx}{\sqrt{9x^2 - 6x + 5}} = \frac{1}{3} \int \frac{3x\ dx}{\sqrt{(3x - 1)^2 + 4}}$$

$$= \frac{1}{3} \int \frac{(3x - 1)\ dx}{\sqrt{(3x - 1)^2 + 4}} + \frac{1}{3} \int \frac{dx}{\sqrt{(3x - 1)^2 + 4}}$$

$$= \frac{1}{3} \sqrt{(3x - 1)^2 + 4} + \frac{1}{9} \ln |(3x - 1) + \sqrt{(3x - 1)^2 + 4}| + C .$$

11. $$\int \frac{dx}{\sqrt{x^2 + 2x}} = \int \frac{dx}{\sqrt{(x + 1)^2 - 1}} = \ln \left| (x + 1) + \sqrt{(x + 1)^2 - 1} \right| + C$$

13. $$\int \frac{x \, dx}{\sqrt{x^2 + 4x + 13}} = \int \frac{x \, dx}{\sqrt{(x + 2)^2 + 9}} = \int \frac{(x + 2) \, dx}{\sqrt{(x + 2)^2 + 9}} - \int \frac{2 \, dx}{\sqrt{(x + 2)^2 + 9}}$$

$$= \sqrt{(x + 2)^2 + 9} - 2\ln \left| (x + 2) + \sqrt{(x + 2)^2 + 9} \right| + C \ .$$

15. $$\int \frac{dx}{\sqrt{x^2 - 2x - 3}} = \int \frac{dx}{\sqrt{(x^2 - 2x + 1) - 4}} = \int \frac{dx}{\sqrt{(x - 1)^2 - 4}}$$

$$= \ln \left| (x - 1) + \sqrt{(x - 1)^2 + 4} \right| + C \ .$$

17.
$$\int \frac{(x+1)\ dx}{\sqrt{2x-x^2}} = \int \frac{(x+1)\ dx}{\sqrt{1-(1-2x+x^2)}} = \int \frac{(x+1)\ dx}{\sqrt{1-(x-1)^2}}$$

$$= \int \frac{(x-1)\ dx}{\sqrt{1-(x-1)^2}} + \int \frac{2\ dx}{\sqrt{1-(x-1)^2}}$$

$$= -\sqrt{1-(x-1)^2} + 2\sin^{-1}(x-1) + C.$$

19.
$$\int \frac{x\ dx}{\sqrt{5+4x-x^2}} = \int \frac{x\ dx}{\sqrt{9-(x-2)^2}} = \int \frac{(x-2)\ dx}{\sqrt{9-(x-2)^2}} + \int \frac{2\ dx}{\sqrt{9-(x-2)^2}}$$

$$= -\sqrt{9-(x-2)^2} + 2\sin^{-1}\left(\frac{x-2}{3}\right) + C.$$

21.
$$\int \frac{(1-x)\ dx}{\sqrt{8+2x-x^2}} = \int \frac{(1-x)\ dx}{\sqrt{9-(x-1)^2}} = \sqrt{9-(x-1)^2} + C.$$

23.
$$\int \frac{x\ dx}{x^2+4x+5} = \int \frac{x\ dx}{(x+2)^2+1} = \int \frac{(x+2)\ dx}{(x+2)^2+1} - \int \frac{2\ dx}{(x+2)^2+1}$$

$$= \frac{1}{2}\ln(x^2+4x+5) - 2\tan^{-1}(x+2)$$

25.
$$\int_a^3 \frac{dx}{(x+1)\sqrt{x^2+2x-3}} = \int_a^3 \frac{dx}{(x+1)\sqrt{(x+1)^2-4}}$$

$$= \frac{1}{2}\sec^{-1}\left|\frac{x+1}{2}\right|\Bigg|_a^3 = \frac{\pi}{6} - \frac{1}{2}\sec^{-1}\left|\frac{a+1}{2}\right|.$$

As $a \to 1^+$, $a+1 \to a^+$, and $\sec^{-1}\left|\frac{a+1}{2}\right| \to 0$. So,

$$\lim_{a\to 1^+} \int_a^3 \frac{dx}{(x+1)\sqrt{x^2+2x-3}} = \frac{\pi}{6}$$

Question: Why must we require $a \to 1^+$?

Article 7.7

1.
$$\frac{5x-13}{(x-3)(x-2)} = \frac{A}{x-3} + \frac{B}{x-2}.$$

By the "cover up" method,

$$A = \frac{5(3)-13}{3-2} = 2 \quad \text{and} \quad B = \frac{5(2)-13}{2-3} = 3.$$

So, $\dfrac{5x-13}{(x-3)(x-2)} = \dfrac{2}{x-3} + \dfrac{3}{x-2}.$

3.
$$\frac{x + 4}{(x + 1)^2} = \frac{A}{x + 1} + \frac{B}{(x + 1)^2}$$

or $(x + 4) = A(x + 1) + B = Ax + (A + B)$.

Equating coefficients of like powers: $A = 1$, $A + B = 4$. So, $B = 3$, and

$$\frac{x + 4}{(x + 1)^2} = \frac{1}{x + 1} + \frac{3}{(x + 1)^2} .$$

5.
$$\frac{x + 1}{x^2(x - 1)} = \frac{A}{x} + \frac{B}{x^2} + \frac{C}{x - 1}$$

$$x + 1 = Ax(x - 1) + B(x - 1) + Cx^2 = (A + C)x^2 + (B - A)x - B .$$

So, $A + C = 0$, $B - A = 1$, and $-B = 1$. Hence, $B = -1$, $A = -2$, $C = 2$, and

$$\frac{x + 1}{x^2(x - 1)} = -\frac{2}{x} - \frac{1}{x^2} + \frac{2}{x - 1} .$$

7.
$$\frac{x^2 + 8}{x^2 - 5x + 6} = 1 + \frac{5x + 2}{x^2 - 5x + 6} = 1 + \frac{5x + 2}{(x - 2)(x - 3)}$$

So, for

$$\frac{5x + 2}{(x - 2)(x - 3)} = \frac{A}{x - 2} + \frac{B}{x - 3} , \qquad A = -12, B = 17.$$

Thus,

$$\frac{x^2 + 8}{x^2 - 5x + 6} = 1 - \frac{12}{x - 2} + \frac{17}{x - 3} .$$

9.
$$\frac{3}{x^2(x^2 + 9)} = \frac{A}{x} + \frac{B}{x^2} + \frac{Cx + D}{x^2 + 9} :$$

$$3 = Ax(x^2 + 9) + B(x^2 + 9) + (Cx + D)x^2$$

$$= (A + C)x^3 + (B + D)x^2 + 9Ax + 9B .$$

So, $A + C = 0$, $B + D = 0$, $9A = 0$, $9B = 3$. Therefore, $A = 0$, $B = 1/3$, $C = 0$, $D = -1/3$, and

$$\frac{3}{x^2(x^2 + 9)} = \frac{1/3}{x^2} - \frac{1/3}{x^2 + 9} .$$

11.
$$\frac{1}{1 - x^2} = \frac{1}{(1 - x)(1 + x)} = -\frac{1}{(x - 1)(x + 1)} = \frac{A}{x - 1} + \frac{B}{x + 1}$$

By the "cover up" method, $A = -1/2$, $B = +1/2$. So,

$$\int_0^{1/2} \frac{dx}{1-x^2} = \frac{1}{2}\int_0^{1/2} \frac{dx}{x-1} - \frac{1}{2}\int_0^{1/2} \frac{dx}{x+1}$$

$$= \left(-\frac{1}{2}\ln|x-1| + \frac{1}{2}\ln|x+1|\right)\Big|_0^{1/2}$$

$$= -\frac{1}{2}\ln\frac{1}{2} + \frac{1}{2}\ln\frac{3}{2} = -\frac{1}{2}\ln\left(\frac{1}{3}\right) = \frac{1}{2}\ln 3.$$

13. $\quad \dfrac{x^3}{x^2+1} = x - \dfrac{x}{x^2+1}$,

so $\quad \displaystyle\int_0^{2\sqrt{2}} \frac{x^3}{x^2+1}\,dx = \int_0^{2\sqrt{2}} x\,dx + \int_0^{2\sqrt{2}} \frac{-x\,dx}{x^2+1}$

$$= \frac{x^2}{2} - \frac{1}{2}\ln(x^2+1)\Big|_0^{2\sqrt{2}} = 4 - \frac{1}{2}\ln 9 = 4 - \ln 3.$$

15. $\quad \dfrac{1}{x-x^2} = \dfrac{1}{x(1-x)} = \dfrac{A}{x} + \dfrac{B}{1-x}$.

So, $1 = A(1-x) + Bx$, or $1 = (B-A)x + A$. Thus, $A = 1$, $B = 1$, and

$$\int_{1/4}^{3/4} \frac{dx}{x-x^2} = \int_{1/4}^{3/4} \frac{dx}{x} - \int_{1/4}^{3/4} \frac{dx}{x-1} = \ln|x| - \ln|x-1|\Big|_{1/4}^{3/4}$$

$$= \ln\frac{3}{4} - \ln\frac{1}{4} - \ln\frac{1}{4} + \ln\frac{3}{4} = 2\ln 3.$$

17. $\quad \dfrac{x+4}{x^2+5x-6} = \dfrac{x-4}{(x+6)(x-1)} = \dfrac{A}{x+6} + \dfrac{B}{x-1}$.

So, $x-4 = A(x-1) + B(x+6)$. Thus, $A+B = 1$ and $-A+6B = -4$. Hence, $B = -3/7$ and $A = 10/7$. Hence,

$$\int \frac{x+4}{x^2+5x-6}\,dx = \frac{10}{7}\int \frac{dx}{x+6} - \frac{3}{7}\int \frac{dx}{x-1}$$

$$= \frac{10}{7}\ln|x+6| - \frac{3}{7}\ln|x-1| + C.$$

19. $\quad \dfrac{3x^2}{x^2+2x+1} = 3 - \dfrac{6x+3}{x^2+2x+1} = 3 - \dfrac{6x+3}{(x+1)^2}$

Now expand

$$\frac{6x+3}{(x+1)^2} = \frac{A}{x+1} + \frac{B}{(x+1)^2}: \quad 6x+3 = A(x+1) + B.$$

Hence, $A = 6$, $B = -3$, and we have

$$\int \frac{3x^2}{x^2 + 2x + 1} \, dx = \int \left(3 - \frac{6}{x + 1} + \frac{3}{(x + 1)^2} \right) dx$$

$$= 3x - \frac{3}{x + 1} - 6 \ln |x + 1| + C.$$

21. $$\frac{x}{x^2 + 4x - 5} = \frac{x}{(x + 5)(x - 1)} = \frac{A}{x + 5} + \frac{B}{x - 1} \; .$$

By the "cover up" method, A = 5/6, B = 1/6, and

$$\int \frac{x \, dx}{x^2 + 4x - 5} = \frac{5}{6} \int \frac{dx}{x + 5} + \frac{1}{6} \int \frac{dx}{x - 1} = \frac{5}{6} \ln |x + 5| + \frac{1}{6} \ln |x - 1| + C.$$

23. $$\frac{x + 1}{x^2 + 4x - 5} = \frac{x + 1}{(x + 5)(x - 1)} = \frac{A}{x + 5} + \frac{B}{x - 1} \; .$$

By the "cover up" method, A = 2/3, B = 1/3, and

$$\int \frac{(x + 1) \, dx}{x^2 + 4x - 5} = \frac{2}{3} \int \frac{dx}{x + 5} + \frac{1}{3} \int \frac{dx}{x - 1} = \frac{2}{3} \ln |x + 5| + \frac{1}{3} \ln |x - 1| + C.$$

25. $$\frac{1}{x(x + 1)^2} = \frac{A}{x} + \frac{B}{x + 1} + \frac{C}{(x + 1)^2} \; ,$$

or $1 = A(x + 1)^2 + B(x + 1)x + Cx = (A + B)x^2 + (2A + B + C)x + A.$
Thus, A = 1, B = -1, C = -1, and

$$\int_1^3 \frac{dx}{x(x + 1)^2} = \int_1^3 \left(\frac{1}{x} - \frac{1}{x + 1} - \frac{1}{(x + 1)^2} \right) dx$$

$$= \ln |x| - \ln |x + 1| + \frac{1}{x + 1} \Big|_1^3 = \ln 3 - \ln 2 - \frac{1}{4} \; .$$

27. $$\frac{x + 3}{2x^3 - 8x} = \frac{x + 3}{2x(x + 2)(x - 2)} = \frac{A}{x} + \frac{B}{x + 2} + \frac{C}{x - 2} \; .$$

By the "cover up" method, A = -3/8, B = 1/16, C = 5/16, and

$$\int \frac{(x + 3) \, dx}{2x^3 - 8x} = \int -\frac{3 \, dx}{8x} + \int \frac{dx}{16(x + 2)} + \int \frac{5 \, dx}{16(x - 2)}$$

$$= -\frac{3}{8} \ln |x| + \frac{1}{16} \ln |x + 2| + \frac{5}{16} \ln |x - 2|.$$

29. $$\int_0^{\sqrt{3}} \frac{5x^2}{x^2 + 1}\, dx = \int_0^{\sqrt{3}} \left(5 - \frac{5}{x^2 + 1}\right)\, dx$$

$$= 5x - 5\, \tan^{-1}(x)\,\Big|_0^{\sqrt{3}} = 5\left(\sqrt{3} - \frac{\pi}{3}\right)$$

31. $$\frac{1}{x(x^2 + x + 1)} = \frac{A}{x} + \frac{Bx + C}{x^2 + x + 1}\,,$$

or $1 = A(x^2 + x + 1) + (Bx + C)x = (A + B)x^2 + (A + C)x + A.$

Then, $A = 1$, $B = -1$, $C = -1$, and

$$\int \frac{dx}{x(x^2 + x + 1)} = \int \frac{dx}{x} - \int \frac{x + 1}{x^2 + x + 1}\, dx = \int \frac{dx}{x} - \int \frac{x + 1}{\left(x + \frac{1}{2}\right)^2 + \frac{3}{4}}\, dx$$

$$= \ln|x| - \frac{1}{2}\ln(x^2 + x + 1) - \frac{1}{\sqrt{3}}\tan^{-1}\left[\frac{2}{\sqrt{3}}\left(x + \frac{1}{2}\right)\right] + C.$$

33. $$\frac{3x^2 + x + 4}{x^3 + x} = \frac{3x^2 + x + 4}{x(x^2 + 1)} = \frac{A}{x} + \frac{Bx + C}{x^2 + 1}\,,$$

or $3x^2 + x + 4 = A(x^2 + 1) + (Bx + C)x = (A + B)x^2 + Cx + A.$

So, $A = 4$, $B = -1$, $C = 1$, and

$$\int \frac{3x^2 + x + 4}{x^3 + x}\, dx = \int \left(\frac{4}{x} - \frac{x - 1}{x^2 + 1}\right)\, dx$$

$$= 4\ln|x| - \frac{1}{2}\ln(x^2 + 1) + \tan^{-1} x + C.$$

35. $$\int \frac{x^3 + 4x^2}{x^2 + 4x + 3}\, dx = \int \left(x - \frac{3x}{x^2 + 4x + 3}\right)\, dx = \int \left(x - \frac{9}{2(x + 3)} + \frac{3}{2(x + 1)}\right)\, dx$$

$$= \frac{x^2}{2} - \frac{9}{2}\ln|x + 3| + \frac{3}{2}\ln|x + 1| + C.$$

37. $$\frac{x^2 + 2x + 1}{\left(x^2 + 1\right)^2} = \frac{Ax + B}{x^2 + 1} + \frac{Cx + D}{\left(x^2 + 1\right)^2}\,,$$

or $x^2 + 2x + 1 = (Ax + B)(x^2 + 1) + Cx + D = Ax^3 + Bx^2 + (A + C)x + (B + D).$

So, $A = 0$, $B = 1$, $C = 2$, $D = 0$, and

$$\int \frac{x^2 + 2x + 1}{\left(x^2 + 1\right)^2}\, dx = \int \left[\frac{1}{x^2 + 1} + \frac{2x}{\left(x^2 + 1\right)^2}\right]\, dx = \tan^{-1} x - \frac{1}{x^2 + 1} + C.$$

39. $$\frac{2x}{(x^2 + 1)(x - 1)^2} = \frac{Ax + B}{x^2 + 1} + \frac{C}{x - 1} + \frac{D}{(x - 1)^2} \, ,$$

or $2x = (Ax + B)(x - 1)^2 + C(x - 1)(x^2 + 1) + D(x^2 + 1)$

$\quad\quad = (A + C)x^3 + (B - 2A - C + D)x^2 + (A - 2B + C)x + (B - C + D).$

So, $B = -1$, $D = 1$, $C = 0$, $A = 0$, and

$$\int \frac{2x}{(x^2 + 1)(x - 1)^2} \, dx = \int \left(\frac{-1}{x^2 + 1} + \frac{1}{(x - 1)^2} \right) dx = -\tan^{-1} x - \frac{1}{x - 1} + C.$$

41. Let $u = e^t$, $du = e^t \, dt$. At $t = 0$. $u = 1$, while at $t = \ln 2$, $u = 2$. Then,

$$\int_0^{\ln 2} \frac{e^t \, dt}{e^{2t} + 3e^t + 2} = \int_1^2 \frac{du}{u^2 + 3u + 2} = \int_1^2 \left[\frac{1}{u + 2} + \frac{1}{u + 1} \right] du$$

$$= \ln |u + 1| - \ln |u + 2| \Big|_1^2 = 2 \ln 3 - 3 \ln 2$$

43. $$\frac{x^4}{\left(x^2 + 1\right)^2} = 1 - \frac{2x^2 + 1}{\left(x^2 + 1\right)^2} \, ,$$

so

$$\int \frac{x^4}{\left(x^2 + 1\right)^2} \, dx = x - \int \frac{2x^2 + 1}{\left(x^2 + 1\right)^2} \, dx \, .$$

Now let $x = \tan u$, $dx = \sec^2 u \, du$. Then,

$$\int \frac{2x^2 + 1}{\left(x^2 + 1\right)^2} \, dx = \int \frac{(2\tan^2 u + 1) \sec^2 u}{\sec^4 u} \, du = \int \frac{2\sec^2 u - 1}{\sec^4 u} \cdot \sec^2 u \, du$$

$$= \int (2 - \cos^2 u) \, du = \int \left(\frac{3}{2} - \frac{1}{2} \cos 2u \right) du$$

$$= \frac{3}{2} u - \frac{1}{4} \sin 2u = \frac{3}{2} \tan^{-1} x - \frac{1}{2} \cdot \frac{x}{x^2 + 1} \, .$$

Consequently,

$$\int \frac{x^4}{\left(x^2 + 1\right)^2} \, dx = x - \frac{3}{2} \tan^{-1} x + \frac{1}{2} \cdot \frac{x}{x^2 + 1} \, .$$

45. If $u = \sqrt{x}$, $x = u^2$ and $dx = 2u \, du$. Then,

$$\int \frac{1 - \sqrt{x}}{1 + \sqrt{x}} \, dx = \int \frac{(1 - u)(2u)}{1 + u} \, du = \int \frac{-2u^2 + 2u}{u + 1} \, du = \int \left(-2u + 4 - \frac{4}{u + 1} \right) du$$

$$= -u^2 + 4u - 4 \ln |u + 1| = 4\sqrt{x} - x - 4 \ln (1 + \sqrt{x}).$$

47. $x = u^6$, $dx = 6u^5$ du, and

$$\int \frac{dx}{\sqrt{x} + \sqrt[3]{x}} = \int \frac{6u^5 \, du}{u^3 + u^2} = \int \frac{6u^3 \, du}{u + 1} = 6 \int \left(u^2 - u + 1 - \frac{1}{u + 1} \right) \, du$$

$$= 6[u^3 - u^2 + u - \ell n \, |u + 1|] + C$$

$$= 6[\sqrt{x} - \sqrt[3]{x} + \sqrt[6]{x} - \ell n \, (1 + \sqrt[6]{x})] + C.$$

49. Integrate by parts with $u = \ell n \, (x^2 + 1)$, $dv = dx$. Then,

$$v = x, \quad du = \frac{2x}{x^2 + 1} \, dx, \quad \text{and}$$

$$\int \ell n \, (x^2 + 1) \, dx = x \, \ell n \, (x^2 + 1) - \int \frac{2x^2}{x^2 + 1} \, dx$$

$$= x \, \ell n \, (x^2 + 1) - \int \left(2 - \frac{2}{x^2 + 1} \right) \, dx$$

$$= x \, \ell n \, (x^2 + 1) - 2x + 2\tan^{-1} x + C.$$

51. a)   If $a = b$, then we have

$$\frac{dx}{(a - x)^2} = k \, dt, \quad \text{or} \quad - \frac{1}{x - a} = kt + C.$$

Since $x = 0$ at $t = 0$, $C = +1/a$, so

$$- \frac{1}{x - a} = kt + \frac{1}{a} = \frac{akt + 1}{a} \, .$$

Therefore,

$$x = \frac{a^2 kt}{akt + 1} \, .$$

b)   If $a \neq b$, we have

$$\frac{dx}{(a - x)(b - x)} = k \, dt.$$

But

$$\frac{1}{(a - x)(b - x)} = \frac{1}{(x - a)(x - b)} = \frac{1}{a - b} \cdot \frac{1}{x - a} + \frac{1}{b - a} \cdot \frac{1}{x - b}$$

by the "cover up" method. Hence,

$$\frac{1}{b - a} \int \left(\frac{1}{x - b} - \frac{1}{x - a} \right) \, dx = \int k \, dt$$

or

$$\frac{1}{b - a}[\ell n \, |x - b| - \ell n \, |x - a|] = kt + C.$$

Since x = 0 at t = 0, C = 1/(b - a) ℓn (b/a) (if a,b > 0). Hence,

$$\ell n \left| \frac{x - b}{x - a} \right| = (b - a)kt + \ell n \frac{b}{a}$$

Article 7.8

1. Let $z = \tan \frac{x}{2}$, $dx = \frac{2\ dz}{1 + z^2}$. Then, $\sin x = \frac{2z}{1 + z^2}$ and

$$\int \frac{dx}{1 + \sin x} = \int \frac{2\ dz}{1 + 2z + z^2} = \int \frac{2\ dz}{(1 + z)^2} = -\frac{2}{1 + z} = \frac{-2}{1 + \tan \frac{x}{2}}$$

Thus,

$$\int_0^\pi \frac{dx}{1 + \sin x} = \frac{-2}{1 + \tan \frac{x}{2}} \Bigg|_0^\pi = 2.$$

3. Let $z = \tan \frac{x}{2}$, $dx = \frac{2\ dz}{1 + z^2}$. Then, $\sin x = \frac{2z}{1 + z^2}$ and

$$\int \frac{dx}{1 - \sin x} = \int \frac{2\ dz}{1 - 2z + z^2} = \int \frac{2\ dz}{(z - 1)^2} = \frac{2}{1 - z} + C = \frac{2}{1 - \tan \frac{x}{2}} + C.$$

5. Let $z = \tan \frac{x}{2}$, $dx = \frac{2\ dz}{1 + z^2}$, $\cos x = \frac{1 - z^2}{1 + z^2}$. Then,

$$\int \frac{\cos x\ dx}{2 - \cos x} = \int \frac{2 - 2z^2}{(z^2 + 1)(z^2 + 3)}\ dz.$$

Now

$$\frac{2 - 2z^2}{(z^2 + 1)(z^2 + 3)} = \frac{Az + B}{z^2 + 1} + \frac{Cz + D}{z^2 + 3},$$

or

$$2 - 2z^2 = (Az + B)(z^2 + 3) + (Cz + D)(z^2 + 1)$$

$$= (A + C)z^2 + (B + D)z^2 + (3A + C)z + (3B + D).$$

Thus, A = 0, B = 2, C = 0, D = -4, and

$$\int \frac{2 - 2z^2}{(z^2 + 1)(z^2 + 3)}\ dz = \int \frac{2\ dz}{z^2 + 1} - \int \frac{4\ dz}{z^2 + 3}$$

$$= 2\tan^{-1} z - \frac{4}{\sqrt{3}} \tan^{-1} \left[\frac{z}{\sqrt{3}} \right] + C$$

$$= x - \frac{4}{\sqrt{3}} \tan^{-1} \left[\frac{1}{\sqrt{3}} \tan \left(\frac{x}{2} \right) \right] + C.$$

7. Let $z = \tan \frac{x}{2}$, $dx = \frac{2\,dz}{1 + z^2}$, $\sin x = \frac{2z}{1 + z^2}$, $\cos x = \frac{1 - z^2}{1 + z^2}$. Then,

$$\int \frac{dx}{\sin x - \cos x} = \int \frac{2\,dz}{z^2 + 2z - 1} = \int \frac{2\,dz}{(z + 1)^2 - 2}.$$

Let $w = z + 1$, $dw = dz$, then,

$$\int \frac{2\,dz}{(z + 1)^2 - 2} = \int \frac{2\,dw}{w^2 - 2} = \frac{1}{\sqrt{2}} \int \left(\frac{1}{w - \sqrt{2}} - \frac{1}{w + \sqrt{2}} \right) dw$$

$$= \frac{1}{\sqrt{2}} \ln \left| \frac{w - \sqrt{2}}{w + \sqrt{2}} \right| + C = \frac{1}{\sqrt{2}} \left| \frac{\tan \frac{x}{2} + 1 - \sqrt{2}}{\tan \frac{x}{2} + 1 + \sqrt{2}} \right| + C.$$

Article 7.9

1. $$\int_0^\infty \frac{dx}{x^2 + 1} = \lim_{b \to \infty} \int_0^b \frac{dx}{x^2 + 1} = \lim_{b \to \infty} \tan^{-1}(b) = \frac{\pi}{2}.$$

3. $$\int_{-1}^1 \frac{dx}{x^{2/3}} = \lim_{b \to 0^-} \int_{-1}^b x^{-2/3}\,dx + \lim_{a \to 0^+} \int_a^1 x^{-2/3}\,dx$$

$$= \lim_{b \to 0^-} (3b^{1/3} + 3) + \lim_{a \to 0^+} (3 - 3a^{1/3}) = 6.$$

5. $$\int_0^4 \frac{dx}{\sqrt{4 - x}} = \lim_{b \to 4^-} \int_0^b \frac{dx}{\sqrt{4 - x}} = \lim_{b \to 4^-} -2\sqrt{4 - b} + 4 = 4$$

7. To find $\int e^{-x} \cos x\,dx$, use integration by parts with $u = e^{-x}$, $dv = \cos x\,dx$. Then, $v = \sin x$, $du = -e^{-x}\,dx$, and $\int e^{-x} \cos x\,dx = e^{-x} \sin x + \int e^{-x} \sin x\,dx$. Now in the last integral let $u = e^{-x}$, $dv = \sin x\,dx$. Then, $v = -\cos x$, $du = -e^{-x}\,dx$, and $\int e^{-x} \sin x\,dx = -e^{-x} \cos x - \int e^{-x} \cos x\,dx$. Hence,

$$\int e^{-x} \cos x\,dx = e^{-x}(\sin x - \cos x) - \int e^{-x} \cos x\,dx$$

or $$\int e^{-x} \cos x\,dx = \frac{e^{-x}}{2}(\sin x - \cos x).$$

Thus,

$$\int_0^\infty e^{-x} \cos x\,dx = \lim_{b \to \infty} \int_0^b e^{-x} \cos x\,dx$$

$$= \lim_{b \to \infty} \left(-e^{-b} \cos b + e^{-b} \sin b + \frac{1}{2} \right) = \frac{1}{2}$$

9. $$\int_{-\infty}^{2}\frac{dx}{4-x} = \lim_{a\to-\infty}\int_{a}^{2}\frac{dx}{4-x} = \lim_{a\to-\infty} -\ln|4-x|\Big|_{a}^{2}$$

$$= \lim_{a\to-\infty}(\ln|4-a| - \ln 2).$$

But this becomes infinite as $a \to -\infty$, so the integral diverges.

11. $$\int_{2}^{\infty}\frac{dx}{x^2-x} = \int_{2}^{\infty}\frac{dx}{\left(x-\frac{1}{2}\right)^2 - \frac{1}{4}} = \int_{3/2}^{\infty}\frac{dw}{w^2-\frac{1}{4}}$$

$$= \int_{3/2}^{\infty}\left[\frac{1}{w-\frac{1}{2}} - \frac{1}{w+\frac{1}{2}}\right]dw \qquad \left(w = x - \frac{1}{2},\ dw = dx\right)$$

$$= \lim_{b\to\infty}\int_{3/2}^{b}\left[\frac{1}{w-\frac{1}{2}} - \frac{1}{w+\frac{1}{2}}\right]dw$$

$$= \lim_{b\to\infty}\left(\ln\left|\frac{b-\frac{1}{2}}{b+\frac{1}{2}}\right| - \ln\left(\frac{1}{2}\right)\right) = \ln 2.$$

13. $$\int_{1}^{\infty}\frac{dx}{\sqrt{x}} = \lim_{b\to\infty}\int_{1}^{b}\frac{dx}{\sqrt{x}} = \lim_{b\to\infty}(2\sqrt{b} - 2) \longrightarrow \infty \quad \text{as } b\to\infty.$$

Hence, the integral is divergent.

15. For $x \geq 1$, $x^3 + 1 \geq x^3$, so $\dfrac{1}{x^3+1} \leq \dfrac{1}{x^3}$. Now,

$$\int_{1}^{\infty}\frac{dx}{x^3} = \lim_{b\to\infty}\int_{1}^{b}\frac{dx}{x^3} = \lim_{b\to\infty}\left(\frac{1}{2} - \frac{1}{2b^2}\right) = \frac{1}{2},$$

so by the domination test,

$$\int_{1}^{\infty}\frac{dx}{x^3+1} \quad \text{converges.}$$

17. $$\int_{0}^{\infty}\frac{dx}{x^3+1} = \lim_{b\to\infty}\int_{0}^{b}\frac{dx}{x^3+1} = \int_{0}^{1}\frac{dx}{x^3+1} + \lim_{b\to\infty}\int_{1}^{b}\frac{dx}{x^3+1}.$$

Now, since

$$\frac{1}{x^3+1} \leq \frac{1}{x^3} \quad \text{and} \quad \int_{1}^{\infty}\frac{dx}{x^3}$$

is convergent, the dominated convergence theorem says

$\lim\limits_{b \to \infty} \int_1^b \dfrac{dx}{x^3 + 1}$ exists and is finite. Since $\int_0^1 \dfrac{dx}{x^3 + 1}$ is a proper integral,

we conclude $\int_0^\infty \dfrac{dx}{x^3 + 1}$ converges.

19. $\displaystyle\int_0^{\pi/2} \tan x \, dx = \lim_{b \to \frac{\pi}{2}^-} \int_0^b \tan x \, dx = \lim_{b \to \frac{\pi}{2}^-} \ln |\sec b|$

which goes to infinity. Hence $\int_0^{\pi/2} \tan x \, dx$ converges.

21. $\displaystyle\int_{-1}^1 \dfrac{dx}{x^{2/5}} = \lim_{b \to 0^-} \int_{-1}^b x^{-0.4} \, dx + \lim_{a \to 0^+} \int_a^1 x^{-0.4} \, dx$

$\displaystyle = \lim_{b \to 0^-} \left(-\dfrac{5}{3} b^{0.6} + \dfrac{5}{3} \right) + \lim_{a \to 0^+} \left(\dfrac{5}{3} - \dfrac{5}{3} a^{0.6} \right) = \dfrac{10}{3} .$

23. $\displaystyle\int_2^\infty \dfrac{dx}{\sqrt{x - 1}} = \lim_{b \to \infty} \int_2^b \dfrac{dx}{\sqrt{x - 1}} = \lim_{b \to \infty} 2\sqrt{b - 1} - 2 ,$

which goes to infinity. Thus, $\int_2^\infty \dfrac{dx}{\sqrt{x - 1}}$ diverges.

25. $\displaystyle\int_0^2 \dfrac{dx}{1 - x^2} = \lim_{b \to 1^-} \int_0^b \dfrac{dx}{1 - x^2} + \lim_{a \to 1^+} \int_a^1 \dfrac{dx}{1 - x^2}$

$\displaystyle = \lim_{b \to 1^-} \dfrac{1}{2} \ln \left| \dfrac{b + 1}{b - 1} \right| - \dfrac{1}{2} \lim_{a \to 1^+} \left| \dfrac{a + 1}{a - 1} \right| .$

But these go to infinity, so $\int_0^2 \dfrac{dx}{1 - x^2}$ diverges.

27. $\displaystyle\int_0^4 \dfrac{dx}{\sqrt{4 - x}} = \lim_{b \to 4^-} \int_0^b \dfrac{dx}{\sqrt{4 - x}} = \lim_{b \to 4^-} (-2\sqrt{4 - b}) + 2 = 2$

29. $\displaystyle\int x^2 e^{-x} \, dx = -x^2 e^{-x} - 2x e^{-x} - 2e^{-x} + C$

(repeated integration by parts, or tabular integration). Since
$\lim\limits_{x \to \infty} x^n e^{-x} = 0$ for all positive integers n,

$\displaystyle\int_0^\infty x^2 e^{-x} \, dx = \lim_{b \to \infty} \int_0^b x^2 e^{-x} \, dx = 2 - \lim_{b \to \infty} (b^2 e^{-b} + 2b e^{-b} + 2e^{-b}) = 2.$

31. Since $2 + \cos x \geq 1$ and $\int_\pi^\infty dx/x$ diverges, $\int_\pi^\infty \frac{2 + \cos x}{x} dx$ diverges.

33.
$$\int_6^\infty \frac{dx}{\sqrt{x + 5}} = \lim_{b \to \infty} \int_6^b \frac{dx}{\sqrt{x + 5}} = \lim_{b \to \infty} 2\sqrt{b + 5} - 2\sqrt{11},$$

which goes to infinity. So, $\int_6^\infty \frac{dx}{\sqrt{x + 5}}$ diverges.

35.
$$\int_2^\infty \frac{dx}{x^2 - 1} = \lim_{b \to \infty} \int_2^b \frac{dx}{x^2 - 1} = \lim_{b \to \infty} \left[\frac{1}{2} \int_2^b \left(\frac{1}{x - 1} - \frac{1}{x + 1} \right) dx \right]$$

$$= \lim_{b \to \infty} \frac{1}{2} \ln \left| \frac{b - 1}{b + 1} \right| - \frac{1}{2} \ln \left(\frac{1}{3} \right) = \frac{1}{2} \ln 3.$$

37. Let $x = e^u$, $dx = e^u du$. Then,

$$\int_2^\infty \frac{dx}{\ln x} = \int_{\ln 2}^\infty \frac{e^u}{u} du.$$

But $e^u/u > 1/u$, and $\int_{\ln 2}^\infty du/u$ diverges. So, $\int_2^\infty dx/\ln x$ diverges.

39.
$$\lim_{x \to \infty} \frac{1/e^x}{1/(e^x - 2^x)} = \lim_{x \to \infty} \left[1 - \left(\frac{2}{e} \right)^x \right] = 1,$$

so the limit comparison test says that $\int_1^\infty dx/(e^x - 2^x)$ converges since $\int_1^\infty e^{-x} dx = 1$.

41.
$$\int_0^\infty \frac{dx}{\sqrt{x + x^4}} = \lim_{\substack{b \to \infty \\ a \to 0^+}} \int_a^b \frac{dx}{\sqrt{x + x^4}}$$

$$= \lim_{a \to 0^+} \int_a^1 \frac{dx}{\sqrt{x + x^4}} + \lim_{b \to \infty} \int_1^b \frac{dx}{\sqrt{x + x^4}}.$$

Now,
$$\frac{1}{\sqrt{x + x^4}} \leq \frac{1}{\sqrt{x}}, \text{ so } \int_0^1 \frac{dx}{\sqrt{x + x^4}} \text{ is finite, since } \int_0^1 \frac{dx}{\sqrt{x}} = 2.$$

Also,
$$\lim_{x \to \infty} \frac{1/x^2}{1/\sqrt{x + x^4}} = \lim_{x \to \infty} \frac{\sqrt{x + x^4}}{x^2} = \lim_{x \to \infty} \sqrt{1 + \frac{1}{x^3}} = 1 \text{ and } \int_1^\infty \frac{dx}{x^2} = 1.$$

So, by the limit comparison test, $\int_1^\infty dx/\sqrt{x + x^4}$ is also finite. Thus, $\int_0^\infty dx/\sqrt{x + x^4}$ converges.

43. $$\int_1^\infty \pi \left(\frac{1}{x}\right)^2 dx = \lim_{b \to \infty} \int_1^b \pi \left(\frac{1}{x}\right)^2 dx = \lim_{b \to \infty} \pi \left(1 - \frac{1}{b}\right) = \pi.$$

45. First, we find $\int dx/x(\ln x)^p$. Let $u = \ln x$, $du = dx/x$. Then,

$$\int \frac{dx}{x(\ln x)^p} = \int \frac{du}{u^p} = \begin{cases} \dfrac{u^{1-p}}{1-p} = \dfrac{(\ln x)^{1-p}}{1-p}, & p \neq 1 \\[3mm] \ln |u| = \ln |\ln x|, & p = 1 \end{cases}$$

a) $$\int_1^2 \frac{dx}{x(\ln x)^p} = \lim_{a \to 1^+} \int_a^2 \frac{dx}{x(\ln x)^p}$$

Assume $p \neq 1$, then

$$\lim_{a \to 1^+} \int_a^2 \frac{dx}{x(\ln x)^p} = \lim_{a \to 1^+} \frac{(\ln 2)^{1-p}}{1-p} - \frac{(\ln a)^{1-p}}{1-p}$$

Now, $\ln a \longrightarrow 0$ as $a \longrightarrow 1^+$, so if $1 - p > 0$ $(p < 1)$ $(\ln a)^{1-p} \longrightarrow 0$ as $a \longrightarrow 1^+$. If $1 - p < 0$ $(p > 1)$, $(\ln a)^{1-p} \longrightarrow \infty$ as $a \longrightarrow 1^+$.

If $p = 1$, then

$$\lim_{a \to 1^+} \int_a^2 \frac{dx}{x \ln x} = \lim_{a \to 1^+} (\ln |\ln 2| - \ln |\ln a|),$$

which goes to infinity.

Conclusion: $\int_1^2 dx/x(\ln x)^p$ converges for $p < 1$.

b) First assume $p \neq 1$. Then,

$$\int_2^\infty \frac{dx}{x(\ln x)^p} = \lim_{b \to \infty} \int_2^b \frac{dx}{x(\ln x)^p} = \lim_{b \to \infty} \frac{(\ln b)^{1-p}}{1-p} + \frac{(\ln 2)^{1-p}}{p-1}.$$

Now, $\ln b \longrightarrow \infty$ as $b \longrightarrow \infty$. So, if $1 - p < 0$ $(p > 1)$, $(\ln b)^{1-p} \longrightarrow 0$ as $b \longrightarrow \infty$ and the integral converges. If $1 - p > 0$ $(p < 1)$, then $(\ln b)^{1-p} \longrightarrow \infty$ as $b \longrightarrow \infty$, and the integral diverges.

If $p = 1$, then

$$\lim_{b \to \infty} \int_2^b \frac{dx}{x \ln x} = \lim_{b \to \infty} \ln (\ln b) - \ln |\ln 2|,$$

which goes to infinity.

Conclusion: $\int_2^\infty dx/x(\ln x)^p$ converges for $p > 1$.

47. a)   Volume $= \pi \int_0^\infty \left(e^{-x}\right)^2 dx = \lim_{b \to \infty} \int_0^b \pi e^{-2x} dx = \frac{\pi}{2} - \lim_{b \to \infty} \frac{\pi}{2} e^{-2b} = \frac{\pi}{2}$.

b)   Volume $= \int_0^\infty 2\pi x e^{-x} dx = \lim_{b \to \infty} (2\pi - 2\pi b e^{-b} - 2\pi e^{-b}) = 2\pi$.

49.   Area $= \int_0^{\pi/2} (\sec x - \tan x) dx = \lim_{b \to \frac{\pi}{2}^-} \int_0^b (\sec x - \tan x) dx$

$$= \lim_{b \to \frac{\pi}{2}^-} \ln |\sec b + \tan b| - \ln |\sec b| = \lim_{b \to \frac{\pi}{2}^-} \ln \left| 1 + \frac{\tan b}{\sec b} \right|.$$

Now,

$$\frac{\tan b}{\sec b} = \frac{\sqrt{\sec^2 b - 1}}{\sec b} = \sqrt{1 - \cos^2 b} \longrightarrow 1 \text{ as } b \longrightarrow \frac{\pi}{2}^-.$$

Thus,

$$\lim_{b \to \frac{\pi}{2}^-} \ln \left| 1 + \frac{\tan b}{\sec b} \right| = \ln 2.$$

Article 7.10

1. $\frac{d}{dx}\left[-\frac{1}{a}\left(\frac{2a - x}{x}\right)^{1/2} \right] = -\frac{1}{2a}\left(\frac{2a - x}{x}\right)^{-1/2} \frac{d}{dx}\left(\frac{2a - x}{x}\right)$

$$= -\frac{1}{2a}\left(\frac{x}{2a - x}\right)^{1/2}\left(-\frac{2a}{x^2}\right) \cdot$$

$$= \frac{1}{x^{3/2}(2a - x)^{1/2}} = \frac{1}{x\sqrt{2ax - x^2}}$$

3. If $u = ax + b$, $x = (u - b)/a$, $dx = 1/a\ du$, then

$$\int \frac{x\ dx}{(ax + b)^2} = \frac{1}{a^2}\int \frac{u - b}{u^2}\ du = \frac{1}{a^2}\left[\int \frac{du}{u} - b\int \frac{du}{u^2}\right]$$

$$= \frac{1}{a^2}\left[\ln |u| + \frac{b}{u}\right] + C = \frac{1}{a^2}\ln |ax + b| + \frac{b}{a^2(ax + b)} + C$$

5. Let $u = x$, $dv = e^{ax} dx$. Then, $v = 1/a\ e^{ax}$, $du = dx$, and

$$\int xe^{ax}\ dx = \frac{xe^{ax}}{a} - \int \frac{e^{ax}}{a}\ dx = \frac{xe^{ax}}{a} - \frac{e^{ax}}{a^2} + C = \frac{e^{ax}}{a^2}(ax - 1) + C.$$

7. By formula 141:

$$\int_0^{\pi/2} \cos^9 x\ dx = \frac{2 \cdot 4 \cdot 6 \cdot 8}{3 \cdot 5 \cdot 7 \cdot 9} = \frac{128}{315}.$$

9. By formula 100 with $n = 1$, $a = 1$:

$$\int x\ \cos^{-1} x\ dx = \frac{x^2}{2} \cos^{-1} x + \frac{1}{2} \int \frac{x^2}{\sqrt{1 - x^2}}\ dx.$$

Now use formula 33:

$$\int x\ \cos^{-1} x\ dx = \frac{x^2}{2} \cos^{-1} x + \frac{1}{4} \sin^{-1} x - \frac{1}{4} x\sqrt{1 - x^2} + C.$$

However, we can simplify this. Suppose $\theta = \sin^{-1} x$. Then, $x = \sin \theta = \cos(\pi/2 - \theta)$, so $\pi/2 - \theta = \cos^{-1} x$. Hence, $\theta = \pi/2 - \cos^{-1} x$. Since $\theta = \sin^{-1} x$, we have the identity $\sin^{-1} x = \pi/2 - \cos^{-1} x$. Therefore,

$$\int x\ \cos^{-1} x\ dx = \left(\frac{x^2}{2} - \frac{1}{4}\right) \cos^{-1} x - \frac{1}{4} x\sqrt{1 - x^2} + A,$$

where $A = C + \pi/8$.

11. By formula 92:

$$\int \sec^4 x\ dx = \frac{\sec^2 x\ \tan x}{3} + \frac{2}{3} \int \sec^2 x\ dx = \frac{\sec^2 x\ \tan x}{3} + \frac{2\tan x}{3} + C.$$

13. By formula 13(a) with $a = 1$, $b = -3$:

$$\int \frac{dx}{x\sqrt{x - 3}} = \frac{2}{\sqrt{3}} \tan^{-1} \sqrt{\frac{x - 3}{3}} + C.$$

15. By formula 19:

$$\int \frac{dx}{\left(9 - x^2\right)^2} = \frac{x}{18(9 - x^2)} + \frac{1}{18} \int \frac{dx}{9 - x^2}$$

By formula 18:

$$\int \frac{dx}{\left(9 - x^2\right)^2} = \frac{x}{18(9 - x^2)} + \frac{1}{108} \ln\left|\frac{x + 3}{x - 3}\right| + C.$$

17. By formula 27:

$$\int \frac{dx}{x^2 \sqrt{7 + x^2}} = - \frac{\sqrt{7 + x^2}}{7x} + C$$

19. By formula 42:

$$\int \frac{\sqrt{x^2 - 2}}{x} \, dx = \sqrt{x^2 - 2} - \sqrt{2} \sec^{-1} \left| \frac{x}{\sqrt{2}} \right| + C.$$

21. By formula 71:

$$\int \frac{dx}{4 + 5\sin 2x} = - \frac{1}{6} \ln \left| \frac{5 + 4\sin 2x + 3\cos 2x}{4 + 5\sin 2x} \right| + C.$$

23. By formula 86:

$$\int \tan^3 \frac{x}{2} \, dx = \tan^2 \frac{x}{2} - \int \tan \frac{x}{2} \, dx$$

By formula 82:

$$\int \tan^3 \frac{x}{2} \, dx = \tan^2 \frac{x}{2} + 2\ln \left| \cos \frac{x}{2} \right| + C.$$

25. By formula 7 with n = 1/2:

$$\int x\sqrt{ax + b} \, dx = \frac{(ax + b)^{3/2}}{a^2} \left[\frac{2(ax + b)}{5} - \frac{2b}{3} \right] + C$$

27. By formula 103:

$$\int 3^{ex} \, dx = \frac{3^{ex}}{e \ln 3} + C.$$

29. By repeated use of formula 105:

$$\int x^6 e^x \, dx = e^x(x^6 - 6x^5 + 30x^4 - 120x^3 + 360x^2 - 720x + 720).$$

So,

$$\int_0^1 x^6 e^x \, dx = 265e - 720.$$

MISCELLANEOUS PROBLEMS

1. $$\int \frac{\cos x}{\sqrt{1+\sin x}} \, dx = 2\sqrt{1+\sin x} + C.$$

3. $$\int \frac{\tan x}{\cos^2 x} \, dx = \int \tan x \, \sec^2 x \, dx = \frac{\tan^2 x}{2} + C.$$

5. $$\int e^{\ln \sqrt{x}} \, dx = \int x^{1/2} \, dx = \frac{2}{3} x^{3/2} + C.$$

7. Let $\tan \theta = x+1$, $dx = \sec^2 \theta \, d\theta$.

$$\int \frac{1}{\sqrt{x^2 + 2x + 2}} \, dx = \int \frac{\sec^2 \theta}{\sqrt{\tan^2 \theta + 1}} \, d\theta = \int \sec \theta \, d\theta = \ln \left| \sec \theta + \tan \theta \right| + C$$

$$= \ln \left| \sqrt{x^2 + 2x + 2} + x + 1 \right| + C.$$

9. $$\int x^2 e^x \, dx = (x^2 - 2x + 2)e^x + C. \quad \text{(See Example 3, Article 7.7.)}$$

11. $$\int \frac{e^t}{1 + e^{2t}} \, dt = \tan^{-1}(e^t) + C.$$

13. Let $u = \sqrt{x}$, $du = dx/2\sqrt{x}$. Then,

$$\int \frac{dx}{(x+1)\sqrt{x}} = \int \frac{2 \, du}{u^2 + 1} = 2 \tan^{-1} u + C = 2 \tan^{-1}(\sqrt{x}) + C.$$

15. $$d\left(1 + t^{5/3}\right)^{5/3} = \frac{5}{3}\left(1 + t^{5/3}\right)^{2/3} \frac{5}{3} t^{2/3} \, dt = \frac{25}{9} t^{2/3} \left(1 + t^{5/3}\right)^{2/3} dt$$

$$\therefore \quad \int t^{2/3}\left(1 + t^{5/3}\right)^{2/3} dt = \frac{9}{25}\left(1 + t^{5/3}\right)^{5/3} + C.$$

17. Let $x = \sqrt{1 + e^t}$, $dt = [2x/(x^2 - 1)] \, dx$.

$$\int \frac{dt}{\sqrt{1 + e^t}} = 2\int \frac{dx}{x^2 - 1} = \int \frac{dx}{x - 1} - \int \frac{dx}{x + 1} = \ln \left| \frac{x-1}{x+1} \right| + C$$

$$= \ln \left| \frac{\sqrt{1 + e^t} - 1}{\sqrt{1 + e^t} + 1} \right| + C.$$

19. $\displaystyle\int \frac{\sin x\, e^{\sec x}}{\cos^2 x}\, dx = \int \tan x\ \sec x\, e^{\sec x}\, dx = e^{\sec x} + C.$

21. $\displaystyle\int \frac{1}{\sqrt{2x - x^2}}\, dx = \int \frac{1}{\sqrt{1 - (x-1)^2}}\, dx = \sin^{-1}(x-1) + C.$

23. $\displaystyle\int \frac{\cos 2t}{(1 + \sin 2t)}\, dt = \frac{1}{2}\ln|1 + \sin 2t| + C$

25. $\displaystyle\int \sqrt{1 + \sin x}\ dx = \int \frac{\sqrt{1 + \sin x}\cdot\sqrt{1 - \sin x}}{\sqrt{1 - \sin x}}\, dx = \int \frac{\cos x}{\sqrt{1 - \sin x}}\, dx$

$$= -2\sqrt{1 - \sin x} + C = \frac{-2\cos x}{\sqrt{1 + \sin x}} + C.$$

27. Let $x = a\cos\theta$, $dx = -a\sin\theta\, d\theta$.

$$\int \frac{1}{\sqrt{(a^2 - x^2)^3}}\, dx = \frac{1}{a^2}\int \frac{-\sin\theta}{\sin^3\theta}\, d\theta = \frac{1}{a^2}\int -\csc^2\theta\ d\theta$$

$$= \frac{1}{a^2}\cot\theta + C = \frac{x}{a^2\sqrt{a^2 - x^2}} + C.$$

29. Let $\cos x = u$, $-\sin x\, dx = du$

$$\int \frac{\sin x\, dx}{\cos^2 x - 5\cos x + 4} = -\int \frac{du}{(u - 4)(u - 1)}$$

$$\frac{-1}{(u - 4)(u - 1)} = \frac{A}{u - 4} + \frac{B}{u - 1} \implies A = -1/3,\ B = 1/3$$

$$\int \frac{\sin x\, dx}{\cos^2 x - 5\cos x + 4} = \int \frac{-1/3\, du}{u - 4} + \int \frac{1/3\, du}{u - 1} = -\frac{1}{3}\ln(u - 4) + \frac{1}{3}\ln(u - 1)$$

$$= \frac{1}{3}\ln\left|\frac{(\cos x - 1)}{(\cos x - 4)}\right| + C.$$

31. Suppose $\displaystyle\frac{1}{x(x+1)\cdots(x+m)} = \frac{A_0}{x} + \frac{A_1}{x+1} + \cdots + \frac{A_m}{x+m}$

To find A_k, use the "cover up" method:

$$A_k = \frac{1}{(-k)(-k+1)\cdots(-1)(1)(2)\cdots(m-k)}$$

$$= \frac{(-1)^k}{k(k-1)(k-2)\cdots(1)(m-k)(m-k-1)\cdots(2)(1)} = \frac{(-1)^k}{k!(m-k)!}$$

(continued)

CHAPTER 7 — Miscellaneous Problems

(This formula holds even for $k = 0, m$ with the convention $0! = 1$. Thus,

$$\frac{1}{x(x+1)\cdots(x+m)} = \sum_{k=0}^{m} \frac{(-1)^k}{k!(m-k)!} \cdot \frac{1}{x+k}$$

and

$$\int \frac{dx}{x(x+1)\cdots(x+m)} = \sum_{k=0}^{m} \frac{(-1)^k}{k!(m-k)!} \ln |x+k| + C.$$

33. Let $\quad \dfrac{1}{y\left(2y^3+1\right)^2} = \dfrac{A}{y} + \dfrac{By^2+Cy+D}{(2y^3+1)} + \dfrac{Ey^2+Fy+G}{\left(2y^3+1\right)^2}$

$4A + 2B = 0$, $2C = 0$, $2D = 0$, $4A + B + E = 0$, $C + F = 0$, $D + G = 0$,

$A = 1 \implies A = 1$, $B = -2$, $C = 0$, $D = 0$, $E = -2$, $F = 0$, $G = 0$

$$\int \frac{dy}{y\left(2y^3+1\right)^2} = \int \frac{dy}{y} - 2\int \frac{y^2\,dy}{2y^3+1} - 2\int \frac{y^2\,dy}{\left(2y^3+1\right)^2}$$

$$= \ln |y| - \frac{1}{3}\ln |2y^3+1| + \frac{1}{3(2y^3+1)} + C$$

$$= \frac{1}{3}\ln \left|\frac{y^3}{2y^3+1}\right| + \frac{1}{3}\left(2y^3+1\right)^{-1} + C.$$

35. $\quad \dfrac{1}{x\left(x^2+1\right)^2} = \dfrac{A}{x} + \dfrac{Bx+C}{x^2+1} + \dfrac{Dx+E}{\left(x^2+1\right)^2}$

$A + B = 0$, $C = 0$, $2A + B + D = 0$, $E + C = 0$, $A = 1 \implies A = 1$, $B = -1$, $C = 0$, $D = -1$, $E = 0$

$$\int \frac{dx}{x\left(x^2+1\right)^2} = \int \frac{dx}{x} - \int \frac{x\,dx}{x^2+1} - \int \frac{x\,dx}{\left(x^2+1\right)^2}$$

$$= \ln |x| - \frac{1}{2}\ln |x^2+1| + \frac{1}{2(x^2+1)} + C$$

$$= \ln \left|\frac{x}{\sqrt{x^2+1}}\right| + \frac{1}{2(x^2+1)} + C$$

$$= \frac{1}{2}\left[\ln x^2 - \ln (1+x^2) - \left(1+x^2\right)^{-1}\right] + C.$$

37. Let $u = e^x - 1$, $dx = [1/(u+1)]\,du$

$$\int \frac{dx}{e^x - 1} = \int \frac{du}{u(u+1)} = -\int \frac{du}{u+1} + \int \frac{du}{u}$$

$$= \ln |u| - \ln |u+1| + C = \ln |e^x - 1| - x + C$$

CHAPTER 7 — Miscellaneous Problems

39. $$\frac{x+1}{x^2(x-1)} = \frac{A}{x} + \frac{B}{x^2} + \frac{C}{x-1} \implies A = -2,\ B = -1,\ C = 2$$

$$\int \frac{(x+1)\,dx}{x^2(x-1)} = -\int \frac{2\,dx}{x} - \int \frac{dx}{x^2} + \int \frac{2\,dx}{x-1} = 2\ln\left|\frac{x-1}{x}\right| + \frac{1}{x} + C$$

41. $$\int \frac{du}{\left(e^u - e^{-u}\right)^2} = \int \frac{du}{4\sinh^2 u} = \frac{1}{4}\int \text{csch}^2 u\ du = -\frac{1}{4}\coth u + C$$

$$= -\frac{1}{4}\cdot\frac{e^u + e^{-u}}{e^u - e^{-u}} + C.$$

43. $$5x^2 + 8x + 5 = 5\left[\left(x+\frac{4}{5}\right)^2 + \left(\frac{3}{5}\right)^2\right] = 5\cdot\frac{9}{25}\left[\left(x+\frac{4}{5}\right)^2 + \left(\frac{3}{5}\right)^2\right]\cdot\left(\frac{5}{3}\right)^2$$

$$= \frac{9}{5}\left[\left(\frac{5}{3}\right)^2\left(x+\frac{4}{5}\right)^2 + 1\right] = \frac{9}{5}\left[\left(\frac{5x+4}{3}\right)^2 + 1\right]$$

$$\int \frac{dx}{5x^2 + 8x + 5} = \frac{1}{5}\int \frac{dx}{[x+(4/5)]^2 + (3/5)^2} = \frac{5}{9}\int \frac{dx}{[(5x+4)/3]^2 + 1}$$

$$= \frac{1}{3}\tan^{-1}\left(\frac{5x+4}{3}\right) + C.$$

45. Let $u = \cos 2x,\ du = -2\sin 2x\,dx;\ dv = e^x\,dx,\ v = e^x.$

$$\int e^x \cos 2x\ dx = e^x \cos 2x + 2\int e^x \sin 2x\ dx;$$

$$\int e^x \sin 2x\ dx = e^x \sin 2x - 2\int e^x \cos 2x\ dx$$

$\therefore \int e^x \cos 2x\ dx = e^x \cos 2x + 2e^x \sin 2x - 4\int e^x \cos 2x\ dx$

$5\int e^x \cos 2x\ dx = e^x(\cos 2x + 2\sin 2x) + C$

$\therefore \int e^x \cos 2x\ dx = \frac{e^x}{5}(\cos 2x + 2\sin 2x) + C$

47. Let $u = (x)^{1/3},\ du = (x^{-2/3}/3)\,dx,\ dx = 3u^2\,du$

$$\int \frac{dx}{x\sqrt{1+(x)^{1/3}}} = 3\int \frac{du}{u(1+u)} = 3\int \frac{du}{u} - 3\int \frac{du}{1+u}$$

$$= 3\ln|u| - 3\ln|1+u| + C = \ln|x| - 3\ln\left|1+(x)^{1/3}\right| + C$$

CHAPTER 7 - Miscellaneous Problems

49. Let $z = \tan\theta$, $dz = \sec^2\theta\ d\theta$

$$\int \frac{z^5}{\sqrt{1+z^2}}\ dz = \int \tan^5\theta\ \sec\theta\ d\theta = \int \left(\sec^2\theta - 1\right)^2 \tan\theta\ \sec\theta\ d\theta$$

$$= \int \sec^4\theta\ \tan\theta\ \sec\theta\ d\theta - 2\int \sec^2\theta\ \tan\theta\ \sec\theta\ d\theta$$

$$+ \int \tan\theta\ \sec\theta\ d\theta$$

$$= \frac{1}{5}\sec^5\theta - \frac{2}{3}\sec^3\theta + \sec\theta + C$$

$$= \frac{1}{5}\left(1+z^2\right)^{5/2} - \frac{2}{3}\left(1+z^2\right)^{3/2} + \left(1+z^2\right)^{1/2} + C$$

$$= \frac{1}{15}\left(1+z^2\right)^{1/2}(3z^4 - 4z^2 + 8) + C.$$

51. Let $u = 1 + x^{4/5}$, $du = (4/5)x^{-1/5}\ dx$. Then,

$$\int \frac{dx}{x^{1/5}\sqrt{1+x^{4/5}}} = \frac{5}{4}\int \frac{du}{\sqrt{u}} = \frac{5}{2}\sqrt{u} + C = \frac{5}{2}\sqrt{1+x^{4/5}} + C.$$

53. Let $u_1 = \sin^{-1}x$, $du_1 = dx/\sqrt{1-x^2}$: $dv_1 = x\ dx$, $v_1 = x^2/2$.

$$\int x\sin^{-1}x\ dx = \frac{x^2}{x}\sin^{-1}x - \int \frac{x^2}{2\sqrt{1-x^2}}\ dx$$

Let $u_2 = x/2$, $du_2 = dx/2$; $dv_2 = -(x/\sqrt{1-x^2})\ dx$, $v_2 = \sqrt{1-x^2}$.

$$-\int \left(\frac{x^2}{2\sqrt{1-x^2}}\right)dx = \frac{x\sqrt{1-x^2}}{2} - \frac{1}{2}\int \sqrt{1-x^2}\ dx$$

Let $\cos\theta = \sqrt{1-x^2}$, $x = \sqrt{1-\cos^2\theta} = \sin\theta$, $dx = \cos\theta\ d\theta$.

$$\int \sqrt{1-x^2}\ dx = \int \cos^2\theta\ d\theta = \frac{1}{2}\int (1+\cos 2\theta)\ d\theta = \frac{\theta}{2} + \frac{1}{4}\sin 2\theta + C_1.$$

$$\therefore \quad \int x\sin^{-1}x\ dx = \frac{x^2}{2}\sin^{-1}x - \int \frac{x^2\ dx}{2\sqrt{1-x^2}}$$

$$= \frac{x^2}{2}\sin^{-1}x + \frac{x}{2}\sqrt{1-x^2} - \frac{1}{2}\int \sqrt{1-x^2}\ dx$$

$$= \frac{x^2}{2}\sin^{-1}x + \frac{x}{2}\sqrt{1-x^2} - \frac{\theta}{4} - \frac{1}{4}\sin\theta\cos\theta + C$$

(continued)

CHAPTER 7 - Miscellaneous Problems

$$= \frac{x^2}{2} \sin^{-1} x + \frac{x}{2} \sqrt{1-x^2} - \frac{1}{4} \sin^{-1} x - \frac{1}{4} x\sqrt{1-x} + C$$

$$= \frac{1}{4}[(2x^2 - 1) \sin^{-1} x + x\sqrt{1-x^2}] + C.$$

55.

$$\int \frac{(x^3+1)\ dx}{x^3 - x} = \int \frac{(x+1)(x^2 - x + 1)\ dx}{x(x+1)(x-1)} = \int dx + \int \frac{dx}{x(x-1)}$$

$$= x - \int \frac{dx}{x} + \int \frac{dx}{x-1} = x - \ln |x| + \ln |x-1| + C$$

$$= x + \ln \left| \frac{x-1}{x} \right| + C.$$

57.

$$\frac{2e^{2x} - e^x}{\sqrt{3e^{2x} - 6e^x - 1}} = \frac{2}{\sqrt{3}} \cdot \frac{[(e^x - 1) + (1/2)]e^x}{\sqrt{\left(e^x - 1\right)^2 - (4/3)}}$$

Let $e^x - 1 = u$, $e^x dx = du$.

$$\int \frac{2e^{2x} - e^x}{\sqrt{3e^{2x} - 6e^x - 1}}\ dx = \frac{2}{\sqrt{3}} \int \frac{[(e^x - 1) + (1/2)]e^x}{\sqrt{\left(e^x - 1\right)^2 - (4/3)}}\ dx = \frac{2}{\sqrt{3}} \int \frac{[u + (1/2)]\ du}{\sqrt{u^2 - (4/3)}}$$

$$= \frac{2}{\sqrt{3}} \int \frac{u\ du}{\sqrt{u^2 - (4/3)}} + \frac{1}{\sqrt{3}} \int \frac{du}{\sqrt{u^2 - (4/3)}}$$

$$= \frac{2}{\sqrt{3}} \sqrt{u^2 - \frac{4}{3}} + \frac{1}{\sqrt{3}} \int \frac{du}{\sqrt{u^2 - (4/3)}}$$

Let $u = (2/\sqrt{3}) \sec \theta$, $du = (2/\sqrt{3}) \sec \theta \tan \theta\ d\theta$, $\tan \theta = (\sqrt{3}/2)\sqrt{u^2 - 4/3}$.

$$\frac{1}{\sqrt{3}} \int \frac{du}{\sqrt{u^2 - (4/3)}} = \frac{1}{\sqrt{3}} \int \frac{(2/\sqrt{3}) \sec \theta \tan \theta\ d\theta}{\sqrt{(4/3) \tan^2 \theta}} = \frac{1}{\sqrt{3}} \int \sec \theta\ d\theta$$

$$= \frac{1}{\sqrt{3}} \ln |\sec \theta + \tan \theta| + C_1$$

$$= \frac{1}{\sqrt{3}} \ln \left| \frac{\sqrt{3}}{2} u + \frac{\sqrt{3}}{2} \sqrt{u^2 - \frac{4}{3}} \right| + C_1$$

$$= \frac{1}{\sqrt{3}} \ln \left| \frac{\sqrt{3}}{2}\left(u + \sqrt{u^2 - \frac{4}{3}}\right) \right| + C_1$$

$$\therefore \quad \int \frac{2e^{2x} - e^x}{\sqrt{3e^{2x} - 6e^x - 1}}\ dx = \frac{2}{3} \sqrt{3e^{2x} - 6e^x - 1} + \frac{1}{\sqrt{3}} \ln \left| (e^x - 1) + \sqrt{e^{2x} - 2e^x - \frac{1}{3}} \right| + C.$$

CHAPTER 7 - Miscellaneous Problems

59. Let $u = 2\sqrt{y^2 + y}$, $du = [(2y+1)/\sqrt{y^2+y}]\,dy$, $u^2 = 4(y^2+y) = 4y^2 + 4y$;
$u^2 + 1 = 4y^2 + 4y + 1 = (2y+1)^2$.

$$\int \frac{dy}{(2y+1)\sqrt{y^2+y}} = \int \frac{(2y+1)\,dy}{(2y+1)^2\sqrt{y^2+1}} = \int \frac{du}{1+u^2}$$

$$= \tan^{-1} u + C = \tan^{-1}(2\sqrt{y^2+y}) + C.$$

61. Let $x = \sin\theta$, $dx = \cos\theta\,d\theta$.

$$\int (1-x^2)^{3/2}\,dx = \int \cos^4\theta\,d\theta = \int \frac{1}{2}(1+\cos 2\theta)^2\,d\theta$$

$$= \frac{3}{8}\theta + \frac{1}{4}\sin 2\theta + \frac{1}{32}\sin 4\theta + C$$

$\theta = \sin^{-1} x$; $\sin 2\theta = 2\sin\theta\cos\theta = 2x\sqrt{1-x^2}$;
$\sin 4\theta = 2\sin 2\theta\cos 2\theta = 4\sin\theta\cos\theta(\cos^2\theta - \sin^2\theta) = 4x\sqrt{1-x^2}(1-2x^2)$

$$\therefore \int (1-x^2)^{3/2}\,dx = \frac{3}{8}\sin^{-1}x + \frac{1}{2}x\sqrt{1-x^2} + \frac{1}{8}x\sqrt{1-x^2}(1-2x^2) + C$$

$$= \frac{3}{8}\sin^{-1}x + \frac{1}{8}(5x - 2x^3)\sqrt{1-x^2} + C.$$

63.
$$\int x\tan^2 x\,dx = \int x(\sec^2 x - 1)\,dx = \int x\sec^2 x\,dx - \int x\,dx$$

Let $u = x$, $du = dx$; $dv = \sec^2 x\,dx$, $v = \tan x$.

$$\int x\sec^2 x\,dx = x\tan x - \int \tan x\,dx = x\tan x + \ln|\cos x| + C_1$$

$$\therefore \int x\tan^2 x\,dx = x\tan x + \ln|\cos x| - (x^2/2) + C.$$

65. Let $u = x$, $du = dx$; $dv = [(1/2) + (1/2)\cos 2x]\,dx$, $v = (x/2) + (1/4)\sin 2x$.

$$\int x\cos^2 x\,dx = x\left(\frac{x}{2} + \frac{1}{4}\sin 2x\right) - \int \left(\frac{x}{2} + \frac{1}{4}\sin 2x\right)\,dx$$

$$= \frac{1}{4}x^2 + \frac{x}{4}\sin 2x + \frac{1}{8}\cos 2x + C.$$

67. Let $u = x$, $du = dx$; $dv = \sin^2 x\,dx$, $v = (x/2) - (\sin 2x/4)$

$$\int x\sin^2 x\,dx = \frac{x^2}{2} - \frac{x}{4}\sin 2x - \int \left(\frac{x}{2} - \frac{1}{4}\sin 2x\right)\,dx$$

$$= \frac{x^2}{4} - \frac{x}{4}\sin 2x - \frac{1}{8}\cos 2x + C$$

69. Let $e^{2u} = x$, $du = (1/2x)\,dx$.

$$\int \frac{du}{e^{4u} + 4e^{2u} + 3} = \int \frac{dx}{2x(x+1)(x+3)} = \int \left[\frac{1}{6x} - \frac{1}{4(x+1)} + \frac{1}{12(x+3)} \right] dx$$

$$= \frac{1}{6} \ln |x| - \frac{1}{4} \ln |x+1| + \frac{1}{12} \ln |x+3| + C$$

$$= \frac{1}{12}[4u - 3 \ln (1 + e^{2u}) + \ln (3 + e^{2u})] + C$$

71. Let $u_1 = (x+1)^2$, $du_1 = 2(x+1)\,dx$; $dv_1 = e^x\,dx$, $v_1 = e^x$.

$$\int (x+1)^2 e^x\,dx = e^x(x+1)^2 - 2\int e^x(x+1)\,dx = e^x(x+1)^2 - 2e^x - 2\int xe^x\,dx$$

Let $u_2 = x$, $du_2 = dx$; $dv_2 = e^x\,dx$, $v_2 = e^x$.

$$-2\int xe^x\,dx = -2xe^x + 2\int e^x\,dx = -2xe^x + 2e^x + C$$

$$\therefore \quad \int (x+1)^2 e^x\,dx = e^x(x+1)^2 - 2xe^x + C = e^x(x^2+1) + C.$$

73.
$$\int \frac{8\,dx}{x^4 + 2x^3} = 8\int \frac{dx}{x^3(x+2)} = \int \left(\frac{4}{x^3} - \frac{2}{x^2} + \frac{1}{x} - \frac{1}{x+2} \right) dx$$

$$= -2x^{-2} + 2x^{-1} + \ln \left| \frac{x}{x+2} \right| + C$$

75. $\cos x = \cos 2(x/2) = \cos^2 (x/2) - \sin^2 (x/2) = 2\cos^2 (x/2) - 1$

$$\sqrt{1 + \cos x} = \sqrt{1 + [2\cos^2 (x/2) - 1]} = \sqrt{2} \cos (x/2)$$

$$\int_0^{\pi/2} \frac{\cos x\,dx}{\sqrt{1 + \cos x}} = \int_0^{\pi/2} \frac{2\cos^2 (x/2) - 1}{\sqrt{2} \cos (x/2)}\,dx$$

$$= \int_0^{\pi/2} \sqrt{2} \cos \frac{x}{2}\,dx - \int_0^{\pi/2} \frac{1}{\sqrt{2}} \sec \frac{x}{2}\,dx$$

$$= 2\sqrt{2} \sin \frac{x}{2} - \frac{2}{\sqrt{2}} \ln \left| \sec \frac{x}{2} + \tan \frac{x}{2} \right| \Bigg]_0^{\pi/2} = 2 - \sqrt{2} \ln |\sqrt{2} + 1|$$

77.
$$\int \frac{du}{\left(e^u + e^{-u} \right)^2} = \int \frac{du}{4 \cosh^2 u} = \frac{1}{4} \int \operatorname{sech}^2 u\,du$$

$$= \frac{1}{4} \tanh^2 u + C = \frac{1}{4} \cdot \frac{(e^u - e^{-u})}{(e^u + e^{-u})} + C.$$

CHAPTER 7 - Miscellaneous Problems

79. Let $x = \tan t$, $dx = \sec^2 t \; dt$.

$$\int \frac{\sec^2 t \; dt}{\sec^2 t - 3 \tan t + 1} = \int \frac{dx}{x^2 - 3x + 2} = \int \left(\frac{1}{x-2} - \frac{1}{x-1} \right) dx$$

$$= \ln |\tan t - 2| - \ln |\tan t - 1| + C$$

$$= \ln \left| \frac{\tan t - 2}{\tan t - 1} \right| + C$$

81. $$\int \frac{dx}{1 + \cos^2 x} = \int \frac{\sec^2 x \; dx}{\sec^2 x + 1} = \int \frac{\sec^2 x \; dx}{\tan^2 x + 2}$$

Let $u = \tan x$, $du = \sec^2 x \; dx$.

$$\int \frac{\sec^2 x \; dx}{\tan^2 x + 2} = \int \frac{du}{u^2 + 2} = \frac{1}{\sqrt{2}} \tan^{-1} \frac{u}{\sqrt{2}} + C = \frac{1}{\sqrt{2}} \tan^{-1} \left(\frac{\tan x}{\sqrt{2}} \right) + C$$

83. Let $u = \ln \sqrt{x^2 + 1}$, $du = [x/(x^2+1)] \; dx$, $dv = dx$, $v = x$.

$$\int \ln \sqrt{x^2 + 1} \; dx = x \ln \sqrt{x^2 + 1} - \int \frac{x^2}{x^2 + 1} \; dx = x \ln \sqrt{x^2 + 1} - \int \left(1 - \frac{1}{x^2 + 1} \right) dx$$

$$= x \ln \sqrt{x^2 + 1} - x + \tan^{-1} x + C$$

85. Let $z = e^{x^2}$, $dz = 2xe^{x^2} \; dx$, $\ln z = x^2$.

$$\int x^3 e^{x^2} \; dx = \int \frac{1}{2} \ln z \; dz$$

Let $u = \ln z$, $du = (1/z) \; dz$, $dv = (1/2) \; dz$, $v = z/2$.

$$\int \frac{1}{2} \ln z \; dz = \frac{z}{2} \ln z - \int \frac{dz}{2} = \frac{z}{2} \ln z - \frac{z}{2} + C = \frac{1}{2} x^2 e^{x^2} - e^{x^2} + C$$

$$\therefore \quad \int x^3 e^{x^2} \; dx = \frac{1}{2}(x^2 - 1) e^{x^2} + C.$$

87. $$\int \frac{\sec^2 x \; dx}{\sqrt{4 - \sec^2 x}} = \int \frac{\sec^2 x \; dx}{\sqrt{3 - \tan^2 x}} = \sin^{-1} \left(\frac{\tan x}{\sqrt{3}} \right) + C$$

89. $$\int \frac{dx}{1 + \sin x} = \int \frac{1 - \sin x}{1 - \sin^2 x} \; dx = \int \frac{1 - \sin x}{\cos^2 x} \; dx$$

$$= \int \sec^2 x \; dx - \int \sec x \; \tan x \; dx = \tan x - \sec x + C.$$

CHAPTER 7 - Miscellanous Problems

91. Let $\sin x = 2z/(1+z^2)$, $dx = 2 \; dz/(1+z^2)$

$$\int \frac{dx}{\sin^3 x} = \int \frac{\left(1+z^2\right)^2}{4z^3} \; dz = \frac{1}{4} \int z^{-3} \; dz + \frac{1}{4} \int z \; dz + \frac{1}{2} \int \frac{dz}{z}$$

$$= -\frac{1}{8} z^{-2} + \frac{1}{8} z^2 + \frac{1}{2} \ln |z| + C$$

$$= \frac{1}{8} \left(\tan^2 \frac{x}{2} - \cot^2 \frac{x}{2} \right) + \frac{1}{2} \ln \left| \tan \frac{x}{2} \right| + C$$

$$= \frac{1}{8} \left(\frac{1-\cos x}{1+\cos x} - \frac{1+\cos x}{1-\cos x} \right) + \frac{1}{4} \ln \left(\frac{1-\cos x}{1+\cos x} \right) + C$$

$$= \frac{-4 \cos x}{8 \sin^2 x} + \frac{1}{4} \ln \left(\frac{1-\cos x}{1+\cos x} \right) + C$$

$$= -\frac{1}{2} \cot x \; \csc x + \frac{1}{4} \ln \left(\frac{1-\cos x}{1+\cos x} \right) + C.$$

93. Let $u_1 = \left(\sin^{-1} x \right)^2$, $du_1 = 2(\sin^{-1} x)(1/\sqrt{1-x^2}) \; dx$; $dv_1 = dx$, $v_1 = x$.

$$\int \left(\sin^{-1} x \right)^2 \; dx = x \left(\sin^{-1} x \right)^2 - 2 \int \frac{x \sin^{-1} x}{\sqrt{1-x^2}} \; dx$$

Let $u_2 = \sin^{-1} x$, $du_2 = (1/\sqrt{1-x^2}) \; dx$; $dv_2 = (x/\sqrt{1-x^2}) \; dx$, $v_2 = -\sqrt{1-x^2}$.

$$\int \left(\sin^{-1} x \right)^2 \; dx = x \left(\sin^{-1} x \right)^2 - 2 \left(-\sin^{-1} x \sqrt{1-x^2} + \int dx \right)$$

$$= x \left(\sin^{-1} x \right)^2 - 2x + 2\sqrt{1-x^2} \; \sin^{-1} x + C$$

95. $$\int \frac{x^3 \; dx}{\left(x^2+1\right)^2} = \int \frac{-x}{\left(x^2+1\right)^2} \; dx + \int \frac{x \; dx}{x^2+1} = \frac{1}{2} \left(x^2+1\right)^{-1} + \frac{1}{2} \ln \left(x^2+1\right) + C$$

97. Let $y = \sqrt{2x+1}$, $x = (y^2-1)/2$, $dx = y \; dy$.

$$\int x\sqrt{2x+1} \; dx = \frac{1}{2} \int (y^4 - y^2) \; dy = \frac{y^5}{10} - \frac{y^3}{6} + C = \frac{1}{15}(3x-1)(2x+1)^{3/2} + C$$

99. Let $u = \ln (x - \sqrt{x^2-1})$, $du = (-1/\sqrt{x^2-1}) \; dx$, $dv = dx$, $v = x$.

$$\int \ln (x - \sqrt{x^2-1}) \; dx = x \ln (x - \sqrt{x^2-1}) + \int \frac{x}{\sqrt{x^2-1}} \; dx$$

$$= x \ln (x - \sqrt{x^2-1}) + \sqrt{x^2-1} + C.$$

CHAPTER 7 – Miscellaneous Problems

101.   Let $u = \tan^{-1} e^x$, $du = [e^x/(1+e^{2x})]\, dx$; $dv = e^{-x}\, dx$, $v = -e^{-x}$.

$$\int e^{-x} \tan^{-1} e^x\, dx = -e^{-x} \tan^{-1} e^x + \int \frac{1}{1+e^{2x}}\, dx$$

Let $t = 1+e^{2x}$, $dt = 2e^{2x}\, dx$.

$$\int \frac{1}{1+e^{2x}}\, dx = \int \frac{1}{2t(t-1)}\, dt = -\frac{1}{2}\int \frac{dt}{t+(1/2)} \int \frac{dt}{t-1}$$

$$= \frac{1}{2}[\ln|t-1| - \ln t] + C = \frac{1}{2}[\ln e^{2x} - \ln(1+e^{2x})] + C$$

$$= x - \frac{1}{2}\ln(1+e^{2x}) + C.$$

∴ $\int e^{-x} \tan^{-1} e^x\, dx = -e^{-x} \tan^{-1} e^x + x - \frac{1}{2}\ln(1+e^{2x}) + C.$

103.   Let $u = \ln(x+\sqrt{x})$, $du = dx/(x+\sqrt{x}) + dx/[2\sqrt{x}(x+\sqrt{x})]$; $dv = dx$, $v = x$.

$$\int \ln(x+\sqrt{x})\, dx = x\ln(x+\sqrt{x}) - \int \frac{x\, dx}{x+\sqrt{x}} - \frac{1}{2}\int \frac{dx}{\sqrt{x}+1}$$

Let $\sqrt{x} = z$, $dx = 2z\, dz$.

$$-\int \frac{x\, dx}{x+\sqrt{x}} - \frac{1}{2}\int \frac{dx}{\sqrt{x}+1} = -\int \frac{2z^2\, dz}{z+1} - \int \frac{z\, dz}{z+1}$$

$$= -2\int z\, dz + 2\int dz - 2\int \frac{dz}{z+1} - \int dz + \int \frac{dz}{z+1}$$

$$= -z^2 + z - \ln|z+1| + C = -x + \sqrt{x} - \ln|1+\sqrt{x}| + C$$

∴ $\int \ln(x+\sqrt{x})\, dx = x\ln(x+\sqrt{x}) - x + \sqrt{x} - \ln|1+\sqrt{x}| + C$

105. $\int \ln(x^2+x)\, dx = \int \ln x\, dx + \int \ln(x+1)\, dx$

Let $u_1 = \ln x$, $du_1 = dx/x$; $dv_1 = dx$, $v_1 = x$.

$$\int \ln x\, dx = x\ln x - \int dx = x\ln x - x + C_1$$

Let $u_2 = \ln(x+1)$, $du_2 = dx/(x+1)$; $dv_2 = dx$, $v_2 = x$.

$$\int \ln(x+1)\, dx = x\ln(x+1) - \int \frac{x\, dx}{x+1} = x\ln(x+1) - \int \left(1 - \frac{1}{x+1}\right) dx$$

$$= (x+1)\ln(x+1) - x + C_2$$

(continued)

CHAPTER 7 - Miscellaneous Problems

$$\therefore \quad \int \ln (x^2 + x)\, dx = (x+1) \ln (x+1) + x \ln x - 2x + C.$$

107. Let $z = \sqrt{x}$, $dx = 2z\, dz$. $\int \cos \sqrt{x}\ dx = 2\int z \cos z\ dz$. Let $u = z$, $du = dz$; $dv = \cos z\ dz$, $v = \sin z$.

$$2 \int z \cos z\ dz = 2z \sin z - 2 \int \sin z\ dz = 2z \sin z + 2 \cos z + C.$$

$$\therefore \quad \int \cos \sqrt{x}\ dx = 2\sqrt{x} \sin \sqrt{x} + 2 \cos \sqrt{x} + C.$$

109. Let $z = \sqrt{x+1}$, $dx = 2z\, dz$. $\int \tan^{-1} \sqrt{x+1}\ dx = \int 2z \tan^{-1} z\ dz$. Let $u = \tan^{-1} z$, $du = dz/(1+z^2)$; $dv = 2z\ dz$, $v = z^2$.

$$\int 2z \tan^{-1} z\ dz = z^2 \tan^{-1} z - \int \frac{z^2}{1+z^2}\ dz = z^2 \tan^{-1} z - \int 1 - \frac{1}{1+z^2}\ dz$$

$$= z^2 \tan^{-1} z - z + \tan^{-1} z + C_1$$

$$= (x+1) \tan^{-1} \sqrt{x+1} - \sqrt{x+1} + \tan^{-1} \sqrt{x+1} + C$$

$$\therefore \quad \int \tan^{-1} \sqrt{x+1}\ dx = (x+2) \tan^{-1} \sqrt{x+1} - \sqrt{x+1} + C$$

111.
$$\int x \sin^2 (2x)\ dx = \int x(1 - \cos^2 2x)\ dx = \int x\left(\frac{1}{2} - \frac{1}{2} \cos 4x\right) dx$$

$$= \frac{x^2}{4} - \frac{1}{2} \int x \cos 4x\ dx$$

Let $u = x$, $du = dx$; $dv = \cos 4x\ dx$, $v = \sin 4x/4$.

$$\int x \sin^2 (2x)\ dx = \frac{x^2}{4} - \frac{x \sin 4x}{8} + \frac{1}{8} \int \sin 4x\ dx$$

$$= \frac{x^2}{4} - \frac{1}{8} x \sin 4x - \frac{1}{32} \cos 4x + C.$$

113. Let $\sqrt{e^{2t} + 1} = u$, $dt = [u/(u^2 - 1)]\, du$.

$$\int \frac{dt}{\sqrt{e^{2t} + 1}} = \int \frac{du}{u^2 - 1} = \frac{1}{2} \int \frac{du}{u - 1} - \frac{1}{2} \int \frac{du}{u + 1} = \frac{1}{2} \ln \left| \frac{u - 1}{u + 1} \right| + C$$

$$= \frac{1}{2} \ln \left| \frac{\sqrt{e^{2t} + 1} - 1}{\sqrt{e^{2t} + 1} + 1} \right| + C$$

(continued)

CHAPTER 7 - Miscellaneous Problems

$$= \frac{1}{2} \ln \left| \frac{(\sqrt{e^{2t}+1} - 1) \cdot (\sqrt{e^{2t}+1} - 1)}{(e^{2t}+1) - 1} \right| + C$$

$$= 2 \cdot \frac{1}{2} \ln \left| \sqrt{e^{2t}+1} - 1 \right| - \frac{1}{2} \ln e^{2t} + C = \ln \left| \sqrt{e^{2t}+1} - 1 \right| - t + C$$

115. Let $a + be^{ct} = z$, $dt = dz/c(z-a)$.

$$\int \frac{dt}{a+be^{ct}} = \frac{1}{c} \int \frac{dz}{(z-a)z} = \frac{1}{ac} \int \frac{dz}{z-a} - \frac{1}{ac} \int \frac{dz}{z} = \frac{1}{ac} \ln \left| \frac{z-a}{z} \right| + C_1$$

$$= \frac{1}{ac} \ln \left| \frac{be^{ct}}{a+be^{ct}} \right| + C_1 = \frac{1}{ac} \ln \left| \frac{e^{ct}}{a+be^{ct}} \right| + C$$

$$= \frac{t}{a} - \frac{1}{ac} \ln (a+be^{ct}) + C.$$

117. $$\frac{1}{9x^4+x^2} = \frac{1}{x^2(9x^2+1)} = \frac{A}{x} + \frac{B}{x^2} + \frac{Cx+D}{9x^2+1} \, .$$

So, $B = 1$, $D = -9$, $A = C = 0$, and

$$\int \frac{dx}{9x^4+x^2} = \int \left(\frac{1}{x^2} - \frac{9}{9x^2+1} \right) dx = \int \left(\frac{1}{x^2} - \frac{1}{x^2+(1/9)} \right) dx$$

$$= -\frac{1}{x} - 3 \tan^{-1}(3x) + C.$$

119. $$\int \ln (2x^2+4) \, dx = \int \ln 2 \, dx + \int \ln (x^2+2) \, dx = x \ln 2 + \int \ln (x^2+2) \, dx$$

Let $u = \ln (x^2+2)$, $du = [2x/(x^2+2)] \, dx$; $dv = dx$, $v = x$.

$$\int \ln (x^2+2) \, dx = x \ln (x^2+2) - \int \frac{2x^2}{x^2+2} \, dx$$

$$= x \ln (x^2+2) - 2 \int dx + 4 \int \frac{1}{x^2+2} \, dx$$

$$= x \ln (x^2+2) - 2x + 2\sqrt{2} \tan^{-1} \frac{x}{\sqrt{2}} + C$$

$$\therefore \int \ln (2x^2+4) \, dx = x \ln 2 + x \ln (x^2+2) - 2x + 2\sqrt{2} \tan^{-1} \frac{x}{\sqrt{2}} + C$$

$$= x \ln (x^2+2) + (\ln 2 - 2)x + 2\sqrt{2} \tan^{-1} \frac{x}{\sqrt{2}} + C.$$

121. $\int \dfrac{1}{x(2 + \ln\ x)}\ dx = \ln\ |2 + \ln\ x| + C$

123. $\int \dfrac{dx}{x^3 + 1} = \dfrac{1}{3} \int \dfrac{dx}{x+1} - \dfrac{1}{3} \int \dfrac{x-2}{x^2 - x + 1}\ dx$

$\quad = \dfrac{1}{3} \ln\ |x+1| - \dfrac{1}{6} \int \dfrac{2x-1}{x^2 - x + 1}\ dx + \dfrac{1}{2} \int \dfrac{dx}{x^2 - x + 1}$

$\quad = \dfrac{1}{3} \ln\ |x+1| - \dfrac{1}{6} \ln\ |x^2 - x + 1| + \dfrac{1}{\sqrt{3}} \tan^{-1} \left(\dfrac{2x-1}{\sqrt{3}} \right) + C$

125. Let $u = \sqrt{x}$, $du = dx/2\sqrt{x}$. Then,

$\int \dfrac{2 \sin \sqrt{x}}{\sqrt{x} \sec \sqrt{x}}\ dx = \int \dfrac{4 \sin u}{\sec u}\ du = \int 4 \sin u \cos u\ du$

$\quad = \int 2 \sin 2u\ du = -\cos 2u + C = -\cos (2\sqrt{x}) + C.$

127. Let $y = \sqrt{x+1}$, $dx = 2y\ dy$. $\int \sin \sqrt{x+1}\ dx = 2\int y \sin y\ dy$. Let $u = y$, $dv = \sin y\ dy$.

$2 \int y \sin y\ dy = -2y \cos y + 2 \int \cos y\ dy = -2y \cos y + 2 \sin y + C$

$\therefore \quad \int \sin \sqrt{x+1}\ dx = -2\sqrt{x+1} \cos \sqrt{x+1} + 2 \sin \sqrt{x+1} + C$

129. $\lim\limits_{n \to \infty} \left(\dfrac{1}{n+1} + \dfrac{1}{n+2} + \cdots + \dfrac{1}{n+n} \right) = \lim\limits_{n \to \infty} \sum\limits_{k=1}^{n} \dfrac{1}{n+k} = \lim\limits_{n \to \infty} \sum\limits_{k=1}^{n} \dfrac{1}{\left(1 + \frac{k}{n}\right)} \cdot \dfrac{1}{n}.$

Let $\Delta x = 1/n$, $x_k = k \Delta x = k/n$, $f(x_k) = 1/(1 + x_k) = 1/(1 + kn)$; $x_0 = 0$ and $x_n = n/n = 1$.

$\lim\limits_{n \to \infty} \sum\limits_{k=1}^{n} \dfrac{1}{1 + (k/n)} \cdot \dfrac{1}{n} = \lim\limits_{\Delta x \to 0} \sum\limits_{k=1}^{n} f(x_k) \Delta x = \int_0^1 \dfrac{dx}{1+x} = \ln\ |1+x| \Big]_0^1 = \ln\ 2$

131.
$$\lim_{n \to \infty} \frac{1}{n}\left(\sin 0 + \sin \frac{\pi}{n} + \sin \frac{2\pi}{n} + \cdots + \sin \frac{(n-1)}{n}\pi\right)$$

$$= \lim_{n \to \infty} \sum_{k=0}^{n-1} \frac{1}{n} \sin \frac{k\pi}{n} = \lim_{\Delta x \to 0} \sum_{k=0}^{n-1} (\sin k\pi\Delta x)\Delta x$$

$$= \int_0^1 \sin \pi x \ dx = -\frac{1}{\pi}\cos \pi x \bigg|_0^1 = \frac{2}{\pi}$$

133.
$$\lim_{n \to \infty}\left(\frac{n}{n^2+0^2} + \frac{n}{n^2+1^2} + \frac{n}{n^2+2^2} + \cdots + \frac{n}{n^2+(n-1)^2}\right)$$

$$= \lim_{n \to \infty} \sum_{k=0}^{n-1} \frac{n}{n^2+k^2} = \lim_{n \to \infty} \sum_{k=0}^{n-1} \frac{1}{n} \cdot \frac{1}{1+(k/n)^2}$$

$$= \lim_{\Delta x \to 0} \frac{\Delta x}{1+(k\Delta x)^2} = \int_0^1 \frac{dx}{1+x^2} = \tan^{-1} x \bigg|_0^1 = \frac{\pi}{4}$$

135.
$$\lim_{n \to \infty} \sum_{k=0}^{n-1} \frac{1}{\sqrt{n^2-k^2}} = \lim_{n \to \infty} \sum_{k=0}^{n-1} \frac{1}{n} \cdot \frac{1}{\sqrt{1-(k/n)^2}} = \lim_{\Delta x \to 0} \sum_{k=0}^{n-1} \frac{\Delta x}{\sqrt{1-(k\Delta x)^2}}$$

$$= \int_0^1 \frac{1}{\sqrt{1-x^2}} \ dx = \sin^{-1} x \bigg|_0^1 = \frac{\pi}{2}$$

137.

Let $u = \ln x$, $du = dx/x$; $dv = dx$, $v = x$.

$$\int_0^1 \ln x \ dx = \lim_{b \to 0} \int_b^1 \ln x \ dx$$

$$= \lim_{b \to 0} [x \ln x - x] \bigg|_b^1 = -1$$

139. Let $u_1 = e^{-x^2}$, $du_1 = -2xe^{-x^2} dx$; $dv_1 = dx$, $v_1 = x$.

$$\int e^{-x^2} dx = xe^{-x^2} + \int 2x^2 e^{-x^2} dx$$

Let $u_2 = u_1$, $du_2 = du_1$, $dv_2 = 2x^2 dx$, $v_2 = (2/3)x^3$.

$$\int 2x^2 e^{-x^2} dx = \frac{2}{3}x^3 e^{-x^2} + \int \frac{4}{3}x^4 e^{-x^2} dx$$

(continued)

CHAPTER 7 - Miscellaneous Problems

Thus as $x \to \infty$, the integral $\int (4/3) x^4 e^{-x^2} \, dx$ decreases exponentially toward zero.

$$\therefore \quad \int e^{-x^2} \, dx = x e^{-x^2} + \frac{2}{3} x^3 e^{-x^2} + \cdots = x e^{-x^2} \left(1 + \frac{2}{3} x^2 + \cdots \right)$$

$$\lim_{h \to 0} \frac{1}{h} \int_2^{2+h} e^{-x^2} \, dx = \lim_{h \to 0} \frac{1}{h} \left[x e^{-x^2} \left(1 + \frac{2}{3} x^2 + \cdots \right) \right]_2^{2+h}$$

$$= \lim_{h \to 0} \frac{1}{h} \left[(2+h) e^{-(2+h)^2} \left(1 + \frac{2}{3}(2+h)^2 + \cdots \right) \right]$$

$$- \lim_{h \to 0} \frac{1}{h} \left[2 e^{-4} \left(1 + \frac{2}{3} \cdot 4 + \cdots \right) \right]$$

$$= \lim_{h \to 0} \frac{1}{h} \left[2 e^{-(2+h)^2} \left(1 + \frac{2}{3}(2+h)^2 + \cdots \right) \right.$$

$$- 2 e^{-4} \left(1 + \frac{2}{3} \cdot 4 + \cdots \right)$$

$$\left. + h e^{-(2+h)^2} \left(1 + \frac{2}{3}(2+h)^2 + \cdots \right) \right] = e^{-4}$$

141. The area of the region is

(area of quarter circle) - (area of isosceles right triangle)

$$= \frac{\pi r^2}{4} - \frac{r^2}{2} = \frac{r^2}{4} (\pi - 2).$$

Moreover, by symmetry, $\bar{x} = \bar{y}$.

$$\bar{x} = \frac{4}{r^2 (\pi - 2)} \int_0^r x \left[\sqrt{r^2 - x^2} - (r - x) \right] \, dx$$

$$= \frac{4}{(\pi - 2) r^2} \left[-\frac{1}{3} \left(r^2 - x^2 \right)^{3/2} - \frac{r x^2}{2} + \frac{x^3}{3} \right] \Bigg|_0^r = \frac{2}{3 (\pi - 2) r^2}$$

143. Let A be the point $(r, 0)$.

$$\int_0^{2\pi r} k r \theta \, d\theta = k r \frac{\theta^2}{2} \Bigg]_0^{2\pi r} = 2\pi^2 k r^2$$

145. $y = e^x$, $y' = e^x$, $ds = \left(1 + e^{2x} \right)^{1/2} \, dx$, $dA = e^x \, dx$, $dV = \pi e^{2x} \, dx$, $dS = 2\pi e^x \, ds$

a) $A = \int_0^\infty e^x \, dx = (e^x) \Big]_0^\infty = \infty$ \qquad b) $V = \int_0^\infty \pi e^{2x} \, dx = \left[\frac{\pi}{e} e^{2x} \right) \Big]_0^\infty = \infty$

c) $\quad S = \int_0^\infty 2\pi e^x \left(1 + e^{2x}\right)^{1/2} dx \geq \int_0^\infty 2\pi e^x \, dx = \infty.$

147.

$$S = 2\pi \int_1^e x \cdot \frac{\sqrt{1+x^2}}{x} \, dx = 2\pi \int_1^e \sqrt{1+x^2} \, dx$$

$$= 2\pi \cdot \frac{1}{2} \left[x\sqrt{1+x^2} + \ln\left(x + \sqrt{1+x^2}\right) \right]_1^e$$

$$= \pi \left(e\sqrt{1+e^2} - \sqrt{2} + \ln\left(e + \sqrt{1+e^2}\right) - \ln\left(1 + \sqrt{2}\right) \right)$$

$$= \pi \cdot (7.29) = 22.89 \approx 23.$$

149.

$$S = 2\pi \int_0^2 y \sqrt{1 + \left(\frac{dy}{dx}\right)^2} \, dx = 2\pi \int_0^2 e^x \sqrt{1+e^x} \, dx$$

Let $e^x = \tan z$, $dx = (\sec^2 z / \tan z) \, dz$; $x = 0$, $z = 0$, $x = 2$, $z = \tan^{-1} e^2$

$$S = 2\pi \int_0^{\tan^{-1} e^2} \sec^3 z \, dz = \pi \left[\frac{\sin z}{\cos^2 z} + \ln(\tan z + \sec z) \right]_0^{\tan^{-1} e^2}$$

$$= \pi \left[e^2 \sqrt{1+e^4} + \ln\left(\sqrt{1+e^4} + e^2\right) - \ln(2.414) \right] = \pi \cdot 55.7 \approx 175$$

151.

$$L = \int_{-\pi/2}^{\pi/2} \sqrt{1 + \sin^2 x} \, dx.$$

Let $\Delta x = \pi/6$, $\Delta x / 3 = \pi/18 = 0.1745$.

$$A_p = 0.1745 \left[\sqrt{1 + \sin^2 \frac{-\pi}{2}} + 4\sqrt{1 + \sin^2 \frac{-\pi}{3}} + 2\sqrt{1 + \sin^2 \frac{\pi}{6}} \right.$$

$$\left. + 4\sqrt{1 + \sin^2(0)} + \cdots + 2\sqrt{1 + \sin^2 \frac{\pi}{6}} + 4\sqrt{1 + \sin^2 \frac{\pi}{3}} + \sqrt{1 + \sin^2 \frac{\pi}{2}} \right]$$

$$= 0.1745 \left[\sqrt{2} + 4\sqrt{1 + \frac{3}{4}} + 2\sqrt{1 + \frac{1}{4}} + 4 + 2\sqrt{1 + \frac{1}{4}} + 4\sqrt{1 + \frac{3}{4}} + \sqrt{2} \right]$$

$$= 3.8187 \approx 3.82$$

153.

$$V = \pi \int_0^1 y^2 \, dx = \pi \int_0^1 \left(\ln \frac{1}{x}\right)^2 \left(-\frac{dx}{x^2}\right) = \pi \left[\frac{1}{x}\left(\ln \frac{1}{x}\right)^2 - \frac{2}{x} \ln \frac{1}{x} + 2 \right]_0^1 = 2\pi$$

CHAPTER 7 - Miscellaneous Problems

155.

$$S = 2 \cdot 2\pi \int_0^{\pi/2} y \sqrt{\left(\frac{dx}{d\theta}\right)^2 + \left(\frac{dy}{d\theta}\right)^2} \, d\theta$$

$$= 4\pi \int_0^{\pi/2} (a \sin^3 \theta) \sqrt{9a^2 \cos^2 \theta \sin^2 \theta} \, d\theta$$

$$= 12\pi a^2 \int_0^{\pi/2} \sin^4 \theta \cos \theta \, d\theta = \frac{12}{5} \pi a^2 \sin^5 \theta \Big]_0^{\pi/2} = \frac{12}{5} \pi a^2$$

157. Let $u = \ln x$, $du = (1/x) \, dx$, $dv = (1/x^2) \, dx$, $v = -1/x$.

$$\int_1^\infty \ln x \, \frac{dx}{x^2} = -\ln x \cdot \frac{1}{x} \Big]_1^\infty + \int_1^\infty \frac{dx}{x^2} = -\frac{\ln x}{x} - \frac{1}{x} \Big]_1^\infty$$

$$\lim_{x \to \infty} \left(\frac{\ln x}{x}\right) = \lim_{x \to \infty} \frac{1/x}{1} = 0 \qquad \therefore \text{ the integral converges.}$$

Compared with the integral

$$\int_1^\infty \frac{dx}{x^{3/2}} \,, \quad \int_1^\infty \ln x \, \frac{dx}{x^2} \le \int_1^\infty \frac{dx}{x^{3/2}}$$

$$\int_1^\infty \frac{dx}{x^{3/2}} = -\frac{1}{2} x^{-1/2} \Big]_1^\infty = \frac{1}{2} \qquad \therefore \text{ the integral converges.}$$

159. (a)

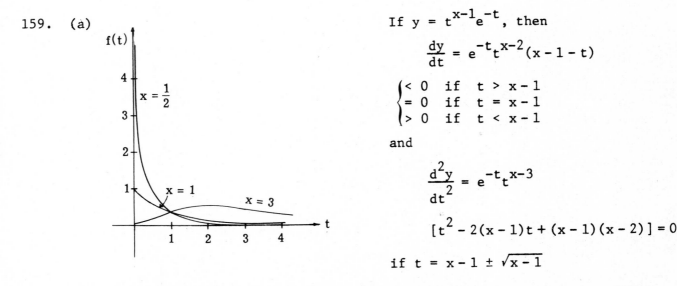

If $y = t^{x-1} e^{-t}$, then

$$\frac{dy}{dt} = e^{-t} t^{x-2} (x - 1 - t)$$

$$\begin{cases} < 0 & \text{if} \quad t > x - 1 \\ = 0 & \text{if} \quad t = x - 1 \\ > 0 & \text{if} \quad t < x - 1 \end{cases}$$

and

$$\frac{d^2 y}{dt^2} = e^{-t} t^{x-3}$$

$$[t^2 - 2(x-1)t + (x-1)(x-2)] = 0$$

if $t = x - 1 \pm \sqrt{x - 1}$

Note: If $0 < x \le 1$, there are no inflection points. If $0 < x < 1$, the graph approaches the y-axis as one asymptote and the t-axis as another, and there are no maximum or minimum points. For $x = 1$, the maximum is at the left end of the curve at $(0,1)$. For $x > 1$, there is a maximum at the point where $t = x - 1$, and a minimum would occur at $(0,0)$ if we add $t = 0$ to the domain, which we do for convenience in sketching. Short tables can be found elsewhere (e.g., Abramowitz and Stegun,

Handbook of Mathematical Functions, 1964, p. 267 et seq.; Jahnke and Emde, *Tables of Functions*, 1945, p. 14). A rough sketch is shown above, for $x = 1/2, 1,$ and 3.

(b) For $0 < x < 1$, the factor t^{x-1} tends to infinity as t approaches 0. Since $e^{-t} < 1$, the graph of $y = t^{x-1}e^{-t}$ lies below the graph of $y = t^{x-1}$. For small values of t, say $0 < t < 1$ and $\varepsilon > 0$, we have

$$\int_{\varepsilon}^{1} t^{x-1}e^{-t}\, dt < \int_{\varepsilon}^{1} t^{x-1}\, dt = \left.\frac{t^{x}}{x}\right]_{\varepsilon}^{1} = \frac{1 - e^{x}}{x} < \frac{1}{x}.$$

$\therefore \displaystyle\lim_{\varepsilon \to 0} \int_{\varepsilon}^{1} t^{x-1}e^{-t}\, dt$ exists and is not greater than $1/x$.

That is, $\int_{0}^{1} t^{x-1}e^{-t}\, dt$ converges if $0 < x < 1$ and its value is less than $1/x$. (For $x = 1/2$, this means that the area between the curve, the y-axis, the x-axis, and the line $x = 1$ is finite and less than 2.) To establish convergence for the integration from 1 to ∞, we use the fact that $t^{x-1} \le C_x e^{0.5t}$ if $x \ge 1$ and $t \ge 1$, where C_x is an appropriate number depending upon x. (The actual numerical value of C_x does not matter; it could be, for example, the maximum value of the function $g(t) = t^{x-1}e^{-0.5t}$, $t \ge 1$.) This maximum occurs at $t = 2(x - 1)$ if $x \ge 3/2$, and at $t = 1$ if $x < 3/2$. Thus for suitable choice of C_x, we can always say that $t^{x-1}e^{-t} \le C_x e^{0.5t}$ for $t \ge 1$ and

$$\int_{1}^{\infty} t^{x-1}e^{-t}\, dt \le \lim_{b \to \infty} C_x \int_{1}^{b} e^{-0.5t}\, dt = 2C_x e^{-0.5}.$$

Since both $\int_{0}^{1} t^{x-1}e^{-t}\, dt$ and $\int_{1}^{\infty} t^{x-1}e^{-t}\, dt$ converge, the integral from 0 to ∞ converges.

(c) If $x = 0$, $t^{x-1}e^{-t} = e^{-t}/t$, and near $t = 0$, $e^{-t} = 1$, so the function behaves like $1/t$. In fact, if $0 < t < 1$, then $1 > e^{-t} > 0.36$ and $e^{-t}/t > 0.36/t$. But $\int_{0}^{1} (0.36/t)\, dt$ diverges because

$$\lim_{\varepsilon \to 0} \int_{\varepsilon}^{1} \frac{1}{t}\, dt = \lim_{\varepsilon \to 0} \ln \frac{1}{\varepsilon} = \infty.$$

$\therefore \displaystyle\lim_{\varepsilon \to 0} \int_{\varepsilon}^{1} t^{-1}e^{-t}\, dt = \infty,$ so $\displaystyle\int_{0}^{1} t^{-1}e^{-t}\, dt$

diverges. For $x < 0$, and $0 < t < 1$, $t^{x} = e^{x \ln t} > 1$, so $t^{x-1} > t^{-1}$ and $\int_{0}^{1} t^{x-1}e^{-t}\, dt$ diverges by comparison with $\int_{0}^{1} t^{-1}e^{-t}\, dt$. Thus, for $x \le 0$, the integral from 0 to 1 diverges, which means that $\int_{0}^{\infty} t^{x-1}e^{-t}\, dt$ also diverges since the integrand is everywhere positive for $t > 0$.

(d) $$\Gamma(x+1) = \int_0^\infty t^x e^{-t}\, dt$$

because $(x+1) - 1 = x$. In this integral, let $u = t^x$, $dv = e^{-t}\, dt$, then $du = xt^{x-1}\, dt$, $v = -e^{-t}$ and

$$\int_0^\infty t^x e^{-t}\, dt = -t^x e^{-t}\Big]_0^\infty + \int_0^\infty xt^{x-1}e^{-t}\, dt = 0 + x\int_0^\infty t^{x-1}e^{-t}\, dt = x\Gamma(x),$$

because $\lim\limits_{t \to \infty} t^x e^{-t} = 0$ and $\lim\limits_{t \to 0^+} t^x e^{-t} = 0$, if $x > 0$.

(e) $$\Gamma(1) = \int_0^\infty t^0 e^{-t}\, dt = -e^{-t}\Big]_0^\infty = -0 + 1 = 1 = 0!$$

Hence, $\Gamma(n) = (n-1)!$ is true for $n = 1$. Suppose it is true that $\Gamma(k) = (k-1)!$ for some positive integer k, then by part (d) above

$$\Gamma(k+1) = k\Gamma(k) = k(k-1)! = k!$$

Therefore, by principle of mathematical induction, $\Gamma(n) = (n-1)!$ is true for all positive integers n.

(f)   Consider any x between 1 and 2. By its definition as a definite integral, $\Gamma(x)$ can be interpreted as the area above the t-axis and under the graph of $y = t^{x-1}e^{-t}$ from $t = 0$ to $t = \infty$. Because $x > 1$, the graph is similar to that shown here. Suppose we wanted to find the area correct to 3 decimal places, we can eliminate the difficulty caused by the infinitely long tail to the right if we locate b such that $t^{x-1}e^{-t} < e^{-0.5t}$, for $t > b$, and $\int_b^\infty e^{-0.5t}\, dt < 0.0001$. The latter requirement is easy to translate into

$$2e^{-b/2} < 0.0001 = 10^{-4}$$

or

$$e^{b/2} > 2 \times 10^4 = 20{,}000.$$

$y = t^{x-1}e^{-t}; x > 1$

Now $e^{10} \approx 22{,}026$, so $b \geq 20$ will work. The condition $t^{x-1} < e^{-0.5t}$ will be satisfied for $1 < x < 2$ if $t < e^{0.5t}$ or $\ln t < 0.5t$ or $\ln t/t < 0.5$. The function $f(t) = \ln t/t$ has derivative $f'(t) = (1 - \ln t)/t^2$ which is zero when $t = e$, positive for $1 < t < e$, and negative for $t > e$. Hence, the maximum value of $f(t)$ is $f(e) = 1/e \approx 0.37$. That is, $(\ln t)/t \leq e^{-1} \approx 0.37 < 0.5$ is true for all $t > 0$. This means that if we compute the area under the original curve $y = t^{x-1}e^{-t}$ for $0 \leq t \leq 20$, the result will differ from $\Gamma(x)$ by less than 1 in the fourth decimal place. Now by using the trapozoidal rule, we can estimate $\int_0^{20} t^{x-1}e^{-t}\, dt$ with an error as small as we wish, say to within 4 units in the fourth decimal place. The result is at least a three-decimal approximation to $\Gamma(x)$. A table of values of the gamma function $\Gamma(n)$ for $1.00 \le n \le 2.00$ is given on page 349 of the <u>Chemical Rubber Company Standard Mathematical Tables</u>, 14th edition. An abbreviated table follows:

x	1.0	1.1	1.2	1.3	1.4	1.5	1.6	1.7	1.8	1.9	2.0
$\Gamma(x)$	1.000	0.951	0.918	0.897	0.887	0.886	0.894	0.909	0.931	0.962	1.000

From this table and the formula, $\Gamma(x+1) = x\Gamma(x)$ or $\Gamma(x) = \Gamma(x+1)/x$, we can compute

$$\Gamma(2.5) = \Gamma(1.5+1) = 1.5\Gamma(1.5) = 1.329$$

$$\Gamma(0.5) = \frac{\Gamma(1.5)}{0.5} = 2\Gamma(1.5) = 1.772$$

and $\Gamma(0.1) = \dfrac{\Gamma(1.1)}{0.1} = 10\Gamma(1.1) = 9.51.$

For a rough sketch of $\Gamma(x)$ for $0 < x \le 3$, the points in the following table are used

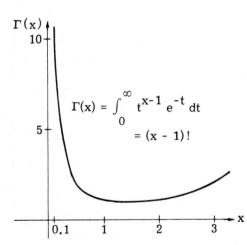

$$\Gamma(x) = \int_0^\infty t^{x-1} e^{-t} dt$$
$$= (x-1)!$$

x	0.1	0.5	1.0	1.5	2.0	2.5	3.0
$\Gamma(x)$	9.5	1.8	1.0	0.9	1.0	1.3	2.0

Remark. Because $\Gamma(x+1) = x!$ for x an integer ≥ 0, and $\Gamma(x+1) = x\Gamma(x)$ for all real x, it is customary to use the gamma function as the definition of the factorial function on the domain $x \ge 0$ to be $x! = \Gamma(x+1)$. Thus $(1.5)! = \Gamma(2.5)$; $(0.6)! = \Gamma(1.6)$, etc.

161. $$\frac{d^2x}{dt^2} = \frac{dv}{dt} = \frac{dv}{dx}\cdot\frac{dx}{dt} = v\frac{dv}{dx}; \qquad v\frac{dv}{dx} = -k^2x \implies \int v\,dv = -k^2\int x\,dx$$

$\therefore \quad \dfrac{v^2}{2} = -\dfrac{k^2x^2}{2} + C_1$

Initial conditions: $x = a$, $dx/dt = v = 0$ when $t = 0$

$\therefore \quad 0 = -\dfrac{k^2a^2}{2} + C_1 \implies C_1 = \dfrac{k^2a^2}{2} \qquad \therefore \quad v^2 = k^2(a^2 - x^2)$

$\therefore \quad v = \dfrac{dx}{dt} = k\sqrt{a^2 - x^2}$

$$\int \frac{1}{\sqrt{a^2 - x^2}} dx = k\int dt \implies \sin^{-1}\frac{x}{a} = kt + C_2$$

CHAPTER 7 – Miscellaneous Problems

Initial condition: $\sin^{-1}(a/a) = 0 + C_2 \implies C_2 = \pi/2$

$$\implies \sin^{-1}(x/a) = kt + (\pi/2)$$

$$x/a = \sin[kt + (\pi/2)] = \cos kt \implies x = a\cos kt$$

163. The formula is easily seen to be correct for $n = 0$. For

$$\int_{-1}^{1} \left(1 - x^2\right)^0 dx = 2 = \frac{2(0!)^2}{1!}$$

since $0! = 1$. To prove it is true in general, we use the method of mathematical induction and the reduction formula

$$\int_{0}^{\pi/2} \cos^m \theta \, d\theta = \left.\frac{\cos^{m-1}\theta \, \sin\theta}{m}\right]_{0}^{\pi/2} + \frac{m-1}{m}\int_{0}^{\pi/2} \cos^{m-2}\theta \, d\theta,$$

Thus, assuming the formula

$$\int_{-1}^{1} \left(1 - x^2\right)^n dx = \frac{2^{2n+1}(n!)^2}{(2n+1)!}$$

is true for the nonnegative integer n, we replace n by $n+1$ and let $x = \sin\theta$ to get

$$\int_{-1}^{1} \left(1 - x^2\right)^{n+1} dx = 2\int_{0}^{\pi/2} \cos^{2n+3}\theta \, d\theta = 2\left(\frac{2n+2}{2n+3}\right)\int_{0}^{\pi/2} \cos^{2n+1}\theta \, d\theta$$

$$= \left(\frac{2n+2}{2n+3}\right)\int_{-1}^{1}\left(1-x^2\right)^n dx = \left(\frac{2n+2}{2n+3}\right)\left(\frac{2^{n+1}(n!)^2}{(2n+1)!}\right)$$

$$= \left(\frac{2(n+1)}{2n+3}\right)\left(\frac{2n+2}{2n+2}\right)\left(\frac{2^{2n+1}(n!)^2}{(2n+1)!}\right) = \frac{2^{2n+3}[(n+1)!]^2}{(2n+3)!},$$

which is the right side of the original formula when n is replaced by $(n+1)$. Q.E.D.

CHAPTER 8

PLANE ANALYTIC GEOMETRY

Article 8.2

1. $\left(x^2 + y^2\right)^{1/2} = y - (-4) \implies x^2 + y^2 = (y + 4)^2 = y^2 + 8y + 16$

    $\implies x^2 = 8y + 16.$ Parabola.

3.

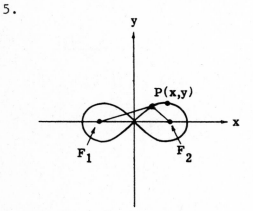

$$\sqrt{(x+2)^2 + (y-1)^2} = \sqrt{(x-2)^2 + (y+3)^2}$$

$$x^2 + 4x + 4 + y^2 - 2y + 1 = x^2 - 4x + 4 + y^2 + 6y + 9$$

$$x - y = 1, \text{ a line}$$

5.

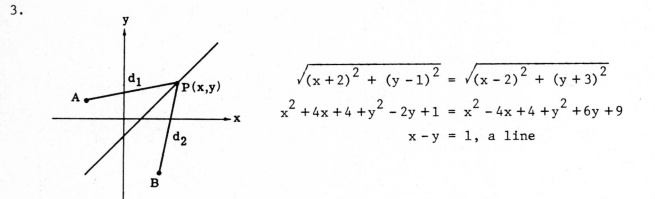

$$\sqrt{(x+2)^2 + y^2} - \sqrt{(x-2)^2 + y^2} = 4$$

$$x^4 + 2x^2y^2 + y^4 + 8y^2 - 8x^2 = 0$$

$$\left(x^2 + y^2\right)^2 + 8(y^2 - x^2) = 0$$

Lemniscate of Bernoulli.

7.

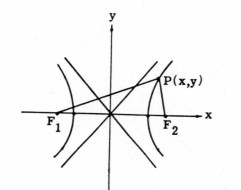

$$x + 2 = 2\sqrt{(x-2)^2 + y^2}$$

$3x^2 + 4y^2 - 20x + 12 = 0$, an ellipse

9.

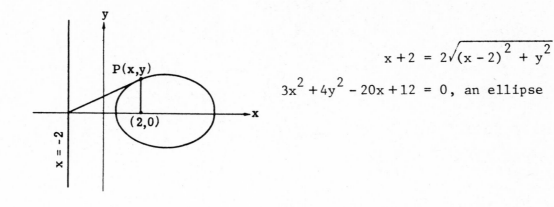

$$\sqrt{(x+3)^2 + y^2} = \sqrt{(x-3)^2 + y^2} + 4$$

$$3x - 4 = 2\sqrt{(x-3)^2 + y^2}$$

$5x^2 - 4y = 20$, hyperbola.

11. $(x-2)^2 + (y-3)^2 = 9$, circle, radius 3, center at $(2,3)$.

13.

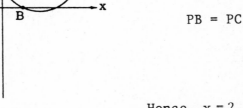

The point (x,y) equidistant from the points A, B and C is at the center of the circle through these points.

$$PA = PB \implies \sqrt{x^2 + (y-1)^2} = \sqrt{(x-1)^2 + y^2}$$

$$\implies x = y$$

$$PB = PC \implies \sqrt{(x-1)^2 + y^2}$$

$$= \sqrt{(x-4)^2 + (y-3)^2}$$

$$\implies x + y = 4.$$

Hence, $x = 2$, $y = 2$. Radius is

$$PA = \sqrt{2^2 + (2-1)^2} = \sqrt{5}$$

Article 8.3

1. $x^2 + (y-2)^2 = 4$
   $x^2 + y^2 - 4y = 0$

3. $(x-3)^2 + (y+4)^2 = 25$
   $x^2 + y^2 - 6x + 8y = 0$

5. $(x+2)^2 + (y+1)^2 < 6$

7. $x^2 + y^2 = 16 \implies (x-0)^2 + (y-0)^2 = 4^2$
   Center = (0,0), radius = 4.

9. $x^2 + y^2 - 2y = 3 \implies x^2 + y^2 - 2y + 1 = 4 \implies x^2 + (y-1)^2 = 4$
   Center = (0,1), radius = 2.

11. $x^2 + 4x + y^2 = 12 \implies x^2 + 4x + 4 + y^2 = 16$
    $(x+2)^2 + y^2 = 4^2 \implies C = (-2,0)$. Radius = 4.

13. $x^2 + y^2 + 2x + 2y = -1 \implies x^2 + 2x + 1 + y^2 + 2y + 1 = 1$
    $(x+1)^2 + (y+1)^2 = 1 \implies C = (-1,-1)$, radius = 1.

15. $2x^2 + 2y^2 + x + y = 0 \implies x^2 + \frac{1}{2}x + \frac{1}{16} + y^2 + \frac{1}{2}y + \frac{1}{16} = \frac{1}{8}$
    $\left(x+\frac{1}{4}\right)^2 + \left(y+\frac{1}{4}\right)^2 = \frac{1}{8}$ $C = \left(-\frac{1}{4}, -\frac{1}{4}\right)$. Radius = $\frac{1}{\sqrt{8}}$

17. $x^2 + y^2 + 2x - 4y + 5 = (x+1)^2 + (y-2)^2 \le 0$ is satisfied only by (-1,2).

19. $(x-2)^2 + (y-2)^2 = r^2$. $(x,y) = (4,5) \implies (4-2)^2 + (5-2)^2 = 13 = r^2$
    $(x-2)^2 + (y-2)^2 = 13 \implies x^2 + y^2 - 4x - 4y = 5$

21. Center is at (a, 2-a)
    $r^2 = (3-a)^2 + (4-2+a)^2 = (2-a)^2 + (2-a+2)^2$
    $9 - 6a + a^2 + 4 + 4a + a^2 = 4 - 4a + a^2 + 16 - 8a + a^2$
    $$10a = 7$$
    $$a = \frac{7}{10}$$
    Center at (7/10, 13/10)
    $$r = \sqrt{\left(3 - \frac{7}{10}\right)^2 + \left(2 + \frac{7}{10}\right)^2} = \sqrt{12.58}. \quad (x-0.7)^2 + (y-1.3)^2 = 12.58$$

23. $x^2 + y^2 + Ax + By + C = 0$

    $(x,y) = (2,3) \implies 4 + 9 + 2A + 3B + C = 0$ (1)
    $(x,y) = (3,2) \implies 9 + 4 + 3A + 2B + C = 0$ (2)
    $(1),(2) \implies A - B = 0$ or $A = B$ (3)
    $(x,y) = (-4,3) \implies 16 + 9 - 4A + 3B + C = 0$ (4)

(3),(4) $\implies$ A = C + 25   (5)

(1) $\implies$ 13 + 5A + (A − 25) = 0 $\implies$ A = B = 2, C = −23

$\qquad x^2 + y^2 + 2x + 2y - 23 = 0$

25. $(x - 1)^2 + (y - 2)^2 = 2.$ $(0.1 - 1)^2 + (3.1 - 2)^2 = 2.02 > 2 \implies$ outside

27. $\qquad x^2 + y^2 = a^2 \implies 2x + 2y\,\dfrac{dy}{dx} = 0 \implies \dfrac{dy}{dx} = -\dfrac{x}{y}\,.$

The slope of the surface of the circle is m = −x/y. The perpendicular has
slope −1/m = y/x, which is the slope of line through the origin and (x,y).
The line through the origin and the perpendicular have the same slope and
both pass through (x,y), so they are the same line. Thus the perpendicular
passes through the origin.

29. Construct the axes so that T is at the origin, and P is on the positive
x-axis. Let the center, C, be at (0,1), so that the radius is 1, and
CM = CN = 1. Let y = PC, and x = PT. We must show that

$\qquad x^2 = (y - 1)(y + 1) = y^2 - 1.$

Since TC = 1, and $\overline{TC}$ is perpendicular to $\overline{TP}$,

$\qquad y^2 = (TC)^2 + (TP)^2 = 1 + x^2\,.$

Article 8.4

1.

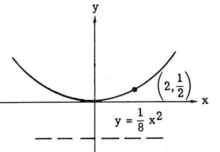

Parabola: $x^2 = 4(2)y = 8y$

Directrix: y = −2; (p = 2)

3. $\qquad p = |V - F| = \left[(0 - 0)^2 + (0 + 4)^2\right]^{1/2} = 4$

Parabola: $x^2 = -4(p)y = -16y$; Opens down.

Directrix: y = 4

5. $\qquad p = |V - F|$

$\qquad\qquad = \left[(-2 + 2)^2 + (3 - 4)^2\right]^{1/2} = 1$

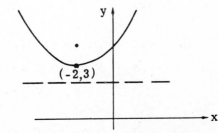

Parabola: $(x + 2)^2 = 4(y - 3)$

Directrix: y = 2

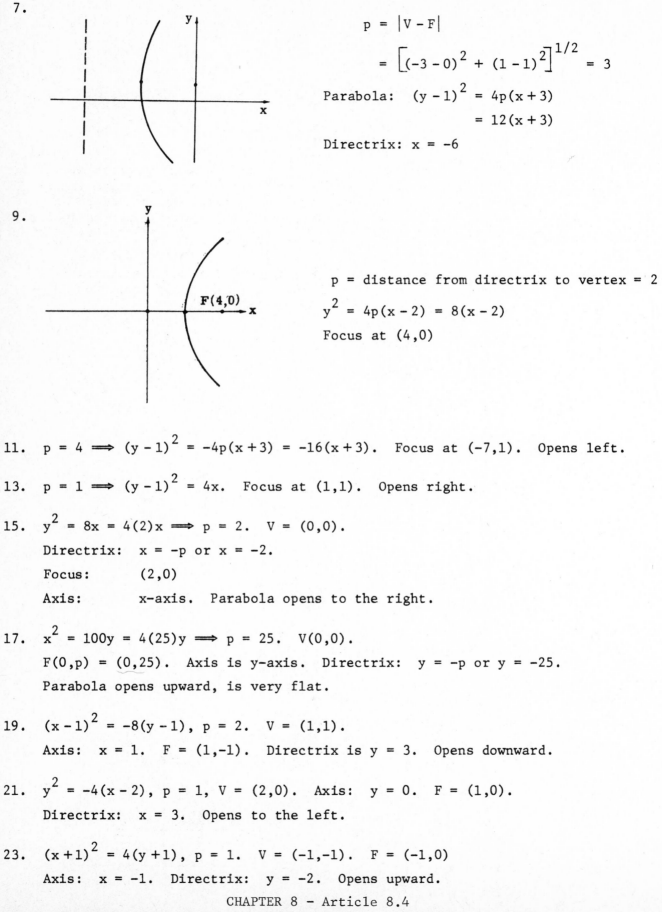

7.

$p = |V - F|$

$\quad = \left[(-3 - 0)^2 + (1 - 1)^2\right]^{1/2} = 3$

Parabola: $(y - 1)^2 = 4p(x + 3)$

$\qquad\qquad\qquad = 12(x + 3)$

Directrix: $x = -6$

9.

p = distance from directrix to vertex = 2

$y^2 = 4p(x - 2) = 8(x - 2)$

Focus at $(4, 0)$

F(4,0)

11. $p = 4 \implies (y - 1)^2 = -4p(x + 3) = -16(x + 3)$. Focus at $(-7, 1)$. Opens left.

13. $p = 1 \implies (y - 1)^2 = 4x$. Focus at $(1, 1)$. Opens right.

15. $y^2 = 8x = 4(2)x \implies p = 2$. $V = (0, 0)$.

Directrix: $x = -p$ or $x = -2$.

Focus: $(2, 0)$

Axis: x-axis. Parabola opens to the right.

17. $x^2 = 100y = 4(25)y \implies p = 25$. $V(0, 0)$.

$F(0, p) = (0, 25)$. Axis is y-axis. Directrix: $y = -p$ or $y = -25$.

Parabola opens upward, is very flat.

19. $(x - 1)^2 = -8(y - 1)$, $p = 2$. $V = (1, 1)$.

Axis: $x = 1$. $F = (1, -1)$. Directrix is $y = 3$. Opens downward.

21. $y^2 = -4(x - 2)$, $p = 1$, $V = (2, 0)$. Axis: $y = 0$. $F = (1, 0)$.

Directrix: $x = 3$. Opens to the left.

23. $(x + 1)^2 = 4(y + 1)$, $p = 1$. $V = (-1, -1)$. $F = (-1, 0)$

Axis: $x = -1$. Directrix: $y = -2$. Opens upward.

CHAPTER 8 - Article 8.4

25. $$(y - 1)^2 = -4\left(\frac{3}{16}\right)(x - 2),$$

$p = 3/16$. $V = (2,1)$, $F = (29/16,1)$.

Axis: $y = 1$. Directrix: $x = 35/16$. Opens to the left.

27. $$y^2 + 4x = 8 \implies (x - 2)^2 = 4\left(\frac{2}{3}\right)(y + 2),$$

$p = 2/3$. $V = (2,-2)$, $F = (2,-4/3)$.

Axis: $x = 2$. Directrix: $y = -8/3$. Opens upward.

29. $x^2 = 8y$ represents an upward opening parabola. Any point (x,y) that satisfies $x^2 < 8y$ lies above this parabola, since it has the same x coordinate and greater y coordinate for some point on the parabola.

31. $$\left[x - \left(\frac{b}{2}\right)\right]^2 = 4p(y - h).$$

$$(x,y) = (0,0) \implies \frac{b^2}{4} = -4ph \implies 4p = \frac{-b^2}{4ph}$$

$$(x,y) = (b,0) \implies \frac{b^2}{4} = 4ph$$

$$\left[x - \left(\frac{b}{2}\right)\right]^2 = \left(\frac{-b^2}{4h}\right)(y - h) \implies b^2 y = 4hx(b - x).$$

33. (1) $y^2 = 4a^2 - 4ax = 4a(a - x)$, $p = a \implies V(0,0)$

(2) $y^2 = 4b^2 + 4bx = 4b(b + x)$, $p = b \implies V(0,0)$.

$$4a(a - x) = 4b(b + x) \implies a^2 - ax = b^2 + bx$$
$$a^2 - b^2 = (a + b)x \implies x = a - b$$
$$y^2 = 4a(a - x) = 4a(a - a + b) = 4ab$$

Thus the points of intersection are $(a-b, \pm 2\sqrt{ab})$

(1) $\implies 2y \, dy = 4a(-dx) \implies \dfrac{dy}{dx} = \dfrac{-2a}{2\sqrt{ab}} = -\sqrt{a/b}$

(2) $\implies 2y \, dy = 4b \, dx \implies \dfrac{dy}{dx} = \dfrac{2b}{2\sqrt{ab}} = \sqrt{b/a}$

Hence, the curves are orthogonal at the points of intersection.

35.

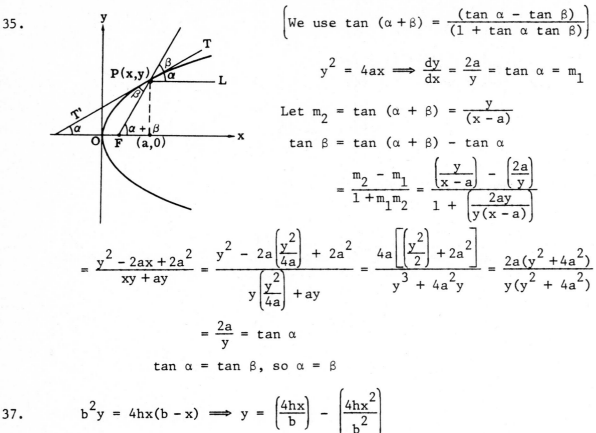

$$\left[\text{We use } \tan(\alpha + \beta) = \frac{(\tan\alpha - \tan\beta)}{(1 + \tan\alpha\,\tan\beta)}\right]$$

$$y^2 = 4ax \implies \frac{dy}{dx} = \frac{2a}{y} = \tan\alpha = m_1$$

$$\text{Let } m_2 = \tan(\alpha + \beta) = \frac{y}{(x - a)}$$

$$\tan\beta = \tan(\alpha + \beta) - \tan\alpha$$

$$= \frac{m_2 - m_1}{1 + m_1 m_2} = \frac{\left(\dfrac{y}{x - a}\right) - \left(\dfrac{2a}{y}\right)}{1 + \left(\dfrac{2ay}{y(x - a)}\right)}$$

$$= \frac{y^2 - 2ax + 2a^2}{xy + ay} = \frac{y^2 - 2a\left(\dfrac{y^2}{4a}\right) + 2a^2}{y\left(\dfrac{y^2}{4a}\right) + ay} = \frac{4a\left[\left(\dfrac{y^2}{2}\right) + 2a^2\right]}{y^3 + 4a^2 y} = \frac{2a(y^2 + 4a^2)}{y(y^2 + 4a^2)}$$

$$= \frac{2a}{y} = \tan\alpha$$

$$\tan\alpha = \tan\beta, \text{ so } \alpha = \beta$$

37.

$$b^2 y = 4hx(b - x) \implies y = \left(\frac{4hx}{b}\right) - \left(\frac{4hx^2}{b^2}\right)$$

$$\int_0^b y\, dx = \int_0^b \left(\frac{4hx}{b} - \frac{4hx^2}{b^2}\right) dx = \frac{4h}{2b}x^2 - \frac{4h}{3b^2}x^3 \Big|_0^b = 2bh - \frac{4}{3}bh = \frac{2}{3}bh$$

39.

$$\frac{dy}{dx} = \frac{wx}{H} \implies dy = \left(\frac{w}{H}\right) x\, dx$$

$$\int dy = \int \left(\frac{w}{H}\right) x\, dx + C \implies y = \frac{w}{2H}x^2 + C$$

One point is $(0,0)$, so $C = 0$. $y = (w/2H)x^2$, which is a parabola.

Article 8.5

1.

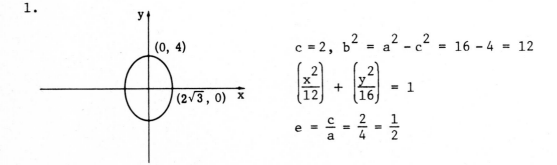

$$c = 2, \quad b^2 = a^2 - c^2 = 16 - 4 = 12$$

$$\left(\frac{x^2}{12}\right) + \left(\frac{y^2}{16}\right) = 1$$

$$e = \frac{c}{a} = \frac{2}{4} = \frac{1}{2}$$

3.

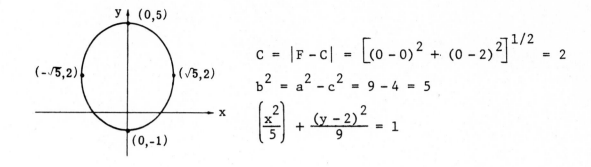

$$C = |F - C| = \left[(0 - 0)^2 + (0 - 2)^2\right]^{1/2} = 2$$

$$b^2 = a^2 - c^2 = 9 - 4 = 5$$

$$\left(\frac{x^2}{5}\right) + \frac{(y - 2)^2}{9} = 1$$

5. $C = 3$, $b^2 = a^2 - c^2 = 10 - 9 = 1$. $\dfrac{(x - 2)^2}{10} + \dfrac{(y - 2)^2}{1} = 1$. $e = \dfrac{3}{\sqrt{10}}$

7. $C = (7,5)$. $c^2 = a^2 - b^2 = 25 - 4 = 21$. $c = \sqrt{21}$
    Foci at $(7, 5 \pm \sqrt{21})$, vertices at $(7, 5 \pm 5)$.

9. $C = (-1,-4)$. $c^2 = a^2 - b^2 = 25 - 9 = 16$. $c = 4$
    Foci at $(-1, -4 \pm c) = (-1, -4 \pm 4)$, vertices at $(-1, -4 \pm 5)$.

11. $$\frac{(x - 3)^2}{4} + \frac{(y - 1)^2}{25} = 1$$

    $C = (3,1)$. $c^2 = a^2 - b^2 = 25 - 4$. $c = \sqrt{21}$
    Foci at $(3, 1 \pm \sqrt{21})$, vertices at $3, 1 \pm 5)$

13. $$x^2 + 10x + 25y^2 = 0 \implies (x + 5)^2 + 25y^2 = 25$$

    $$\frac{(x + 5)^2}{25} + y^2 = 1$$

    $C = (-5,0)$. $c^2 = a^2 - b^2 = 25 - 1 = 24$. $c = 2\sqrt{6}$
    Foci at $(-5 \pm 2\sqrt{6}, 0)$, vertices at $(-5 \pm 5, 0)$

15. $$(x^2 - 4x + 4) + 9(y^2 + 2y + 1) = 9 \implies \frac{(x - 2)^2}{9} + (y + 1)^2 = 1$$

    $C = (2,-1)$. $c^2 = a^2 - b^2 = 9 - 1 = 8$. $c = 2\sqrt{2}$
    Foci at $(2 \pm c, -1) = (2 \pm 2\sqrt{2}, -1)$, vertices at $(2 \pm a, -1) = (2 \pm 3, -1)$

17. $$4(x^2 - 4x + 4) + (y^2 + 4y + y) = 4$$

    $$(x - 2)^2 + \frac{(y + 2)^2}{4} = 1$$

    $C = (2,-2)$. $c^2 = a^2 - b^2 = 4 - 1 = 3$. $c = \sqrt{3}$
    Foci at $(2, -2 \pm \sqrt{3})$, vertices at $(2, -2 \pm 2)$

19. $9(x^2 + 2x + 1) + 16(y^2 - 6y + 9) = 144;$ $\dfrac{(x+1)^2}{16} + \dfrac{(y-3)^2}{9} = 1$

$C = (-1,3).$ $c^2 = a^2 - b^2 = 16 - 9 = 7.$ $c = \sqrt{7}$

Foci at $(-1 \pm \sqrt{7}, 3)$, vertices at $(-1 \pm 4, 3)$

21. a) $\dfrac{x^2}{4} + \dfrac{y^2}{9} = 1$ b) $\dfrac{x^2}{36} + \dfrac{y^2}{16} = 1$

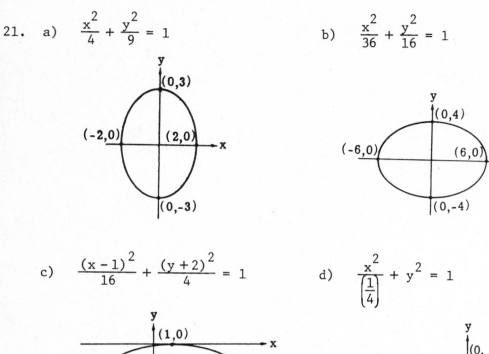

c) $\dfrac{(x-1)^2}{16} + \dfrac{(y+2)^2}{4} = 1$ d) $\dfrac{x^2}{\left(\frac{1}{4}\right)} + y^2 = 1$

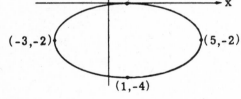

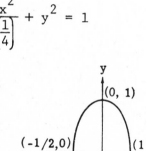

e) $\dfrac{(x-2)^2}{9} + \dfrac{(y+3)^2}{16} = 1$

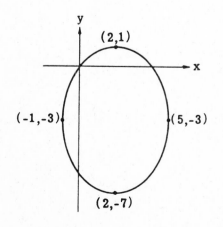

23. $c^2 = 16 - 7 = 9.$ y axis is major axis. $e = \dfrac{c}{a} = \dfrac{3}{4}$

Directrix: $a/e = 16/3$. center at $(0,0)$. So, $y = 16/3$ and $y = -16/3$.

25.
$$V = 2\int_0^a \pi y^2\,dx = 2\pi\int_0^a \left(b^2 - \frac{b^2}{a^2}x^2\right)dx = 2\pi b^2\left[x - \frac{x^3}{3a^2}\right]_0^a = \frac{4}{3}\pi ab^2.$$

27.

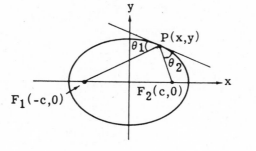

Let F_1 and F_2 be the two foci and P any point on the ellipse. A ray from F_1 to P will be reflected through F_2 if and only if the angles θ_1 and θ_2 formed between $\overline{F_1P}$ and $\overline{F_2P}$ each with the tangent line are equal. Thus, we must show that $\theta_1 = \theta_2$.

Let P' be any other point on the tangent line. Since P' is outside of the ellipse and P on the ellipse, we have

$$F_1P' + F_2P' > 2a = F_1P + F_2P.$$

This proves that P is the point which minimizes the sum of distances from F_1 and F_2 to a point on the tangent line. A general principle is that the rays to a point on a line which minimize the sum of distances form equal angles with the line. The two such angles are θ_1 and θ_2, so the principle confirms that $\theta_1 = \theta_2$.

(The first step in proving the principle is to draw a mirror image of one of the points on the opposite side of the line, and note that finding the shortest distance between one of the points and the image of the other is equivalent to finding the shortest sum of distances from one point to the line and back to the other point. A similar triangles argument completes the proof.)

29. $x = h + (a/e) = 9$ is the directrix [where the center is at $(h,0)$]

$h + c = 4$ is the x-coordinate of the focus.

$$\left(\frac{a}{e}\right) - ae = 5 \implies 9a - 4a = 30,\ a = 6,\ h = 0$$

$C = (0,0)$. $b^2 = a^2 - c^2 = 36 - 16 = 20$.

$$\left(\frac{x^2}{36}\right) + \left(\frac{y^2}{20}\right) = 1$$

31.
$$\left(\frac{x^2}{a^2}\right) + \left(\frac{y^2}{b^2}\right) = 1 \implies \frac{dy}{dx} = \frac{-b^2 x_1}{a^2 y_1} = \frac{y - y_1}{x - x_1} \implies \left(\frac{xx_1}{a^2}\right) + \left(\frac{yy_1}{b^2}\right) = 1$$

33.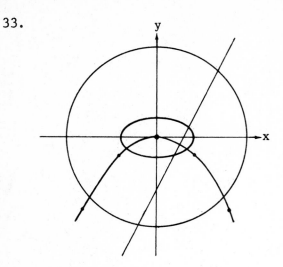

The equation is satisfied whenever any term is zero.

$$x^2 + 4y = 0 \quad \text{(parabola)}$$
$$2x - y - 3 = 0 \quad \text{(straight line)}$$
$$x^2 + y^2 - 25 = 0 \quad \text{(circle)}$$
$$x^2 + 4y^2 - 4 = 0 \quad \text{(ellipse)}$$

35. $e = c/a = 4/5$. Let $a = 5$, $c = 4$, we get $b = \sqrt{a^2 - c^2} = 3$. Using center at $(0,0)$, we have

$$\frac{x^2}{25} + \frac{y^2}{9} = 1. \quad \text{In general,} \quad \frac{(x-h)^2}{25r} + \frac{(y-k)^2}{9r} = 1.$$

37. a) From the text (p. 527),

$$a = 18.09, \quad b = 4.56, \quad e = 0.97, \quad c = ae = 17.54$$

If the sun is at a focus, then the center is at $(c,0)$. Thus the equation is

$$\frac{(x - 17.5)^2}{(18.09)^2} + \frac{y^2}{(4.56)^2} = 1.$$

b) Minimum distance = $a - c$ = 0.5842 AU
$$= (0.5842) \times (9.26 \times 10^7) = 5.41 \times 10^7 \text{ miles}$$
$$\text{(because 1 AU} = 9.26 \times 10^7 \text{ mile)}$$

c) Maximum distance = $a + c$ = 35.60 AU = 3.30×10^9 miles

$$\tfrac{1}{2}(\text{period}) = \tfrac{1}{2}(76) = 38.$$

At newest point to sun in January 1986, at farthest point in January 1948 (1986 – 38).

d) T = 76 years = $76 \times 365 + (76/4)$ days.
$$\frac{T^2}{D^3} = 1.66 \times 10^{-9}, \text{ so } D = 1.668 \times 10^9 \text{ mi} = 18.01 \text{ AU}$$

39. a) When the position of the ring is fixed, the loop has constant length. The sum of the distances of the pegs to the loop is constant, and this defines an ellipse.

CHAPTER 8 – Article 8.5

b)   Minimum energy occurs when the ring is at the lowest point of the ellipse, and this is exactly between the two foci. The foci are symmetric about the vertical line through the ring.

<u>Article 8.6</u>

1.   a) $\left(\dfrac{x^2}{9}\right) - \left(\dfrac{y^2}{16}\right) = 1$ b) $\left(\dfrac{x^2}{16}\right) - \left(\dfrac{y^2}{9}\right) = 1$

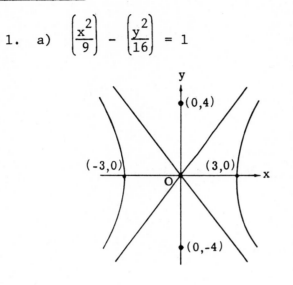
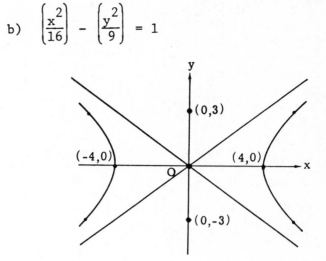

   c) $\left(\dfrac{y^2}{9}\right) - \left(\dfrac{x^2}{16}\right) = 1$ d) $\left(\dfrac{y^2}{16}\right) - \left(\dfrac{x^2}{9}\right) = 1$

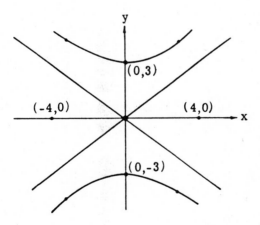
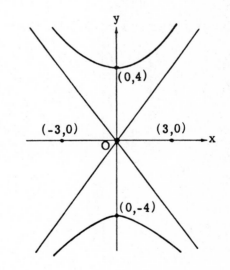

3.

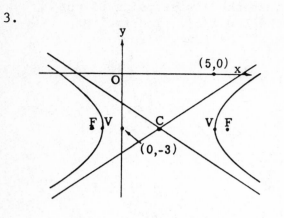

$$4(x-2)^2 - 9(y+3)^2 = 36$$

$$\implies \frac{(x-2)^2}{9} - \frac{(y+3)^2}{4} = 1$$

Center: $(2,-3)$. $a = 3$, $b = 2$

$$c^2 = a^2 + b^2 = 13, \quad c = \sqrt{13}$$

Foci at $(2 \pm c, -3) = (2 \pm \sqrt{13}, -3)$

Asym: $2(x-2) = \pm 3(y+3)$

Vertices: $(2 \pm a, -3) = (2 \pm 3, -3)$

5. $$5(x^2 + 4x + 4) - 4(y^2 - 2y + 1) = 20 \implies \left[\frac{(x+2)^2}{4}\right] - \left[\frac{(y-1)^2}{5}\right] = 1$$

Center: $(-2,1)$. $a = 2$, $b = \sqrt{5}$, $c = \sqrt{a^2 + b^2} = 3$

Foci: $(-2 \pm 3, 1)$. Vertices: $(-2 \pm 2, 1)$. Asym: $\sqrt{5}(x+2) = \pm 2(y-1)$.

Curves open to left and right.

7. $$x^2 - 4y^2 - 4x = 0 \implies \left[\frac{(x-2)^2}{4}\right] - y^2 = 1$$

Center: $(2,0)$. $a = 2$, $b = 1$, $c = \sqrt{a^2 + b^2} = \sqrt{5}$

Foci: $(2 \pm \sqrt{5}, 0)$. Vertices: $(2 \pm 2, 0)$. Asym: $(x-2) = \pm 2y$

Curves open to right and left.

9. $$\left[\frac{x^2}{a^2}\right] - \left[\frac{y^2}{b^2}\right] = 1 \implies \left[\frac{2x}{a^2}\right] - \left[\frac{2y}{b^2}\right]\frac{dy}{dx} = 0 \implies \frac{dy}{dx} = \frac{x_1 b^2}{y_1 a^2} = \frac{\Delta y}{\Delta x} = \frac{y_1 - y}{x_1 - x}$$

$$0 = \left[\frac{x_1^2}{a^2} - \frac{xx_1}{a^2}\right] - \left[\frac{y_1^2}{b^2} - \frac{yy_1}{b^2}\right] = -\frac{xx_1}{a^2} + \frac{yy_1}{b^2} + 1$$

$$\implies b^2 xx_1 - a^2 yy_1 = a^2 b^2.$$

11. $$\left[\frac{x^2}{(9-c)}\right] + \left[\frac{y^2}{(5-c)}\right] = 1$$

a) If $c < 5$, then since $9 - c > 0$, and $5 - c > 0$, we can let $a^2 = 9 - c$, $b^2 = 5 - c$, and get the equation

$$\left[\frac{x^2}{a^2}\right] + \left[\frac{y^2}{b^2}\right] = 1,$$

for an ellipse.

b) When $5 < c < 9$, we can let $a^2 = 9 - c > 0$, and $-b^2 = 5 - c < 0$. The equation,

$$\frac{x^2}{a^2} - \frac{y^2}{b^2} = 1$$

is of an hyperbola.

c) When $c > 9$, we make the same substitution as in (b) and get

$$\left(\frac{x^2}{a^2}\right) + \left(\frac{y^2}{b^2}\right) = -1,$$

which has no real points.

The foci of the ellipse are at $(\pm h, 0)$, where

$$h^2 = a^2 - b^2 = (9 - c) - (5 - c) = 4, \quad h = 2.$$

The foci of the hyperbola are at $(\pm k, 0)$, where

$$k^2 = a^2 + b^2 = (9 - c) + (c - 5) = 4, \quad k = 2.$$

13. $PF = ePD.$ $\sqrt{(x-1)^2 + (y+3)^2} = \left(\frac{3}{2}\right)|y - 2|$

$x^2 - 2x + 1 + y^2 + 6y + 9 = \left(\frac{9}{4}\right)(y^2 - 4y + 4)$

$4x^2 - 5y^2 - 8x + 60y + 4 = 0$

15. a) $|PA - PB| = 980$ ft/ms $\times 1400$ ms $= 1.372 \times 10^6$ ft $= 259.8$ mi

The boat is on hyperbola with $a = 259.8/2 = 130$ mi, $c = AB/2$, and foci at A and B.

17. (A sketch is necessary to follow this argument.) The foci are F_1 and F_2, as in Fig. 8.33. T_1 and T_2 are on the tangent line to the hyperbola, and R_1 and R_2 are on the tangent line to the ellipse. C is on the ray from F_2 through P. We prove the curves are perpendicular by showing that the tangents are perpendicular. This amounts to showing that

$$\angle R_1 P T_2 = \angle R_1 P T_1 \quad \text{or} \quad \angle R_1 PC + \angle CPT_2 = \angle F_1 PR_1 + \angle F_1 PT_1$$

From Problem 16, $\angle CPT_2 = \angle F_1 PT_1$ (in that problem we assumed these angles were equal and showed that F_2, P, and C were colinear).

$\angle R_1 PC = \angle F_2 PR_2$, because they are opposite vertical.

$\angle F_2 PR_2 = \angle F_1 PR_1$, because the tangent to an ellipse forms equal angles with the rays through the foci. (See Prob. 27, Sec. 8-5.)

It follows that

$$\angle R_1 PC = \angle F_1 PR \text{ and } \angle R_1 PC + \angle CPT_2 = \angle F_1 PR_1 + \angle F_1 PT_1 \, .$$

Article 8.7

1. $A = C = 0$, $B = 1$, $\cot 2\alpha = \dfrac{(A - C)}{B} = 0$, $\alpha = \dfrac{\pi}{4}$

$x = \dfrac{(x' - y')}{\sqrt{2}}$, $y = \dfrac{(x' + y')}{\sqrt{2}}$, $xy = \dfrac{(x'^2 - y'^2)}{2} = 2$ or $x'^2 - y'^2 = 4$.

3. $A = -1$, $B = 2$, $C = \sqrt{3}$, $\cot 2\alpha = \dfrac{(A - B)}{C} = -\sqrt{3}$, $\alpha = \dfrac{5\pi}{12} = 75°$

$\cos \alpha = \left[\dfrac{(2 + \sqrt{3})}{2}\right]^{1/2}$, $\sin \alpha = \left[\dfrac{(2 - \sqrt{3})}{2}\right]^{1/2}$,

$x = (x' \cos \alpha - y' \sin \alpha)$, $y = x' \sin \alpha + y' \cos \alpha$

$2y^2 + \sqrt{3}xy - x^2 = \left[\dfrac{1}{2} - \sqrt{3}\right]x'^2 + \left[\dfrac{1}{2} + \sqrt{3}\right]y'^2 = 2$

5. $A = C = 1$, $B = -3$, $\cot 2\alpha = \dfrac{(A - C)}{B} = 0$, $\alpha = \dfrac{\pi}{4}$

$x = \dfrac{(x' - y')}{\sqrt{2}}$, $y = \dfrac{(x' + y')}{\sqrt{2}}$

$x^2 - 3xy + y^2 = x'^2 + y'^2 - \left[\dfrac{3}{2}\right]x'^2 + \left[\dfrac{3}{2}\right]y'^2 = 5$

$5y'^2 - x'^2 = 10$

7. $A = 14$, $B = 16$, $C = 2$, $\cot 2\alpha = \dfrac{(A - C)}{B} = \dfrac{3}{4}$

Let $x = \cos 2\alpha$, $\dfrac{3}{4} = \cot 2\alpha = \dfrac{\cos 2\alpha}{\cos 2\alpha} = \dfrac{x}{\sqrt{1 - x^2}}$

$\dfrac{x^2}{1 - x^2} = \dfrac{9}{16} \implies 25x^2 = 9$, $x = \dfrac{3}{5}$.

$\cos^2 \alpha = \dfrac{(1 + \cos 2\alpha)}{2} = \dfrac{(1 + 3/5)}{2} = \dfrac{4}{5}$. $\cos \alpha = \dfrac{2}{\sqrt{5}}$, $\sin \alpha = \dfrac{1}{\sqrt{5}}$.

9. $x = (x' \cos \alpha - y' \sin \alpha)$, $y = (x' \sin \alpha + y' \cos \alpha)$

$r^2 = x^2 + y^2 = (x' \cos \alpha - y' \sin \alpha)^2 + (x' \sin \alpha + y' \cos \alpha)^2$

$\qquad = x'^2(\sin^2 \alpha + \cos^2 \alpha) + y'^2(\sin^2 \alpha + \cos^2 \alpha) = x'^2 + y'^2$.

Therefore, $x'^2 + y'^2 = r^2$.

11. $B'^2 = [B(\cos^2 \alpha - \sin^2 \alpha) + 2(C - A)\sin \alpha \cos \alpha]^2$

$= B^2(\sin^4 \alpha + \cos^4 \alpha - 2\cos^2 \alpha \sin^2 \alpha) + (4C^2 + 4A^2 - 8AC)\sin^2 \alpha \cos^2 \alpha$

$+ (4BC - 4AB)(\sin \alpha \cos^3 \alpha - \sin^3 \alpha \cos \alpha)$

$4A'C' = 4(A\cos^2 \alpha + B\sin \alpha \cos \alpha + C\sin^2 \alpha)$

$\times (A\sin^2 \alpha - B\sin \alpha \cos \alpha + C\cos^2 \alpha)$

$= 4(A^2 - B^2 + C^2)\sin^2 \alpha \cos^2 \alpha - 4AB\sin \alpha \cos^3 \alpha + AB\sin^3 \alpha \cos \alpha$

$+ 4AC\cos^4 \alpha + 4AC\sin^4 \alpha + 4BC\sin \alpha \cos^3 \alpha - 4BC\sin^3 \alpha \cos \alpha$

$B'^2 - 4A'C' = B^2(\sin^4 \alpha + \cos^4 \alpha + 2\sin^2 \alpha \cos^2 \alpha)$

$- 8AC\sin^2 \alpha \cos^2 \alpha - 4AC\cos^4 \alpha - 4AC\sin^4 \alpha$

$= B^2\left(\sin^2 \alpha + \cos^2 \alpha\right)^2 - 4AC(\sin^2 \alpha + \cos^2 \alpha) = B^2 - 4AC.$

13. $A = 1$, $B = 2$, $C = 1$.

$A' = A\cos^2 \alpha + B\cos \alpha \sin \alpha + C\sin^2 \alpha$

$= (\cos \alpha + \sin \alpha)^2 = 0 \implies \cos \alpha = -\sin \alpha,$

$\alpha = \dfrac{3\pi}{4}, \quad \cos \alpha = \dfrac{-1}{\sqrt{2}}, \quad \sin \alpha = \dfrac{1}{\sqrt{2}}$

$x = -\dfrac{(x' + y')}{\sqrt{2}}, \quad y = \dfrac{(x' - y')}{\sqrt{2}}$

$x^2 + 2xy + y^2 = 1 \implies (x + y)^2 = 1 \qquad x + y = \pm 1$

$-\dfrac{(x' + y')}{\sqrt{2}} + \dfrac{(x' - y')}{\sqrt{2}} = \pm 1 \implies y' = \pm \dfrac{1}{\sqrt{2}}$

Article 8.8

1. $A = 1$, $B = 0$, $C = -1$, $B^2 - 4AC = 4 > 0$. Hyperbola

3. $A = 0$, $B = 0$, $C = 1$, $B^2 - 4AC = 0$. Parabola

5. $A = 1$, $B = 0$, $C = 4$, $B^2 - 4AC = -16 < 0$. Ellipse.

7. $A = 2$, $B = 4$, $C = -1$, $B^2 - 4AC = 24 > 0$. Hyperbola.

9. $A = 1$, $B = 0$, $C = 1$, $B^2 - 4AC = -4 < 0$. Ellipse (circle)

11. $A = 3$, $B = 6$, $C = 3$, $B^2 - 4AC = 0$. Parabola

13. $A = 2$, $B = 0$, $C = 3$, $B^2 - 4AC = -24 < 0$. Ellipse.

15. A = 25, B = 0, C = -4, $B^2 - 4AC = 400 < 0$. Hyperbola.

17. A = 3, B = 12, C = 12, $B^2 - 4AC = 0$. Parabola.

19. $D' = D \cos \alpha + E \sin \alpha$, $E' = -D \sin \alpha + E \cos \alpha$

$$D'^2 + E'^2 = (D^2 \cos^2 \alpha + 2DE \sin \alpha \cos \alpha + E^2 \sin^2 \alpha)$$
$$+ (D^2 \sin^2 \alpha - 2DE \sin \alpha \cos \alpha + E^2 \cos^2 \alpha)$$
$$= D^2(\cos^2 \alpha + \sin^2 \alpha) + E^2(\sin^2 \alpha + \cos^2 \alpha) = D^2 + E^2$$

Article 8.9

1.

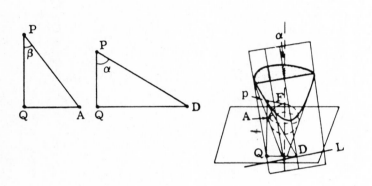

PA = PF

PQ = PA cos β

PQ = PD cos α

PA cos β = PD cos α

$$\Rightarrow \frac{PA}{PD} = \frac{\cos \alpha}{\cos \beta}$$

Since $\alpha = \beta$, PA/PD = PF/PD = $\cos \alpha / \cos \beta$ = 1. Hence, e = 1, and we have a parabola.

3. With a circle, α = 90°. The plane perpendicular to the cone's axis and the plane of the circle are the same, so their intersection does not form a unique line, and we cannot locate D.

MISCELLANEOUS PROBLEMS

1. Let $f(x,y) = 0$ be the equation of the curve C. Let $x' = kx$ and $y' = ky$. Then $x = k^{-1}x'$ and $y = k^{-1}y'$. Therefore, $f(k^{-1}x', k^{-1}y') = 0$ is the locus of P'(x',y'). In our case, $f(x,y) = 0$ is $x^2 + xy + y^2 = 3$. Hence,

$$f(k^{-1}x', k^{-1}y') = 0 \text{ is } k^{-2}(x')^2 + k^{-2}x'y' + k^{-2}(y')^2 = 3$$
or
$$(x')^2 + x'y' + (y')^2 = 3k^2.$$

Since the primes were used merely to distinguish between points P and P', we have

$$x^2 + xy + y^2 = 3k^2.$$

3. (a) x < 3:
        anywhere for x < 3,
        and all y

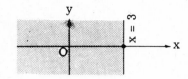

(b) x < y

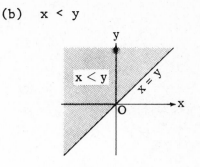

(c) $x^2 < y$

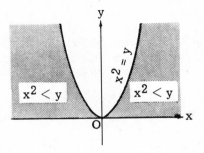

(d) $x^2 + y^2 > 4$

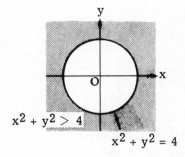

(e) $x^2 + xy + y^2 < 3 \implies \dfrac{x'^2}{2} + \dfrac{y'^2}{6} < 1,$

where $x' = (x+y)/\sqrt{2}$, $y' = (y-x)/\sqrt{2}$. The region is inside the
ellipse $x^2/2 + y^2/6 = 1$, rotated 45° counter clockwise.

(f) $x^2 + xy + y^2 > 3$

is the region outside of the rotated ellipse in (e).

5. $xy = 2 \implies \dfrac{dy}{dx} = -\dfrac{y}{x} = m_1;$ $x^2 - y^2 = 3 \implies \dfrac{dy}{dx} = \dfrac{x}{y} = m_2$

$m_1 = -\dfrac{1}{m_2} \implies$ curves are orthogonal.

7.

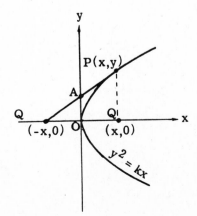

$y^2 = kx;$ $\dfrac{dy}{dx} = \dfrac{k}{2y}.$ Equation of $\overline{PQ}$

$y - y_1 = \dfrac{k}{2y}(x - x_1)$

At Q, $y_1 = 0$, $2y^2 = k(x - x_1)$

$2y^2 - kx = -kx_1$

$2kx - kx = -kx_1$

$kx = -kx_1 \implies x_1 = -x \implies |QO| = |OQ'|.$

Since $\triangle QOA \sim \triangle QQ'P$, $\overline{PQ}$ is bisected by y-axis
at A(0,y/2).

9.

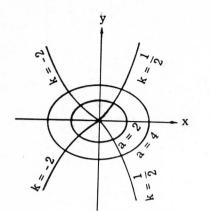

$$2x^2 + 3y^2 = a^2$$

$$\frac{dy}{dx} = -\frac{2x}{3y} = m_1$$

$$ky^2 = x^3$$

$$\frac{dy}{dx} = \frac{3x^2}{2ky} = \frac{3y}{2x} = m_2$$

$m_1 = -1/m_2 \implies$ orthogonal, and is independent of a and k, $a \neq 0$, $k \neq 0$.

11.

Let $Q(x,y)$ be on the line $x + 2y - 5 = 0$ and let $P(u,v)$ be on the ray $\overline{OQ}$ so that for some $t > 0$, $u = tx$, $v = ty$.

The condition that P and Q be symmetric with respect to the given circle is

$$\sqrt{x^2 + y^2} \sqrt{(tx)^2 + (ty)^2} = 4$$

or

$$t(x^2 + y^2) = 4.$$

Therefore,

$$u = tx = \frac{4x}{(x^2 + y^2)},$$

$$v = ty = \frac{4y}{(x^2 + y^2)}$$

with $x + 2y = 5$ or $x = 5 - 2y$. In terms of y, the parametric equations for the locus of $P(u,v)$ are

$$u = \frac{4(5 - 2y)}{[(5 - 2y)^2 + y^2]}, \qquad v = \frac{4y}{[(5 - 2y)^2 + y^2]}$$

To get a cartesian equation for u and v, we eliminate y between these two equations. By division, we get

$$\frac{u}{v} = \frac{(5 - 2y)}{y} = \left(\frac{5}{y}\right) - 2 \quad \text{or} \quad \frac{5}{y} = 2 + \left(\frac{u}{v}\right) = \frac{(2v + u)}{v}.$$

Therefore,

$$y = \frac{5v}{(2v + u)} \quad \text{and} \quad x = 5 - 2y = y\left(\frac{u}{v}\right) = \frac{5u}{(2v + u)}.$$

Substituting into the equation for v, we get

$$v = \frac{4y}{[(5 - 2y)^2 + y^2]} = \frac{20v(2v + u)}{[25(u^2 + v^2)]} = \frac{4v(2v + u)}{[5(u^2 + v^2)]}$$

or

$$u^2 + v^2 = \left(\frac{4}{5}\right)(2v + u).$$

CHAPTER 8 - Miscellaneous Problems

The locus is the circle

$$\left[u^2 - \left(\frac{4}{5}\right)u + \left(\frac{4}{25}\right)\right] + \left[v^2 - \left(\frac{8}{5}\right)v + \left(\frac{16}{25}\right)\right] = \frac{20}{25}$$

with center $(2/5, 4/5)$ and radius $2\sqrt{5}/5$.

13. Suppose a chord has endpoints (x_1, y_1) and (x_2, y_2) and has slope m. We show that $(x_1 + x_2)/2 = 2pm$. This proves that the midpoint of each chord lies on the line $x = 2pm$. Since the endpoints lie on the curve $y = (1/4p)x^2$, and the slope of the chord is m, we have

$$m = \frac{(y_2 - y_1)}{(x_2 - x_1)} = \frac{\left[\left(\frac{1}{4p}\right)x_2^2 - \left(\frac{1}{4p}\right)x_1^2\right]}{(x_2 - x_1)} = \frac{\left(\frac{1}{4p}\right)[(x_2 + x_1)(x_2 - x_1)]}{(x_2 - x_1)} = \frac{(x_1 + x_2)}{4p}$$

Hence,

$$\frac{(x_1 + x_2)}{2} = 2pm.$$

15.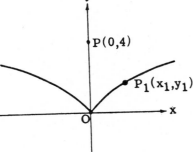

$$x^2 = y^3 \implies \frac{dx}{dy} = \frac{3y^2}{2x}$$

$$d = \overline{PP_1} = \sqrt{x_1^2 + (y_1 - 4)^2}$$

$$\frac{d}{dx}(d) = \frac{d}{dx}\left[x_1^2 + (y_1 - 4)^2\right]^{1/2}$$

$$= \frac{x_1\left(\frac{dx}{dy}\right) + (y_1 - 4)}{\sqrt{x_1^2 + (y_1 - 4)^2}}$$

$$= \frac{3y_1^2 + 2(y_1 - 4)}{2\sqrt{x_1^2 + (y_1 - 4)^2}} = 0$$

$$3y_1^2 + 2y_1 - 8 = 0 \implies y_1 = -2 \pm \sqrt{4 + \frac{96}{6}} = -2 \pm \frac{10}{6} = -2 \quad \text{or} \quad \frac{4}{3}$$

-2 is not on the curve.

$$x^2 = \left(\frac{4}{3}\right)^3 = \pm\left(\frac{8}{9}\right)\sqrt{3}. \qquad \therefore P_1\left[\pm\left(\frac{8}{9}\right)\sqrt{3}, \frac{4}{3}\right]$$

17.

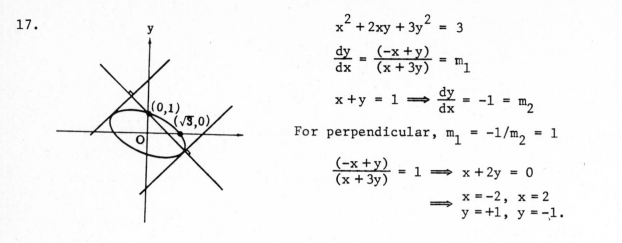

$$x^2 + 2xy + 3y^2 = 3$$

$$\frac{dy}{dx} = \frac{(-x+y)}{(x+3y)} = m_1$$

$$x+y = 1 \implies \frac{dy}{dx} = -1 = m_2$$

For perpendicular, $m_1 = -1/m_2 = 1$

$$\frac{(-x+y)}{(x+3y)} = 1 \implies x+2y = 0$$

$$\implies \begin{array}{l} x = -2, \ x = 2 \\ y = +1, \ y = -1. \end{array}$$

19. a) $(x+y)(x^2+y^2-1) = 0 \implies x+y = 0$ or $x^2+y^2 = 1$ or both.

Sketch is a circle and a line.

b) Let $x+y = \frac{1}{k}$, $x^2+y^2 = 1+k$, $k \geq -1$, $k \neq 0$.
Then,

$$(x+y)(x^2+y^2-1) = \left(\frac{1}{k}\right)(k) = 1 .$$

Using the rotation,

$$x' = \frac{(x+y)}{\sqrt{2}} = \frac{1}{(\sqrt{2}k)} , \quad y' = \frac{(y-x)}{\sqrt{2}} , \quad x'^2 + y'^2 = k+1 ,$$

we get

$$y' = \left[k+1 - \tfrac{1}{2}k^2\right]^{1/2} = \left[\frac{1}{(\sqrt{2}x')} + 1 - x'^2\right]^{1/2} .$$

Note that in the case of (b), the curve is the intersection of the circle with radius $\sqrt{k+1}$, and the line $x+y = k$, and k is between -1 and 4. This curve has asymptote $y = -x$.

21.

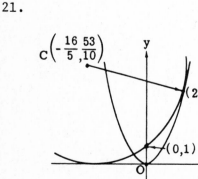

$$(x-h)^2 + (y-k)^2 = r^2$$

At $(0,1)$, $h^2 + 1 - 2k + k^2 = r^2$

At $(2,4)$, $4 - 4h + h^2 + 16 - 8k + k^2 = r^2$

$$\implies 4h + 6k - 19 = 0$$

$$\frac{dy}{dx} = \frac{(-x-h)}{(y-k)} = 2x$$

At $(2,4)$, $\frac{(-2-h)}{(4-k)} = 4 \implies h + 4k - 18 = 0$

$\therefore$ $k = 53/10$; $h = -16/5$

CHAPTER 8 - Miscellaneous Problems

23. $\qquad \overline{PF} = 4 \times 10^7$ miles $= \overline{PD}$

$x = (4 \times 10^7) \cos 60° = 2 \times 10^7$ miles

$y = (4 \times 10^7) \sin 60° = 2\sqrt{3} \times 10^7$ miles

$\overline{PD} = \overline{AF} + \overline{FF'} \implies \overline{FF'} = 2 \times 10^7$ miles

$\therefore \overline{OF} = \frac{1}{2}\overline{FF'} = 10^7$ miles = nearest distance.

25. If d_1 and d_2 are the distances respectively to the line $x = 3$ and the point $(4,0)$ from the point (x,y), then $d_1 = d_2$ implies

$$d_1^2 = (x-3)^2 = d_2^2 = (x-4)^2 + y^2 \implies y^2 = 2x - 7, \text{ a parabola.}$$

27.

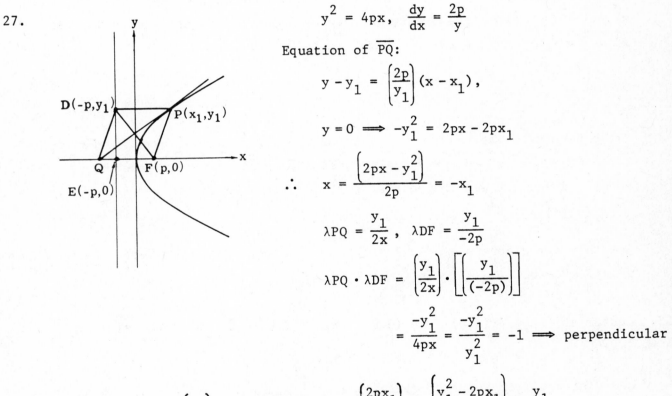

$y^2 = 4px, \quad \dfrac{dy}{dx} = \dfrac{2p}{y}$

Equation of $\overline{PQ}$:

$$y - y_1 = \left(\frac{2p}{y_1}\right)(x - x_1),$$

$$y = 0 \implies -y_1^2 = 2px - 2px_1$$

$$\therefore \quad x = \frac{\left(2px - y_1^2\right)}{2p} = -x_1$$

$$\lambda PQ = \frac{y_1}{2x}, \quad \lambda DF = \frac{y_1}{-2p}$$

$$\lambda PQ \cdot \lambda DF = \left(\frac{y_1}{2x}\right) \cdot \left[\left[\frac{y_1}{(-2p)}\right]\right]$$

$$= \frac{-y_1^2}{4px} = \frac{-y_1^2}{y_1^2} = -1 \implies \text{perpendicular}$$

At $x = 0$, $y - y_1 = \left(\dfrac{2p}{y_1}\right)(-x_1) \implies y = y_1 - \left(\dfrac{2px_1}{y_1}\right) = \dfrac{\left(y_1^2 - 2px_1\right)}{y_1} = \dfrac{y_1}{2}$

Equation of $\overline{DF}$:

$$y - 0 = \left(\frac{y_1}{-2p}\right)(x - p)$$

at $x = 0$, $y = y_1/2$. $\therefore \overline{DF}$ and $\overline{PQ}$ bisect each other and figure is a rhombus.

29. Let the parabola be $y = kx^2$, $p = (x_1, y_1)$. Suppose the tangent meets the
y-axis at $T = (0, -y_2)$. The vertex, $V = (0,0)$. The distance from P to the
tangent at the vertex is the distance from P to the x-axis, which is y_1.
Thus we must show that $y_1 = VT = y_2$. The slope of the parabola is $y' = 2kx_1$,
and the slope of the tangent is $[y_1 - (-y_2)]/(x_1 - 0)$. Hence,

$$\frac{y_1 + y_2}{x_1} = 2kx_1 \implies y_1 + y_2 = 2kx_1^2 = 2y_1 .$$

Thus, $y_1 = y_2$.

31.
$$x^4 - \left(y^2 - 9\right)^2 = 1 \implies [x^2 - (y^2 - 9)][x^2 + (y^2 - 9)] = 1.$$

$$(x^2 - y^2 + 9)(x^2 + y^2 - 9) = 1 \tag{1}$$

Note thet if $x^2 + y^2 < 9$, then $x^2 + y^2 - 9 < 0$, and (since $9 - y^2 > 0$),
$x^2 - y^2 + 9 > 0$ hence the product cannot be 1, and no points inside the circle
$x^2 + y^2 = 9$ are on the curve. Let

$$x^2 + y^2 = 9 + k \tag{2}$$

$$x^2 - y^2 = \frac{1}{k} - 9 . \tag{3}$$

These permit (1) to hold. As $x \to \pm\infty$ or $y \to \pm\infty$, $9 + k \to \infty$, by (2). Hence,
$1/k \to 0$. Thus, $x^2 - y^2 + 9 = 1/k \to 0$, by (3). Using similar reasoning
$x^2 + y^2 - 9 \to 0$. Thus, the points on (1) converge to the curve

$$x^4 - \left(y^2 - 9\right)^2 = (x^2 - y^2 + 9)(x^2 + y^2 - 9) = 0,$$

for this curve consists of all points on either $x^2 + y^2 - 9 = 0$ or
$x^2 - y^2 + 9 = 0$.

33. Center at $[(1,0) + (5,0)]/2 = (3,0)$. Since $V = (0,0)$ is on the ellipse,

$$2a = F_1 V + F_2 V = 1 + 5 = 6 \implies a = 3.$$

c = distance from focus to center = 2. $b^2 = a^2 - c^2 = 9 - 4 = 5$. Equation is

$$\left[\frac{(x-3)^2}{9}\right] + \left[\frac{y^2}{5}\right] = 1.$$

35.
$$\left[\frac{x^2}{a^2}\right] + \left[\frac{y^2}{b^2}\right] = 1 \implies a^2 y^2 = a^2 b^2 - b^2 x^2 \tag{1}$$

Suppose the ellipse meets the rectangle in the positive orthant at (x,y).
Then,

$$\text{area} = 4xy \tag{2}$$

Let

$$A = \left(\frac{area}{4}\right)^2 a^2 = x^2 y^2 a^2 .$$

To minimize A is to minimize the area.

$$A = x^2(y^2 a^2) = x^2(a^2 b^2 - b^2 x^2) = a^2 b^2 x^2 - b^2 x^4 \qquad \text{by (1)}$$

$$\frac{dA}{dx} = 2a^2 b^2 x - 4b^2 x^3 = 0 \implies x^2 = \frac{a^2}{2} .$$

By (1), $y^2 = b^2/2$. Hence, area $= 4xy = 2ab$.

37. $$Ax^2 + Bxy + Cy^2 + Dx + Ey + F = 0$$

(a) Symmetric with respect to the origin $\implies F(x,y) = F(-x,-y)$
$$\implies D = E = 0;$$

$A = 1$ and $x^2 + Bxy + Cy^2 + F = 0$

(b) passes through $P(1,0) \implies x = 1, y = 0 \implies F = -A \implies F = -1$

(c) $$\frac{dy}{dx} = \frac{(-2Ax + By)}{(2Cy + Bx)}$$

at $(-2,1)$,

$$\frac{dy}{dx} = 0 \implies B = 4A \implies B = 4$$

$$\therefore x^2 + 4xy + Cy^2 - 1 = 0$$

$$x = -2, y = 1, 4 + 4(-2)(1) + C - 1 = 0 \implies C = 5$$

$$\therefore x^2 + 4xy + 5y^2 - 1 = 0$$

39. We make two changes of coordinates, first a parallel shift, then a rotation.

$$1 - xy - x - y + 1 - 1 = (x-1)(y-1) - 1 = x'y' - 1 ,$$

where $x' = x - 1$, $y' = y - 1$.

Next, we show that this curve, $x'y' = 2$, is a hyperbola of the form $x''^2 - y''^2 = 4$. Let

$$x'' = \frac{(x' + y')}{\sqrt{2}} , \quad y'' = \frac{(y' - x')}{\sqrt{2}} .$$

Then,

$$x''^2 - y''^2 = \frac{(x'^2 + 2x'y' + y'^2)}{2} - \frac{(x'^2 - 2x'y' + y'^2)}{2} = 2x'y' = 4.$$

41. Using equation (6) on page 540,

$$A = C = D = 0, B = -2, E = \sqrt{2}, F = -3.$$

CHAPTER 8 – Miscellaneous Problems

$$\cot 2\alpha = \frac{(A-C)}{B} = 0 \implies \alpha = \frac{\pi}{4},$$

$$A' = -1, \; B' = 0, \; C' = 1, \; D' = 1, \; E' = 1, \; F' = -3.$$

The new equation is

$$y^2 - x^2 + x + y - 3 = 0.$$

Completing the square,

$$\left(y + \frac{1}{2}\right)^2 - \left(x - \frac{1}{2}\right)^2 = 3, \quad a^2 = b^2 = 3, \quad c^2 = a^2 + b^2 = 6, \quad e = \frac{c}{a} = \frac{\sqrt{6}}{\sqrt{3}} = \sqrt{2}$$

Center at $(1/2, 1/2)$. Curve is hyperbola. Opening upward and downward.

43. The region is the square formed by the four lines:

     1) $x + y = 1$ 2) $x - y = 1$ 3) $-x + y = 1$ 4) $-x - y = 1$

The vertices of the square are $(0, \pm 1)$ and $(\pm 1, 0)$, so the area is 2.

45. a) $\qquad (x^2 - 4x + 4) + 4(y^2 + 2y + 1) = 1 + 4 + 4 = 9.$

$$\left[\frac{(x-2)^2}{9}\right] + \left[\frac{(y+1)^2}{9/4}\right] = 1.$$

$a = 3$, $b = 3/2$, $c^2 = a^2 - b^2 = 9 - 9/4 = 27/4$, $c = 3\sqrt{3}/2$. Center at $(2, -1)$.
Foci at $(2 \pm c, -1) = (2 \pm 3\sqrt{3}/2, -1)$. $\quad e = c/a = \sqrt{3}/2$

    b) $\qquad 3(x^2 + 4x + 4) - \left[y^2 + 3y + \left(\frac{9}{4}\right)\right] = \frac{39}{4}$

$$\frac{12(x+2)^2}{39} - \frac{4(y + (3/2))^2}{39} = 1. \qquad a^2 = \frac{39}{12}, \quad b^2 = \frac{39}{4}$$

Center at $(-2, -3/2)$. $\quad c = \sqrt{a^2 + b^2} = \sqrt{13}$
Foci at $(-2 \pm c, -3/2) = (-2 \pm \sqrt{13}, -3/2)$.
Asymptotes: $\pm a(y - k) = b(x - h) \implies \pm(y + 3/2) = \sqrt{3}(x + 2)$.

47. a) $\qquad (9x^2 + 4y^2 - 36)(4x^2 + 9y^2 - 36) = 0 \implies 9x^2 + 4y^2 = 36$
$$\text{or } 4x^2 + 9y^2 = 36$$

Graph is two ellipses.

    b) Let $9x^2 + 4y^2 = 36 + k$, $\quad 4x^2 + 9y^2 = 36 + \frac{1}{k}$.

Then, $(9x^2 + 4y^2 - 36)(4x^2 + 9y^2 - 36) = k\left(\frac{1}{k}\right) = 1.$

Subtracting, we find

$$5x^2 - 5y^2 = k - \frac{1}{k}.$$

CHAPTER 8 - Miscellaneous Problems

The graph is the intersection of the two ellipses, as k varies. The graph is bounded, because if $4(x^2 + y^2) > 37$, then

$$9x^2 + 4y^2 - 36 \geq 4(x^2 + y^2) - 36 > 1$$

and

$$4x^2 + 9y^2 - 36 \geq 4(x^2 + y^2) - 36 > 1,$$

so the product is greater than one.

49. $xy = 1$. Let $x = \dfrac{(x' - y')}{\sqrt{2}}$, $y = \dfrac{(x' \overset{+}{-} y')}{\sqrt{2}}$

$$1 = xy = \frac{(x' - y')(x' + y')}{2} = \frac{(x'^2 - y'^2)}{2} .$$

$a = b = \sqrt{2}$, $c = \sqrt{a^2 + b^2} = 2$, $e = \dfrac{c}{a} = \dfrac{2}{\sqrt{2}} = \sqrt{2}$

51. The asymptotes are the x and y axes. Suppose the tangent meets the y-axis at $(0,c)$ and the x-axis at $(d,0)$. We must show that the area $= cd/2 = 2a^2$.

$$xy = a^2 \implies \frac{dy}{dx} = -\frac{y}{x} .$$

The chord from $(0,c)$ to (x,y) has slope

$$\frac{(c - y)}{(0 - x)} = -\frac{y}{x} \implies c - y = y \implies c = 2y .$$

The chord from (x,y) to $(d,0)$ has slope

$$\frac{(y - 0)}{(x - d)} = -\frac{y}{x} \implies x - d = -x \implies d = 2x .$$

Thus,

$$\text{area} = \frac{cd}{2} = \frac{(2y)(2x)}{2} = 2a^2 .$$

53. $\qquad 7x^2 - 8xy + y^2 = 9$, $\qquad A = 7$, $B = -8$, $C = 1$.

$$\cot 2\alpha = \frac{(A - C)}{B} = -\frac{3}{4} , \qquad \cos 2\alpha = -\frac{3}{\sqrt{3^2 + 4^2}} = -\frac{3}{5}$$

$$\cos^2 \alpha = \frac{(1 + \cos 2\alpha)}{2} = \frac{1}{5} , \qquad \sin^2 \alpha = 1 - \cos^2 \alpha = \frac{4}{5}$$

$$\sin \alpha = \frac{2}{\sqrt{5}} , \qquad \cos \alpha = -\frac{1}{\sqrt{5}} .$$

Using Equation (6) on page 540, $A' = -1$, $C' = 9$. So, $9y'^2 - x'^2 = 9$, where

$$x = -\frac{(x' + 2y')}{\sqrt{5}} , \qquad y = \frac{(2x' - y')}{\sqrt{5}} .$$

55. a)
$$x^2 + 12y^2 - 6x - 48y + 9 = 0$$

$$(x^2 - 6x + 9) + 12(y^2 - 4y + 4) = 48$$

$$\left[\frac{(x-3)^2}{48}\right] + \left[\frac{(y-2)^2}{4}\right] = 1$$

$$C(3,2), \quad c = \sqrt{a^2 - b^2} = \sqrt{48 - 4} = \sqrt{44} = 2\sqrt{11}$$

$$e = \frac{c}{a} = \frac{2\sqrt{11}}{2\sqrt{12}} = \sqrt{\frac{11}{12}}$$

b) $x^2 - 6x - 12y = 0 \implies (x-3)^2 = 12y.$ Vertex at (3,0)

c) Part (a) is a nearly circular ellipse that passes through (3,0). Part (b) is a parabola with vertex at (3,0).

57.

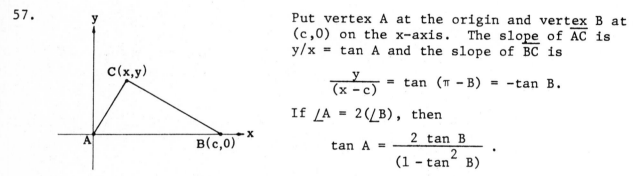

Put vertex A at the origin and vertex B at $(c,0)$ on the x-axis. The slope of $\overline{AC}$ is $y/x = \tan A$ and the slope of $\overline{BC}$ is

$$\frac{y}{(x-c)} = \tan(\pi - B) = -\tan B.$$

If $\angle A = 2(\angle B)$, then

$$\tan A = \frac{2 \tan B}{(1 - \tan^2 B)}.$$

So, we have

$$\frac{y}{x} = \frac{2\left(\frac{y}{(c-x)}\right)}{1 - \left(\frac{y}{(c-x)}\right)^2} = \frac{2y(c-x)}{(c-x)^2 - y^2}$$

or

$$(c-x)^2 - y^2 = 2x(c-x), \quad 3x^2 - 4cx - y^2 + c^2 = 0$$

The locus is thus part of the hyperbola $3x^2 - 4cx - y^2 + c^2 = 0$, whose center is at $(2c/3, 0)$. From simple geometry, since $\angle A$ is greater than $\angle B$, C must lie to the left of the perpendicular bisector of $\overline{AB}$. This means that C cannot be on the right-hand branch of the hyperbola and the locus is just the left-hand branch.

59.
$$x = x' \cos \alpha - y' \sin \alpha, \quad dx = \cos \alpha \, dx' - \sin \alpha \, dy'$$

$$(dx)^2 = \cos^2 \alpha \, (dx')^2 + \sin^2 \alpha \, (dy')^2 - 2 \cos \alpha \sin \alpha \, (dx')(dy')$$

$$y = x' \sin \alpha + y' \cos \alpha, \quad dy = \sin \alpha \, dx' + \cos \alpha \, dy'$$

$$(dy)^2 = \sin^2 \alpha \, (dx')^2 + \cos^2 \alpha \, (dy')^2 + 2 \cos \alpha \sin \alpha \, (dx')(dy')$$

$$(dx)^2 + (dy)^2 = (\sin^2 \alpha + \cos^2 \alpha)(dx')^2 + (\sin^2 \alpha + \cos^2 \alpha)(dy')^2$$

$$= (dx')^2 + (dy')^2$$

CHAPTER 8 - Miscellaneous Problems

61. a) $n = 1$, $x^2 + y^2 = a^2$, a perfect circle.

b) $n = 2$, $x^4 + y^4 = a^4$, squarish circle.

c) $n = 100$, similar to $n = \infty$. When $n = \infty$, region is perfect square...

<u>region is perfect square</u> with vertices at $(\pm a, \pm a)$. To see this, note that the equation can be written as $(x/a)^{2n} + (4/a)^{2n} = 1$. If n is large, and $-1 < (y/a) < 1$, then $(y/a)^{2n}$ is near zero, so $(x/a)^{2n}$ must be near 1, that is, x is near a.

Chapter 9

HYPERBOLIC FUNCTIONS

<u>Article 9.2</u>

1.
$$x^2 - y^2 = \left(\frac{e^u + e^{-u}}{2}\right)^2 - \left(\frac{e^u - e^{-u}}{2}\right)^2$$

$$= \frac{e^{2u} + 2 + e^{-2u}}{4} - \frac{e^{2u} - 2 + e^{-2u}}{4} = 1$$

So, $(x,y) = (-\cosh u, \sinh u)$ satisfies the equation of the hyperbola. Now, since

$$e^u > 0 \text{ for all } u, \quad -\cosh u = \frac{-e^u - e^{-u}}{2} < 0 \text{ for all } u.$$

So, $x < 0$, and the point $(-\cosh u, \sinh u)$, therefore, lies on the left branch of the hyperbola.

3.
$$\sinh (x + y) = \frac{e^{x+y} - e^{-x-y}}{2} = \left(\frac{e^x e^y - e^{-x} e^y + e^x e^{-y} - e^{-x} e^{-y}}{4}\right)$$

$$+ \left(\frac{e^x e^y + e^{-x} e^y - e^x e^{-y} - e^{-x} e^{-y}}{4}\right)$$

$$= \left[\frac{e^y}{2}\left(\frac{e^x - e^{-x}}{2}\right) + \frac{e^{-y}}{2}\left(\frac{e^x - e^{-x}}{2}\right)\right]$$

$$+ \left[\frac{e^y}{2}\left(\frac{e^x + e^{-x}}{2}\right) - \frac{e^{-y}}{2}\left(\frac{e^x + e^{-x}}{2}\right)\right]$$

$$= \left(\frac{e^x - e^{-x}}{2}\right)\left(\frac{e^y + e^{-y}}{2}\right) + \left(\frac{e^x + e^{-x}}{2}\right)\left(\frac{e^y - e^{-y}}{2}\right)$$

$$= \sinh x \cosh y + \cosh x \sinh y.$$

$$\cosh (x + y) = \frac{e^{x+y} + e^{-x-y}}{2} = \left(\frac{e^x e^y + e^{-x} e^y + e^x e^{-y} + e^{-x} e^{-y}}{4}\right)$$

$$+ \left(\frac{e^x e^y - e^{-x} e^y - e^x e^{-y} + e^{-x} e^{-y}}{4}\right)$$

$$= \left[\frac{e^y}{2}\left(\frac{e^x + e^{-x}}{2}\right) + \frac{e^{-y}}{2}\left(\frac{e^x + e^{-x}}{2}\right)\right]$$

$$+ \left[\frac{e^y}{2}\left(\frac{e^x - e^{-x}}{2}\right) - \frac{e^{-y}}{2}\left(\frac{e^x - e^{-x}}{2}\right)\right]$$

$$= \left(\frac{e^x + e^{-x}}{2}\right)\left(\frac{e^y + e^{-y}}{2}\right) + \left(\frac{e^x - e^{-x}}{2}\right)\left(\frac{e^y - e^{-y}}{2}\right)$$

$$= \cosh x \cosh y + \sinh x \sinh y.$$

5. We need the slope of the tangent line. Since $x^2 - y^2 = 1$, we differentiate implicitly: $2x - 2y(dy/dx) = 0$, so $dy/dx = x/y$. At $(x_1, y_1) = (\cosh u, \sinh u)$, the tangent line L is given by

$$y - \sinh u = \frac{\cosh u}{\sinh u}(x - \cosh u).$$

To find the x intercept, set $y = 0$:

$$-\sinh u = \frac{\cosh u}{\sinh u}(x - \cosh u),$$

or

$$x = \cosh u - \frac{\sinh^2 u}{\cosh u} = \frac{\cosh^2 u - \sinh^2 u}{\cosh u} = \frac{1}{\cosh u} = \text{sech } u.$$

Therefore, L cuts the x-axis at $(\text{sech } u, 0)$. To find the y-intercept, set $x = 0$:

$$y - \sinh u = \frac{\cosh u}{\sinh u}(-\cosh u)$$

or

$$y = \sinh u - \frac{\cosh^2 u}{\sinh u} = \frac{\sinh^2 u - \cosh^2 u}{\sinh u} = \frac{-1}{\sinh u} = -\text{csch } u.$$

Therefore, L cuts the y-axis at $(0, -\text{csch } u)$.

7.
$$r = \left(\cosh^2 u + \sinh^2 u\right)^{1/2} = \left[\left(\frac{e^u + e^{-u}}{2}\right)^2 + \left(\frac{e^u - e^{-u}}{2}\right)^2\right]^{1/2}$$

$$= \left[\frac{e^{2u} + 2 + e^{-2u}}{4} + \frac{e^{2u} - 2 + e^{-2u}}{4}\right]^{1/2} = \left[\frac{e^{2u} + e^{-2u}}{2}\right]^{1/2} = \sqrt{\cosh(2u)}$$

9. Since $-\pi/2 < \theta < \pi/2$, $\cos\theta > 0$. Then,

$$\sec\theta = \sqrt{1 + \tan^2\theta} = \sqrt{1 + \sinh^2 x} = \cosh x$$

So,

$$\cos\theta = \text{sech } x, \quad \sin\theta = \tan\theta\cos\theta = \sinh x\,\text{sech } x = \frac{\sinh x}{\cosh x} = \tanh x;$$

$$\cot\theta = \frac{1}{\tan\theta} = \frac{1}{\sinh x} = \text{csch } x;$$

and

$$\csc\theta = \frac{1}{\sin\theta} = \frac{1}{\tanh x} = \coth x.$$

Summary table:

$\sin\theta = \tanh x$	$\cot\theta = \text{csch } x$
$\cos\theta = \text{sech } x$	$\sec\theta = \cosh x$
$\tan\theta = \sinh x$	$\csc\theta = \coth x$

Article 9.3

1. a) If
$$y = \coth u = \frac{\cosh u}{\sinh u},$$
then
$$\frac{dy}{dx} = \frac{(\sinh u)\, d(\cosh u)/dx - (\cosh u)\, d(\sinh u)/dx}{\sinh^2 u}$$

$$= \frac{\sinh^2 u \, \frac{du}{dx} - \cosh^2 u \, \frac{du}{dx}}{\sinh^2 u} = \frac{-1}{\sinh^2 u} \cdot \frac{dy}{dx} = -\text{csch}^2 u \, \frac{du}{dx}$$

b) If
$$y = \text{sech } u = \frac{1}{\cosh u},$$
then
$$\frac{dy}{dx} = \frac{-1}{\cosh^2 u} \cdot \frac{d(\cosh u)}{dx} = \frac{-\sinh u}{\cosh^2 u} \cdot \frac{du}{dx}$$

$$= -\left(\frac{\sinh u}{\cosh u}\right)\left(\frac{1}{\cosh u}\right)\frac{du}{dx} = -\tanh u \, \text{sech } u \, \frac{du}{dx}.$$

c) If
$$y = \text{csch } u = \frac{1}{\sinh u},$$
then
$$\frac{dy}{dx} = \frac{-1}{\sinh^2 u} \cdot \frac{d(\sinh u)}{dx} = \frac{-\cosh u}{\sinh^2 u} \cdot \frac{du}{dx}$$

$$= -\left(\frac{\cosh u}{\sinh u}\right)\left(\frac{1}{\sinh u}\right)\frac{du}{dx} = -\coth u \, \text{csch } u \, \frac{du}{dx}.$$

3. $y = \cosh^2 5x$, $\frac{dy}{dx} = (5)(2\cosh 5x)(\sinh 5x) = 10 \cosh 5x \sinh 5x$

5. $y = \tanh 2x$, $\frac{dy}{dx} = 2 \, \text{sech}^2 \, 2x$

7. $y = \text{sech}^3 x$, $\frac{dy}{dx} = (3 \, \text{sech}^2 \, x)(-\text{sech } x \tanh x) = -3 \, \text{sech}^3 x \tanh x$

9. Implicitly differentiate:
$$(\cosh y)\frac{dy}{dx} = \sec^2 x$$

But
$$\cosh y = \sqrt{1 + \sinh^2 y} = \sqrt{1 + \tan^2 x} = \sec x$$

So, $dy/dx = \sec x$.

11. $y = \operatorname{csch}^2 x,\ \dfrac{dy}{dx} = (2\ \operatorname{csch} x)(-\operatorname{csch} x \coth x) = -2\ \operatorname{csch}^2 x \coth x.$

13. $y = x - \dfrac{1}{4} \coth (4x),$

so

$$\dfrac{dy}{dx} = 1 - \dfrac{1}{4}(-4\ \operatorname{csch}^2 4x) = 1 + \operatorname{csch}^2 4x = 1 + \dfrac{1}{\sinh^2 4x} = \dfrac{1 + \sinh^2 4x}{\sinh^2 4x}$$

$$= \dfrac{\cosh^2 4x}{\sinh^2 4x} = \coth^2 4x.$$

15. $y = x^4 \sinh x,$

so

$$\dfrac{dy}{dx} = 4x^3 \sinh x + x^4\ \dfrac{d(\sinh x)}{dx} = 4x^3 \sinh x + x^4 \cosh x.$$

17. $y = x \sinh x - \cosh x,$

so

$$\dfrac{dy}{dx} = (\sinh x + x \cosh x) - \sinh x = x \cosh x.$$

19. $\displaystyle \int \tanh x\ dx = \int \dfrac{\sinh x}{\cosh x}\ dx = \int \dfrac{d(\cosh x)}{\cosh x} = \ln(\cosh x) + C.$

21. $\displaystyle \int \dfrac{4\ dx}{\left(e^x + e^{-x}\right)^2} = \int \left[\dfrac{2}{e^x + e^{-x}}\right]^2 dx = \int \dfrac{dx}{\cosh^2 x} = \int \operatorname{sech}^2 x\ dx = \tanh x + C.$

23. $\displaystyle \int \tanh^2 x\ dx = \int \dfrac{\sinh^2 x}{\cosh^2 x}\ dx = \int \dfrac{\cosh^2 x - 1}{\cosh^2 x}\ dx$

$$= \int (1 - \operatorname{sech}^2 x)\ dx = x - \tanh x + C.$$

25. $\displaystyle \int \cosh^2 3x\ dx = \int \dfrac{1 + \cosh 6x}{2}\ dx = \dfrac{x}{2} + \dfrac{\sinh 6x}{12} + C.$

27. $\displaystyle \int \cosh^2 5x\ dx = \int \dfrac{1 + \cosh 10x}{2}\ dx = \dfrac{x}{2} + \dfrac{\sinh 10x}{20} + C.$

29. $\displaystyle \int \cosh^3 x = \int \cosh^2 x \cosh x\ dx = \int (1 + \sinh^2 x) \cosh x\ dx$

$$= \sinh x + \dfrac{\sinh^3 x}{3} + C.$$

31. $$\int \operatorname{csch} x \, dx = \int \frac{(\operatorname{csch} x)(\operatorname{csch} x + \coth x) \, dx}{\operatorname{csch} x + \coth x}$$

$$= \int \frac{-d(\operatorname{csch} x + \coth x)}{\operatorname{csch} x + \coth x} = -\ln \left| \operatorname{csch} x + \coth x \right| + C.$$

33. $$\int \sinh^4 3x \, dx = \frac{\sinh 12x}{96} - \frac{\sinh 6x}{12} + \frac{3x}{8} \, .$$

35. $$\int e^{3x} \cosh 2x \, dx = \frac{e^{3x}}{5}(3 \cosh 2x - 2 \sinh 2x) + C$$

37.

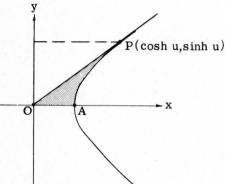

The line OP is given by $y = x \tanh u$.
So, the area is given by:

$$\text{area} = \int_0^{\sinh u} [\sqrt{y^2 + 1} - y \coth u] \, dy$$

With the help of a table of integrals, we find:

$$\text{area} = \frac{y}{2} \sqrt{y^2 + 1} + \frac{1}{2} \log \left| y + \sqrt{y^2 + 1} \right|$$

$$- \frac{y^2}{2} \coth u \Big|_0^{\sinh}$$

$$= \frac{\sinh u \cosh u}{2}$$

$$+ \frac{1}{2} \ln \left| \sinh u + \cosh u \right|$$

$$- \frac{\sinh u \cosh u}{2}$$

$$= \frac{1}{2} \ln \left| \sinh u + \cosh u \right| = \frac{1}{2} \ln (x + y)$$

$$= \frac{1}{2} \ln \left(\frac{e^u - e^{-u}}{2} + \frac{e^u + e^{-u}}{2} \right) = \frac{1}{2} \ln (e^u) = \frac{u}{2} \, .$$

39. We will use Newton's method. Let

$$F(x) = \cosh x - \frac{x}{2} - 1 \, .$$

Since $F(0) = 0$, the curves $y = \cosh x$ and $y = x/2 + 1$ intersect at $(0,1)$.

Newton's method says that if we guess x_n as a root of $F(x)$, then a (hopefully better) guess is

$$x_{n+1} = x_n - \frac{F(x_n)}{F'(x_n)} = x_n - \frac{2 \cosh x_n - x_n + 2}{2 \sinh x_n - 1} \, .$$

Then, guess $x_1 = 1$, so

$$x_1 = 1 \qquad F(x_1) = 0.0431$$
$$x_2 = 0.9362 \qquad F(x_2) = 0.0031$$
$$x_3 = 0.9309 \qquad F(x_3) = 0.00002$$

So, the curves intersect approximately at $(0.930, 1.465)$

Article 9.4

1. Let $e^y = w$. Then, $x = 1/2(w - 1/w)$, or $2wx = w^2 - 1$, so $w^2 - 2xw - 1 = 0$. Then, by the quadratic formula,

$$w = \frac{2x + \sqrt{4x^2 + 4}}{2} = x + \sqrt{x^2 + 1}$$

(We must choose the plus sign in front of the square root to guarantee that $w > 0$.) So,

$$y = \ln w = \ln(x + \sqrt{x^2 + 1}), \quad \text{or} \quad \sinh^{-1} x = \ln(x + \sqrt{x^2 + 1})$$

3. Let $y = \sinh^{-1} u$. Then, $u = \sinh y$, so

$$\frac{du}{dx} = \frac{d(\sinh y)}{dx} = \cosh y \frac{dy}{dx} = \sqrt{1 + \sinh^2 y}\frac{dy}{dx} = \sqrt{1 + u^2}\frac{dy}{dx}.$$

Therefore,

$$\frac{dy}{dx} = \frac{1}{\sqrt{1 + u^2}} \cdot \frac{du}{dx}.$$

5. Let $y = \text{sech}^{-1} u$, $u = \text{sech } y$. Then,

$$\frac{du}{dx} = \frac{d(\text{sech } y)}{dx} = -\text{sech } y \tanh y \frac{dy}{dx} = -(\text{sech } y)\sqrt{1 - \tanh^2 y}\frac{dy}{dx}$$

$$= -u\sqrt{1 - u^2}\frac{dy}{dx}.$$

So, $\frac{dy}{dx} = -\frac{1}{u\sqrt{1 - u^2}} \cdot \frac{du}{dx}.$

7. $y = \tanh^{-1}(\cos x)$, $\frac{dy}{dx} = \left(\frac{1}{1 - (\cos x)^2}\right)(-\sin x) = \frac{-\sin x}{\sin^2 x} = -\csc x.$

9. $y = \coth^{-1}(\sec x)$, $\frac{dy}{dx} = \left(\frac{1}{1 - (\sec x)^2}\right)(\sec x \tan x) = \frac{\sec x \tan x}{-\tan^2 x} = -\csc x.$

11. $y = \cosh^{-1}(x^2)$, $\dfrac{dy}{dx} = \left(\dfrac{1}{\sqrt{\left(x^2\right)^2 - 1}}\right)(2x) = \dfrac{2x}{\sqrt{x^4 - 1}}$

13. $y = \operatorname{csch}^{-1}(\tan x)$, $\dfrac{dy}{dx} = \dfrac{-1}{|\tan x|\ \sqrt{1 + \tan^2 x}}(\sec^2 x) = \dfrac{-\sec^2 x}{|\tan x \sec x|}$

$$= -\left|\dfrac{\sec x}{\tan x}\right| = -|\csc x|.$$

15. $y = \cosh^{-1}\sqrt{x+1}$, $\dfrac{dy}{dx} = \left(\dfrac{1}{\left[(\sqrt{x+1})^2 - 1\right]^{1/2}}\right)\left(\dfrac{1}{2\sqrt{x+1}}\right) = \dfrac{1}{2\sqrt{x^2 + x}}$.

17. $y = \coth^{-1}(\sin x)$, $\dfrac{dy}{dx} = \left(\dfrac{1}{1 - \sin^2 x}\right)(\cos x) = \dfrac{\cos x}{\cos^2 x} = \sec x.$

19. $\qquad y = \dfrac{x}{2}\sqrt{x^2 - 1} - \dfrac{1}{2}\cosh^{-1} x,$

so

$$\dfrac{dy}{dx} = \left(\dfrac{1}{2}\sqrt{x^2 - 1} - \dfrac{x^2}{2\sqrt{x^2 - 1}}\right) - \dfrac{1}{2\sqrt{x^2 - 1}} = \dfrac{-1}{\sqrt{x^2 - 1}}$$

21. Let $x = \dfrac{y}{2}$, $dx = \dfrac{1}{2}\,dy$. Then,

$$\int \dfrac{dx}{\sqrt{1 + 4x^2}} = \dfrac{1}{2}\int \dfrac{dy}{\sqrt{1 + y^2}} = \dfrac{1}{2}\sinh^{-1} y + C = \dfrac{1}{2}\sinh^{-1}(2x) + C.$$

23. $\qquad \displaystyle\int_0^{0.5} \dfrac{dx}{1 - x^2} = \tanh^{-1} x \Big|_0^{0.5} = \dfrac{1}{2}\ln\left(\dfrac{1+x}{1-x}\right)\Big|_0^{0.5} = \dfrac{1}{2}\ln 3.$

25. Let $x = 2y$, $dx = 2\,dy$. Then,

$$\int \dfrac{dx}{x\sqrt{4 + x^2}} = \int \dfrac{2\,dy}{2y\sqrt{4 + 4y^2}} = \dfrac{1}{2}\int \dfrac{dy}{y\sqrt{y^2 + 1}} = -\dfrac{1}{2}\operatorname{csch} y + C = -\dfrac{1}{2}\operatorname{csch}\left(\dfrac{x}{2}\right) + C.$$

27. By integration by parts: let $u = \sinh^{-1} x$, $dv = dx$. Then,

$$v = x, \quad du = \dfrac{dx}{\sqrt{1 + x^2}} ,$$

and

$$\int \sinh^{-1} x \, dx = x \sinh^{-1} x - \int \frac{x \, dx}{\sqrt{1+x^2}} = x \sinh^{-1} x - \sqrt{1+x^2} + C.$$

29. Let $u = \sec x$, $du = \sec x \tan x \, dx$. Then,

$$\int \frac{du}{\sqrt{u^2-1}} = \int \frac{\sec x \tan x \, dx}{\tan x} = \int \sec x \, dx$$

$$= \ln |\sec x + \tan x| + C = \ln |u + \sqrt{u^2-1}| + C.$$

31. Let $u = \tan x$, $du = \sec^2 x \, dx$. Then,

$$\int \frac{du}{u\sqrt{u^2+1}} = \int \frac{\sec^2 x \, dx}{\tan x \sec x} = \int \csc x \, dx = -\ln |\csc x + \cot x| + C$$

$$= -\ln \left| \frac{1}{u} + \frac{\sqrt{u^2+1}}{u} \right| + C.$$

33. $\int \frac{\sqrt{x^2-25}}{x} \, dx = \sqrt{x^2-a^2} - 5 \sec^{-1} \left(\frac{x}{5} \right) + C.$

35. $\int \frac{x^2 \, dx}{\sqrt{x^2-25}} = \frac{x}{2} \sqrt{x^2-25} + \frac{25}{2} \ln (x + \sqrt{x^2-25})$

37. $\int \sqrt{x^2-4} \, dx = \frac{1}{2}[x\sqrt{x^2-4} - 4 \ln (x + \sqrt{x^2-4})] + C$

39. We have $\dfrac{dv}{mg-kv^2} = \dfrac{dt}{m}$, so $\displaystyle\int \dfrac{dv}{gm-kv^2} = \dfrac{t}{m} + C$. Let $v = \sqrt{\dfrac{gm}{k}}\, w$, $dv = \sqrt{\dfrac{gm}{k}}\, dw$. Then,

$$\int \frac{dv}{gm-kv^2} = \frac{1}{gm} \sqrt{\frac{gm}{k}} \int \frac{dw}{1-w^2} = \frac{1}{\sqrt{gmk}} \tanh^{-1} w = \frac{1}{\sqrt{gmk}} \tanh^{-1} \left(\sqrt{\frac{k}{gm}}\, v \right).$$

Therefore,

$$\tanh^{-1} \left(\sqrt{\frac{k}{gm}}\, v \right) = \sqrt{\frac{gk}{m}}\, t + C'.$$

To determine C', use the condition $v = 0$ at $t = 0$. Thus, $C' = 0$, and

$$\tanh^{-1} \left(\sqrt{\frac{k}{gm}}\, v \right) = \sqrt{\frac{gk}{m}}\, t.$$

Solving for v, we find

$$v = \sqrt{\frac{gm}{k}} \tanh \left(\sqrt{\frac{gk}{m}}\, t \right).$$

As $x \to \infty$, $\tanh x \to 1$, so $v \to \sqrt{gm/k}$ as $t \to \infty$.

Article 9.5

1. $$\frac{dy}{dx} = \sinh\left(\frac{x}{a}\right),$$

   so

   $$ds = \sqrt{1 + \left(\frac{dy}{dx}\right)^2}\, dx = \sqrt{1 + \sinh^2\left(\frac{x}{a}\right)}\, dx = \cosh\left(\frac{x}{a}\right) dx.$$

   $$\text{Length} = \int_0^{x_1} \cosh\left(\frac{x}{a}\right) dx = a \sinh\left(\frac{x}{a}\right)\Big|_0^{x_1} = a \sinh\left(\frac{x_1}{a}\right).$$

3. $$\text{Surface area} = \int_0^{x_1} 2\pi y\, ds = \int_0^{x_1} \left[2\pi a \cosh\left(\frac{x}{a}\right)\right] \sqrt{1 + \sinh^2\left(\frac{x}{a}\right)}\, dx$$

   $$= \int_0^{x_1} 2\pi a \cosh^2\left(\frac{x}{a}\right) dx = \pi a \int_0^{x_1} \left[\cosh\left(\frac{2x}{a}\right) + 1\right] dx$$

   $$= \pi a \left[x + \frac{a}{2}\sinh\left(\frac{2x}{a}\right)\right]\Big|_0^{x_1} = \pi a \left[x_1 + \frac{a}{2}\sinh\left(\frac{2a}{x_1}\right)\right].$$

5. $$\text{Volume} = \int_0^{x_1} \pi y^2\, dx = \pi a^2 \int_0^{x_1} \cosh^2\left(\frac{x}{a}\right) dx$$

   $$= \frac{\pi a^2}{2} \int_0^{x_1} \left[1 + \cosh\left(\frac{2x}{a}\right)\right] dx = \frac{\pi a^2}{2}\left[x + \frac{a}{2}\sinh\left(\frac{2x}{a}\right)\right]\Big|_0^{x_1}$$

   $$= \frac{\pi a^2}{2}\left[x_1 + \frac{a}{2}\sinh\left(\frac{2x_1}{a}\right)\right] = \frac{\pi a^2}{2}\left[x_1 + a \sinh\left(\frac{x_1}{a}\right)\cosh\left(\frac{x_1}{a}\right)\right]$$

   $$= \frac{\pi a^2}{2}\left[x_1 + \frac{1}{a}y_1\sqrt{y_1^2 - 1}\right] \quad [\text{since } y_1 = a \cosh(x_1/a)]$$

7.

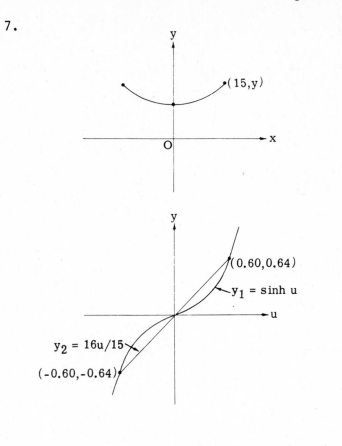

a) $\qquad x = a \sinh^{-1}\left(\dfrac{s}{a}\right)$,

so $\quad 15 = a \sinh^{-1}\left(\dfrac{16}{a}\right)$.

Let $u = \dfrac{15}{a}$, then

$$\dfrac{15}{a} = \sinh^{-1}\left(\dfrac{16}{15} \cdot \dfrac{15}{a}\right)$$

or $\quad \dfrac{16}{15} u = \sinh u$.

b) $\qquad \sinh 0.6 \approx \left(\dfrac{16}{15}\right)(0.6)$

$$\approx 0.64$$

c) From (a) and (b), $u \approx 0.6$. Since $u = 15/a$, we have

$$a = \dfrac{15}{0.6} = 25 \text{ ft}$$

so $\quad y = 25 \cosh\left(\dfrac{x}{25}\right)$.

For $x = 0$, $y = 25$; for $x = 15$, $y = 25 \cosh (0.6) \approx 29.6$. Thus, the dip is approximately 4.6 ft.

d) $\quad T = wy = 2\left[25 \cosh\left(\dfrac{0}{25}\right)\right] = 50 \text{ lbs.}$

MISCELLANEOUS PROBLEMS

1. $\qquad \cosh^2 x + \sinh^2 x = \left(\dfrac{e^x + e^{-x}}{2}\right)^2 + \left(\dfrac{e^x - e^{-x}}{2}\right)^2$

$$= \left(\dfrac{e^{2x} + 2 + e^{-2x}}{4}\right) + \left(\dfrac{e^{2x} - 2 + e^{-2x}}{4}\right)$$

$$= \dfrac{e^{2x} + e^{-2x}}{2} = \cosh 2x.$$

3.

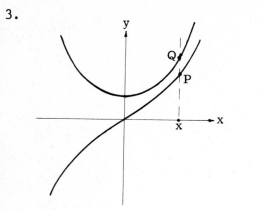

$$PQ = \cosh x - \sinh x$$

$$= \frac{e^x + e^{-x}}{2} - \frac{e^x - e^{-x}}{2} = e^{-x}$$

So,

$$\lim_{x \to \infty} PQ = \lim_{x \to \infty} e^{-x} = 0$$

5. $\quad \sinh x = \dfrac{1}{\text{csch } x} = -\dfrac{40}{9},$

so

$\cosh x = \sqrt{1 + \sinh^2 x} = \dfrac{41}{9}$ and $\tanh x = \dfrac{\sinh x}{\cosh x} = -\dfrac{40}{41}$.

7.

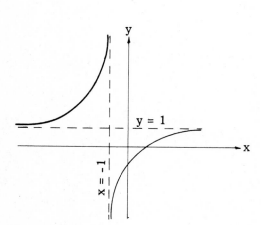

The radius of the semicircle is 1. So,

$$RP^2 + PQ^2 = RQ^2 = 1.$$

Since

$$RP = \tanh x,$$

$$PQ = \sqrt{1 - \tanh^2 x} = \text{sech } x.$$

9.

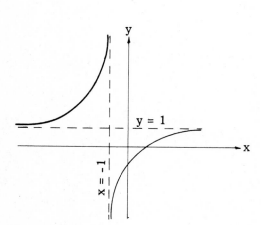

First, we simplify

$$y = \tanh\left(\frac{1}{2} \ln x\right) = \tanh(\ln \sqrt{x})$$

$$= \frac{e^{\ln \sqrt{x}} - e^{-\ln \sqrt{x}}}{e^{\ln \sqrt{x}} + e^{-\ln \sqrt{x}}} = \frac{\sqrt{x} - 1/\sqrt{x}}{\sqrt{x} + 1/\sqrt{x}}$$

$$= \frac{x - 1}{x + 1} \ .$$

So, we see there is a vertical asymptote at $x = -1$ (the denominator can not be zero).

Now, solve for x in terms of y

$$(x + 1)y = x - 1, \text{ or } x = \frac{1 + y}{1 - y} \ .$$

Therefore, there is also a horizontal asymptote at $y = 1$.

CHAPTER 9 – Miscellaneous Problems

11.

y	$\cosh x$	$\sinh x$	$\cos x$	$\sin x$
$\dfrac{dy}{dx}$	$\sinh x$	$\cosh x$	$-\sin x$	$\cos x$
$\dfrac{d^2y}{dx^2}$	$\cosh x$	$\sinh x$	$-\cos x$	$-\sin x$
$\dfrac{d^3y}{dx^3}$	$\sinh x$	$\cosh x$	$\sin x$	$-\cos x$
$\dfrac{d^4y}{dx^4}$	$\cosh x$	$\sinh x$	$\cos x$	$\sin x$

In each case, $d^4y/dx^4 = y$.

13. By implicit differentiation: if $\tan x = \tanh^2 y$, then

$$\sec^2 x = 2 \tanh y \, \operatorname{sech}^2 y \, \frac{dy}{dx},$$

so $\dfrac{dy}{dx} = \dfrac{\sec^2 x}{2 \tanh y \, \operatorname{sech}^2 y} = \dfrac{\sec^2 x}{2 \tanh y (1 - \tanh^2 y)} = \dfrac{\sec^2 x}{2\sqrt{\tan x}(1 - \tan x)}$

15. Implicitly differentiate $\sinh y = \sec x$. Then,

$$\cosh y \, \frac{dy}{dx} = \sec x \tan x,$$

or $\dfrac{dy}{dx} = \dfrac{\sec x \tan x}{\cosh y} = \dfrac{\sec x \tan x}{\sqrt{1 + \sinh^2 y}} = \dfrac{\sec x \tan x}{\sqrt{1 + \sec^2 x}}$

17. First, we simplify: since

$$\tanh u = \frac{e^u - e^{-u}}{e^u + e^{-u}}, \quad y = \tanh(\ln x) = \frac{x - 1/x}{x + 1/x} = \frac{x^2 - 1}{x^2 + 1}.$$

So, $\dfrac{dy}{dx} = \dfrac{(x^2 + 1)(2x) - (x^2 - 1)(2x)}{\left(x^2 + 1\right)^2} = \dfrac{4x}{\left(x^2 + 1\right)^2}$

19. $y = \sinh(\tan^{-1} e^{3x})$,

so $\dfrac{dy}{dx} = [\cosh(\tan^{-1} e^{3x})]\left[\dfrac{1}{1 + \left(e^{3x}\right)^2}\right][3e^{3x}]$

$\qquad = [\cosh(\tan^{-1} e^{3x})]\left[\dfrac{3e^{3x}}{1 + e^{6x}}\right].$

21. By implicit differentiation:

$$2y \frac{dy}{dx} + \left(\cosh y + x \sinh y \frac{dy}{dx}\right) + 2 \sinh x \cosh x = 0,$$

or $(2y + x \sinh y) \frac{dy}{dx} = -2 \sinh x \cosh x - \cosh y.$

Hence,

$$\frac{dy}{dx} = -\frac{2 \sinh x \cosh x + \cosh y}{2y + x \sinh y} = -\frac{\sinh 2x + \cosh y}{2y + x \sinh y}.$$

23.
$$\int \frac{\cosh \theta}{\sinh \theta + \cosh \theta} d\theta = \int \frac{\cosh \theta}{\sinh \theta + \cosh \theta} \cdot \frac{\cosh \theta - \sinh \theta}{\cosh \theta - \sinh \theta} d\theta$$

$$= \int (\cosh^2 \theta - \cosh \theta \sinh \theta) \, d\theta$$

$$= \int \frac{1 + \cosh 2\theta}{2} d\theta - \frac{\sinh^2 \theta}{2} = \frac{\theta}{2} + \frac{\sinh 2\theta}{4} - \frac{\sinh^2 \theta}{2} + C.$$

25. Using integration by parts, let $u = \sinh 2x$, $dv = e^x dx$. Then,

$$v = e^x, \quad du = 2 \cosh 2x \, dx,$$

and $\int e^x \sinh 2x \, dx = e^x \sinh 2x - 2 \int e^x \cosh 2x \, dx.$

Now, in the last integral, let $u = \cosh 2x$, $dv = e^x dx$. Then,

$$v = e^x, \quad du = 2 \sinh 2x \, dx,$$

and $\int e^x \cosh 2x \, dx = e^x \cosh 2x - 2 \int e^x \sinh 2x \, dx.$

Therefore,

$$\int e^x \sinh 2x \, dx = e^x \sinh 2x - 2e^x \cosh 2x + 4 \int e^x \sinh 2x \, dx$$

or $\int e^x \sinh 2x \, dx = \frac{e^x}{3}(2 \cosh 2x - \sinh 2x).$

27.
$$\int_0^1 \frac{dx}{4 - x^2} = \frac{1}{4} \int_0^1 \left(\frac{1}{2 - x} + \frac{1}{2 + x}\right) dx = \frac{1}{4} \ln \left|\frac{2 + x}{2 - x}\right|\Big|_0^1 = \frac{1}{4} \ln 3.$$

29. Let $u = e^t$, $du = e^t dt$. Then,

$$\int \frac{e^t \, dt}{\sqrt{1 + e^{2t}}} = \int \frac{du}{\sqrt{1 + u^2}} = \sinh^{-1} u + C = \sinh^{-1} (e^t) + C.$$

CHAPTER 9 - Miscellaneous Problems

31. Let $u = \tan \theta$, $du = \sec^2 \theta \, d\theta$. Then,

$$\int \frac{\sec^2 \theta \, d\theta}{\sqrt{\tan^2 \theta - 1}} = \int \frac{du}{\sqrt{u^2 - 1}} = \cosh^{-1} u + C = \cosh^{-1} (\tan \theta) + C.$$

33.

$y = \tan \left((\pi/2) \tanh x \right)$

Let $g(x) = (\pi/2) \tanh x.$

35. The arc length from the lowest point $(x = 0)$ to an arbitrary value of x is

$$L(x) = \int_0^x \sqrt{1 + \sinh^2 \left(\frac{t}{a}\right)} \, dt = \int_0^x \cosh \left(\frac{t}{a}\right) dt = a \sinh \left(\frac{t}{a}\right) \Big|_0^x$$

$$= a \sinh \left(\frac{x}{a}\right) \equiv s$$

Now, $\dfrac{dy}{dx} = \sinh \left(\dfrac{x}{a}\right) = \dfrac{1}{a} \left[a \sinh \left(\dfrac{x}{a}\right) \right] = \dfrac{s}{a}$.

37.

$$\int_1^x \left[\frac{1}{\sqrt{1+t^2}} - \frac{1}{t} \right] dt = \sinh^{-1} t - \ln t \Big|_1^x = \sinh^{-1} x - \ln x - \sinh^{-1} 1$$

$$= \ln (x + \sqrt{x^2 + 1}) - \ln x - \ln (1 + \sqrt{2})$$

Thus,

$$\lim_{x \to \infty} \int_1^x \left[\frac{1}{\sqrt{1+t^2}} - \frac{1}{t} \right] dt = \lim_{x \to \infty} \ln \left[\frac{x + \sqrt{x^2 + 1}}{x} \right] - \ln (1 + \sqrt{2})$$

$$= \ln 2 - \ln (1 + \sqrt{2})$$

(Compare with Problem 36.)

CHAPTER 10

POLAR COORDINATES

Article 10.1

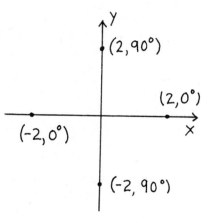

1. a) $x = r \cos \theta = 2 \cos 90° = 0$, $y = r \sin \theta = 2 \sin 90° = 2$. All polar
       coordinates: $(2, 90 + 360n)$ where n is any integer, and $(-2, 270 + 360n)$,
       where n can again be any integer.

    b) $x = 2 \cos 0° = 2$, $y = 2 \sin 0° = 0$. All polar coordinates: $(2, 360n)$ and
       $(-2, 180 + 360n)$, where n is any integer.

    c) $x = -2 \cos 90° = 0$, $y = -2 \sin 90° = -2$. All polar coordinates:
       $(-2, 90 + 360n)$ and $(2, 270 + 360n)$, where n can be any integer.

    d) $x = -2 \cos 0° = -2$, $y = -2 \sin 0° = 0$. All polar coordinates: $(-2, 360n)$
       and $(2, 180 + 360n)$, where n is any integer.

3. $r = 2$

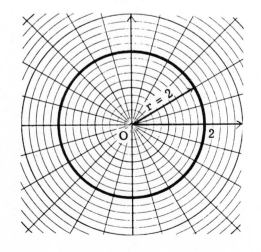

5.

    r > 1

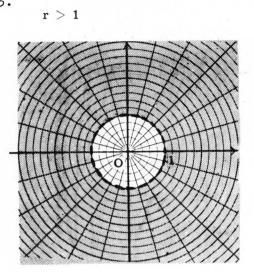

7.

    0° ≤ θ ≤ 30°, r ≥ 0

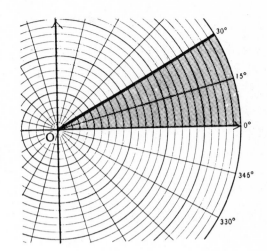

9.   θ = 60°, -1 ≤ r ≤ 3

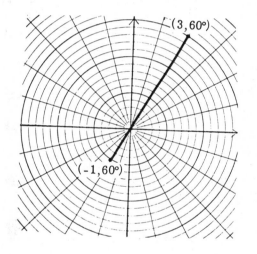

11.

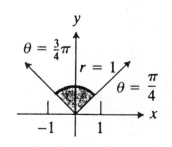

13.

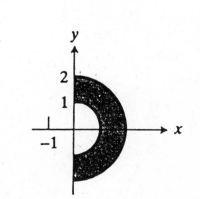

15. Since $x = r \cos \theta$, the equation becomes $x = 2$.

17. $y = r \sin \theta$, so $y = 4$.

19. $y = r \sin \theta$, so $y = 0$.

21. $x = r \cos \theta$, $y = r \sin \theta$, so the equation becomes $x + y = 1$.

23. Since $x^2 + y^2 = r^2$, we have $x^2 + y^2 = 1$. The graph is a circle with center at the origin and radius 1.

25. Since $x = r \cos \theta$ and $y = r \sin \theta$, we have $y = e^x$.

27. We may rewrite the equation as $r(\sin \theta - 2 \cos \theta) = 5$, or $r \sin \theta - 2(r \cos \theta) = 5$, so that $y - 2x = 5$ or $y = 2x + 5$.

29. We rewrite this as $r = \dfrac{4 \sin \theta}{\cos \theta} \cdot \dfrac{1}{\cos \theta}$, or $r \cos \theta = \dfrac{4r \sin \theta}{r \cos \theta}$. Thus, $x = \dfrac{4y}{x}$, or $y = \dfrac{1}{4}x^2$.

31. $x = r \cos \theta$, so $r \cos \theta = 7$.

33. $x = r \cos \theta$, $y = r \sin \theta$, so $r \cos \theta = r \sin \theta$.

35. $r^2 = 4$ or $r = 2$.

37. $\dfrac{r^2 \cos^2 \theta}{9} + \dfrac{r^2 \sin^2 \theta}{4} = 1$.

39. $r^2 \sin^2 \theta = 4r \cos \theta$.

41. $\cos(\theta - 60°) = \cos \theta \cos 60° + \sin \theta \sin 60° = \dfrac{1}{2} \cos \theta + \dfrac{\sqrt{3}}{2} \sin \theta$. So, $\dfrac{1}{2}r \cos \theta + \dfrac{\sqrt{3}}{2}r \sin \theta = 3$, or $x + \sqrt{3}y = 6$.

43. $\sin(45° - \theta) = \sin 45° \cos \theta - \cos 45° \sin \theta = \dfrac{1}{\sqrt{2}}(\cos \theta - \sin \theta)$. Thus, $\dfrac{1}{\sqrt{2}}(r \cos \theta - r \sin \theta) = \sqrt{2}$, or $x - y = 2$, so $y = x - 2$.

45. $\sin 2\left(\dfrac{3\pi}{4}\right) = \sin \dfrac{3\pi}{2} = -1$, so $2 \sin 2\theta = -2$. Therefore, we try an equivalent pair of coordinates for which r is negative, such as $\left(-2, \dfrac{3\pi}{4} - \pi\right) = \left(-2, -\dfrac{\pi}{4}\right)$. Then $\sin 2\theta = -1$ and $r = \sin 2\theta$, as claimed.

47. We note that $\cos(\theta + \pi) = \cos \theta \cos \pi - \sin \theta \sin \pi = -\cos \theta$. Suppose (r_0, θ_0) satisfies $r = \cos \theta - 1$. Then $\cos(\theta_0 + \pi) - 1 = \cos \theta_0 - 1$ $= -(1 + \cos \theta_0)$, so the point $(-r_0, \theta_0 + \pi)$ satisfies $r = \cos \theta + 1$. But the points (r_0, θ_0) and $(-r_0, \theta_0 + \pi)$ are the same point. Now suppose (r_0, θ_0) satisfies $r = \cos \theta + 1$. Then $\cos(\theta_0 + \pi) + 1 = -\cos \theta_0 + 1$ $= -(\cos \theta_0 - 1)$, so $(-r_0, \theta_0 + \pi)$ satisfies $r = \cos \theta - 1$. Therefore, the two graphs consist exactly of the same points, so the two curves are the same.

49. $r = a \sin \theta$, $r = a \cos \theta$: $a \sin \theta = a \cos \theta$, or $\tan \theta = 1$. Thus, $\theta = \dfrac{\pi}{4}$ or $\dfrac{5\pi}{4}$, so the points of intersection we have found are $\left(\dfrac{a}{\sqrt{2}}, \dfrac{\pi}{4}\right)$ and $\left(\dfrac{a}{\sqrt{2}}, \dfrac{5\pi}{4}\right)$.

51. We use the identity $\cos 2\theta = 2\cos^2\theta - 1$. $a(1 + \sin\theta) = 2a\cos\theta$. Square both sides: $1 + 2\sin\theta + \sin^2\theta = 4\cos^2\theta = 4(1 - \sin^2\theta)$, or $5\sin^2\theta + 2\sin\theta - 3 = 0$, or $(5\sin\theta - 3)(\sin\theta + 1) = 0$. $\sin\theta + 1 = 0$: $\theta = \frac{3\pi}{2}$, $r = a(1 + \sin\theta) = 0$, so we have the origin. (We need to be careful and check that the origin lies on the graph of $r = 2a\cos\theta$; this is so for $\theta = \frac{\pi}{2}$.) $\sin\theta = \frac{3}{5}$: $\theta \approx 37°$, $143°$. However, if we go back to the original equations, we find $\theta \approx 143°$ is not a root. (It is an extraneous root produced by squaring.) Therefore, at $\theta \approx 37°$, $1 + \sin\theta = \frac{8}{5}$, and $r = \frac{8a}{5}$. Thus, the point of intersection is $(\frac{8a}{5}, 37°)$.

53. a) $-4a^2\cos\theta = a^2(1 - \cos\theta)^2$: $\cos^2\theta + 2\cos\theta + 1 = 0$, $(\cos\theta + 1)^2 = 0$, $\cos\theta = -1$, $\theta = \pi$. Then $r = a(1 - \cos\theta) = 2a$, so $(2a, \pi)$ is a point of intersection.

   b) The origin satisfies $r = a(1 + \cos\theta)$ since $\cos 0 = 1$; the origin also satisfies $r = 4a^2\cos\theta$ since $\cos\frac{\pi}{2} = 0$. Therefore, the origin lies on the intersection.

Article 10.2

1. $x = r\cos\theta$, $y = r\sin\theta$; $3x + 4y = 5$ becomes $r(3\cos\theta + 4\sin\theta) = 5$.

3. $r = a(1 + \cos\theta) = a\big(1 + \cos(-\theta)\big)$ symmetric with respect to x-axis.
   $-1 \leq \cos\theta \leq 1 \Rightarrow 0 \leq r \leq 2a$
   When $\theta = 0$, $r = 2a$ and as θ increases r decreases.

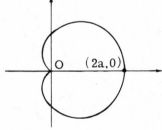

5. $r = a\sin 2\theta$; $-r = a\sin 2(-\theta)$, symmetric with respect to y-axis.
   $-r = a\sin 2(\pi - \theta)$, symmetric with respect to x-axis.
   $-1 \leq \sin 2\theta \leq 1 \Rightarrow -1 \leq r \leq 1$
   On the closed interval $[0, \pi/4]$, r increases from 0 to a and on the closed interval $[\pi/4, \pi/2]$, r decreases from a to 0.

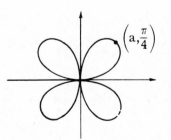

Lemniscate of Bernoulli

7. r = a(2 + sin θ) = a[2 + sin(π - θ)]
    symmetric with respect to
    y-axis.
    On the closed interval [-π/2,
    π/2], r increases from a to
    3a.

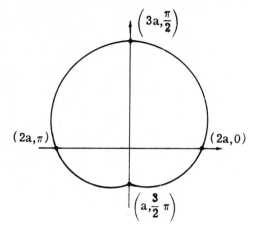

9. r = θ (Spiral of Archimedes)
    -r = -θ, symmetric with respect to
    y-axis.

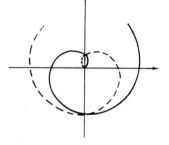

11.

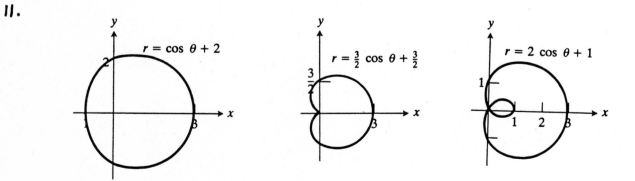

13. $1 = 2 \sin 2\theta$: $\sin 2\theta = \frac{1}{2}$, so $2\theta = \frac{\pi}{6}, \frac{5\pi}{6}, \frac{13\pi}{6}, \frac{17\pi}{6}$ (other values of θ lead to duplication of points already given). So, there are four distinct points of intersection: $(1, \pi/12)$, $(1, 5\pi/12)$, $(1, 13\pi/12)$, and $(1, 17\pi/12)$.

15.

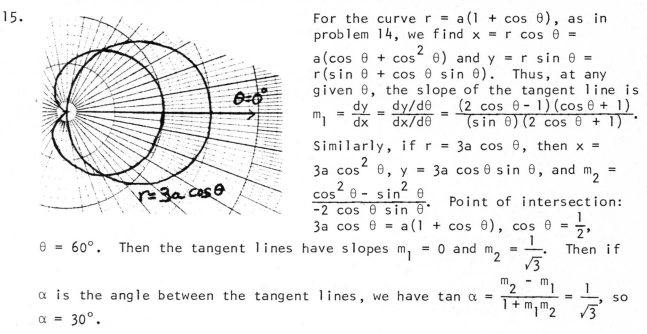

For the curve $r = a(1 + \cos \theta)$, as in problem 14, we find $x = r \cos \theta = a(\cos \theta + \cos^2 \theta)$ and $y = r \sin \theta = r(\sin \theta + \cos \theta \sin \theta)$. Thus, at any given θ, the slope of the tangent line is $m_1 = \frac{dy}{dx} = \frac{dy/d\theta}{dx/d\theta} = \frac{(2 \cos \theta - 1)(\cos \theta + 1)}{(\sin \theta)(2 \cos \theta + 1)}$.

Similarly, if $r = 3a \cos \theta$, then $x = 3a \cos^2 \theta$, $y = 3a \cos \theta \sin \theta$, and $m_2 = \frac{\cos^2 \theta - \sin^2 \theta}{-2 \cos \theta \sin \theta}$. Point of intersection: $3a \cos \theta = a(1 + \cos \theta)$, $\cos \theta = \frac{1}{2}$,

$\theta = 60°$. Then the tangent lines have slopes $m_1 = 0$ and $m_2 = \frac{1}{\sqrt{3}}$. Then if

α is the angle between the tangent lines, we have $\tan \alpha = \frac{m_2 - m_1}{1 + m_1 m_2} = \frac{1}{\sqrt{3}}$, so $\alpha = 30°$.

Article 10.3

1. Multiplying both sides by r: $r^2 = 4r \cos \theta$, or $x^2 + y^2 = 4x$, or $(x - 2)^2 + y^2 = 4$. Graph: circle with center $(2, 0)$ and radius 2.

3. Multiplying both sides by r: $r^2 = -2r \cos \theta$, or $x^2 + y^2 = -2x$, or $(x + 1)^2 + y^2 = 1$ (completing the square). Graph: circle with center $(-1, 0)$ and radius 1.

5.

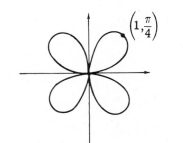

$r = \sin 2\theta = 2 \sin \theta \cos \theta$, so $r^3 = 2(r \cos \theta)(r \sin \theta)$, or $(x^2 + y^2)^{3/2} = 2xy$.

7.

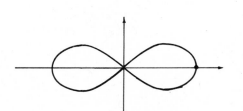

$r^2 = 8 \cos 2\theta = 8(\cos^2 \theta - \sin^2 \theta)$, so $r^4 = 8(r \cos \theta)^2 - 8(r \sin \theta)^2$, or $(x^2 + y^2)^2 = 8(x^2 - y^2)$.

9. $r = 8 - 16 \cos \theta$, or $r^2 = 8r - 16r \cos \theta$,
so $x^2 + y^2 = 8(x^2 + y^2)^{1/2} - 16x$.

11. $r = 4(1 - \cos \theta)$ or
$r^2 = 4r - 4(r \cos \theta)$, so
$x^2 + x^2 = 4(x^2 + y^2)^{1/2} - 4x$.

13. $3r - 6(r \cos \theta) = 12$, or
$3(x^2 + y^2)^{1/2} - 6x = 12$. To sketch the
graph, we note that
$r = \dfrac{12}{3 - 6 \cos \theta} = \dfrac{4}{1 - 2 \cos \theta}$; this is the
equation of the hyperbola.

15. a) $\cos(\theta + 45°) = \cos \theta \cos 45° - \sin \theta \sin 45°$. Thus $r = 2 \cos(\theta + 45°)$ or
$r^2 = 2r \cos(\theta + 45°) = \sqrt{2}(r \cos \theta - r \sin \theta)$, so we have $x^2 + y^2$
$= \sqrt{2}x - \sqrt{2}y$, or $(x - \dfrac{\sqrt{2}}{2})^2 + (y - \dfrac{\sqrt{2}}{2})^2 = 1$. Graph: circle with center
$\left(\dfrac{1}{\sqrt{2}}, -\dfrac{1}{\sqrt{2}} \right)$ and radius 1.

   b) $r = 4 \csc(\theta - 30°)$ or $r \sin(\theta - 30°) = 4$. But $r \sin(\theta° - 30°)$
$= r[\sin \theta \cos 30° - \cos \theta \sin 30°]$, so we have $\dfrac{\sqrt{3}}{2}x - \dfrac{1}{2}y = 4$, whose graph
is a line.

   c) $r = 5 \sec(60° - \theta)$, or $r \cos(60° - \theta) = 5$. But $\cos(60° - \theta)$
$= \cos 60° \cos \theta + \sin 60° \sin \theta = \dfrac{1}{2}\cos \theta + \dfrac{\sqrt{3}}{2} \sin \theta$, so we have

$a)$

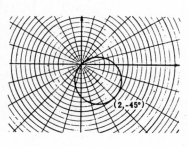

$b)$

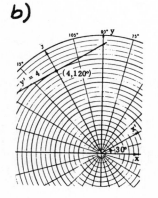

$c)$

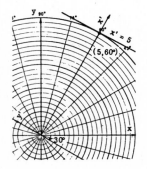

continued

CHAPTER 10 - Article 10.3

$\frac{1}{2}x + \frac{\sqrt{3}}{2}y = 5$, or $x + \sqrt{3}y = 10$, whose graph is a line.

d) $r = 3\sin(\theta + 30°)$, so $r^2 = 3r\sin(\theta + 30°)$. But $\sin(\theta + 30°)$

$= \sin\theta\cos 30° + \cos\theta\sin 30° = \frac{\sqrt{3}}{2}\sin\theta + \frac{1}{2}\cos\theta$, so we have

$x^2 + y^2 = \frac{1}{2}x + \frac{\sqrt{3}}{2}y$, or $(x^2 - \frac{1}{2}x) + (y^2 - \frac{\sqrt{3}}{2}y) = 0$. Completing the square

we find $(x - \frac{1}{4})^2 + (y - \frac{\sqrt{3}}{4})^2 = \frac{1}{4}$, whose graph is the circle with center

$(\frac{1}{4}, \frac{\sqrt{3}}{4})$ and radius $\frac{1}{2}$.

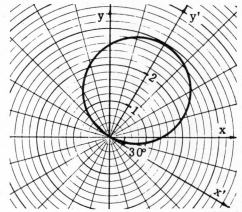

c) $r = a + a\cos(\theta - 30°)$

$= a + \frac{a\sqrt{3}}{2}\cos\theta + \frac{a}{2}\sin\theta$, so

$r^2 = ar + \frac{\sqrt{3}}{2}a(r\cos\theta) + \frac{a}{2}(r\sin\theta)$, or

$x^2 + y^2 = a(x^2 + y^2)^{1/2} + \frac{\sqrt{3}}{2}ax + \frac{ay}{2}$.

Graph: Let $\theta' = \theta - 30°$; then

$r' = a(1 + \cos\theta')$. The graph is a
cardioid rotated 30°.

c)

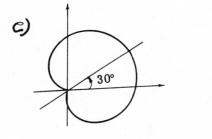

f) The region is the area inside the
cardioid $r = 2(1 - \cos\theta)$.

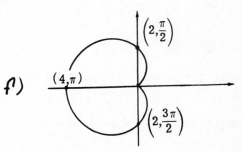

17. (e) $r = \theta$. Except for (c) and (i), which represent conics, this is the only equation for which r is unbounded.

19. (f) $r^2 = \cos 2\theta$. This matches the lemniscate in figure 10.26 of the text on page 584.

21. (h) $r = 1 - \sin\theta$. The equation of the cardioid will either be (g) or (h), and since $r = 0$ at $\theta = 90°$, the correct choice is (h).

23. (j) $r^2 = \sin 2\theta$. This is the lemniscate of problem 19 rotated by 45°, so we replace θ by $\theta - 45°$ to find the new equation: $r^2 = \cos 2(\theta - 45°)$ $= \cos(2\theta - 90°) = \cos(90° - 2\theta) = \sin 2\theta$.

25. a) $r = 2\cos\theta \Rightarrow r^2 = 2r\cos\theta \Rightarrow x^2 + y^2 = 2x$ or $(x - 1)^2 + y^2 = 1$.
    Also, $r = \sec\theta \Rightarrow r\cos\theta = 1$ or $x = 1$.

    b) Intersection: $(1, 1)$, which has polar coordinates $(\sqrt{2}, 45°)$ or $(\sqrt{2}, 315°)$.

27. Since the distance from the directrix to the origin is k = 4 and since the eccentricity of a parabola is e = 1, the equation of the parabola is
$$r = \frac{ke}{1 - e \cos \theta} = \frac{4}{1 - \cos \theta}.$$

29.

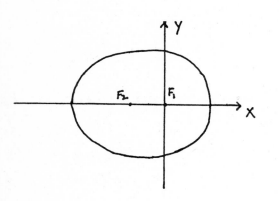

$r = \dfrac{-8}{-\cos \theta}$ is the equation of the ellipse (k < 0 since the given directrix lies to the right of the focus). The vertices of the ellipse are at $(-\frac{8}{3}, 0°) = (\frac{8}{3}, 180°)$ and $(-\frac{8}{5}, 180°) = (\frac{8}{5}, 0°)$. Thus, 2a = $\frac{8}{5} - (-\frac{8}{3}) = \frac{64}{15}$, and the distance between the two foci is 2ae = $\frac{16}{15}$. Since F_2 lies to the left of $F_1 (0, 0°)$, F_2 is at $(-\frac{16}{15}, 0°) = (\frac{16}{15}, 180°)$.

Remark: It may be noted that the equation we have found for the ellipse, $r = \dfrac{8}{4 - \cos \theta}$, does not match that offered in the appendix of the textbook, namely $r = \dfrac{8}{4 + \cos \theta}$. The two equations, however, give the same ellipse.

To see this, we recall that the points (r, θ) and (-r, θ + π) are the same. Consequently, we may replace r by -r and θ by θ + π in any equation without changing the equation. We do this to our equation $r = \dfrac{8}{4 - \cos \theta}$ and obtain $-r = \dfrac{-8}{4 - \cos (\theta + \pi)}$. Since cos(θ + π) = -cos θ, this is the same as $r = \dfrac{8}{4 + \cos \theta}$. In general, if we have a focus F_1 of our conic at the origin and a vertical directrix k units to the <u>right</u> of F_1, then the equation of the conic is $r = \dfrac{ke}{1 + e \cos \theta}$.

31. Since $r^2 = 2a^2 \cos 2\theta$, $\dfrac{dr}{d\theta} = \dfrac{-2a^2 \sin 2\theta}{r}$. Then, $\dfrac{dx}{d\theta} = \dfrac{d}{d\theta}(r \cos \theta)$

$= -r \sin \theta + (\cos \theta)\dfrac{dr}{d\theta}$ and $\dfrac{dy}{d\theta} = \dfrac{d}{d\theta}(r \sin \theta) = r \cos \theta + (\sin \theta)\dfrac{dr}{d\theta}$.

Therefore, $\dfrac{dy}{dx} = \dfrac{dy/d\theta}{dx/d\theta} = \dfrac{r \cos \theta + (\sin \theta)(dr/d\theta)}{-r \sin \theta + (\cos \theta)(dr/d\theta)}$

$= \dfrac{r^2 \cos \theta - 2a^2 \sin \theta \sin 2\theta}{-r^2 \sin \theta - 2a^2 \cos \theta \, 2\theta} = \dfrac{-2a^2(\cos \theta \cos 2\theta - \sin \theta \sin 2\theta)}{2a^2(\sin \theta \cos 2\theta + \cos \theta \sin 2\theta)}$

$= -\dfrac{\cos 3\theta}{\sin 3\theta} = -\cot 3\theta$, where we have used (in order of appearance) the expressions for $\dfrac{dr}{d\theta}$, r, and the trigonometric identities for cos(a + b) and sin(a + b). The lemniscate passes through the origin when r = 0; this happens when cos 2θ = 0, or θ = $\frac{\pi}{4}$, $\frac{3\pi}{4}$, $\frac{5\pi}{4}$, or $\frac{7\pi}{4}$. At θ = $\frac{\pi}{4}$ and $\frac{5\pi}{4}$, -cot 3θ = 1, so the tangent line has slope 1. At θ = $\frac{3\pi}{4}$ and $\frac{7\pi}{4}$, -cot 3θ = -1, so the tangent line has slope -1.

CHAPTER 10 - Article 10.3

33. a) The vertices of the hyperbola are at $(\frac{ke}{1-e}, 0°)$ and $(\frac{ke}{1+e}, 180°)$

$= (\frac{-ke}{1+e}, 0°)$. Then the distance a satisfies: $2a = \left| \frac{ke}{e-1} - \frac{ke}{e+1} \right|$

$= \left| \frac{2ke}{e^2-1} \right| = \frac{2ke}{e^2-1}$, since $e > 1$. Thus, $ke = a(e^2-1)$, and the

equation of the hyperbola can be written as $r = \frac{a(e^2-1)}{1-e\cos\theta}$.

b) We look for the angles θ_0 such that $\lim_{\theta \to \theta_0} r = \infty$. This happens when

$\lim_{\theta \to \theta_0} (1 - e\cos\theta) = 0$, or $1 - e\cos\theta_0 = 0$. Thus, $\cos\theta_0 = \frac{1}{e}$, so

$\sin\theta_0 = \frac{\pm\sqrt{e^2-1}}{e}$ and $\tan\theta_0 = \pm\sqrt{e^2-1}$. Therefore, the asymptotes have

slope $\pm\sqrt{e^2-1}$.

35. Completing the square: $x^2 + (y-a)^2 = a^2$. Graph: circle with center

(0, a) and radius a. Polar equation: $r^2 - 2ar\sin\theta = 0$, or

$r = 2a\sin\theta$.

37. Graph: straight line of Figure 10.22 in text. Polar equation:
$(r\cos\theta)\cos\alpha + (r\sin\theta)\sin\alpha = p$, or $r\cos(\theta - \alpha) = p$, or
$r = p\sec(\theta - \alpha)$.

39. Polar equation: $r^4 = r^2\cos^2\theta - r^2\sin^2\theta$, or $r^2 = \cos^2\theta - \sin^2\theta$
$= \cos 2\theta$. Graph: lemniscate as in Figure 10.26 of the text.

Article 10.4

1. Area $= \frac{1}{2}\int_0^{2\pi} a^2(1 + \cos\theta)^2 \, d\theta = \frac{a^2}{2}\int_0^{2\pi} (1 + 2\cos\theta + \cos^2\theta)d\theta$

$= \frac{a^2}{a}\int_0^{2\pi} (1 + 2\cos\theta + \frac{1 + \cos 2\theta}{2})d\theta = \frac{a^2}{2}(\frac{3\theta}{2} + 2\sin\theta + \frac{\sin 2\theta}{4})\Big|_0^{2\pi} = \frac{3\pi a^2}{2}$.

3.

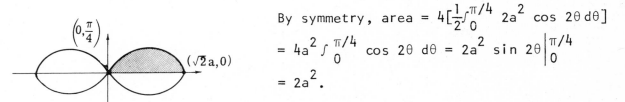

By symmetry, area $= 4[\frac{1}{2}\int_0^{\pi/4} 2a^2\cos 2\theta \, d\theta]$

$= 4a^2\int_0^{\pi/4} \cos 2\theta \, d\theta = 2a^2\sin 2\theta\Big|_0^{\pi/4}$

$= 2a^2$.

5. Area $= \frac{1}{2}\int_0^{2\pi} a^2(2 + \cos\theta)^2 \, d\theta = \frac{a^2}{a}\int_0^{2\pi} (4 + 4\cos\theta + \cos^2\theta)d\theta$

$= \frac{a^2}{2}\int_0^{2\pi} (4 + 4\cos\theta + \frac{1 + \cos 2\theta}{2})d\theta = \frac{9\pi a^2}{2}$.

7.

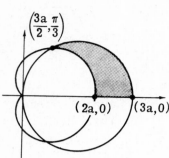

$r_1 = 3a \cos \theta, \; r_2 = a(1 + \cos \theta)$.

Intersection: $3a \cos\theta = a + a \cos \theta$,

so $\cos \theta = \dfrac{1}{2}$. Then $\theta = \dfrac{\pi}{3}$ or $-\dfrac{\pi}{3}$, so the area

is $A = \dfrac{1}{2} \int_{-\pi/3}^{\pi/3} (r_1^2 - r_2^2) d\theta$

$= \dfrac{a^2}{2} \int [9 \cos^2 \theta - (1 + \cos \theta)^2] d\theta$

$= \dfrac{a^2}{2} \int [8 \cos^2 \theta - 2 \cos \theta - 1] d\theta$

$= \dfrac{a^2}{2} \int [8(\dfrac{1 + \cos 2\theta}{2}) - 2 \cos \theta - 1] d\theta$

$= \dfrac{a^2}{2} (3\theta + 2 \sin 2\theta - 2 \sin \theta) \Big|_{-\pi/3}^{\pi/3} = \pi a^2$.

9.

Points of intersection: $2a \sin \theta = a$,

or $\sin \theta = \dfrac{1}{2}$. Then $\theta = \dfrac{\pi}{6}$ or $\dfrac{5\pi}{6}$. By

symmetry:

Area $= 2[\dfrac{1}{2} \int_0^{\pi/6} 4a^2 \sin^2 \theta \, d\theta + \dfrac{1}{2}\int_{\pi/6}^{\pi/2} a^2 d\theta]$

$= \dfrac{\pi a^2}{3} + 4a^2 \int_0^{\pi/6} (\dfrac{1 - \cos 2\theta}{2}) d\theta$

$= \dfrac{\pi a^2}{3} + 4a^2 (\dfrac{\theta}{2} - \dfrac{\sin 2\theta}{4}) \Big|_0^{\pi/6}$

$= (\dfrac{2\pi}{3} - \dfrac{\sqrt{3}}{2}) a^2$.

11. Area $= \dfrac{1}{2} \int_{-\pi/4}^{\pi/4} \cos^2 2\theta \, d\theta = \dfrac{1}{2} \int_{-\pi/4}^{\pi/4} (\dfrac{1 + \cos 4\theta}{2}) d\theta = \dfrac{\theta}{4} + \dfrac{\sin 4\theta}{16} \Big|_{-\pi/4}^{\pi/4} = \dfrac{\pi}{8}$.

13. Area $= \dfrac{1}{2} \int_0^{2\pi} (4 + 2 \cos \theta)^2 d\theta = \int_0^{2\pi} (8 + 8 \cos \theta + 2 \cos^2 \theta) d\theta$

$= 16\pi + \int_0^{2\pi} (1 + \cos 2\theta) d\theta = 18\pi$.

15. a) By symmetry, the area inside the larger loop is

$A = 2[\dfrac{1}{2} \int_0^{2\pi/3} (1 + 2 \cos \theta)^2 d\theta] = \int_0^{2\pi/3} (1 + 4 \cos \theta + 4 \cos^2 \theta) d\theta$

$= \int_0^{2\pi/3} [1 + 4 \cos \theta + (2 + 2 \cos 2\theta)] d\theta = (3\theta + 4 \sin \theta + \sin 2\theta) \Big|_0^{2\pi/3}$

$= 2\pi + \dfrac{3\sqrt{3}}{2}$.

b) Since the area inside the smaller loop is $\pi - \dfrac{3\sqrt{3}}{2}$ (found in example 2),

the area we are looking for is $(2\pi + \dfrac{3\sqrt{3}}{2}) - (\pi - \dfrac{3\sqrt{3}}{2}) = \pi + 3\sqrt{3}$.

17. (See figure 10.35 on page 589 of the text.) By symmetry, $\bar{y} = 0$; from problem 1, the area of the cardioid is $3\pi a^2/2$. $\int_0^{2\pi} (\frac{2r}{3} \cos \theta)(\frac{1}{2} r^2 \, d\theta)$

$= \frac{a^3}{3} \int_0^{2\pi} \cos \theta (1 + \cos \theta)^3 \, d\theta = \frac{a^3}{3} \int_0^{2\pi} (\cos \theta + 3 \cos^2 \theta + 3 \cos^3 \theta + \cos^4 \theta) d\theta.$

Now $\int (\cos \theta + 3 \cos^3 \theta) d\theta = \int (4 - 3 \sin^2 \theta) \cos \theta \, d\theta = 4 \sin \theta - \sin^3 \theta,$

$\int 3 \cos^2 \theta \, d\theta = \int \frac{3(1 + \cos 2\theta)}{2} \, d\theta = \frac{3\theta}{2} + \frac{3 \sin 2\theta}{4},$ and $\int \cos^4 \theta \, d\theta$

$= \int (\frac{1 + \cos 2\theta}{2})^2 d\theta = \frac{1}{4} \int (1 + 2 \cos 2\theta + \cos^2 2\theta) d\theta$

$= \frac{1}{4} \int (\frac{3}{2} + 2 \cos 2\theta + \frac{\cos 4\theta}{2}) d\theta = \frac{3\theta}{8} + \frac{\sin 2\theta}{4} + \frac{\sin 4\theta}{32}.$ Hence,

$\int_0^{2\pi} \cos \theta (1 + \cos \theta)^3 \, d\theta = [4 \sin \theta - \sin^3 \theta + \frac{15\theta}{8} + \sin 2\theta + \frac{\sin 4\theta}{32}]\Big|_0^{2\pi}$

$= \frac{15\pi}{4}.$ Hence $\bar{x} = \frac{5\pi a^3/4}{3\pi a^2/2} = \frac{5a}{6}.$ Center of gravity: $(\frac{5a}{6}, 0).$

19. a) r is constant, so $dr = 0$. Circumference $= \int_0^{2\pi} a\theta d = 2\pi a.$

b) $dr = -a \cos \theta \, d\theta$, so $r^2 \, d\theta^2 + dr^2 = a^2 (\cos^2 \theta + \sin^2 \theta) d\theta^2 = a^2 \, d\theta^2.$ Thus, $ds = a \, d\theta$ and circumference $= \int_0^{2\pi} a \, d\theta = 2\pi a.$

c) $dr = a \cos \theta \, d\theta$, so $ds^2 = a^2 (\sin^2 \theta + \cos^2 \theta) d\theta^2 = a^2 \, d\theta^2.$ Hence $ds = a \, d\theta$ and the circumference is again $\int_0^{2\pi} a \, d\theta = 2\pi a.$

21. To sketch the curve, note that $\sin^2 (\theta/2) = \frac{1 - \cos \theta}{2}$, so $r = \frac{a}{2}(1 - \cos \theta).$ The graph is therefore a cardioid; see figure 10.11 on page 578 of the text. Using the form $r = \frac{a}{2}(1 - \cos \theta)$, $dr = \frac{a}{2} \sin \theta \, d\theta.$ Hence, $ds^2 = r^2 \, d\theta^2 + dr^2 = \frac{a^2}{4}[(1 - \cos \theta)^2 + \sin^2 \theta] d\theta^2 = a^2 (\frac{1 - \cos \theta}{2}) d\theta^2$ $= a^2 \sin^2 (\theta/2) d\theta^2.$ Hence, $ds = a|\sin(\theta/2)| d\theta$; we note that $\sin(\theta/2) \geq 0$ for $0 \leq \theta \leq 2\pi.$ So, the required length is $L = \int_0^\pi ds = \int_0^\pi a \sin(\frac{\theta}{2}) d\theta$ $= -2a \cos (\theta/2)\Big|_0^\pi = 2a.$

23. $dr = \sin^2 (\theta/3) \cos (\theta/3) d\theta$, so $ds^2 = a^2 [\sin^6 (\theta/3) + \sin^4 (\theta/3) \cos^2 (\theta/3)] d\theta^2$ $= a^2 \sin^4 (\theta/3) d\theta^2.$ Hence, $L = a \int_0^\pi \sin^2 (\theta/3) d\theta = a \int_0^\pi \frac{1 - \cos(2\theta/3)}{2} d\theta$

$= a[\frac{\theta}{2} - \frac{3 \sin(2\theta/3)}{4}]\Big|_0^\pi = (\frac{4\pi - 3\sqrt{3}}{8})a.$

25. $dS = 2\pi y \, ds$ where $y = r \sin \theta.$ $ds^2 = r^2 \, d\theta^2 + dr^2$, $2r\frac{dr}{d\theta} = -4a^2 \sin 2\theta.$ Hence, $dr = \frac{-2a^2 \sin 2\theta}{r} d\theta$, so $dr^2 = \frac{4a^4 \sin^2 2\theta}{2a^2 \cos 2\theta} d^2$, and

$ds^2 = 2a^2 [\cos 2\theta + \frac{\sin^2 2\theta}{\cos 2\theta}] d\theta^2 = \frac{2a^2}{\cos 2\theta} d\theta^2.$ Hence, $ds = \frac{a\sqrt{2}}{\sqrt{\cos 2\theta}} d\theta.$

By symmetry, the surface area is $2 \int_0^{\pi/4} dS = 4\pi \int_0^{\pi/4} r \sin \theta \, ds$

$= 4\pi \int_0^{\pi/4} (a\sqrt{2} \sqrt{\cos 2\theta})(\sin \theta)\left(\frac{a\sqrt{2}}{\sqrt{\cos 2\theta}} d\theta\right) = 8\pi a^2 \int_0^{\pi/4} \sin \theta \, d\theta$

$= 4\pi a^2 (2 - \sqrt{2}).$

CHAPTER 10 - Article 10.4

27. $dS = 2\pi y \, ds$. $dr = -\sin \theta \, d\theta$, so $ds^2 = r^2 \, d\theta^2 + dr^2$
$= [(1 + \cos \theta)^2 + \sin^2 \theta]d\theta^2 = [2 + 2 \cos \theta]d\theta^2 = 4 \cos^2(\theta/2)d\theta$. Hence,
$ds = 2 \cos(\theta/2)d\theta$ (since $0 \le \theta \le \pi/2$, we do not need absolute values here).
Now $dS = 2\pi(r \sin \theta)ds = 2\pi(1 + \cos \theta)(\sin \theta)ds$
$= 4\pi(1 + \cos \theta)(\sin \theta)\cos(\theta/2)d\theta$. But since $1 + \cos \theta = 2 \cos^2(\theta/2)$ and
$\sin \theta = 2 \sin(\theta/2)\cos(\theta/2)$, $dS = 16\pi \cos^4(\theta/2)\sin(\theta/2)d\theta$. Hence,
$S = \int_0^{\pi/2} 16\pi \cos^4(\theta/2)\sin(\theta/2)d$
$= \dfrac{-32\pi \cos^5(\theta/2)}{5}\Big|_0^{\pi/2} = \dfrac{4\pi(8 - \sqrt{2})}{5}$.

Miscellaneous Problems, Chapter 10

1. $r = a\,\theta$

   (Spiral of Archimedes)

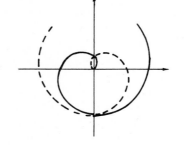

2. $r = a(1 + \cos 2\theta)$

   $= 2a \cos^2 \theta$

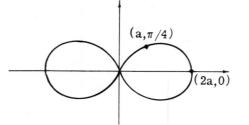

3. (a) $r = a \sec \theta \Rightarrow x = a$

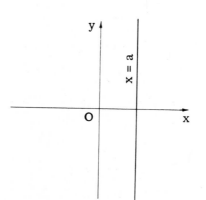

(b) $r = a \csc \theta. \Rightarrow y = a$

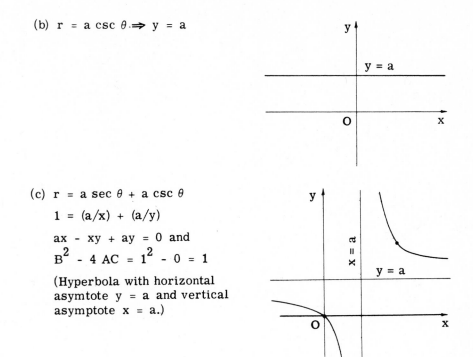

(c) $r = a \sec \theta + a \csc \theta$

$1 = (a/x) + (a/y)$

$ax - xy + ay = 0$ and

$B^2 - 4AC = 1^2 - 0 = 1$

(Hyperbola with horizontal asymtote $y = a$ and vertical asymptote $x = a$.)

5. $r^2 + 2r(\cos \theta + \sin \theta) = 7$

$x^2 + 2x + 1 + y^2 + 2y + 1 = 9$

$(x + 1)^2 + (y + 1)^2 = 9$

Circle with center at $(-1, -1)$ and radius 3.

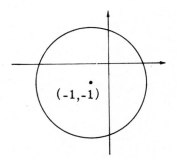

7. $r \cos (\theta/2) = a$

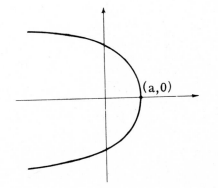

CHAPTER 10 - Miscellaneous Problems

9. $r^2 = 2a^2 \sin 2\theta$

    $r^2 = 4a^2 \sin \theta \cos \theta$

    Symmetric about origin

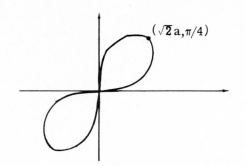

11. (a) $r = \cos 2\theta$

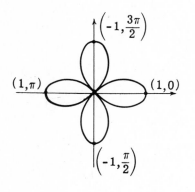

    (b) $r^2 = \cos 2\theta$

    Symmetric about origin
    Symmetric about x-axis

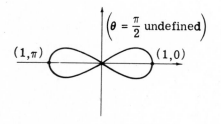

13. (a) $r = 2/(1 - \cos \theta)$

    $e = 1, \ k = 2$

    Focus: (0, 0)

    Directrix: x = -2

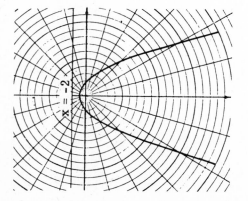

13. (b) $r = 2/(1 + \sin \theta)$

$\quad = 2/\left[1 - \cos\left(\theta + (\pi/2)\right)\right]$

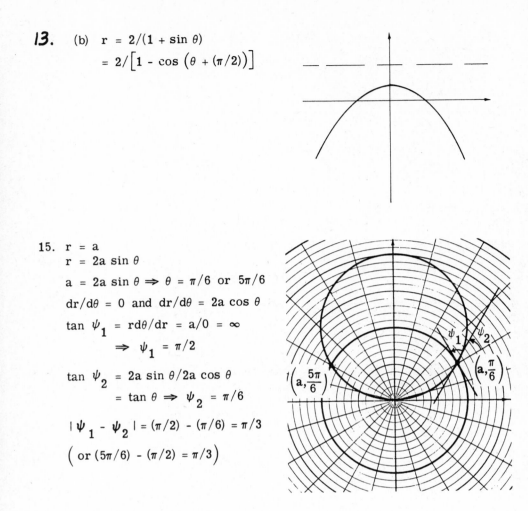

15. $r = a$

$r = 2a \sin \theta$

$a = 2a \sin \theta \Rightarrow \theta = \pi/6$ or $5\pi/6$

$dr/d\theta = 0$ and $dr/d\theta = 2a \cos \theta$

$\tan \psi_1 = rd\theta/dr = a/0 = \infty$

$\qquad \Rightarrow \psi_1 = \pi/2$

$\tan \psi_2 = 2a \sin \theta/2a \cos \theta$

$\qquad = \tan \theta \Rightarrow \psi_2 = \pi/6$

$|\psi_1 - \psi_2| = (\pi/2) - (\pi/6) = \pi/3$

$\left(\text{or } (5\pi/6) - (\pi/2) = \pi/3\right)$

17. $r = a \sec \theta$ and $r = 2a \sin \theta$

$1 = 2 \sin \theta \sec \theta = 2 \tan \theta$

$\theta = \pi/4 \qquad r = \sqrt{2}\, a$

For $r = a \sec \theta$,

$\tan \psi_1 = a \sec \theta/a \sec \theta \tan \theta = \cot \theta$

For $r = 2a \sin \theta$,

$\tan \psi_2 = 2a \sin \theta/2a \cos \theta = \tan \theta$,

$\psi_2 = \pi/4$ and $\psi_1 = \pi/4$

Hence the angle between the curves is $(\pi/4) - (\pi/4) = 0$.

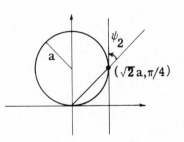

19. $r_1 = a(1 + \cos 2\theta)$

$dr_1 = -2a \sin 2\theta \, d\theta$

$r_2 = a \cos 2\theta$

$dr_2 = -2a \sin 2\theta \, d\theta$

$a(1 + \cos 2\theta) = a \cos 2\theta$

$\Rightarrow a = 0$

But $a \neq 0$ $\therefore$ if $r \neq 0$, intersections occur π radians out of phase, and

$a(1 + \cos 2\theta) = -a \cos 2(\theta + \pi) \Rightarrow \cos 2\theta = -1/2.$

$2\theta = (2\pi/3) + 2n\pi$, $\theta_1 = \pi/3$, $\theta_2 = 4\pi/3$; $\theta_1 = 4\pi/3$, $\theta_2 = \pi/3$ or

$2\theta = (4\pi/3) + 2n\pi$, $\theta_1 = 2\pi/3$, $\theta_2 = 5\pi/3$; $\theta_1 = 5\pi/3$, $\theta_2 = 2\pi/3$ and

$r_1 = -r_2 = a/2.$ Also $\pi/4$ when $r = 0$

$\tan \psi_1 = \dfrac{a/2}{-2a \sin (2\pi/3)} = -\dfrac{\sqrt{3}}{6}$, $\tan \psi_2 = \dfrac{-a/2}{-2a \sin (2\pi/3)} = \dfrac{\sqrt{3}}{6}$ or

$\tan \psi_1 = \dfrac{a/2}{-2a \sin (4\pi/3)} = -\tan \psi_2.$

$\therefore \psi_1 = \psi_2 = \tan^{-1}(\sqrt{3}/6)$ and the angle at which the curves intersect is $\psi_1 + \psi_2 = 2 \tan^{-1}(\sqrt{3}/6).$

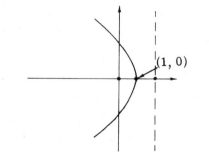

21.

$r = k/(1 - \cos \theta)$

Here $k = 2$, $\theta = (\theta - \pi)$,

$\cos (\theta - \pi) = -\cos \theta$

$\therefore r = 2/(1 + \cos \theta)$

23. $\cos \angle POQ = PO/QO = r/2a$

But $\cos \angle POQ = \pi - \theta$

$\cos (\pi - \theta) = -\cos \theta = r/2a$

$\therefore r = -2a \cos \theta$

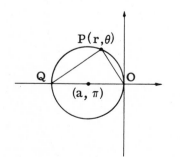

25. $r = ke/(1 - e \cos \theta), \ e < 1$

Center at $(1, 0)$, $a = 3$,

$c = 1 \Rightarrow e = c/a = 1/3$

$k = (a/e) - ae = 3/(1/3) - 3 \cdot 1/3 = 8$

$\therefore \ r = \dfrac{8/3}{1 - (1/3) \cos \theta} = \dfrac{8}{3 - \cos \theta}$

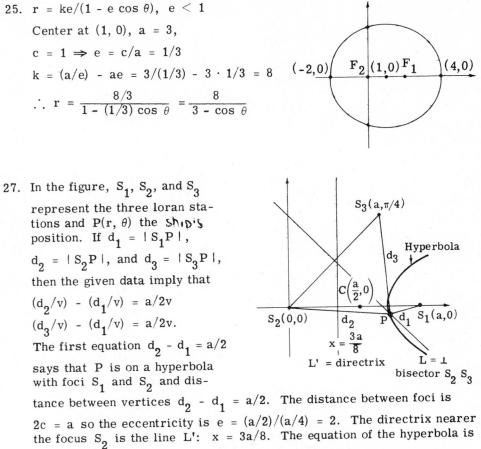

27. In the figure, S_1, S_2, and S_3 represent the three loran stations and $P(r, \theta)$ the ship's position. If $d_1 = |S_1 P|$, $d_2 = |S_2 P|$, and $d_3 = |S_3 P|$, then the given data imply that

$(d_2/v) - (d_1/v) = a/2v$

$(d_3/v) - (d_1/v) = a/2v$.

The first equation $d_2 - d_1 = a/2$ says that P is on a hyperbola with foci S_1 and S_2 and dis-

tance between vertices $d_2 - d_1 = a/2$. The distance between foci is $2c = a$ so the eccentricity is $e = (a/2)/(a/4) = 2$. The directrix nearer the focus S_2 is the line L': $x = 3a/8$. The equation of the hyperbola is $r = 2[r \cos \theta - (3a/8)]$ or $r = 3a/(8 \cos \theta - 4)$ (1). The point P is also on the perpendicular bisector L of $S_2 S_3$, so $r \cos [\theta - (\pi/4)] = a/2$

(2). We need to solve the Equations (1) and (2) simultaneously. The equation $3a/(8 \cos \theta - 4) = a/\{2 \cos [\theta - (\pi/4)]\}$ leads to $3[\cos \theta \cos (\pi/4) + \sin \theta \sin (\pi/4)] = 4 \cos \theta - 2$ or $1.879 \cos \theta - 2.121 \sin \theta = 2$ (3). Consider $\cos (\pi/4) = \sin (\pi/4) = 0.707$.

Now let $C \sin \phi = 2.121$, $C \cos \phi = 1.879$, with $C^2 = (1.879)^2 + (2.121)^2$ $= 8.029$, $C = 2.834$, $\sin \phi = 2.121/C = 0.7485 \Rightarrow \phi = 0.846$ (radian) (4). Eq. (3) becomes $C(\cos \phi \cos \theta - \sin \phi \sin \theta) = 2$ or $\cos (\theta + \phi) = 2/C$ $= 0.7058$ (5). The possible solutions of Eq. (5) are $\theta + \phi = 0.787$ or $\theta + \phi = -0.787$. Substituting $\phi = 0.846$ from the first of these, we get $\theta = -0.059$ (radian) $\approx -3° 22' 50''$ and $r = a/\{2 \cos [\theta - (\pi/4)]\} = a/1.3288$ $\approx 0.75a$.

(The solution $\theta + \phi = -0.787$ leads to $\theta = -1.633$ and corresponds to an intersection of L with the left-hand branch of the hyperbola, Equation (1) nearer to S_2 than to S_1. This does not correspond to $d_2 - d_1 > 0$.)

CHAPTER 10 - Miscellaneous Problems

29.
$$r = a \cos^3 (\theta/3)$$

$$dr = -a \cos^2 (\theta/3) \sin (\theta/3) \, d\theta$$

$$ds = \sqrt{(dr)^2 + (rd\,\theta)^2}$$

$$= \sqrt{a^2 \cos^4 (\theta/3) \sin^2 (\theta/3) + a^2 \cos^6 (\theta/3)} \, d\theta$$

$$= \sqrt{a^2 \cos^4 (\theta/3) \left[\sin^2 (\theta/3) + \cos^2 (\theta/3) \right]} \, d\theta = a \cos^2 (\theta/3) \, d\theta$$

In computing the perimeter or the total length of the closed curve, care must be exercised that the curve changes from a at 0 to 0 at $3\pi/2$, and then from 0 at $3\pi/2$ to -a at 3π (see the figure). Thus

$$P = 2 \int_0^{3\pi/2} a \cos^2 (\theta/3) \, d\theta$$

$$= 2a \int_0^{3\pi/2} (1/2)[1 + \cos (2\theta/3)] \, d\theta$$

$$= a \int_0^{3\pi/2} d\theta + a \int_0^{3\pi/2} \cos (2\theta/3) \, d\theta$$

$$= a\,\theta \Big]_0^{3\pi/2} + (3a/2) \sin (2\theta/3) \Big]_0^{3\pi/2} = 3\pi a/2.$$

31.
$$r_1 = 2a \cos^2 (\theta/2)$$
$$= 2a[(1/2)(1 + \cos \theta)]$$
$$= a(1 + \cos \theta)$$
$$r_2 = 2a \sin^2 (\theta/2)$$
$$= a(1 - \cos \theta)$$

$$A = 4 \cdot (1/2) \int_0^{\pi/2} r_2^2 \, d\theta$$

$$= 2 \int_0^{\pi/2} a^2 (1 - \cos \theta)^2 \, d\theta$$

$$= 2a^2 \int_0^{\pi/2} (1 - 2 \cos \theta + \cos^2 \theta) \, d\theta$$

$$= 2a^2 \int_0^{\pi/2} d\theta - 4a^2 \int_0^{\pi/2} \cos \theta \, d\theta + 2a^2 \int_0^{\pi/2} (1/2)(1 + \cos 2\theta) \, d\theta$$

$$= 3a^2 \int_0^{\pi/2} d\theta - 4a^2 \int_0^{\pi/2} \cos \theta \, d\theta + 2a^2 \int_0^{\pi/2} \cos 2\theta \, d\theta$$

$$= 3a^2 \theta \Big]_0^{\pi/2} - 4a^2 \sin \theta \Big]_0^{\pi/2} + a^2 \sin 2\theta \Big]_0^{\pi/2} = (3\pi a^2/2) - 4a^2$$

$$= (3\pi - 8) a^2/2$$

CHAPTER 10 - Miscellaneous Problems

33. $r = a(2 - \cos \theta)$

$$A = 2 \cdot (1/2) \int_0^\pi a^2 (2 - \cos \theta)^2 d\theta$$

$$= a^2 \int_0^\pi (4 - 4 \cos \theta + \cos^2 \theta) d\theta$$

$$- 4a^2 \int_0^\pi d\theta - 4a^2 \int_0^\pi \cos \theta \, d\theta$$

$$+ a^2 \int_0^\pi (1/2)(1 + \cos 2\theta) d\theta$$

$$= 4a^2 \int_0^\pi d\theta - 4a^2 \int_0^\pi \cos \theta \, d\theta + (a^2/2) \int_0^\pi d\theta + (a^2/2) \int_0^\pi \cos 2\theta \, d\theta$$

$$= (9/2)a^2 \theta \Big]_0^\pi - 4a^2 \sin \theta \Big]_0^\pi + (a^2/4) \sin 2\theta \Big]_0^\pi = 9\pi a^2/2$$

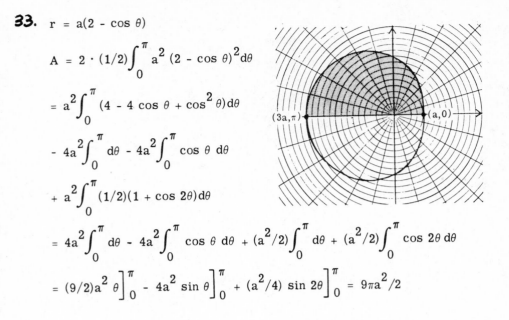

35. $r = 2a \cos \theta$

$$A = 2 \cdot (1/2) \int_0^{\pi/2} 4a^2 \cos^2 \theta \, d\theta$$

$$= 4a^2 \int_0^{\pi/2} (1/2)(1 + \cos 2\theta) d\theta$$

$$= 2a^2 \int_0^{\pi/2} d\theta$$

$$+ 2a^2 \int_0^{\pi/2} \cos 2\theta \, d\theta$$

$$= 2a^2 \theta \Big]_0^{\pi/2} + a^2 \sin 2\theta \Big]_0^{\pi/2}$$

$$= \pi a^2$$

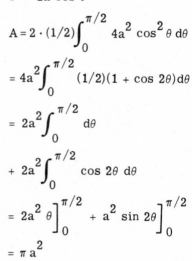

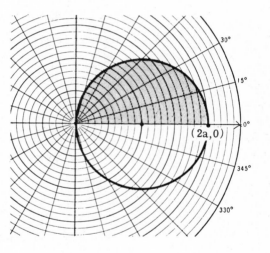

37.

$$r^2 = 2a^2 \sin 3\theta$$

$$A = 12 \cdot (1/2) \int_0^{\pi/6} 2a^2 \sin 3\theta \, d\theta$$

$$= 12a^2 \int_0^{\pi/6} \sin 3\theta \, d\theta$$

$$= -4a^2 \cos 3\theta \Big]_0^{\pi/6}$$

$$= 4a^2$$

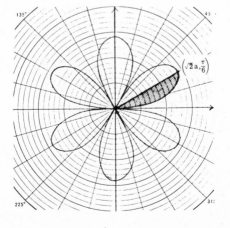

CHAPTER 10 — Miscellaneous Problems

39. $r_1 = a(1 + \sin\theta)$

$r_2 = a\sin\theta$

$A_1 = 2 \cdot a^2/2 \displaystyle\int_{-\pi/2}^{\pi/2} (1 + \sin\theta)^2 \, d\theta$

$= a^2 \displaystyle\int_{-\pi/2}^{\pi/2} (1 + 2\sin\theta + \sin^2\theta) \, d\theta$

$= a^2 \displaystyle\int_{-\pi/2}^{\pi/2} d\theta + 2a^2 \displaystyle\int_{-\pi/2}^{\pi/2} \sin\theta \, d\theta$

$\qquad + (a^2/2) \displaystyle\int_{-\pi/2}^{\pi/2} (1 - \cos 2\theta) \, d\theta$

$= a^2 \displaystyle\int_{-\pi/2}^{\pi/2} d\theta + (a^2/2) \displaystyle\int_{-\pi/2}^{\pi/2} d\theta + 2a^2 \displaystyle\int_{-\pi/2}^{\pi/2} \sin\theta \, d\theta - (a^2/2) \displaystyle\int_{-\pi/2}^{\pi/2} \cos 2\theta \, d\theta$

$= (3a^2/2)\, \theta \Big]_{-\pi/2}^{\pi/2} - 2a^2 \cos\theta \Big]_{-\pi/2}^{\pi/2} - (a^2/4)\sin 2\theta \Big]_{-\pi/2}^{\pi/2} = 3\pi a^2/2$

$A_2 = \pi(a/2)^2 = \pi a^2/4 \quad \therefore \; A = A_1 - A_2 = \big((3/2) - (1/4)\big)\pi a^2$

$= 5\pi a^2/4$

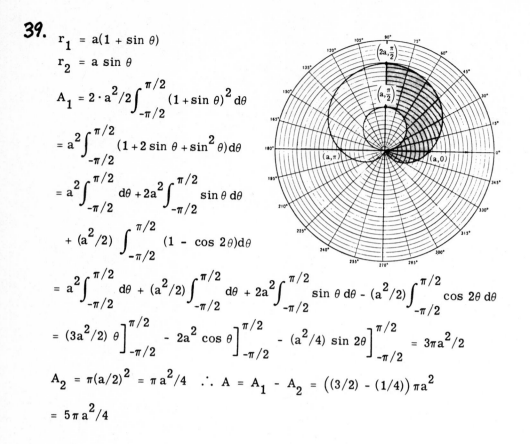

41. $\dfrac{dr}{d\theta} = \sin^3(\theta/4)\cos(\theta/4)$, so $\tan\psi = \dfrac{\sin^4(\theta/4)}{\sin^3(\theta/4)\cos(\theta/4)} = \tan(\theta/4)$.

43.

45.

47. Using the results of problem 44: $\quad \tan\psi_1 = \dfrac{a(1 + \cos\theta)^{-1}}{a\sin\theta(1 + \cos\theta)^{-2}}$

$= \dfrac{1 + \cos\theta}{\sin\theta}$, $\tan\psi_2 = \dfrac{b(1 - \cos\theta)^{-1}}{-b\sin\theta(1 - \cos\theta)^{-2}} = -\dfrac{1 - \cos\theta}{\sin\theta}$. Then,

$1 + \tan\psi_1 \tan\psi_2 = 1 - \dfrac{1 - \cos^2\theta}{\sin^2\theta} = 0$, so $\beta = \pi/2$ and the curves intersect orthogonally.

CHAPTER 10 - Miscellaneous Problems

49. Intersections: $3 \sec \theta = 4 + 4 \cos \theta$, or $4 \cos^2 \theta + 4 \cos \theta - 3 = 0$, or $(2 \cos \theta - 1)(2 \cos \theta + 3) = 0$. Intersection occurs when $2 \cos - 1 = 0$, or $\cos \theta = \frac{1}{2}$, so $\theta = \frac{\pi}{3}$ or $\frac{5\pi}{3}$. Now, if $r = a(1 - \cos \theta)$, then

$x = a(\cos \theta - \cos^2 \theta)$ and $y = a(\sin \theta - \sin \theta \cos \theta)$, so $\frac{dy}{dx} = \frac{dy/d\theta}{dx/d\theta}$

$= \frac{\cos \theta - \cos^2 \theta + \sin^2 \theta}{-\sin \theta + 2 \cos \theta \sin \theta}$. Moreover, the line $r = 3 \sec \theta$ is $r \cos \theta = 3$ or $x = 3$, which makes an angle of $\pi/2$ with the x-axis. At both $\theta = \pi/3$ and $\theta = 5\pi/3$, $2 \cos \theta - 1 = 0$, so the denominator in our expression for $dy/dx = 0$. Since the numerator is nonzero at these points, we conclude that the tangent line is vertical and that the tangent line is, therefore, $r = 3 \sec \theta$.

51. We will use equation (6) from problem 44. $\tan \psi_1 = \dfrac{(1 - \cos \theta)^{-1}}{-\sin \theta (1 - \cos \theta)^{-2}}$

$= \dfrac{1 - \cos \theta}{-\sin \theta}$; $\tan \psi_2 = \dfrac{3(1 + \cos \theta)^{-1}}{3 \sin \theta (1 + \cos \theta)^{-2}} = \dfrac{1 + \cos \theta}{\sin \theta}$. At $\theta = \pi/3$,

$\tan_1 = -\dfrac{1}{\sqrt{3}}$ and $\tan_2 = \sqrt{3}$, so $1 + \tan \psi_1 \tan \psi_2 = 0$. Thus, the curves intersect orthogonally at $(2, \pi/3)$, so the angle between their tangents is $\pi/2$.

53. a) Referring to figure 10.37 and equation (5), we are given that $\psi = \alpha$.
Therefore, $\tan \alpha = \dfrac{r}{dr/d\theta}$, or $(\tan \alpha)\dfrac{dr}{d\theta} = r$, so $\dfrac{dr}{r} = \dfrac{1}{\tan \alpha} d\theta$.
Integrating both sides: $\ln r = \dfrac{\theta}{\tan \alpha} + C$, or $r = Ae^{\theta/\tan \alpha}$, where A is
a (positive) constant. Therefore, the area is $\frac{1}{2}\int_{\theta_1}^{\theta_2} r^2 \, d\theta$

$= \frac{1}{2}\int_{\theta_1}^{\theta_2} A^2 e^{2\theta/\tan \alpha} \, d\theta = \dfrac{A^2 \tan \alpha}{4} e^{2\theta/\tan \alpha} \Big|_{\theta_1}^{\theta_2}$

$= (\dfrac{\tan \alpha}{4})(r_2^2 - r_1^2)$.

b) Since $\dfrac{dr}{d\theta} = \dfrac{r}{\tan \alpha}$, $dr^2 = \dfrac{r^2}{\tan^2 \alpha} \, d\theta^2$, so $ds^2 = r^2 \, d\theta^2 + dr^2$

$= r^2(\dfrac{1 + \tan^2 \alpha}{\tan^2 \alpha}) d\theta^2 = r^2 \dfrac{\sec^2 \alpha}{\tan^2 \alpha} d\theta^2$. Hence, $ds = r \dfrac{\sec \alpha}{\tan \alpha} d\theta$, so the

length is $\int_{\theta_1}^{\theta_2} (\dfrac{\sec \alpha}{\tan \alpha}) Ae^{\theta/\tan \alpha} d\theta = (\sec \alpha)(Ae^{\theta/\tan \alpha}) \Big|_{\theta_1}^{\theta_2} = (\sec \alpha)(r_2 - r_1)$.

CHAPTER 11

SEQUENCES AND INFINITE SERIES

Article 11.1

1. $x_0 = 1$, $x_1 = x_0 + (\frac{1}{2})^{0+1} = \frac{3}{2}$; $x_2 = x_1 + (\frac{1}{2})^{1+1} = \frac{7}{4}$. Continuing we get
$\frac{3}{2}, \frac{7}{4}, \frac{15}{8}, \frac{31}{16}, \frac{63}{32}, \frac{127}{64}$.

3. $x_0 = 2$; $x_{n+1} = x_n/2$. $x_1 = x_0/2 = 1$; $x_2 = x_1/2 = 1/2$; $x_3 = x_2/2 = 1/4$.
Continuing we get $1, \frac{1}{2}, \frac{1}{4}, \frac{1}{8}, \frac{1}{16}, \frac{1}{32}$.

5. $x_0 = x_1 = 1$. $x_{n+2} = x_n + x_{n+1}$. $x_2 = x_0 + x_1 = 2$. $x_3 = x_1 + x_2 = 3$,
$x_4 = x_2 + x_3 = 5$, $x_5 = x_3 + x_4 = 8$, $x_6 = x_4 + x_5 = 13$.

7. $x_{n+1} = x_n + \sin x_n - x_n^2$. $x_0 = 1$, $x_1 = 0.841$, $\ldots$, $x_7 = 0.876726171$.
If $x_{n+1} = x_n$, then $x_{n+1} - x_n = \sin x_n - x_n^2 = 0$, or $\sin x_n = x_n^2$, and we
have solved the problem.

Article 11.2

1. $\lim_{n \to \infty} \frac{1-n}{2} = \lim_{n \to \infty} \frac{-1}{2n} = 0$.

3. $\lim_{n \to \infty} (\frac{1}{3})^n = 0$.

5. $\lim_{n \to \infty} \frac{(-1)^{n+1}}{2n-1} = 0$.

7. $\{\cos(\frac{n\pi}{2})\}$ diverges.

9. $\lim_{n \to \infty} \frac{(-1)^{n-1}}{\sqrt{n}} = 0$.

11. $\lim\limits_{n\to\infty} \dfrac{1}{10^n} = 0.$

13. $\lim\limits_{n\to\infty} (1 + \dfrac{(-1)^n}{n}) = 1.$

15. $\{(-1)^n(1 - \dfrac{1}{n})\}$ diverges.

17. $\lim\limits_{n\to\infty} \dfrac{2n + 1}{1 - 3n} = -\dfrac{2}{3}.$

19. $\lim\limits_{n\to\infty} \sqrt{\dfrac{2n}{n + 1}} = \lim\limits_{n\to\infty} \sqrt{\dfrac{2}{1 + (1/n)}} = \sqrt{\dfrac{2}{1 + 0}} = \sqrt{2}.$

21. $\sin(n\pi) = 0$, for all integers n, so $\lim\limits_{n\to\infty} \sin(n\pi) = 0.$

23. $\{n\pi \cos n\pi\} = \{(-1)^n n\pi\}$ diverges.

25. $\lim\limits_{n\to\infty} \dfrac{n^2}{(n + 1)^2} = \lim\limits_{n\to\infty} \dfrac{2n}{2(n + 1)} = 1.$

27. $\lim\limits_{n\to\infty} \dfrac{1 - 5n^4}{n^4 + 8n^3} = \lim\limits_{n\to\infty} \dfrac{(1/n)^4 - 5}{1 - (8/n)} = \dfrac{0 - 5}{1 - 0} = -5.$

29. $\lim\limits_{n\to\infty} \tanh(n) = \lim\limits_{n\to\infty} \dfrac{e^n - e^{-n}}{e^n + \epsilon^{-n}} = \lim\limits_{n\to\infty} \dfrac{1 - e^{-2n}}{1 + e^{-2n}} = 1.$

31. $\lim\limits_{n\to\infty} \dfrac{2(n + 1) + 1}{2n + 1} = 1.$

33. $\lim\limits_{n\to\infty} 5 = 5.$

35. $\lim\limits_{n\to\infty} (0.5)^n = \lim\limits_{n\to\infty} (\dfrac{1}{2})^n = 0.$

37. $0 \le \dfrac{n^n}{(n + 1)^{n+1}} \le \dfrac{1}{n + 1} \to \lim\limits_{n\to\infty} \dfrac{n^n}{(n + 1)^{n+1}} = 0.$

39. $\lim\limits_{n\to\infty} \sqrt{2 - 1/n} = \sqrt{2}.$

41. (We use: $d/dx(a^x) = (\ln a)a^x$.)

$\lim\limits_{n\to\infty} \dfrac{3^n}{n^3} = \lim\limits_{n\to\infty} \dfrac{\ln 3 \cdot 3^n}{3n^2} = \lim\limits_{n\to\infty} \dfrac{(\ln 3)^2 \cdot 3^n}{6n} = \lim\limits_{n\to\infty} \dfrac{(\ln 3)^3 3^n}{6} = \infty.$

43. $\lim\limits_{n\to\infty} (\ln n - \ln(n + 1)) = \lim\limits_{n\to\infty} \ln(\dfrac{n}{n + 1}) = \ln(1) = 0.$

45. $\lim\limits_{n\to\infty} \dfrac{n^2 - 2n + 1}{n - 1} = \lim\limits_{n\to\infty} \dfrac{(n - 1)(n - 1)}{n - 1} = \lim\limits_{n\to\infty} (n - 1) = \infty.$

47. $\lim\limits_{n\to\infty} (-\frac{1}{2})^n = 0.$

49. $\lim\limits_{n\to\infty} \arctan(n) = \pi/2.$

51. $\lim\limits_{n\to\infty} \dfrac{\sin(1/n)}{(1/n)} = 1.$

53. $\lim\limits_{n\to\infty} (\dfrac{n}{2n-1} \cdot \dfrac{\sin(1/n)}{(1/n)}) = \dfrac{1}{2}.$

55. $0 \le \dfrac{n!}{n^n} = (\dfrac{1}{n} \cdot \dfrac{2}{n} \cdot \dfrac{3}{n} \cdot \ldots \cdot \dfrac{n}{n}) \le \dfrac{1}{n} \to \lim\limits_{n\to\infty} \dfrac{n!}{n^n} = 0.$

57. $\lim\limits_{n\to\infty} x^{1/n} = x^0 = 1$, whenever $x > 0$. If we begin with 10 or 0.1, and
repeatedly strike the square-root key, then eventually the display will
show 1.0. It does not matter which positive number we begin with.

59. $\lim\limits_{n\to\infty} nf(1/n) = \lim\limits_{n\to\infty} \dfrac{f(1/n)}{(1/n)} = \lim\limits_{n\to\infty} \dfrac{f'(1/n)(-1/n^2)}{(-1/n^2)} = f'(0).$

61. $f(x) = e^x - 1$, $f(0) = 0$, $f'(0) = 1 \to \lim\limits_{n\to\infty} n(e^{1/n} - 1) = 1.$

63. By definition, either a_n converges to a limit L, or a_n diverges. Suppose
    a_n converges to L. Let $\varepsilon = 1 > 0$. Then for some N_1, larger than n_0,
    $n > N_1$ implies (1) $|a_n - L| < 1$. (2) $a_n = f(n)$. Since $\lim\limits_{n\to\infty} f(x) = +\infty$, there
    is a number N_2 such that $x > N_2$ implies $f(x) > L + 2$.
    Let N be the larger of N_1 and N_2. $n > N$ implies (4) $|a_n - L| = |f(n) - L| < 1$
    from (1) and (2), (5) $|f(n) - L| > 2$ from (3). Since (4) and (5) cannot
    both be true, a_n cannot converge. Hence a_n diverges. (Note: If $f(x)$
    diverges, then a_n may converge. $f(x) = \sin(\pi x)$ is an example. $a_n = 0$ for
    all n, so it converges, but $f(x)$ does not converge.)

65. Suppose $\lim\limits_{n\to\infty} a_n = L$, $f(x)$ is continuous at L and $\varepsilon > 0$ is given. Then
    there exists $\delta > 0$ such that $|f(x) - f(L)| < \varepsilon$ for all x for which
    $0 < |x - L| < \delta$. Since $\lim\limits_{n\to\infty} a_n = L$, there exists N such that for all $n \ge N$,
    $|a_n - L| < \delta$ and hence $|f(a_n) - f(L)| < \varepsilon$, as required.

Article 11.3

1. $\lim\limits_{n\to\infty} \dfrac{1 + \ln n}{n} = \lim\limits_{n\to\infty} (\dfrac{1}{n} + \dfrac{\ln n}{n}) = 0 + 0 = 0.$

3. $\lim\limits_{n\to\infty} \dfrac{(-4)^n}{n!} = 0$, by formula 6, for $x = -4$.

5. $\lim_{n\to\infty} (0.5)^n = 0$, by formula 4, for $x = 0.5$

7. $\lim_{n\to\infty} (1 + \frac{7}{n})^n = e^7$, by formula 5, for $x = 7$.

9. $\lim_{n\to\infty} \frac{\ln(n+1)}{n} = \lim_{n\to\infty} \frac{1/(n+1)}{1} = 0$, by L'Hôpital's rule.

11. $\lim_{n\to\infty} (\frac{n!}{10^{6n}}) = \lim_{n\to\infty} \{\frac{(10^6)^n}{n!}\}^{-1} = \infty$.

13. $\lim_{n\to\infty} (n)^{1/2n} = (\lim_{n\to\infty} n^{1/n})^{1/2} = 1$.

15. $\lim_{n\to\infty} \frac{1}{3^{2n-1}} = 0$.

17. $\lim_{n\to\infty} (\frac{n}{n+1}) = \lim_{n\to\infty} [(1 + \frac{-1}{n+1})^{n+1}]^{(\frac{n}{n+1})} = \lim_{n\to\infty} (e^{-1})(1 - \frac{1}{n+1}) = e^{-1}$.

19. $0 < \frac{\ln(2n+1)}{n} \leq \frac{\ln(3n)}{n} = \frac{\ln 3}{n} + \frac{\ln n}{n}$. Hence, $\lim_{n\to\infty} \frac{\ln(2n+1)}{n} = 0$.

21. $(2n+1)^{1/n} > (n)^{1/n} \to 1$, as $n \to \infty$. $(2n+1)^{1/n} \leq (3n)^{1/n} = (3^{1/n})(n^{1/n}) \to 1$, as $n \to \infty$. Hence $\lim_{n\to\infty} (2n+1)^{1/n} = 1$.

$\lim_{n\to\infty} (\frac{x^n}{2n+1})^{1/n} = \lim_{n\to\infty} \frac{x}{(2n+1)^{1/n}} = \frac{x}{1} = x$.

23. $\lim_{n\to\infty} (n^2 + n)^{1/n} \leq \lim_{n\to\infty} (2n^2)^{1/n} = \lim_{n\to\infty} (2^{1/n})(n^{1/n})^2 = 1$.

$\lim_{n\to\infty} (n^2 + n)^{1/n} \geq \lim_{n\to\infty} n^{1/n} = 1$. Hence, $(n^2 + n)^{1/n} \to 1$.

25. $\lim_{n\to\infty} (\frac{3}{n})^{1/n} = \lim_{n\to\infty} \frac{(3^{1/n})}{(n^{1/n})} = 1$.

27. $\lim_{n\to\infty} (1 + \frac{-1}{n})^n = e^{-1}$ by formula 5, for $x = -1$.

29. $\lim_{n\to\infty} \frac{\ln(n^2)}{n} = \lim_{n\to\infty} \frac{2\ln(n)}{n} = 0$.

31. As $n \to \infty$, $n^{1/n} \to 1$, and $\ln n \to \infty$. Hence $\lim_{n\to\infty} \frac{\ln n}{n^{1/n}} = \infty$.

33. $\lim_{n\to\infty} [\int_1^n \frac{dx}{x^p}] = \lim_{n\to\infty} [\frac{x^{(1-p)}}{(1-p)}]_1^n = \frac{1}{p-1} + \lim_{n\to\infty} (\frac{1}{n})^{p-1} = \frac{1}{p-1}$.

35. $N \geq 9124$

37. Pre-proof computations: $c > 0$. Assume $0 < \varepsilon < 1$. We need $\dfrac{1}{n^c} < \varepsilon$.

$n^c > \dfrac{1}{\varepsilon}$, $n > \left(\dfrac{1}{\varepsilon}\right)^c = \varepsilon^{-c}$. Proof: Let $\varepsilon > 0$ be given. Let $N = \varepsilon^{-c}$. Then for $n > N$, $0 < \dfrac{1}{n^c} < \varepsilon$, so $\dfrac{1}{n^c}$ converges to zero.

Article 11.4

1. Geometric series. $s_n = 2\left(\dfrac{1 - (1/3)^{n+1}}{1 - (1/3)}\right)$; $\lim\limits_{n \to \infty} s_n = 3$.

3. Geometric series. $s_n = \left(\dfrac{1 - (1/e)^{n+1}}{1 - (1/e)}\right)$; $\lim\limits_{n \to \infty} s_n = \dfrac{1}{1 - (1/e)} = \dfrac{1}{e - 1}$.

5. $r = -2$, $a_0 = 1$. $s_n = \left(\dfrac{1 - (-2)^{n+1}}{1 - (-2)}\right) = \dfrac{1}{3}(1 - (-2)^{n+1})$; $\lim\limits_{n \to \infty} s_n$ does not exist.

7. $s_n = (\ln 1 - \ln 2) + (\ln 2 - \ln 3) + \ldots + (\ln n - \ln(n + 1)) = -\ln(n + 1)$; $\lim\limits_{n \to \infty} s_n$ does not exist.

9. $\displaystyle\sum_{n=1}^{\infty} \dfrac{1}{(n + 1)(n + 2)} = \sum_{n=-2}^{\infty} \dfrac{1}{(n + 4)(n + 5)} = \sum_{n=0}^{\infty} \dfrac{1}{(n + 2)(n + 3)}$

$= \displaystyle\sum_{n=5}^{\infty} \dfrac{1}{(n - 3)(n - 2)}$.

11. $S = \dfrac{1}{1 - (1/4)} = \dfrac{4}{3}$

13. $S = \dfrac{(7/4)}{1 - (1/4)} = \dfrac{7}{3}$

15. $\displaystyle\sum_{n=0}^{\infty} \left(\dfrac{5}{2^n} + \dfrac{1}{3^n}\right) = 5 \sum_{n=0}^{\infty} \dfrac{1}{2^n} + \sum_{n=0}^{\infty} \dfrac{1}{3^n} = 5\left(\dfrac{1}{1 - 1/2}\right) + \left(\dfrac{1}{1 - 1/3}\right) = \dfrac{23}{2}$.

17. $\displaystyle\sum_{n=0}^{\infty} \left(\dfrac{2}{5}\right)^n = \dfrac{1}{1 - 2/5} = \dfrac{5}{3}$.

19. $\displaystyle\sum_{n=1}^{\infty} \left[\dfrac{1}{4n - 3} - \dfrac{1}{4n + 1}\right] = \left(1 - \dfrac{1}{5}\right) + \left(\dfrac{1}{3} - \dfrac{1}{9}\right) + \ldots + \left(\dfrac{1}{4n - 3} - \dfrac{1}{4n + 1}\right) + \ldots = 1$.

21. $\displaystyle\sum_{n=3}^{\infty} \left(\dfrac{1}{4n - 3} - \dfrac{1}{4n + 1}\right) = \left(\dfrac{1}{9} - \dfrac{1}{13}\right) + \left(\dfrac{1}{13} - \dfrac{1}{17}\right) + \ldots = \dfrac{1}{9}$.

23. a) $0.234234, \ldots = 0.234(1 + 10^{-3} + 10^{-6} + \ldots) = 0.234\left(\dfrac{1}{1 - (0.001)}\right)$

   $= 234/999$.

   b) If the decimal form of m is $m = 0.a_1 a_2 \ldots a_n a_1 a_2 \ldots a_n \ldots$, then

   $m = a_1 a_2 \ldots a_n \left(\dfrac{1}{10^n} + \dfrac{1}{10^{2n}} + \dfrac{1}{10^{3n}} + \ldots\right)$

$$= a_1 a_2 \cdots a_n \left(\frac{1/10^n}{1 - (1/10^n)} \right) = \frac{a_1 a_2 \cdots a_n}{10^n - 1}.$$ m is expressed as the ratio of

two integers so the answer is yes.

25. $\displaystyle\sum_{n=0}^{\infty} \left(\frac{1}{\sqrt{2}}\right)^n = \frac{1}{1 - 1/\sqrt{2}} = \frac{\sqrt{2}}{\sqrt{2} - 1}$

27. $\displaystyle\sum_{n=1}^{\infty} (-1)^{n+1} \frac{3}{2^n} = \frac{3/2}{1 - (-1/2)} = 1.$

29. $\displaystyle\sum_{n=0}^{\infty} \cos n\pi = \sum_{n=0}^{\infty} (-1)^n$ diverges because $\displaystyle\lim_{n\to\infty} (-1)^n \neq 0.$

31. $\displaystyle\sum_{n=0}^{\infty} e^{-2n} = \frac{1}{1 - e^{-2}} = \frac{e^2}{e^2 - 1}$

33. $\displaystyle\sum_{n=1}^{\infty} (-1)^{n+1} n$ diverges because $\displaystyle\lim_{n\to\infty} (-1)^{n+1} n \neq 0.$

35. $\displaystyle\sum_{n=0}^{\infty} \frac{2^n - 1}{3^n} = \sum_{n=0}^{\infty} \left(\frac{2}{3}\right)^n - \sum_{n=0}^{\infty} \left(\frac{1}{3}\right)^n = \frac{1}{1 - 2/3} - \frac{1}{1 - 1/3} = \frac{3}{2}.$

37. $\displaystyle\sum_{n=0}^{\infty} \frac{n!}{(1000)^n}$ diverges because $\displaystyle\lim_{n\to\infty} \frac{n!}{(1000)^n} > \frac{(1000)!}{(1000)^{1000}} > 0.$

39. $\displaystyle\sum_{n=0}^{\infty} (-x)^n = \frac{1}{1 - (-x)} = \frac{1}{1 + x}$ for $|x| < 1$ and $a = 1$, $r = -x.$

41. The area of each square is one-half of the area of the preceeding square. So the sum of areas $= 4 + 4(1/2) + 4(1/2) + \cdots = \dfrac{4}{1 - 1/2} = 8.$

43. $\displaystyle\sum_{n=1}^{\infty} n$, and $\displaystyle\sum_{n=1}^{\infty} (-n)$ both diverge but the term-by-term sequence is zero, which converges.

45. Let $a_n = 2^{-n+1}$, $b_n = 2^{-n+1}$. Then $\displaystyle\sum_{n=1}^{\infty} a_n = 2$, $\displaystyle\sum_{n=1}^{\infty} b_n = 2$ but $\displaystyle\sum_{n=1}^{\infty} (a_n/b_n) = \sum_{n=1}^{\infty} (1) = \infty.$

47. If $\displaystyle\sum_{n=1}^{\infty} a_n$ converges and $a_n \neq 0$, then $\displaystyle\lim_{n\to\infty} a_n = 0$, $\displaystyle\lim_{n\to\infty} \frac{1}{a_n} \neq 0$ and hence $\displaystyle\sum_{n=1}^{\infty} \left(\frac{1}{a_n}\right)$ diverges.

Article 11.5

1. Converges. Geometric series with $r = 1/10 < 1.$

3. $\displaystyle\sum_{n=1}^{\infty} \frac{\sin^2 n}{2^n} \leq \sum_{n=1}^{\infty} \frac{1}{2^n} = 1.$ Hence $\displaystyle\sum_{n=1}^{\infty} \frac{\sin^2 n}{2^n}$ converges.

5. $\sum\limits_{n=1}^{\infty} \dfrac{n^3}{2^n}$ converges by the ratio test: $\lim\limits_{n\to\infty} \left(\dfrac{(n+1)^3}{2^{n+1}} \cdot \dfrac{2^n}{n^3}\right) = \dfrac{1}{2} < 1.$

7. $\sum\limits_{n=1}^{\infty} \dfrac{\ln n}{n}$ diverges by the integral test: $\int_2^{\infty} \left(\dfrac{\ln x}{x}\right) dx = \left[\dfrac{(\ln x)^2}{2}\right]_2^{\infty} = \infty.$

9. $\sum\limits_{n=1}^{\infty} \left(\dfrac{2}{3}\right)^n$ converges to 2, being a geometric series with $a = 2/3$ and $r = 2/3.$

11. $\infty = \sum\limits_{n=1}^{\infty} \dfrac{1}{n+1} < \sum\limits_{n=1}^{\infty} \dfrac{1}{1 + \ln n}$ and hence $\sum\limits_{n=1}^{\infty} \dfrac{1}{1 + \ln n}$ diverges.

13. $\sum\limits_{n=1}^{\infty} \dfrac{2^n}{n+1} \geq \sum\limits_{n=1}^{\infty} \dfrac{1}{n+1} = \infty.$ Therefore $\sum\limits_{n=1}^{\infty} \dfrac{2^n}{n+1}$ diverges.

15. $\sum\limits_{n=1}^{\infty} \dfrac{-n^2}{2^n}$ converges by the ratio test: $\lim\limits_{n\to\infty} \left[\dfrac{-(n+1)^2}{2^{n+1}} \cdot \dfrac{2^n}{-n^2}\right] = \dfrac{1}{2} < 1.$

17. $\sum\limits_{n=1}^{\infty} \dfrac{1}{\sqrt{n^3 + 2}} \leq \sum\limits_{n=1}^{\infty} \dfrac{1}{n^{3/2}} < \infty$ by p-series and comparison tests.

19. $\sum\limits_{n=1}^{\infty} \dfrac{(n+1)(n+2)}{n!}$ converges by the ratio test:

$\lim\limits_{n\to\infty} \dfrac{(n+2)(n+3)}{(n+1)!} \cdot \dfrac{n!}{(n+1)(n+2)} = 0.$

21. $\infty = \sum\limits_{n=1}^{\infty} \dfrac{1}{n+1} < \sum\limits_{n=1}^{\infty} \dfrac{n}{n^2 + 1}$ and hence $\sum\limits_{n=1}^{\infty} \dfrac{n}{n^2 + 1}$ diverges.

23. $\sum\limits_{n=1}^{\infty} \left(1 + \dfrac{1}{n}\right)^n$ diverges since $\lim\limits_{n\to\infty} \left(1 + \dfrac{1}{n}\right)^n = e \neq 0.$

25. $\sum\limits_{n=1}^{\infty} \dfrac{(n+3)!}{3!\,n!\,3^n}$ converges by the ratio test:

$\lim\limits_{n\to\infty} \dfrac{(n+4)!\,3!\,n!\,3^n}{3!\,(n+1)!\,3^{n+1}\,(n+3)!} = \dfrac{1}{3} < 1.$

27. $\sum\limits_{n=1}^{\infty} \dfrac{1}{(2n+1)!}$ converges by the ratio test: $\lim\limits_{n\to\infty} \dfrac{(2n+1)!}{(2n+3)!} = 0 < 1.$

29. $\sum\limits_{n=1}^{\infty} \dfrac{n!}{n^n}$ converges by the ratio test:

$\lim\limits_{n\to\infty} \left(\dfrac{(n+1)!}{(n+1)^{n+1}} \cdot \dfrac{n^n}{n!}\right) = \lim\limits_{n\to\infty} \left(\dfrac{n}{n+1}\right)^n = \lim\limits_{n\to\infty} \left(1 + \dfrac{1}{n+1}\right)^n = e^{-1} < 1.$

31. $2n - 1 < 2n$, so $1/(2n-1) > 1/(2n).$

$\sum\limits_{n=1}^{\infty} \dfrac{1}{2n-1} > \sum\limits_{n=1}^{\infty} \dfrac{1}{2n} = \dfrac{1}{2} \sum\limits_{n=1}^{\infty} \dfrac{1}{n} = \dfrac{1}{2}(\infty) = \infty.$

33. By the ratio test, $\sum\limits_{n=1}^{\infty} \dfrac{x^{2n+1}}{n^2}$ converges if $x^2 < 1$ or $|x| < 1.$ For $|x| = 1,$ the series converges by the p-series test. For $|x| > 1,$ the series diverges.

CHAPTER 11 - Article 11.5

35. $\displaystyle\sum_{n=1}^{\infty} \frac{1}{n^2} \approx \sum_{n=1}^{10} \frac{1}{n^2} + \frac{1}{10} \approx 1.649767731.$

37. $\displaystyle\sum_{n=1}^{\infty} \frac{1}{n^4} \approx \sum_{n=1}^{27} \frac{1}{n^4} + \frac{n^{-3}}{3} \approx 1.082324151.$

39. $\displaystyle\sum_{n=1}^{\infty} \frac{1}{nx} = \frac{1}{x}\sum_{n=1}^{\infty} \left(\frac{1}{n}\right)$ and $\displaystyle\sum_{n=1}^{\infty}\frac{1}{n}$ diverges. So, $\displaystyle\sum_{n=1}^{\infty}\frac{1}{nx}$ diverges.

41. Since Σa_n converges, $\displaystyle\lim_{n\to\infty} a_n = 0$. So for some N, $n \geq N$ implies $a_n \leq 1$ and

$a_n b_n \leq b_n$. $\displaystyle\sum_{n=1}^{\infty} a_n b_n = \sum_{n=1}^{N} a_n b_n + \sum_{n=N+1}^{\infty} a_n b_n \leq \sum_{n=1}^{N} a_n b_n + \sum_{n=N+1}^{\infty} b_n < \infty,$

because Σb_n converges.

43. Let $\{a_n\}$ be a decreasing sequence of positive terms that converges to 0.

Let $A_n = \displaystyle\sum_{k=1}^{n} a_k$ and $B_n = \displaystyle\sum_{k=1}^{n} 2^k a_{2^k}$.

(i) $B_n = (2a_2) + (4a_4) + (8a_8) + \ldots + (2^n a_{2^n})$

$= (2a_2 + (2a_4 + 2a_4) + (4a_8 + 4a_8) + \ldots + (2^{n-1}a_{2^n} + 2^{n-1}a_{2^n})$

$\leq (2a_1) + (2a_2 + 2a_3) + (2a_4 + \ldots + 2a_n) + \ldots$

$+ (2a_{2^{n-1}} + 2a_{2^{n-1}+1} + \ldots + 2a_{2^n-1}) = 2A_{2^n-1} \leq \displaystyle\sum_{k=1}^{\infty} a_k$

Therefore, if $\displaystyle\sum_{k=1}^{n} a_k$ converges, then $\displaystyle\sum_{k=1}^{n} 2^k a_{2^k}$ converges.

(ii) $A_n = a_1 + a_2 + a_3 + \ldots + a_n = a_1 + (a_2 + a_3) + (a_4 + a_5 + a_6 + a_7)$

$+ (a_8 + a_9 + \ldots + a_{15}) + \ldots + a_n \leq a_1 + B_n \leq a_1 + \displaystyle\sum_{k=1}^{\infty} 2^k a_{2^k}$

Therefore, if $\displaystyle\sum_{k=1}^{\infty} 2^k a_{2^k}$ converges, then $\displaystyle\sum_{k=1}^{\infty} a_n$ converges.

45. a) Let $f(x) = \frac{1}{x}$. From inequality (8) we have

$\ln(n) < \ln(n + 1) = \displaystyle\int_1^{n+1} \frac{1}{x}\, dx \leq 1 + \frac{1}{2} + \frac{1}{3} + \ldots + \frac{1}{n}$

$\leq 1 + \displaystyle\int_1^n \frac{1}{x}\, dx = 1 + \ln n$ or $0 < 1 + \frac{1}{2} + \frac{1}{3} + \ldots + \frac{1}{n} - \ln(n) < 1.$

b) Let $a_n = 1 + \frac{1}{2} + \frac{1}{3} + \ldots + \frac{1}{n} - \ln(n)$. Then $a_{n+1} - a_n$

$= \frac{1}{n+1} - (\ln(n + 1) - \ln n)$. From the graph of $f(x) = 1/x$ it is clear

that $\frac{1}{n+1} \cdot 1 < \displaystyle\int_n^{n+1} \frac{1}{x}\, dx = \ln(n + 1) - \ln(n)$ or

$\frac{1}{n+1} - (\ln(n + 1) - \ln n) < 0$. Therefore, the sequence $\{a_n\}$ is a

decreasing sequence that is bounded below by 0. Let $\gamma = \displaystyle\lim_{n\to\infty} a_n$. Then γ

is the Euler's constant.

Article 11.6

1. $\sum\limits_{n=1}^{\infty} \dfrac{1}{n^2}$ converges absolutely by the p-series test for $p = 2$.

3. $\sum\limits_{n=1}^{\infty} \left| \dfrac{1-n}{n^2} \right| = \sum\limits_{n=1}^{\infty} \left(\dfrac{1}{n} - \dfrac{1}{n^2} \right) \leq \sum\limits_{n=1}^{\infty} \dfrac{1}{2n} = \infty$. The series diverges.

5. $\sum\limits_{n=1}^{\infty} \dfrac{1}{(n+1)^2} \leq \sum\limits_{n=1}^{\infty} \dfrac{1}{n^2} < \infty$. The series converges absolutely.

7. $\sum\limits_{n=1}^{\infty} \dfrac{\cos n\pi}{n^{3/2}} = \sum\limits_{n=1}^{\infty} \dfrac{1}{n^{3/2}} < \infty$. The series converges by the p-series test for $p = 3/2$.

9. $\sum\limits_{n=0}^{\infty} \left| \dfrac{(-1)^n}{(2n)!} \right| = \sum\limits_{n=0}^{\infty} \dfrac{1}{(2n)!}$ converges by the ratio test.

11. $\sum\limits_{n=2}^{\infty} \left| \dfrac{(-1)^n n}{n+1} \right| = \sum\limits_{n=2}^{\infty} \dfrac{n}{n+1} \geq \sum\limits_{n=2}^{\infty} \dfrac{1}{n} = \infty$.

13. $\sum\limits_{n=1}^{\infty} (5)^{-n} = \dfrac{1}{1 - 1/5} = 5/4$.

15. $\sum\limits_{n=1}^{\infty} \left| \dfrac{(-100)^n}{n!} \right| = \sum\limits_{n=1}^{\infty} \dfrac{(100)^n}{n!}$ converges by the ratio test.

17. $\sum\limits_{n=1}^{\infty} \left| \dfrac{2-n}{n^3} \right| \leq \sum\limits_{n=1}^{\infty} \dfrac{1}{n^2} < \infty$.

19. If $\sum\limits_{n=1}^{\infty} |a_n|$ converges, then $\sum\limits_{n=1}^{\infty} a_n$ converges by theorem 1. If follows that if $\sum\limits_{n=1}^{\infty} a_n$ diverges then $\sum\limits_{n=1}^{\infty} |a_n|$ also diverges.

21. Let $A = \sum\limits_{n=1}^{\infty} |a_n|$, $B = \sum\limits_{n=1}^{\infty} |b_n|$.

a) Since for all n, $|a_n + b_n| \leq |a_n| + |b_n|$, $\sum\limits_{n=1}^{\infty} |a_n + b_n| \leq \sum\limits_{n=1}^{\infty} |a_n| + |b_n|$ $= A + B$. $\Sigma|a_n + b_n|$ is a bounded increasing function, so it converges.

b) For all n, $|a_n - b_n| \leq |a_n| + |b_n|$, so by the same reasoning, $\Sigma|a_n - b_n|$ converges.

c) Since $\sum\limits_{n=1}^{\infty} |a_n|$ converges, theorem 1 says $\sum\limits_{n=1}^{\infty} a_n$ converges. Hence, $\sum\limits_{n=1}^{\infty} ka_n = k \sum\limits_{n=1}^{\infty} a_n$ also converges.

23. For all n, $|b_n| \leq |a_n|$, so $\sum\limits_{n=1}^{\infty} |b_n| \leq \sum\limits_{n=1}^{\infty} |a_n| < \infty$. $\Sigma|b_n|$ converges. By theorem 1, Σb_n converges. The second statement follows if we use $(-a_n)$ for a_n.

25. $a_k b_{n-k} = \dfrac{1}{2^k 3^{n-k}} \le \dfrac{1}{2^n}$. $C_n = \sum\limits_{k=0}^{\infty} a_k b_{n-k} \le \dfrac{n+1}{2^n}$. $\lim\limits_{n\to\infty} C_{n+1}/C_n = \dfrac{1}{2} < 1$, so by

the ratio test $\sum\limits_{n=1}^{\infty} C_n$ converges. Since $|C_n| = C_n$, it converges absolutely.

Article 11.7

1. $\{1/n^2\}$ is a decreasing sequence, so the series converges.

3. $S_n = \sum\limits_{k=1}^{n} (-1)^{n+1}$ is 1 for n odd and 0 for n even. Hence the series diverges.

5. $\{(\sqrt{n} + 1)/(n + 1)\}$ is a decreasing sequence, so the series converges.

7. $\sum\limits_{n=1}^{\infty} n^{-3/2}$ converges, so $\sum\limits_{n=1}^{\infty} n^{-3/2}(-1)^{n+1}$ converges.

9. $\{\ln(1 + \frac{1}{n})\}$ is a decreasing sequence, so by Leibniz's Theorem, the series converges.

11. $\sum\limits_{n=1}^{\infty} (-1)^{n+1} (0.1)^n$ converges absolutely.

13. $\sum\limits_{n=1}^{\infty} (-1)^{n+1} \dfrac{n}{n^3 + 1}$ converges absolutely.

15. $\sum\limits_{n=1}^{\infty} (-1)^n \dfrac{1}{n + 3}$ converges conditionally.

17. $\sum\limits_{n=1}^{\infty} (-1)^{n+1} \dfrac{3 + n}{5 + n}$ diverges since $\lim\limits_{n\to\infty} \dfrac{3 + n}{5 + n} = 1 \ne 0$.

19. $\sum\limits_{n=1}^{\infty} (-1)^{n+1} \dfrac{1 + n}{n^2}$ converges conditionally.

21. $\sum\limits_{n=1}^{\infty} n^2 (\frac{2}{3})^n$ converges absolutely by the ratio test.

23. $\sum\limits_{n=1}^{\infty} (-1)^n \dfrac{\arctan(n)}{n^2 + 1}$ converges absolutely since $-\pi/2 < \arctan n < \pi/2$.

25. $\sum\limits_{n=1}^{\infty} [\frac{1}{n} - \frac{1}{2n}] = \sum\limits_{n=1}^{\infty} \frac{1}{2n} = \infty$. The series diverges.

27. $\sqrt{n+1} - \sqrt{n} = \dfrac{\sqrt{n+1} + \sqrt{n}}{\sqrt{n+1} + \sqrt{n}} \cdot (\sqrt{n+1} - \sqrt{n}) = \dfrac{n + 1 - n}{\sqrt{n+1} + \sqrt{n}} = \dfrac{1}{\sqrt{n+1} + \sqrt{n}} > \dfrac{1}{3\sqrt{n}}$.

$\{\dfrac{1}{\sqrt{n+1} + \sqrt{n}}\}$ is a decreasing sequence that converges to zero, so by Leibniz's theorem, $\sum\limits_{n=1}^{\infty} (-1)^{n+1} (\sqrt{n+1} - \sqrt{n})$ converges. Since $\sum\limits_{n=1}^{\infty} \dfrac{1}{3\sqrt{n}}$ diverges, the convergence is conditional.

29. $a_5 = 1/5$, Error $\leq 1/5$.

31. $a_5 = (10^{-10})/5 = 2 \times 10^{-11}$. Error $\leq 2 \times 10^{-11}$.

33. $\cos 1 \approx 1 - \frac{1}{2!} + \frac{1}{4!} + \frac{1}{6!} + \frac{1}{8!} \approx 0.540302579$.

35. a) Condition 2. $(a_n \geq a_{n+1})$ fails for all odd n.

   b) Sum is $\sum\limits_{n=1}^{\infty} (\frac{1}{3^n} - \frac{1}{2^n}) = \sum\limits_{n=1}^{\infty} \frac{1}{3^n} - \sum\limits_{n=1}^{\infty} \frac{1}{2^n} = \frac{1}{2} - 1 = -\frac{1}{2}$.

37. $\sum\limits_{k=1}^{\infty} b_k = \sum\limits_{k=1}^{\infty} b_k + (b_{n+1} + b_{n+2}) + (b_{n+3} + b_{n+4}) + \ldots$ If $\sum\limits_{k=1}^{\infty} b_k$ is an

   alternating convergent series approximated by $\sum\limits_{k=1}^{n} b_k$, then the remainder

   has the same sign as b_{n+1} because each $(b_{n+k} + b_{n+k+1})$ has the same sign

   as b_{n+1}.

39. Let $a_n = b_n = (-1)^{n+1}/\sqrt{n}$. The sequence $\{1/\sqrt{n}\}$ is decreasing and converges

   to zero, so $\sum\limits_{n=i}^{\infty} a_n$ and $\sum\limits_{n=1}^{\infty} b_n$ converge. However, $\sum\limits_{n=1}^{\infty} a_n b_n = \sum\limits_{n=1}^{\infty} 1/n$

   diverges.

41. The first five pairs of (a_n, s_n) are $(-\frac{1}{2}, -\frac{1}{2})$, $(1, \frac{1}{2})$, $(-\frac{1}{4}, \frac{1}{4})$, $(-\frac{1}{6}, \frac{1}{12})$,

   $(-\frac{1}{8}, -\frac{1}{24})$. The convergence to $s_n = -\frac{1}{2}$ is exceedingly slow.

Miscellaneous Problems, Chapter 11

1. $\sum_{k=2}^{n} \ln(1 - 1/k^2) = \sum_{k=2}^{n} [\ln(1 - 1/k) + \ln(1 + 1/k)]$

   $= \sum_{k=2}^{n} [\ln(k - 1) - \ln k + \ln(k + 1) - \ln k] = \ln(n + 1) - \ln n - \ln 2$

   $= \ln((n + 1)/2n) = \ln(1/2 + 1/2n) \to \ln(1/2)$ as $n \to \infty$.

   $\therefore \sum_{2}^{\infty} \ln(1 - 1/k^2) = -\ln 2$.

3. $\sum_{1}^{\infty} (x_{k+1} - x_k) = \lim\limits_{n\to\infty} \sum_{1}^{n} (x_{k+1} - x_k) = \lim\limits_{n\to\infty} (x_{n+1} - x_1) = \lim\limits_{n\to\infty} x_{n+1} - x_1$.

   Therefore, the series and the sequence either both converge or both diverge.

5. $\frac{1}{1 - x} = \frac{1/x}{\frac{1}{x} - 1} = -1/x (\frac{1}{1 - (\frac{1}{x})}) = -\frac{1}{x}(1 + (\frac{1}{x}) + (\frac{1}{x})^2 + \ldots) = -\frac{1}{x} - \frac{1}{x^2} - \frac{1}{x^3} - \ldots$

   The expression is valid because $(1/x) < 1$.

7. $\lim\limits_{n\to\infty} (e^n - e^{-n})/(e^n + e^{-n}) = 1 \neq 0$. Series diverges.

9. $\sum_1^\infty n/[2(n + 1)(n + 2)] = \sum_1^\infty 1/[2(1 + 1/n)(n + 2)] \le \sum_1^\infty 1/[4(n + 2)] = \infty$.

11. $\sum_1^\infty 1/[n(\ln n)^2]$ converges, since $\int_2^\infty [(1/n)dn]/(\ln n)^2 = -1/(\ln n)\big|_2^\infty < \infty$.

13. $\sum_1^\infty n/(10^3 n^2 + 1) \ge \sum_1^\infty 1/[2(10)^3 n] = \infty$.

15. $\sum_1^\infty 1/[n\sqrt{n^2 + 1}] \le \sum_1^\infty 1/n^2 < \infty$ by p-series test p = 2.

17. $\dfrac{1 \cdot 3 \cdot 5 \ldots (2n - 1)}{2 \cdot 4 \ldots (2n)} \ge \dfrac{1 \cdot 2 \cdot 4 \ldots (2n - 2)}{2 \cdot 4 \ldots (2n)} = \dfrac{1}{2n}$. The series diverges because $\sum 1/n$ diverges.

19. $a_n = (n + 1)/n!$ $\lim_{n \to \infty} a_{n+1}/a_n = \lim_{n \to \infty} \left(\dfrac{n + 2}{n + 1}\right)\left(\dfrac{1}{n + 1}\right) = 0$. Series converges by ratio test.

21. a) $\sum_1 (a_n)/n = a_1 + a_2/2 + a_3/3 + \ldots$

$= a_1 + (1/2)a_2 + (1/3 + 1/4)a_4 + (1/5 + 1/6 + 1/7 + 1/8)a_8$

$+ (1/9 + 1/10 + 1/11 + \ldots + 1/16)a_{16} + \ldots \ge 1/2(a_2 + a_4 + a_8 + \ldots)$

$= \infty$.

b) Let $a_n = 1/(\ln (n))$, n = 2, 3, 4, ... Then,

(i) $a_2 \ge a_3 \ge a_4 \ge \ldots$

(ii) $a_2 + a_4 + a_8 + a_{16} + \ldots = 1/\ln(2) + 1/\ln(2^2) + 1/\ln 2^3 + \ldots$

$+ 1/(n \ln 2) + \ldots = 1/(\ln 2)(1 + 1/2 + 1/3 + \ldots) = \infty$.

By a), the series $\sum_1^\infty a_n/n = \sum_1^\infty 1/(n \ln n)$ diverges.

23. $\sum_2^\infty 1/[n(\ln n)^k]$ converges since $\int_2^\infty 1/[n(\ln n)^k]dn = \int_2^\infty (\ln n)^{-k} d(\ln n)$

$= [(\ln n)^{-k+1}]/(-k + 1)\big|_2^\infty < \infty$ for k > 1.

CHAPTER 12 POWER SERIES

Article 12.1

1.

n	$f^{(n)}(x)$	$f^{(n)}(0)$
0	e^{-x}	1
1	$-e^{-x}$	-1
2	e^{-x}	1
3	$-e^{-x}$	-1
4	e^{-x}	1

So, $f_3(x) = 1 - x + \dfrac{x^2}{2} - \dfrac{x^3}{6}$,

$f_4(x) = 1 - x + \dfrac{x^2}{2} - \dfrac{x^3}{6} + \dfrac{x^4}{24}$

3.

n	$f^{(n)}(x)$	$f^{(n)}(0)$
0	$\cos x$	1
1	$-\sin x$	0
2	$-\cos x$	-1
3	$\sin x$	0
4	$\cos x$	1

So, $f_3(x) = 1 - \dfrac{x^2}{2}$, $f_4(x) = 1 - \dfrac{x^2}{2} + \dfrac{x^4}{24}$

5.

n	$f^{(n)}(x)$	$f^{(n)}(0)$
0	$\sinh x$	0
1	$\cosh x$	1
2	$\sinh x$	0
3	$\cosh x$	1
4	$\sinh x$	0

So, $f_3(x) = x + \dfrac{x^3}{6}$, $f_4(x) = x + \dfrac{x^3}{6}$

7.
n	$f^n(x)$	$f^n(0)$
0	x^4-2x+1	1
1	$4x^3-2$	-2
2	$12x^2$	0
3	$24x$	0
4	24	24

So, $f_3(x) = 1 - 2x$,

$$f_4(x) = 1 - 2x + \frac{24x^4}{4!} = x^4 - 2x + 1$$

9.
n	$f^{(n)}(x)$	$f^{(n)}(0)$
0	x^2-2x+1	1
1	$2x-2$	-2
2	2	2
3	0	0
4	0	0

So, $f_3(x) = 1 - 2x + \frac{2x^2}{2!} = x^2 - 2x + 1$,

$$f_4(x) = x^2 - 2x + 1$$

11. Any polynomial is its own Maclaurin series. So, the Maclaurin series of x^2 is $x^2 = 0 + 0x + x^2 + \ldots$

13. By the binomial theorem,

$$(1+x)^{3/2} = 1^{3/2} + \frac{3}{2}(1)^{1/2}x + \left(\frac{3}{2}\right)\left(\frac{1}{2}\right)(1)^{-1/2} x^2$$

$$+ \left(\frac{3}{2}\right)\left(\frac{1}{2}\right)\left(-\frac{1}{2}\right)\left(1\right)^{-3/2} x^3 + \ldots$$

$$= 1 + \frac{3}{2}x + \frac{3}{4}x^2 - \frac{3}{8}x^3 + \ldots$$

15.

n	$f^n(x)$	$f^n(10)$
0	e^x	e^{10}
1	e^x	e^{10}
2	e^x	e^{10}
3	e^x	e^{10}

So, $e^x = e^{10} + e^{10}(x - 10)$

$+ \dfrac{e^{10}}{2} (x - 10)^2 + \dfrac{e^{10}}{6} (x - 10)^3 + \ldots$

17.

n	$f^n(x)$	$f^{(n)}(1)$
0	$\ln x$	0
1	x^{-1}	1
2	$-x^{-2}$	-1
3	$2x^{-3}$	2
4	$-6x^{-4}$	-6
5	$24x^{-5}$	24

In general, $f^{(n)}(1) = (-1)^{n-1} (n-1)!$

for $n \geq 1$, $f^{(0)}(1) = 0$.

So, $\ln x = \displaystyle\sum_{n=0}^{\infty} \dfrac{f^{(n)}(1)}{n!}(x-1)^n = \sum_{n=1}^{\infty} \dfrac{(-1)^{n-1}}{n}(x-1)^n$

$= (x - 1) - \dfrac{1}{2}(x - 1)^2 + \dfrac{1}{3}(x - 1)^3 + \ldots$

19.

n	$f^{(n)}(x)$	$f^{(n)}(-1)$
0	x^{-1}	-1
1	$-x^{-2}$	-1
2	$2x^{-3}$	-2
3	$-6x^{-4}$	-6

$\dfrac{1}{x} = -1 - (x + 1) - (x + 1)^2 - (x + 1)^3 - \ldots$

$= - \displaystyle\sum_{n=0}^{\infty} (x+1)^n$

CHAPTER 12 − Article 12.1

21.

n	$f^{(n)}(x)$	$f^{(n)}(\pi/4)$
0	$\tan x$	1
1	$\sec^2 x$	2
2	$2 \sec^2 x \tan x$	4

So, $\tan x = 1 + 2(x - \pi/4) + 2(x - \pi/4)^2 + \ldots$

23. $(1 + x)^{1/2} = 1 + \frac{1}{2}x - \frac{1}{4}x^2 + \frac{3}{8}x^3 - \frac{15}{16} x^4 + \ldots$

So, $\sqrt{1.02} = 1 + 0.01 - \frac{1}{4}(0.02)^2 + \frac{3}{8}(0.02)^3 - + \ldots$

$= 1.001 - 0.0001 + 0.000003$

Hence, $\sqrt{1.02} = 1.001$ with error ≤ 0.0001 (by virtue of the alternating series).

Article 12.2

1. $e^{x/2} = 1 + \left(\frac{x}{2}\right) + \frac{1}{2!}\left(\frac{x}{2}\right)^2 + \frac{1}{3!}\left(\frac{x}{2}\right)^3 + \frac{1}{4!}\left(\frac{x}{2}\right)^4 + \ldots$

$= 1 + \frac{x}{2} + \frac{x^2}{8} + \frac{x^3}{48} + \frac{x^4}{384} + \ldots$

3. $5 \cos(x/\pi) = 5\left[1 - \frac{(x/\pi)^2}{2!} + \frac{(x/\pi)^4}{4!} - \frac{(x/\pi)^6}{6!} + - \ldots\right]$

$= 5 - \frac{5x^2}{2\pi^2} + \frac{5x^4}{24\pi^4} - \frac{x^6}{144} + - \ldots$

5. $\frac{x^2}{2} - 1 + \cos x = \frac{x^2}{2!} - 1$

$+ \left(1 - \frac{x^2}{2!} + \frac{x^4}{4!} - \frac{x^6}{6!} + \frac{x^8}{8!} - + \ldots\right)$

$= \frac{x^4}{4!} - \frac{x^6}{6!} + \frac{x^8}{8!} - + \ldots$

7. a) $\cos(-x) = 1 - \dfrac{(-x)^2}{2!} + \dfrac{(-x)^4}{4!} - \dfrac{(-x)^6}{6!} + -$

$= 1 - \dfrac{x^2}{2!} + \dfrac{x^4}{4!} - \dfrac{x^6}{6!} + - \ldots = \cos x$

b) $\sin(-x) = (-x) - \dfrac{(-x)^3}{3!} + \dfrac{(-x)^5}{5!} - \dfrac{(-x)^7}{7!} + - \ldots$

$= -x + \dfrac{x^3}{3!} - \dfrac{x^5}{5!} + \dfrac{x^7}{7!} - + \ldots$

$= -\left(x - \dfrac{x^3}{3!} + \dfrac{x^5}{5!} - \dfrac{x^7}{7!} + - \ldots\right) = -\sin x$

9.

n	$f^{(n)}(x)$	$f^{(n)}(0)$
0	$\dfrac{1}{1+x}$	1
1	$\dfrac{-1}{(1+x)^2}$	-1
2	$\dfrac{2}{(1+x)^3}$	2

Hence, $\dfrac{1}{1+x} = 1 - x + \dfrac{2}{2!}x^2 + R_n(x,a)$

$= 1 - x + x^2 + R_2(x,0).$

11.

n	$f^{(n)}(x)$	$f^{(n)}(0)$
0	$(1+x)^{1/2}$	1
1	$\frac{1}{2}(1+x)^{-1/2}$	$\frac{1}{2}$
2	$-\frac{1}{4}(1+x)^{-3/2}$	$-\frac{1}{4}$

So, $\sqrt{1+x} = 1 + \dfrac{1}{2}x - \dfrac{1}{8}x^2 + R_2(x,0)$

13. Because $x - x^3/6$ is the fourth degree Taylor polynomial of $\sin x$ (as well as the third degree polynomial), the error is no greater than $|x|^5/5!$. So, we want

$\dfrac{|x|^5}{120} < 5 \times 10^{-4}$, or $|x|^5 < 0.06$, or $|x| < 0.5697$.

15. Since x is the Taylor polynomial of degree 2 for $\sin x$, the remainder term is

$R_2(x,0) = \frac{f^{(3)}(c)}{3!} x3 = \frac{-\cos c}{3!} x^3$

Thus, $|R_2(x,0)| = \frac{\cos c}{6}|x|^3$.

Since $|\cos c| \le 1$ and $|x| < 10^{-3}$,

$|R_2(x,0)| < \frac{10^{-9}}{6} \simeq 1.67 \times 10^{-10}$.

Now, $\cos c > 0$ for $-10^{-3} < c < 10^{-3}$. So, if $x < 0$, $R_2(x,0) > 0$, so $x < \sin x$; if $x > 0$, $R_2(x,0) < 0$, so $x > \sin x$. Thus, $x < \sin x$ for $-10^{-3} < x < 0$.

17. We have $|f^{(n+1)}(t)| = e^t < e^{0.1}$ for $|t| < 0.1$; so, by the Remainder Estimation Theorem,

$|R_2(x,0)| < \frac{e^{0.1}}{2!}(0.1)^2 \simeq 5.53 \times 10^{-3}$.

19. $R_4(x,0) = \frac{f^{(5)}(c)}{5!} x^5 = \frac{\sinh(c)}{120} x^5$.

For $|x|$, $|c| < 0.5$, $|\sinh c| < 1.128$; hence,

$|R_4(x,0)| < \frac{1.128}{120}(0.5)^5 \simeq 2.94 \times 10^{-4}$.

21. The Maclaurin series for $(f+g)$ is

$(f+g)(x) = \sum_{n=0}^{\infty} \frac{(f+g)^{(n)}(0)}{n!} x^n$

$= \sum_{n=0}^{\infty} \frac{f^{(n)}(0) + g^{(n)}(0)}{n!} x^n$.

23. a) $e^{i\pi} = \cos \pi + i \sin \pi = -1 + 0i$

b) $e^{i\pi/4} = \cos(\pi/4) + i \sin(\pi/4) = \frac{\sqrt{2}}{2} + \frac{\sqrt{2}}{2} i$

c) $e^{-i\pi/2} = \cos(-\pi/2) + i \sin(-\pi/2) = 0 - 1i$

d) $e^{i\pi} \cdot e^{-i\pi/2} = (-1 + 0i)(0 - 1i) = 0 + 1i$

25. Use $(x + y)^3 = x^3 + 3x^2y + 3xy^2 + y^3$, $(x-y)^3 = x^3 - 3x^2y + 3xy^2 - y^3$.

Then $\cos^3 \theta = \left(\frac{e^{i\theta} + e^{-i\theta}}{2}\right)^3$

$$= \frac{e^{i(3\theta)} + 3e^{i\theta} + 3e^{-i\theta} + e^{i(-3\theta)}}{8}$$

$$= \frac{1}{4} \; \frac{e^{(3\theta)i} + e^{(-3\theta)i}}{2} \; + \frac{3}{4} \; \frac{e^{i\theta} + e^{-i\theta}}{2}$$

$$= \frac{1}{4} \cos 3\theta + \frac{3}{4} \cos\theta$$

Similarly, $\sin^3\theta = \left(\dfrac{e^{i\theta} - e^{-i\theta}}{2i} \right)^3$

$$= \frac{e^{3i\theta} - 3e^{i\theta} + 3e^{-i\theta} - e^{-3\theta i}}{-8i}$$

$(i^3 = i^2 i = -i)$

$$= -\frac{1}{4} \; \frac{e^{i(3\theta)} - e^{-i(3\theta)}}{2i} \; + \frac{3}{4} \; \frac{e^{i\theta} - e^{-i\theta}}{2i}$$

$$= -\frac{1}{4} \sin 3\theta + \frac{3}{4} \sin\theta$$

27. $\displaystyle\int e^{(a+bi)x}\, dx = \int e^{ax}(\cos bx + i \sin bx)\, dx$

$$= \int e^{ax} \cos bx\, dx + i \int e^{ax} \sin bx\, dx$$

$$\frac{a - ib}{a^2 + b^2} \; e^{(a+bi)x}$$

$$= \frac{(a-ib)\, e^{ax}(\cos bx + i \sin bx)}{a^2 + b^2}$$

$$= \frac{e^{ax}(a \cos bx + ai \sin bx - bi \cos bx - i^2 b \sin bx)}{a^2 + b^2}$$

$$= \frac{e^{ax}(a \cos bx + b \sin bx)}{a^2 + b^2}$$

$$+ i \; \frac{e^{ax}(a \sin bx - b \cos bx)}{a^2 + b^2}$$

Since two complex numbers are equal if and only if their real parts and imaginary parts are equal, we have

$$\int e^{ax} \cos bx\, dx = \frac{e^{ax}}{a^2 + b^2} (a \cos bx + b \sin bx) + c_1,$$

$$\int e^{ax} \sin bx\, dx = \frac{e^{ax}}{a^2 + b^2} (a \sin bx - b \cos ax) + c_2.$$

Article 12.3

1. $31^\circ = 31\left(\frac{\pi}{180}\right) \simeq 0.5411$ radians. So,

$$\cos(31^\circ) \simeq 1 - \frac{(0.5411)^2}{2!} + \frac{(0.5411)^4}{4!}$$

$$- \frac{(0.5411)^6}{6!} + - \ldots$$

$$\simeq 1 - \frac{(0.5411)^2}{2} + \frac{(0.5411)^4}{24} = 0.8572.$$

The remainder is

$$R_5(x,0) = \frac{\cos^{(6)}(c)}{6!} x^6 \text{ for some } 0 < c < 0.5411.$$

But $\dfrac{d^6}{dx^6}(\cos x) = -\cos x$

So, $|R_5(x,0)| < \dfrac{1}{6!}(0.5411)^6 \simeq 3.5\times10^{-5}$

3. Since 6.3 is close to 2π, $\sin 6.3 = \sin(6.3 - 2\pi)$

$$= \sin 0.0168 \simeq 0.0168 - \frac{(0.0168)^3}{3!} + - \ldots \simeq 0.0168;$$

$$|R_2(0.0168,0)| \leq \frac{(0.0168)^3}{6} \simeq 7.9\times10^{-7}$$

5. $\ln(1.25) \simeq 0.25 - \dfrac{(0.25)^2}{2} + \dfrac{(0.25)^3}{3} - \dfrac{(0.25)^4}{4}$

$$= 0.223; \quad |R_4(0.25,0)| \leq \frac{(0.25)^5}{5} = 0.0002$$

7. $\ln(1+y) = \sum\limits_{n=1}^{\infty} \dfrac{(-1)^{n-1} y^n}{n}$; let $y = 2x$, so

$$\ln(1+2x) = \sum\limits_{n=1}^{\infty} \frac{(-1)^{n-1} 2^n x^n}{n} \ .$$

Since the series for ln $(1 + y)$ converges for $-1 < y \leq 1$, the series for ln $(1 + 2x)$ converges for $-1 < 2x \leq 1$ or $-1/2 < x \leq 1/2$.

9. $$\int_0^{0.1} \frac{\sin x}{x}\, dx = \int_0^{0.1} \frac{1}{x}\left(x - \frac{x^3}{6} + \frac{x^5}{120} - + \ldots\right) dx$$

$$= \int_0^{0.1} \left(1 - \frac{x^2}{6} + \frac{x^4}{120} - + \ldots\right) dx$$

$$= x - \frac{x^3}{18} + \frac{x^5}{600} - + \ldots \Bigg|_0^{0.1} = 0.1$$

with error $|E| < (0.1)^3/18 = 0.0002$ (since we have an alternating series).

11. We wish to show that $|a \cosh(x/a) - (a + x^2/2a)|$ is less than $0.003\,|a|$ for $|x/a| < 1/3$. Now,

$$a \cosh(x/a) = a\left[1 + \frac{(x/a)^2}{2!} + \frac{(x/a)^4}{4!} + \ldots\right],$$

so the Taylor series for $f(x) = a \cosh(x/a) - (a + x^2/2a)$, is

$$a\left[\frac{(x/a)^4}{4!} + \frac{(x/a)^6}{6!}\right].$$

But this is approximately 0 with error

$$|R_3(x/a, 0)| \leq \frac{|a|\,(1/3)^4}{24} = 0.0005|a|.$$

13. This is the series for $\ln(1 + x)$ with $x = 1/2$, so the sum is $\ln(3/2) = \ln 3 - \ln 2$.

15. $\tan^{-1}(1/18) = \tan^{-1}(0.0556)$

$$= 0.0556 - \frac{(0.0556)^3}{3} \ldots$$

$$= 0.0556 - 0.0001 + \ldots \simeq 0.0555$$

$\tan^{-1}(1/57) = \tan^{-1}(0.0175) \simeq 0.0175$ with

error $|E| < \dfrac{(0.0175)^3}{3} \simeq 1.8 \times 10^{-6}$

$\tan^{-1}(1/239) = \tan^{-1}(0.0042) \simeq 0.0042$ with

error $|E| < \dfrac{(0.0042)^3}{3} \simeq 2.5 \times 10^{-8}$

Hence, $\pi = 48(0.0555) + 32(0.0175) - 20(0.0042) \simeq 3.1400$

CHAPTER 12 – Article 12.3

17. $a_0 = 1$ $\qquad\qquad$ $b_0 = 0.707107$

$a_1 = 0.853553$ $\qquad$ $b_1 = 0.840896$

$a_2 = 0.847225$ $\qquad$ $b_2 = 0.847201$

$a_3 = 0.847213$ $\qquad$ $b_3 = 0.847213$

$a_0^2 - b_0^2 = 0.5$

$a_1^2 - b_1^2 = 0.0147$

$a_2^2 - b_2^2 = 0.000041$

$a_3^2 - b_3^2 = 0$ to six dicimal places

$c_3 = \dfrac{2.871079}{0.994633} = 2.886571.$

19. From the formula for the partial sums of the geometric series, we have:

$$1 + (t^2) + (t^2)^2 + \ldots + (t^2)^n$$

$$= \frac{1 - (t^2)^{n+1}}{1 - (t^2)} ,$$

or $\dfrac{1}{1-t^2} = 1 + t^2 + t^4 \ldots + t^{2n} + \dfrac{t^{2n+2}}{1-t^2} .$

(N.B.: This is an __identity__, not a power series expansion.)
Thus,

$$\int_0^x \frac{dt}{1-t^2} = \int_0^x \left[1 + t^2 + \ldots + t^{2n} + \frac{t^{2n+2}}{1-t^2} \right] dt$$

$$= \int_0^x dt + \int_0^x t^2\, dt + \ldots + \int_0^x t^{2n} dt + \int_0^x \frac{t^{2n+2}}{1-t^2}\, dt$$

$$= x + \frac{x^3}{3} + \ldots + \frac{x^{2n+1}}{2n+1} + R. \quad \text{Since}$$

$$\int_0^x \frac{dt}{1-t^2} = \tanh^{-1}(t) \Big|_0^x = \tanh^{-1}(x), \text{ the result follows.}$$

21. a) $\dfrac{d}{dx} \left(\dfrac{1}{1-x} \right)$

$$= \frac{d}{dx} \left(1 + x + x^2 + x^3 + \ldots + x^n + \frac{x^{n+1}}{1-x} \right)$$

or $\dfrac{1}{(1-x)^2} = 1 + 2x + 3x^2 + \ldots + nx^{n-1}$

$+ \left(\dfrac{x^{n+1} + (n+1)(1-x)x^n}{(1-x)^2} \right)$

$= 1 + 2x + 3x^2 + \ldots + nx^{n-1} + R.$

b) For $|x| < 1$, clearly $x^{n+1} \to 0$ as $n \to \infty$. We will show that $(n+1)\,x^n \to \infty$ as $n \to \infty$. Simply substituting ∞ for n will not do, since we then have an indeterminate expression $\infty \cdot 0$. So we apply l'Hopital's rule to $f(n) = (n+1)x^n$ (Note: $\underline{n}$ is the independent variable; x is $\underline{\text{held fixed}}$ subject only to $|x| < 1$.)

$f(n) = (n + 1)x^n = \dfrac{n + 1}{1/x^n};$

$\lim_{n \to \infty} f(n) = \lim_{n \to \infty} \dfrac{1}{-n/x^{n+1}} = \lim_{n \to \infty} \dfrac{-x^{n+1}}{n} = 0$

So, $\lim_{n \to \infty} R = 0.$

c) Call the expected number of throws E:

$E = \sum_{n=1}^{\infty} nq^{n-1}\,p = p \cdot \sum_{n=1}^{\infty} nq^{n-1} = \dfrac{p}{(1-q)^2}$

by part (a). Since $p = \dfrac{1}{6}$, $q = \dfrac{5}{6}$, $E = 6$.

d) $\sum_{n=1}^{\infty} np^{n-1}\,q = q \cdot \sum_{n=1}^{\infty} np^{n-1} = \dfrac{q}{(1-p)^2}.$

Article 12.4

1. $\lim_{h \to 0} \dfrac{\sin h}{h} = \lim_{h \to 0} \dfrac{h - h^3/3! + h^5/5! - + \ldots}{h}$

$= \lim_{h \to 0} \dfrac{h(1 - h^2/3! + h^4/5! - + \ldots)}{h}$

$= \lim_{h \to 0} 1 - \dfrac{h^2}{3!} + \dfrac{h^4}{5!} - + \ldots = 1.$

3. $\lim\limits_{t\to 0} \dfrac{1 - \cos t - t^2/2}{t^4}$

$= \lim\limits_{t\to 0} \dfrac{(1 - t^2/2) - (1 - t^2/2 + t^4/24 - t^6/720 + - \ldots)}{t^4}$

$= \lim\limits_{t\to 0} \dfrac{-t^4/24 + t^6/270 + - \ldots}{t^4}$

$= \lim\limits_{t\to 0} \dfrac{t^4 \, (-1/24 + t^2/270 - + \ldots)}{t^4}$

$= \lim\limits_{t\to 0} - \dfrac{1}{24} + \dfrac{t^2}{720} - + \ldots = - \dfrac{1}{24}$.

5. $\lim\limits_{x\to 0} \dfrac{x^2}{1 - \cosh x}$

$= \lim\limits_{x\to 0} \dfrac{x^2}{1 - (1 + x^2/2! + x^4/4! + x^6/6! + \ldots)}$

$= \lim\limits_{x\to 0} \dfrac{x^2}{-x^2/2 - x^4/24 - x^6/720 - \ldots}$

$= \lim\limits_{x\to 0} \dfrac{x^2}{x^2 \, (-1/2 - x^2/24 - x^4/720 - \ldots)}$

$= \lim\limits_{x\to 0} \dfrac{1}{-1/2 - x^2/24 - x^4/720 - \ldots} = -2$

7. $\lim\limits_{x\to 0} \dfrac{1 - \cos x}{\sin x}$

$= \lim\limits_{x\to 0} \dfrac{1 - (1 - x^2/2 + x^4/24 - + \ldots)}{x - x^3/6 + x^5/120 - + \ldots}$

$= \lim\limits_{x\to 0} \dfrac{x^2/2 - x^4/24 + - \ldots}{x - x^3/6 + x^5/120 - + \ldots}$

$= \lim\limits_{x\to 0} \dfrac{x^2 \, (1 + 2 - x^2/24 + - \ldots)}{x \, (1 - x^2/6 + x^4/120 - + \ldots)} = 0$

9. $\lim\limits_{z\to 0} \dfrac{\sin(z^2) - \sinh(z^2)}{z^6}$

$= \lim\limits_{z\to 0} \dfrac{(z^2 - (z^2)^3/6 + (z^2)^5/120 - + \ldots)}{z^6}$

$- \dfrac{(z^2 + (z^2)^3/6 + (z^2)^5/120 + \ldots)}{z^6}$

CHAPTER 12 - Article 12.4

$$= \lim_{Z \to 0} \frac{-Z^6/3 + \text{(higher powers of Z)}}{Z^6}$$

$$= \lim_{Z \to 0} -\frac{1}{3} + \text{(powers of Z)} = -\frac{1}{3} .$$

11. $$\lim_{x \to 0} \frac{\sin x - x + x^3/6}{x^5}$$

$$= \lim_{x \to 0} \frac{(x - x^3/6 + x^5/120 - x^7/5040 + - \ldots) - x + x^3/6}{x^5}$$

$$= \lim_{x \to 0} \frac{x^5/120 - x^7/5040 + - \ldots}{x^5} = \frac{1}{120}$$

13. $$\lim_{x \to 0} \frac{x - \tan^{-1} x}{x^3}$$

$$= \lim_{x \to 0} \frac{x - (x - x^3/3 + x^5/5 - x^7/7 + - \ldots)}{x^3}$$

$$= \lim_{x \to 0} \frac{x^3/3 - x^5/5 + x^7/7 - + \ldots}{x^3} = \frac{1}{3} .$$

15. $$\lim_{x \to \infty} x^2 (e^{-1/x^2} - 1)$$

$$= \lim_{x \to \infty} x^2 \left[\left(1 - \frac{1}{x^2} + \frac{1}{2} \left(\frac{1}{x^2} \right)^2 + \ldots \right) - 1 \right]$$

$$= \lim_{x \to \infty} x^2 \left(- \frac{1}{x^2} + \frac{1}{2x^4} - \frac{1}{6x^6} + - \ldots \right) = -1.$$

17. $$\lim_{x \to 0} \frac{\tan 3x}{x}$$

$$= \lim_{x \to 0} \frac{(3x) + (3x)^3/3 + 2(3x)^5/15 + \ldots}{x}$$

$$= \lim_{x \to 0} 3 + 9x^2 + 162x^4/15 + \ldots = 3.$$

19. For $x > 0$, all the terms in the power series

$$e^x = \sum_{n=0}^{\infty} \frac{x^n}{n!} \quad \text{are positive.}$$

Therefore, $e^x > 1 + x + x^2/2! + \ldots + x^k/k!$, for any positive integer K. In particular, $e^x > 1 + x + x^2/2! + \ldots + x^{101}/101!$.

CHAPTER 12 - Article 12.4

Therefore, $0 < \dfrac{x^{100}}{e^x} < \dfrac{x^{100}}{1 + x + \ldots + x^{101}/101!}$.

Consequently,

$$0 \le \lim_{x \to \infty} \frac{x^{100}}{e^x} \le \lim_{x \to \infty} \frac{x^{100}}{1 + x + \ldots + x^{101}/101!} = 0.$$

Hence, $\displaystyle\lim_{x \to \infty} \frac{x^{100}}{e^x} = 0$.

Remark: This can easily be modified to prove that

$\displaystyle\lim_{x \to \infty} \frac{x^N}{e^x} = 0$ for any positive integer N.

21. a) For $x \ge 0$, we have the inequality $e^{x^2} \ge 1$.

Therefore, $\displaystyle\int_0^x e^{t^2}\, dt \ge \int_0^x dt = x$, so

$\displaystyle\int_0^x e^{t^2}\, dt$ diverges as $x \to \infty$.

b) $\displaystyle\lim_{x \to \infty} x \int_0^x e^{t^2 - x^2}\, dt = \lim_{x \to \infty} x \int_0^x e^{t^2} e^{-x^2}\, dt$

$\displaystyle = \lim_{x \to \infty} x e^{-x^2} \int_0^x e^{t^2}\, dt$

$\displaystyle = \lim_{x \to \infty} \frac{x \int_0^x e^{t^2}\, dt}{e^{x^2}}.$

$\displaystyle = \lim_{x \to \infty} \frac{\int_0^x e^{t^2}\, dt + x e^{x^2}}{2x e^{x^2}}$ (by l'Hopital's rule)

$\displaystyle = \lim_{x \to \infty} \frac{2e^{x^2} + 2x^2 e^{x^2}}{2e^{x^2} + 4x^2 e^{x^2}}$ (by l'Hopital's rule)

$\displaystyle = \lim_{x \to \infty} \frac{2x^2 + 2}{4x^2 + 2} = \frac{1}{2}$

23.

x	actual sin x	sin x $\simeq$ x	sin x $\simeq 6x/6+x^2$
± 1.0	± 0.8415	± 1.0	± 0.8571
± 0.1	± 0.0998	± 0.1	± 0.998
± 0.01	± 0.0100	± 0.01	± 0.0100

For x = ± 1.0, ± 0.1, the approximation

$$\sin x \simeq \frac{6x}{6 + x^2} \text{ is better.}$$

Article 12.5

1. By the root test:

$$\sqrt[n]{|a_n x^n|} = \sqrt[n]{|x|^n} = |x| = \zeta.$$

So, $\zeta < 1$ if $|x| < 1$, or $-1 < x < 1$.
Now we check the endpoints: at $x = 1$, the series is

$$\sum_{n=0}^{\infty} 1^n = 1 + 1 + \dots, \text{ which diverges;}$$

at $x = -1$, the series is

$$\sum_{n=0}^{\infty} (-1)^n = 1 - 1 + 1 - 1 + 1 - 1 + - \dots,$$

which diverges since the partial sums alternate between 1 and 0.
Therefore, the interval of absolute convergence is $-1 < x < 1$.

3. $\sqrt[n]{|a_n x^n|} = \sqrt[n]{\dfrac{n|x|^n}{2^n}} = \left(\dfrac{1}{2}\right)(n^{1/n}) \, |x|.$ Now,

$$\lim_{n \to \infty} n^{1/n} = 1, \text{ so } \zeta = \lim_{n \to \infty} \sqrt[n]{|a_n x^n|} = \frac{|x|}{2} < 1$$

if $|x| < 2$. So, $-2 < x < 2$ is the interval of absolute
convergence, by the root rest. We must still check the endpoints
of this interval.

At $x = 2$, the series is $\displaystyle\sum_{n=1}^{\infty} \frac{(n)(2^n)}{2^n}$

$= 1 + 2 + 3 + \dots,$ which diverges. At $x = -2$, the series is

$\displaystyle\sum_{n=1}^{\infty} \frac{n(-2)^n}{2^n} = -1 + 2 - 3 + 4 - + \dots,$ which also diverges.

5. Let C_k denote the kth term of the series; then

$$C_k = \frac{(-1)^k \, x^{2k+1}}{(2k + 1)!} . \text{ So,}$$

$$\left| \frac{c_{k+1}}{c_k} \right| = \left| \frac{(-1)^{k+1} \, x^{2k+3} \, / \, (2k+3)!}{(-1)^k \, x^{2k+1} \, / \, (2k+1)!} \right|$$

$$= \left| \frac{(-1)^{k+1} \, x^{2k+3}}{(2k+3)!} \cdot \frac{(2k+1)!}{(-1)^k \, x^{2k+1}} \right|$$

$$= \left| \frac{(-1)^{k+1} \, x^{2k+3}}{(2k+3)(2k+2)\left[(2k+1)!\right]} \cdot \frac{(2k+1)!}{(-1)^k \, x^{2k+1}} \right|$$

$$= \left| \frac{-x^2}{(2k+2)(2k+3)} \right|$$

Thus, $\lim\limits_{k \to \infty} \left| \dfrac{c_{k+1}}{c_k} \right| = 0$ for all x.

So, the series converges absolutely for $-\infty < x < \infty$.

Remark: The given series is the Taylor series for sin x about x = 0.

7. By the ratio test, $\lim\limits_{n \to \infty} \left| \dfrac{c_{n+1} \, (x+2)^{n+1}}{c_n (x+2)^n} \right|$

$$= \lim_{n \to \infty} \left| \frac{\dfrac{(n+1)^2 \, (x+2)^{n+1}}{2^{n+1}}}{\dfrac{n^2 \, (x+2)^n}{2^n}} \right|$$

$$= \lim_{n \to \infty} \left| \frac{(n+1)^2 \, (x+2)^{n+1}}{2^{n+1}} \cdot \frac{2^n}{n^2 \, (x+2)^n} \right|$$

$$= \lim_{n \to \infty} \left| \left(\frac{n+1}{n}\right)^2 \cdot \frac{(x+2)}{2} \right| = \frac{|x+2|}{2} \lim_{n \to \infty} \left(\frac{n+1}{n}\right)^2$$

$= \dfrac{|x+2|}{2}$. Hence, the series converges absolutely

for $\dfrac{|x+2|}{2} < 1$, or $-4 < x < 0$.

At $x = -4$, the series is $\sum\limits_{n=0}^{\infty} n^2 \cdot \dfrac{(-2)^n}{2^n}$

$$= \sum_{n=0}^{\infty} (-1)^n \, n^2 = -1 + 4 - 9 + 16 - + \dots \; ,$$

which diverges; at $x = 0$, the series is

$\sum\limits_{n=0}^{\infty} n^2 \cdot \dfrac{(2)^n}{2^n} = \sum\limits_{n=0}^{\infty} n^2$, which also diverges.

9. By the ratio test,

$$\lim_{n\to\infty} \left| \frac{(n+1) - \text{th term}}{n - \text{th term}} \right|$$

$$= \lim_{n\to\infty} \left| \frac{\dfrac{(-1)^{n+1} x^{2(n+1)+1}}{2(n+1) + 1}}{\dfrac{(-1)^n x^{2n+1}}{2n+1}} \right|$$

$$= \lim_{n\to\infty} \left| \frac{-x^2 (2n+1)}{2n+3} \right|$$

$$= |x|^2 \cdot \lim_{n\to\infty} \left| \frac{2n+1}{2n+3} \right| = |x|^2.$$

So, the series converges absolutely for $|x|^2 < 1$, or $|x| < 1$, or $-1 < x < 1$.

Check endpoints: At $x = -1$, the series is

$$\sum_{n=0}^{\infty} (-1)^n \frac{(-1)^{2n+1}}{2n+1}$$

$$\sum_{n=0}^{\infty} \frac{(-1)^{n+1}}{2n+1} = 1 + \frac{1}{3} - \frac{1}{5} + \frac{1}{7} - \frac{1}{9} + - \cdots$$

which converges (but not absolutely); for $x = 1$, the series is

$$\sum_{n=0}^{\infty} \frac{(-1)^n}{2n+1} = 1 - \frac{1}{3} + \frac{1}{5} - \frac{1}{7} + - \cdots, \text{ which converges.}$$

11. Here, neither the root test nor the ratio test helps. However, we observe that

$$\left| \frac{\cos nx}{2^n} \right| \le \frac{1}{2^n} \quad (\text{since } |\cos \theta| \le 1), \text{ and}$$

$$\sum_{n=0}^{\infty} \frac{1}{2^n} \text{ converges. Therefore, the given series converges}$$

absolutely for all x.

13. By the ratio test, $\displaystyle\lim_{n\to\infty} \left| \frac{c_{n+1} x^{n+1}}{c_n x^n} \right|$

$$= \lim_{n\to\infty} \left| \frac{\dfrac{e^{n+1} x^{n+1}}{n+2}}{\dfrac{e^n x^n}{n+1}} \right|$$

$$= \lim_{n\to\infty} \left| \frac{e^{n+1} x^{n+1}}{n+2} \cdot \frac{n+1}{e^n x^n} \right|$$

$$= \lim_{n \to \infty} \left| \frac{(n+1)ex}{n+2} \right|$$

$$= e|x| \lim_{n \to \infty} \left| \frac{n+1}{n+2} \right| = e \ |x|.$$

Thus, the series converges absolutely for $e|x| < 1$, or $-1/e < x < 1/e$. Now for the endpoints: as $x = 1/e$, the series is

$$\sum_{n=0}^{\infty} \frac{1}{n+1} \ , \text{ which diverges; at } x = -1/e, \text{ the series is}$$

$$\sum_{n=0}^{\infty} \frac{(-1)^n}{n+1} = 1 - \frac{1}{2} + \frac{1}{3} - \frac{1}{4} + - \ \dots, \text{ which converges.}$$

15. By the root test, $\lim_{n \to \infty} \sqrt[n]{c_n \ x^n}$

$$= \lim_{n \to \infty} \sqrt[n]{n^n \ x^n} = \lim_{n \to \infty} |nx| = \infty \ \underline{\text{unless}} \ x = 0.$$

So, the series $\sum_{n=1}^{\infty} n^n \ x^n$ converges only for $x = 0$.

17. By the ratio test, $\lim_{n \to \infty} \left| \frac{c_{n+1}(x-2)^{n+1}}{c_n(x-2)^n} \right|$

$$= \lim_{n \to \infty} \left| \frac{(-2)^{n+1} \ (n+2) \ (x-2)^{n+1}}{(-2)^n \ (n+1) \ (x-2)^n} \right|$$

$$= \lim_{n \to \infty} \left| \frac{-2(x-2) \ (n+2)}{n+1} \right| = 2|x-2| \lim_{n \to \infty} \left| \frac{n+2}{n+1} \right| = 2 \ |x-2|.$$

Therefore, the series converges absolutely for $2|x-2|<1$, or $7/2 < x < 9/2$.

At $x = \frac{7}{2}$, the series is $\sum_{n=1}^{\infty} (n+1)$, which diverges;

at $x = \frac{9}{2}$, the series is $\sum_{n=1}^{\infty} (-1)^n \ (n+1)$,

which also diverges.

19. By the root test: $\lim_{n \to \infty} \left[\left\{ \frac{x^2-1}{2} \right\}^n \right]^{1/n} = \left| \frac{x^2-1}{2} \right|$

So, the series converges absolutely for $| \ x^2-1| < 2$, or $-1 < x^2 < 3$. Since $x^2 \geq 0$ for all x, this inequality is the same as $x^2 < 3$, or $-\sqrt{3} < x < \sqrt{3}$.

At the endpoints $x = \pm\sqrt{3}$, $\frac{x^2-1}{2} = 1$, so

$$\sum_{n=0}^{\infty} \left[\frac{x^2-1}{2} \right]^n \text{ diverges at } x = \pm\sqrt{3} \ .$$

21. We recognize $e^w = \sum\limits_{n=0}^{\infty} \frac{w^n}{n!}$; the series

$\sum\limits_{n=0}^{\infty} \frac{(3x+6)^n}{n!}$ is this series with $w = 3x+6$. So,

$\sum\limits_{n=0}^{\infty} \frac{(3x+6)^n}{n!} = e^{3x+6}$

23. a) $\cos x = \sum\limits_{n=0}^{\infty} \frac{(-1)^n x^{2n}}{(2n)!}$; since the series converges

absolutely for all x, we may differentiate term by term:

$\frac{d}{dx} (\cos x) = \sum\limits_{n=1}^{\infty} \frac{(-1)^n 2n \, x^{2n-1}}{(2n)!}$

(The term for n=0 in the series for cos x is constant, so we drop n=0 in the summation of the term by term derivatives.

But $\sum\limits_{n=1}^{\infty} \frac{(-1)^n (2n) \, x^{2n-1}}{(2n)!}$

$= \sum\limits_{m=0}^{\infty} \frac{(-1)^{m+1} (2m+2) \, x^{2m+1}}{(2m+2)!}$

(letting n=m+1) $= (-1) \sum\limits_{m=0}^{\infty} \frac{(-1)^m x^{2m+1}}{(2m+1)!} = -\sin x,$

since $\sin x = \sum\limits_{m=0}^{\infty} \frac{(-1)^m x^{2m+1}}{(2m+1)!}$

b) Again, the series for cos t converges uniformly for all x, so we may integrate term by term:

$\int_0^x \cos t \, dt = \int_0^x \left[\sum\limits_{n=0}^{\infty} \frac{(-1)^n t^{2n}}{(2n)!} \right] dt$

$= \sum\limits_{n=0}^{\infty} \left[\int_0^x \frac{(-1)^n t^{2n}}{(2n)!} \, dt \right]$

$= \sum\limits_{n=0}^{\infty} \left[\frac{(-1)^n t^{2n+1}}{(2n+1) (2n)!} \right]_0^x$

$= \sum\limits_{n=0}^{\infty} \frac{(-1)^n x^{2n+1}}{(2n+1)!} = \sin x.$

c) We will show directly that if

$$y = \sum_{n=0}^{\infty} \frac{x^n}{n!} = e^x, \text{ then } \frac{dy}{dx} = y.$$

Since the series converges absolutely for all x, we may differentiate term by term:

$$\frac{dy}{dx} = \frac{d}{dx}\left[\sum_{n=0}^{\infty} \frac{x^n}{n!}\right] = \sum_{n=0}^{\infty} \frac{d}{dx}\left[\frac{x^n}{n!}\right]$$

$$= \sum_{n=1}^{\infty} \frac{nx^{n-1}}{n!} \text{ (Why does the term for n=0 not appear?)}$$

$$= \sum_{n=1}^{\infty} \frac{x^{n-1}}{(n-1)!} = \sum_{m=0}^{\infty} \frac{x^m}{m!} \text{ (where m=n-1)} = e^x = y.$$

25. Since $\frac{1}{1-x^2} = \sum_{n=0}^{\infty} x^{2n}$ converges uniformly for

-1 < x < 1, we may differentiate term by term for
-1 < x < 1:

$$\frac{2x}{(1-x^2)^2} = \frac{d}{dx}\left[\frac{1}{1-x^2}\right] = \frac{d}{dx}\left[\sum_{n=0}^{\infty} x^{2n}\right]$$

$$= \sum_{n=0}^{\infty} \frac{d}{dx}(x^{2n})$$

$$= \sum_{n=1}^{\infty} 2nx^{2n-1} \text{ (again please note the bottom index)}$$

$$= \sum_{m=0}^{\infty} 2(m+1)x^{2m+1} \text{ (where m=n-1)}.$$

Hence, $\frac{2x}{(1-x^2)^2} = \sum_{m=0}^{\infty} 2(m+1)x^{2m+1}.$ or

$$\frac{x}{(1-x^2)^2} = \sum_{m=0}^{\infty} (m+1)x^{2m+1}, \text{ valid for } -1 < x < 1.$$

27. $\displaystyle\int_{0}^{0.2} \sin x^2 \, dx$

$$= \int_{0}^{0.2}\left[\sum_{n=0}^{\infty} \frac{(-1)^n (x^2)^{2n+1}}{(2n+1)!}\right] dx$$

$$= \sum_{n=0}^{\infty} \frac{(-1)^n}{(2n+1)!}\left[\int_{0}^{0.2} x^{4n+2} \, dx\right]$$

$$= \sum_{n=0}^{\infty} \frac{(-1)^n}{(2n+1)!} \left[\frac{x^{4n+3}}{4n+3} \right]_0^{0.2}$$

$$= \sum_{n=0}^{\infty} \frac{(-1)^n (0.2)^{4n+3}}{(4n+3)(2n+1)!} = \frac{(0.2)^3}{3} - \frac{(0.2)^7}{42} + \frac{(0.2)^{11}}{1320}$$

$$- + \dots \simeq \frac{(0.2)^3}{3} \simeq 0.0027 \text{ with error}$$

$$|E| < \frac{(0.2)^7}{42} \simeq 3 \times 10^{-5} < 0.00$$

since the series is alternating.

29. $$\int_0^{0.1} x^2 e^{-x^2} dx = \int_0^{0.1} x^2 \left[\sum_{n=0}^{\infty} \frac{(-1)^n x^{2n}}{n!} \right] dx$$

$$= \sum_{n=0}^{\infty} \frac{(-1)^n}{n!} \left[\int_0^{0.1} x^{2n+2} dx \right]$$

$$= \sum_{n=0}^{\infty} \frac{(-1)^n (0.1)^{2n+3}}{(2n+3)(n!)} = \frac{(0.1)^3}{3} - \frac{(0.1)^5}{5} + - \dots$$

$$\simeq \frac{(0.1)^3}{3} \simeq 0.0003 \text{ with error } |E| < \frac{(0.1)^5}{5} = 2 \times 10^{-6}.$$

31. $$1 - e^{-x} = 1 - \sum_{n=0}^{\infty} \frac{(-1)^n x^n}{n!} = 1 + \sum_{n=0}^{\infty} \frac{(-1)^{n+1} x^n}{n!}$$

$$= \sum_{n}^{\infty} \frac{(-1)^{n+1} x^n}{n!} \text{ , so } \frac{1-e^{-x}}{x} = \sum_{n=1}^{\infty} \frac{(-1)^{n+1} x^{n-1}}{n!}$$

$$= \sum_{m=0}^{\infty} \frac{(-1)^{m+2} x^m}{(m+1)!} = \sum_{m=0}^{\infty} \frac{(-1)^m x^m}{(m+1)!} \text{ , where } m = n-1$$

Hence, $$\int_0^{0.4} \frac{1-e^{-x}}{x} dx = \int_0^{0.4} \left[\sum_{m=0}^{\infty} \frac{(-1)^m x^m}{(m+1)!} \right] dx$$

$$= \sum_{m=0}^{\infty} \frac{(-1)^m}{(m+1)!} \left[\int_0^{0.4} x^m dx \right]$$

$$= \sum_{m=0}^{\infty} \frac{(-1)^m (0.4)^{m+1}}{(m+1)(m+1)!} = 0.4 - \frac{(0.4)^2}{4} + \frac{(0.4)^3}{18}$$

$$- \frac{(0.4)^4}{96} + - \dots \simeq 0.4 - \frac{(0.4)^2}{4} + \frac{(0.4)^3}{18} \simeq 0.3636$$

with error $$|E| < \frac{(0.4)^4}{96} \simeq 0.0003 < 0.001.$$

CHAPTER 12 - Article 12.5

33. $[1+x^4]^{-1/2} = 1 - \frac{1}{2}x^4 + - \ldots,$ so

$$\int_0^{0.1} \frac{dx}{\sqrt{1+x^4}} = \int_0^{0.1} (1 - \frac{1}{2}x^4 + - \ldots)\, dx$$

$$= \int_0^{0.1} dx - \frac{1}{2}\int_0^{0.1} x^4\, dx + - \ldots$$

$$= 0.1 - \frac{(0.1)^5}{10} + - \ldots$$

$$= 0.1 \text{ with error } |E| < \frac{(0.1)^5}{10} = 1\times10^{-6}$$

35. $\dfrac{1}{\sqrt{1+t^2}} = (1+t^2)^{-1/2} = 1 - \frac{1}{2}t^2 + \dfrac{(-1/2)(-3/2)}{2}\, t^4 - + \ldots$

$$= 1 - \frac{t^2}{2} + \frac{3t^4}{8} - + \ldots$$

Therefore, $\sinh^{-1}x = \displaystyle\int_0^x \frac{dt}{\sqrt{1+t^2}}$

$$= \int_0^x \left\{ 1 - \frac{t^2}{2} + \frac{3t^4}{8} - + \ldots \right\} dt$$

$$= x - \frac{x^3}{6} + \frac{3x^5}{40} - + \ldots$$

b) $\sinh^{-1} 0.25 = 0.25 - \dfrac{(0.25)^3}{6} + \dfrac{3(0.25)^5}{40} - + \ldots$

$$= 0.2474 \text{ with error } |E| < \frac{3(0.25)^5}{40} \simeq 7.3\times10^{-5}$$

37. Problem 15 of this section is an example of a series which converges only for x = 0.

39. Our goal is to show that for any value of x between -r and r, the series converges <u>absolutely</u> (we only know that it converges; it may converge conditionally.) So, suppose we are given a definite value of x, which we will call a: x=a. Because -r < a < r, we can find a (positive) number b such that -r < -b < a < b < r:

By our original assumptions, the series converges for x=b. By theorem 1, the series converges absolutely for $|x| < b$; since $-b < a < b$, the series converges absolutely for x=a. Now, the only restriction on a was that $-r < a < r$. Therefore, the series converges absolutely for $-r < x < r$.

41. $e^x \sin x = \left\{ 1 + x + \dfrac{x^2}{2} + \dfrac{x^3}{6} + \dfrac{x^4}{24} + \dfrac{x^5}{120} + \ldots \right\} \cdot$

$\cdot \left\{ x - \dfrac{x^3}{6} + \dfrac{x^5}{120} - + \ldots \right\}$

$= x \left(1 + x + \dfrac{x^2}{2} + \dfrac{x^3}{6} + \dfrac{x^4}{24} + \ldots \right)$

$\quad - \dfrac{x^3}{6} \left(1 + x + \dfrac{x^2}{2} + \ldots \right) + \dfrac{x^5}{120} \left(1 + \ldots \right)$

$= \left(x + x^2 + \dfrac{x^3}{2} + \dfrac{x^4}{6} + \dfrac{x^5}{24} + \ldots \right)$

$\quad - \left(\dfrac{x^3}{6} + \dfrac{x^4}{6} + \dfrac{x^5}{12} + \ldots \right) + \left(\dfrac{x^5}{120} + \ldots \right)$

$= x + x^2 + \dfrac{x^3}{3} - \dfrac{x^5}{30} - \ldots$

To check the answer as suggested:

$(1 + i)^2 = 1 + 2i + i^2 = 1 + 2i - 1 = 2i$

$(1 + i)^3 = (1 + i)^2(1 + i) = 2i(1 + i) = -2 + 2i$

$(1 + i)^4 = (2i)^2 = -4$

$(1 + i)^5 = (1 + i)^4(1 + i) = -4(1 + i) = -4 - 4i$

Therefore, $e^{(1+i)x} = \sum\limits_{n=0}^{\infty} \dfrac{(1+i)^n \, x^n}{n!}$

$= 1 + (1+i)x + \dfrac{2ix^2}{2} + \dfrac{(-2+2i)x^3}{6} \dfrac{-4x^4}{4!} + \dfrac{(-4-4i)x^5}{5!}$

The imaginary part of this series is

$x + x^2 + \dfrac{x^3}{3} - \dfrac{x^5}{30} \ldots$, as before.

Because both the series for e^x and that for $\sin x$ converge absolutely for all x, we expect the series for $e^x \sin x$ to converge absolutely for all x by the theorem given at the end of example 11.

CHAPTER 12 - Article 12.5

43. Since $\sec 0 = 1$, $\displaystyle\int_0^x \tan t\, dt$

$$= \ln \sec x \bigg|_0^x = \ln \sec x. \quad \text{However,}$$

$$\tan t = t + \frac{t^3}{3} + \frac{2t^5}{15} + \dots, \text{ so}$$

$$\ln \sec x = \int_0^x \left[t + \frac{t^3}{3} + \frac{2t^5}{15} + \dots \right]\, dt$$

$$= \int_0^x t\, dt + \frac{1}{3}\int_0^x t^3\, dt + \frac{2}{15}\int_0^x t^5\, dt + \dots$$

$$= \frac{x^2}{2} + \frac{x^4}{12} + \frac{x^6}{45} + \dots$$

45.

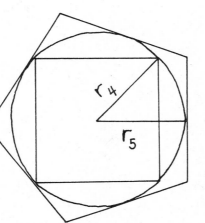

a) Circle 2 passes through the vertices of a triangle (3-gon) and is inscribed in a square (4-gon); circle 3 passes through the vertiaces of a square (regular 4-gon) and is inscribed in a regular pentagon (5-gon). In general, circle n passes through the vertices of a regular (n+1)-gon and is inscribed in a regular (n+2)-gon. Now, because

$$\alpha = \frac{1}{2}\left[\frac{2\pi}{n+1}\right] = \frac{\pi}{n+1}, \text{ we have: } \frac{r_{n-1}}{r_n} = \cos\alpha,$$

or $r_n = r_{n-1} \sec\alpha = r_{n-1} \sec\left[\dfrac{\pi}{n+1}\right]$

so, $r_2 = r, \sec \pi/3$, $r_3 = r_2 \sec \pi/4$, ...

b) By induction on n: <u>Case</u> n=2:

$\ln r_2 = \ln (r_1 \sec \pi/3) = \ln (r_1) + \ln \sec \pi/3$.

Suppose now $\ln (r_{k-1}) = \ln r_1 + \displaystyle\sum_{p=3}^{(k-1)+1} \ln \sec (\pi/p)$

$$= \ln r_1 + \sum_{p=3}^{k} \ln \sec(\pi/p). \quad \text{Then, since}$$

$$r_k = r_{k-1} \sec(\pi/(k+1)), \ln r_k = \ln r_{k-1} + \ln \sec(\pi/k+1)$$

$$= \left[\ln r_1 + \sum_{p=3}^{k} \ln \sec(\pi/p) \right] + \ln \sec \left[\frac{\pi}{k+1} \right]$$

$$= \ln r_1 + \sum_{p=3}^{k+1} \ln \sec(\pi/p).$$

c) $\quad \lim\limits_{n\to\infty} \dfrac{\ln \sec(\pi/n)}{1/n^2} = \lim\limits_{n\to\infty} n^2 \ln \sec(\pi/n)$

$$= \lim_{n\to\infty} n^2 \left[\frac{\pi^2}{2n^2} + \frac{\pi^4}{12n^4} + \text{(higher powers of } 1/n) \right]$$

$= \pi^2/2. \quad$ Since $\quad \sum\limits_{n=3}^{\infty} \dfrac{1}{n^2} \quad$ converges,

so does $\sum\limits_{n=3}^{\infty} \ln \sec(\pi/n)$. Therefore, r_n tends to a

finite limit as $n\to\infty$.

CHAPTER 12

Miscellaneous Problems

1. a) Since $\dfrac{1}{1+x} = \dfrac{1}{1-(-x)} = 1 - x + x^2 - x^3 + - \ldots,$

$$\frac{x^2}{1+x} = x^2 - x^3 + x^4 - x^5 + x^6 - x^7 + - \ldots$$

b) Since the geometric series $\dfrac{1}{1+x} = 1 - x + x^2 - + \ldots$

converges only for $|x| < 1$, the above series expansion should not
converge for $x = 2$.

3. $e^y = 1 + y + \dfrac{y^2}{2} + \dfrac{y^3}{6} + \dfrac{y^4}{24} + \ldots$

$$y = \sin x = x - \frac{x^3}{6} + \ldots$$

$$y^2 = \left[x^2 - \frac{x^3}{6} + \ldots \right]^2 = x^2 - \frac{x^4}{3} + \ldots$$

$$y^3 = \left[x - \frac{x^3}{6} + \ldots\right]^3 = x^3 + \ldots$$

$$y^4 = \left[x - \frac{x^3}{6} + \ldots\right]^4 = x^4 + \ldots$$

where "..." means that we have ignored powers of x higher than x^4. Thus, $e^y = e^{\sin x}$

$$= 1 + x + \frac{x^2}{2} - \frac{x^4}{8} + \ldots$$

Note: Although the problem asks only for the first four terms, we had to keep powers of x through x^4 because the coefficient of x^3 in the series for $e^{\sin x}$ is zero.

5. a) Since $y = 1 - \cos x = \frac{x^2}{2} - \frac{x^4}{24} + \frac{x^6}{720} - + \ldots,$

any power of y higher than y^3 will start with a power of x higher than x^6 [$y = x^3 [1/2 - \ldots]$].

Now $\ln(1 - y) = - y - \frac{y^2}{2} - \frac{y^3}{3} - \ldots;$

$$y^2 = x^4 \left[\frac{1}{2} - \frac{x^2}{24} + \ldots\right]\left[\frac{1}{2} - \frac{x^2}{24} + \ldots\right]$$

$$= x^4 \left[\frac{1}{4} - \frac{x^2}{24} + \ldots\right]$$

$$= \frac{x^4}{4} - \frac{x^6}{24} + \ldots; \quad y^3 = \frac{x^6}{8} + \ldots. \text{ So,}$$

$$\ln(\cos x) = - \frac{x^2}{2} - \frac{x^4}{12} - \frac{x^6}{45} - \ldots$$

b) $\displaystyle\int_0^{0.1} \ln(\cos x)\, dx = \int_0^{0.1} \left[- \frac{x^2}{2} - \frac{x^4}{12} - \frac{x^6}{45} - \ldots\right] dx$

$$= - \frac{(0.1)^3}{6} - \frac{(0.1)^5}{60} - \frac{(0.1)^7}{315} - \ldots \simeq -0.00017$$

7. $\displaystyle\int_0^1 e^{-x^2}\, dx =$

$$\int_0^1 \left[1 - x^2 + \frac{x^4}{2} - \frac{x^6}{6} + \frac{x^8}{24} - \frac{x^{10}}{120} + \frac{x^{12}}{720} - \frac{x^{14}}{5040} + - \ldots\right] dx$$

$$= 1 - \frac{1}{3} + \frac{1}{10} - \frac{1}{42} + \frac{1}{216} - \frac{1}{1320} + - \ldots \simeq 0.747$$

CHAPTER 12 - Miscellaneous Problems

with error $|E| < \frac{1}{1320} \simeq 7.6 \times 10^{-5}$

9.

n	$f^{(n)}(x)$	$f^{(n)}(2)$
0	$(1-x)^{-1}$	-1
1	$(1-x)^{-2}$	1
2	$2(1-x)^{-3}$	-2
3	$6(1-x)^{-4}$	6
4	$24(1-x)^{-5}$	-24

The general trend is that $f^{(n)}(x) = n!(1-x)^{n+1}$, so $f^{(n)}(2) = (-1)^{n+1}n!$ Therefore,

$$f(x) = \sum_{n=0}^{\infty} \frac{f^{(n)}(2)}{n!}(x-2)^n = \sum_{n=0}^{\infty} (-1)^{n+1}(x-2)^n.$$

By the ratio test,
$$\lim_{n \to \infty} \left| \frac{a_{n+1} (x-2)^{n+1}}{a_n (x-2)^n} \right| = |(x-2)|,$$

so the series converges for $|x-2| < 1$, or $1 < x < 3$.

11.

n	$f^{(n)}(x)$	$f^{(n)}(\pi/3)$
0	$\cos x$	1/2
1	$-\sin x$	$-\sqrt{3}/2$
2	$-\cos x$	-1/2
3	$\sin x$	$\sqrt{3}/2$
4	$\cos x$	1/2

so, $\cos x = \frac{1}{2}\left[1 - \frac{\sqrt{3}(x-\pi/3)}{2} - \frac{(x-\pi/3)^2}{4} + \frac{\sqrt{3}(x-\pi/3)^3}{12} + \ldots \right]$

13. To develop a Taylor series about 0 for $f(x)$, we note that $f'(0) = g(0) = 0$, $f''(0) = (g)'(0) = f(0) = 1$, $f'''(0) = (f')''(0) = (g')'(0) = (f')(0) = g(0) = 0$. In general, if n is even, $f^{(n)}(x) = f(x)$, while if n is odd, $f^{(n)}(x) = g(x)$. Thus, the Taylor series about 0 for f is

$$f(x) = 1 + \frac{x^2}{2!} + \frac{x^4}{4!} + \frac{x^6}{6!} + \ldots \text{, and } f(1) \simeq 1.543$$

CHAPTER 12 - Miscellaneous Problems

(the error $R_7(1)$ is bounded in absolute value by $1/8! \approx 2.5 \times 10^{-5}$).

Note: from the power series we see $f(x) = \cosh x$, $g(x) = \sinh x$.

15.

n	$f^{(n)}(x)$	$f^{(n)}(0)$
0	e^{e^x}	e
1	$e^x \cdot e^{e^x} = e^{(x+e^x)}$	e
2	$(1+e^x) \cdot e^{x+e^x}$	$2e$
3	$(1+3e^x+e^{2x}) \cdot e^{x+e^x}$	$5e$

So, $e^{e^x} = e\left[1 + x + x^2 + \dfrac{5x^3}{6} + \dots\right]$

17. Because the binomial series for $(1+x)^{1/3}$ alternates after the first term and because the next term after $x/3$ is $-x^2/9$, the error E is negative and $|E| < (0.1)^2/9 \approx 0.0011$.

19. We will find $\lim\limits_{x \to 0} \ln\left\{\left[\dfrac{\sin x}{x}\right]^{1/x^2}\right\}$

$= \lim\limits_{x \to 0} \dfrac{1}{x^2} \ln\left[\dfrac{\sin x}{x}\right]$. Now $\ln\left[\dfrac{\sin x}{x}\right]$

$= \ln\left[1 - \dfrac{x^2}{6} + \dfrac{x^4}{120} - + \dots\right]$

$= \ln\left\{1 - x^2\left[\dfrac{1}{6} - \dfrac{x^2}{120} + \dots\right]\right\}$

$= -x^2\left[\dfrac{1}{6} - \dfrac{x^2}{120} + \dots\right] - \left\{x^2\left[\dfrac{1}{6} - \dfrac{x^2}{120} + \dots\right]\right\}^2 - \dots$

$\left\{\text{substituting } y = x^2\left[\dfrac{1}{6} - \dfrac{x^2}{120} + \dots\right]\right.$

$\left.\text{in the series for } \ln(1-y)\right\} = -\dfrac{x^2}{6} - \dfrac{7x^4}{360} - \dots$

So, $\lim\limits_{x \to 0} \dfrac{1}{x^2} \ln\left[\dfrac{\sin x}{x}\right] = -\dfrac{1}{6}$, whence

$\lim\limits_{x \to 0}\left[\dfrac{\sin x}{x}\right]^{1/x^2} = e^{-1/6}$.

21. By the ratio test, $\lim\limits_{n \to \infty}\left|\dfrac{a_{n+1}(x+2)^{n+1}}{a_n(x+2)^n}\right|$

$= \lim\limits_{n \to \infty}\left|\dfrac{(x+2)^{n+1}}{3^{n+1}(n+1)} \cdot \dfrac{n\, 3^n}{(x+2)^n}\right|$

$= \lim\limits_{n\to\infty} \left| \dfrac{n}{n+1} \cdot \dfrac{(x+2)}{3} \right| = \dfrac{|x+2|}{3}$. So, the

series converges absolutely for $\dfrac{|x+2|}{3} < 1$,

or $-5 < x < 1$. At $x = 1$, the series is

$1 + 1 + \dfrac{1}{2} + \dfrac{1}{3} + \dots$, which diverges; at $x = -5$,

the series is $1 - 1 + \dfrac{1}{2} - \dfrac{1}{3} + - \dots$, which converges.

23. By the root test, $\zeta = \lim\limits_{n\to\infty} \sqrt[n]{(x/n)^n} = |x/n| = 0$

for all x. So, the series converges absolutely for all x.

25. By the ratio test,

$$\lim\limits_{n\to\infty} \left| \dfrac{n+2}{2n+3} \dfrac{(x-3)^{n+1}}{2^{n+1}} \cdot \dfrac{2^n}{(x-3)^n} \dfrac{2n+1}{n+1} \right|$$

$$= \lim\limits_{n\to\infty} \left| \dfrac{n+2}{n+1} \cdot \dfrac{2n+1}{2n+3} \cdot \dfrac{(x-3)}{2} \right| = \dfrac{|x-3|}{2} . \text{ So, the series}$$

converges absolutely for $|x-3| < 2$, or $1 < x < 5$. At the endpoints: since

$\lim\limits_{n\to\infty} \dfrac{n+1}{2n+1} = \dfrac{1}{2}$, the series diverges at both endpoints.

27. By the ratio test:

$$\lim\limits_{n\to\infty} \left| \dfrac{(-1)^n (x-1)^{n+1}}{(n+1)^2} \cdot \dfrac{n^2}{(-1)^{n-1} (x-1)^n} \right|$$

$$= \lim\limits_{n\to\infty} \left| \dfrac{-n^2 (x-1)}{(n+1)^2} \right| = |x-1|. \text{ So, the series converges}$$

absolutely for $|x-1| < 1$, or $0 < x < 2$. At $x=0$, the series
is $\sum\limits_{n=1}^{\infty} \dfrac{(-1)^{n-1} (-1)^n}{n^2} = -\sum\limits_{n=1}^{\infty} \dfrac{1}{n^2}$

which converges; at $x=2$, the series is

$\sum\limits_{n=1}^{\infty} \dfrac{(-1)^{n-1}}{n^2}$, which converges (absolutely).

So, the series converges absolutely for $0 \le x \le 2$.

CHAPTER 12 – Miscellaneous Problems

29. $\lim\limits_{n\to\infty} \left| \dfrac{(x-2)^{3n+3}}{(n+1)!} \cdot \dfrac{n!}{(x-2)^{3n}} \right|$

$= \lim\limits_{n\to\infty} \left| \dfrac{(x-1)^3}{n+1} \right| = 0.$

So, the series converges for all x.

31. By the ratio test, we have:

$\lim\limits_{n\to\infty} \left| \dfrac{1}{n+1}\left[\dfrac{x-1}{x}\right]^{n+1} \cdot n\left[\dfrac{x}{x-1}\right]^{n} \right|$

$= \lim\limits_{n\to\infty} \left| \dfrac{n}{n+1}\cdot\dfrac{x-1}{x} \right| = \left| \dfrac{x-1}{x} \right|.$

Hence the series converges for $-1 < \dfrac{x-1}{x} < 1.$

To make this last inequality neater, consider:
Case I: x > 0. Then we want -x < x-1 < x; x-1 < x for all x,
and -x < x-1 ⟹ 2x > 1 ⟹ x > 1/2.
Case II: x < 0. Then we want -x > x-1 > x. But x-1 > x is
impossible.
Hence, the series converges absolutely for x > 1/2. At x=1/2,
the series is

$\sum\limits_{n=1}^{\infty} \dfrac{(-1)^n}{n}$, which converges.

Hence, the series converges for x ≥ 1/2.

33. If $a_n > 0, \dfrac{1}{1+a_n} < 1$, so $0 < \dfrac{a_n}{1+a_n} < a_n.$

Therefore $\sum\limits_{n=1}^{\infty} \dfrac{a_n}{1+a_n}$ converges,

by the comparison test.

35. If $-1 < a_n \le 1$ (so that the series converges), we have

$\ln(1+a_n) = a_n - \dfrac{a_n^2}{2} + \dfrac{a_n^3}{3} - + \dots$ However,

$(-1)^{k+1}\dfrac{a_n^k}{k} \le |a_n|^k$ for k = 1, 2, 3, ...; thus,

$|\ln(1+a_n)| \le |a_n| + |a_n|^2 + |a_n|^3 + \dots = \dfrac{|a_n|}{1-|a_n|}.$

CHAPTER 12 - Miscellaneous Problems

Now, since $\sum_{n=1}^{\infty} |a_n|$ converges, we can find some N

such that $|a_n| \leq \frac{1}{2}$ if $n \geq N$. But then $1 - |a_n| \geq \frac{1}{2}$,

so, $\frac{1}{1-|a_n|} \leq 2$ if $n \geq N$. So, $0 < \frac{|a_n|}{1-|a_n|} \leq 2|a_n|$

if $n \geq N$; since the series $\sum_{n=1}^{\infty} 2|a_n|$ converges,

the comparison test yields convergence of $\sum_{n=1}^{\infty} \frac{|a_n|}{1-|a_n|}$;

since $|\ln(1 + |a_n|)| \leq \frac{|a_n|}{1-|a_n|}$, the comparison test

yields absolute convergence of $\sum_{n=1}^{\infty} \ln(1+a_n)$. Thus the product

$\prod_{n=1}^{\infty} (1+a_n)$ converges.

37. $\dfrac{\tan^{-1}x}{1-x} = (x - \dfrac{x^3}{3} + \dfrac{x^5}{5} - + \dots)$

$(1 + x + x^2 + x^3 + x^4 + x^5 + \dots)$

$= x + x^2 + \dfrac{2x^3}{3} + \dfrac{2x^4}{3} + \dfrac{13x^5}{15} + \dots$